Lecture Notes in Computer Science 14488

Founding Editors

Gerhard Goos
Juris Hartmanis

The series Lecture Notes in Computer Science (LNCS), including its subseries Lecture Notes in Artificial Intelligence (LNAI) and Lecture Notes in Bioinformatics (LNBI), has established itself as a medium for the publication of new developments in computer science and information technology research, teaching, and education.

LNCS enjoys close cooperation with the computer science R & D community, the series counts many renowned academics among its volume editors and paper authors, and collaborates with prestigious societies. Its mission is to serve this international community by providing an invaluable service, mainly focused on the publication of conference and workshop proceedings and postproceedings. LNCS commenced publication in 1973.

Zahir Tari · Keqiu Li · Hongyi Wu
Editors

Algorithms and Architectures for Parallel Processing

23rd International Conference, ICA3PP 2023
Tianjin, China, October 20–22, 2023
Proceedings, Part II

 Springer

Editors
Zahir Tari
Royal Melbourne Institute of Technology
Melbourne, VIC, Australia

Keqiu Li
Tianjin University
Tianjin, China

Hongyi Wu
University of Arizona
Tucson, AZ, USA

ISSN 0302-9743 ISSN 1611-3349 (electronic)
Lecture Notes in Computer Science
ISBN 978-981-97-0800-0 ISBN 978-981-97-0801-7 (eBook)
https://doi.org/10.1007/978-981-97-0801-7

This Springer imprint is published by the registered company Springer Nature Singapore Pte Ltd.
The registered company address is: 152 Beach Road, #21-01/04 Gateway East, Singapore 189721, Singapore

Paper in this product is recyclable.

Preface

On behalf of the Conference Committee, we welcome you to the proceedings of the 2023 International Conference on Algorithms and Architectures for Parallel Processing (ICA3PP 2023), which was held in Tianjin, China from October 20–22, 2023. ICA3PP2023 was the 23rd in this series of conferences (started in 1995) that are devoted to algorithms and architectures for parallel processing. ICA3PP is now recognized as the main regular international event that covers the many dimensions of parallel algorithms and architectures, encompassing fundamental theoretical approaches, practical experimental projects, and commercial components and systems. This conference provides a forum for academics and practitioners from countries around the world to exchange ideas for improving the efficiency, performance, reliability, security, and interoperability of computing systems and applications.

A successful conference would not be possible without the high-quality contributions made by the authors. This year, ICA3PP received a total of 503 submissions from authors in 21 countries and regions. Based on rigorous peer reviews by the Program Committee members and reviewers, 193 high-quality papers were accepted to be included in the conference proceedings and submitted for EI indexing. In addition to the contributed papers, six distinguished scholars, Lixin Gao, Baochun Li, Laurence T. Yang, Kun Tan, Ahmed Louri, and Hai Jin, were invited to give keynote lectures, providing us with the recent developments in diversified areas in algorithms and architectures for parallel processing and applications.

We would like to take this opportunity to express our sincere gratitude to the Program Committee members and 165 reviewers for their dedicated and professional service. We highly appreciate the twelve track chairs, Dezun Dong, Patrick P. C. Lee, Meng Shen, Ruidong Li, Li Chen, Wei Bao, Jun Li, Hang Qiu, Ang Li, Wei Yang, Yu Yang, and Zhibin Yu, for their hard work in promoting this conference and organizing the reviews for the papers submitted to their tracks. We are so grateful to the publication chairs, Heng Qi, Yulei Wu, Deze Zeng, and the publication assistants for their tedious work in editing the conference proceedings. We must also say "thank you" to all the volunteers who helped us at various stages of this conference. Moreover, we were so honored to have many renowned scholars be part of this conference. Finally, we would like to thank all speakers, authors, and participants for their great contribution to and support for the success of ICA3PP 2023!

October 2023

Jean-Luc Gaudiot
Hong Shen
Gudula Rünger
Zahir Tari
Keqiu Li
Hongyi Wu
Tian Wang

Organization

General Chairs

Jean-Luc Gaudiot   University of California, Irvine, USA
Hong Shen   University of Adelaide, Australia
Gudula Rünger   Chemnitz University of Technology, Germany

Program Chairs

Zahir Tari   Royal Melbourne Institute of Technology,
     Australia
Keqiu Li   Tianjin University, China
Hongyi Wu   University of Arizona, USA

Program Vice-chair

Wenxin Li   Tianjin University, China

Publicity Chairs

Hai Wang   Northwest University, China
Milos Stojmenovic   Singidunum University, Serbia
Chaofeng Zhang   Advanced Institute of Industrial Technology,
     Japan
Hao Wang   Louisiana State University, USA

Publication Chairs

Heng Qi   Dalian University of Technology, China
Yulei Wu   University of Exeter, UK
Deze Zeng   China University of Geosciences (Wuhan), China

Workshop Chairs

Laiping Zhao   Tianjin University, China
Pengfei Wang   Dalian University of Technology, China

Local Organization Chairs

Xiulong Liu   Tianjin University, China
Yitao Hu   Tianjin University, China

Web Chair

Chen Chen   Shanghai Jiao Tong University, China

Registration Chairs

Xinyu Tong   Tianjin University, China
Chaokun Zhang   Tianjin University, China

Steering Committee Chairs

Yang Xiang (Chair)   Swinburne University of Technology, Australia
Weijia Jia   Beijing Normal University and UIC, China
Yi Pan   Georgia State University, USA
Laurence T. Yang   St. Francis Xavier University, Canada
Wanlei Zhou   City University of Macau, China

Program Committee

Track 1: Parallel and Distributed Architectures

Dezun Dong (Chair)   National University of Defense Technology,
                                  China
Chao Wang   University of Science and Technology of China,
                                  China
Chentao Wu   Shanghai Jiao Tong University, China

Chi Lin	Dalian University of Technology, China
Deze Zeng	China University of Geosciences, China
En Shao	Institute of Computing Technology, Chinese Academy of Sciences, China
Fei Lei	National University of Defense Technology, China
Haikun Liu	Huazhong University of Science and Technology, China
Hailong Yang	Beihang University, China
Junlong Zhou	Nanjing University of Science and Technology, China
Kejiang Ye	Shenzhen Institute of Advanced Technology, Chinese Academy of Sciences, China
Lei Wang	National University of Defense Technology, China
Massimo Cafaro	University of Salento, Italy
Massimo Torquati	University of Pisa, Italy
Mengying Zhao	Shandong University, China
Roman Wyrzykowski	Czestochowa University of Technology, Poland
Rui Wang	Beihang University, China
Sheng Ma	National University of Defense Technology, China
Songwen Pei	University of Shanghai for Science and Technology, China
Susumu Matsumae	Saga University, Japan
Weihua Zhang	Fudan University, China
Weixing Ji	Beijing Institute of Technology, China
Xiaoli Gong	Nankai University, China
Youyou Lu	Tsinghua University, China
Yu Zhang	Huazhong University of Science and Technology, China
Zichen Xu	Nanchang University, China

Track 2: Software Systems and Programming Models

Patrick P. C. Lee (Chair)	Chinese University of Hong Kong, China
Erci Xu	Ohio State University, USA
Xiaolu Li	Huazhong University of Science and Technology, China
Shujie Han	Peking University, China
Mi Zhang	Institute of Computing Technology, Chinese Academy of Sciences, China

Jing Gong	KTH Royal Institute of Technology, Sweden
Radu Prodan	University of Klagenfurt, Austria
Wei Wang	Beijing Jiaotong University, China
Himansu Das	KIIT Deemed to be University, India
Rong Gu	Nanjing University, China
Yongkun Li	University of Science and Technology of China, China
Ladjel Bellatreche	National Engineering School for Mechanics and Aerotechnics, France

Track 3: Distributed and Network-Based Computing

Meng Shen (Chair)	Beijing Institute of Technology, China
Ruidong Li (Chair)	Kanazawa University, Japan
Bin Wu	Institute of Information Engineering, China
Chao Li	Beijing Jiaotong University, China
Chaokun Zhang	Tianjin University, China
Chuan Zhang	Beijing Institute of Technology, China
Chunpeng Ge	National University of Defense Technology, China
Fuliang Li	Northeastern University, China
Fuyuan Song	Nanjing University of Information Science and Technology, China
Gaopeng Gou	Institute of Information Engineering, China
Guangwu Hu	Shenzhen Institute of Information Technology, China
Guo Chen	Hunan University, China
Guozhu Meng	Chinese Academy of Sciences, China
Han Zhao	Shanghai Jiao Tong University, China
Hai Xue	University of Shanghai for Science and Technology, China
Haiping Huang	Nanjing University of Posts and Telecommunications, China
Hongwei Zhang	Tianjin University of Technology, China
Ioanna Kantzavelou	University of West Attica, Greece
Jiawen Kang	Guangdong University of Technology, China
Jie Li	Northeastern University, China
Jingwei Li	University of Electronic Science and Technology of China, China
Jinwen Xi	Beijing Zhongguancun Laboratory, China
Jun Liu	Tsinghua University, China

Kaiping Xue	University of Science and Technology of China, China
Laurent Lefevre	National Institute for Research in Digital Science and Technology, France
Lanju Kong	Shandong University, China
Lei Zhang	Henan University, China
Li Duan	Beijing Jiaotong University, China
Lin He	Tsinghua University, China
Lingling Wang	Qingdao University of Science and Technology, China
Lingjun Pu	Nankai University, China
Liu Yuling	Institute of Information Engineering, China
Meng Li	Hefei University of Technology, China
Minghui Xu	Shandong University, China
Minyu Feng	Southwest University, China
Ning Hu	Guangzhou University, China
Pengfei Liu	University of Electronic Science and Technology of China, China
Qi Li	Beijing University of Posts and Telecommunications, China
Qian Wang	Beijing University of Technology, China
Raymond Yep	University of Macau, China
Shaojing Fu	National University of Defense Technology, China
Shenglin Zhang	Nankai University, China
Shu Yang	Shenzhen University, China
Shuai Gao	Beijing Jiaotong University, China
Su Yao	Tsinghua University, China
Tao Yin	Beijing Zhongguancun Laboratory, China
Tingwen Liu	Institute of Information Engineering, China
Tong Wu	Beijing Institute of Technology, China
Wei Quan	Beijing Jiaotong University, China
Weihao Cui	Shanghai Jiao Tong University, China
Xiang Zhang	Nanjing University of Information Science and Technology, China
Xiangyu Kong	Dalian University of Technology, China
Xiangyun Tang	Minzu University of China, China
Xiaobo Ma	Xi'an Jiaotong University, China
Xiaofeng Hou	Shanghai Jiao Tong University, China
Xiaoyong Tang	Changsha University of Science and Technology, China
Xuezhou Ye	Dalian University of Technology, China
Yaoling Ding	Beijing Institute of Technology, China

Yi Zhao	Tsinghua University, China
Yifei Zhu	Shanghai Jiao Tong University, China
Yilei Xiao	Dalian University of Technology, China
Yiran Zhang	Beijing University of Posts and Telecommunications, China
Yizhi Zhou	Dalian University of Technology, China
Yongqian Sun	Nankai University, China
Yuchao Zhang	Beijing University of Posts and Telecommunications, China
Zhaoteng Yan	Institute of Information Engineering, China
Zhaoyan Shen	Shandong University, China
Zhen Ling	Southeast University, China
Zhiquan Liu	Jinan University, China
Zijun Li	Shanghai Jiao Tong University, China

Track 4: Big Data and Its Applications

Li Chen (Chair)	University of Louisiana at Lafayette, USA
Alfredo Cuzzocrea	University of Calabria, Italy
Heng Qi	Dalian University of Technology, China
Marc Frincu	Nottingham Trent University, UK
Mingwu Zhang	Hubei University of Technology, China
Qianhong Wu	Beihang University, China
Qiong Huang	South China Agricultural University, China
Rongxing Lu	University of New Brunswick, Canada
Shuo Yu	Dalian University of Technology, China
Weizhi Meng	Technical University of Denmark, Denmark
Wenbin Pei	Dalian University of Technology, China
Xiaoyi Tao	Dalian Maritime University, China
Xin Xie	Tianjin University, China
Yong Yu	Shaanxi Normal University, China
Yuan Cao	Ocean University of China, China
Zhiyang Li	Dalian Maritime University, China

Track 5: Parallel and Distributed Algorithms

Wei Bao (Chair)	University of Sydney, Australia
Jun Li (Chair)	City University of New York, USA
Dong Yuan	University of Sydney, Australia
Francesco Palmieri	University of Salerno, Italy

George Bosilca	University of Tennessee, USA
Humayun Kabir	Microsoft, USA
Jaya Prakash Champati	IMDEA Networks Institute, Spain
Peter Kropf	University of Neuchâtel, Switzerland
Pedro Soto	CUNY Graduate Center, USA
Wenjuan Li	Hong Kong Polytechnic University, China
Xiaojie Zhang	Hunan University of Technology and Business, China
Chuang Hu	Wuhan University, China

Track 6: Applications of Parallel and Distributed Computing

Hang Qiu (Chair)	Waymo, USA
Ang Li (Chair)	Qualcomm, USA
Daniel Andresen	Kansas State University, USA
Di Wu	University of Central Florida, USA
Fawad Ahmad	Rochester Institute of Technology, USA
Haonan Lu	University at Buffalo, USA
Silvio Barra	University of Naples Federico II, Italy
Weitian Tong	Georgia Southern University, USA
Xu Zhang	University of Exeter, UK
Yitao Hu	Tianjin University, China
Zhixin Zhao	Tianjin University, China

Track 7: Service Dependability and Security in Distributed and Parallel Systems

Wei Yang (Chair)	University of Texas at Dallas, USA
Dezhi Ran	Peking University, China
Hanlin Chen	Purdue University, USA
Jun Shao	Zhejiang Gongshang University, China
Jinguang Han	Southeast University, China
Mirazul Haque	University of Texas at Dallas, USA
Simin Chen	University of Texas at Dallas, USA
Wenyu Wang	University of Illinois at Urbana-Champaign, USA
Yitao Hu	Tianjin University, China
Yueming Wu	Nanyang Technological University, Singapore
Zhengkai Wu	University of Illinois at Urbana-Champaign, USA
Zhiqiang Li	University of Nebraska, USA
Zhixin Zhao	Tianjin University, China

Ze Zhang   University of Michigan/Cruise, USA
Ravishka Rathnasuriya   University of Texas at Dallas, USA

Track 8: Internet of Things and Cyber-Physical-Social Computing

Yu Yang (Chair)   Lehigh University, USA
Qun Song   Delft University of Technology, The Netherlands
Chenhan Xu   University at Buffalo, USA
Mahbubur Rahman   City University of New York, USA
Guang Wang   Florida State University, USA
Houcine Hassan   Universitat Politècnica de València, Spain
Hua Huang   UC Merced, USA
Junlong Zhou   Nanjing University of Science and Technology,
                                  China
Letian Zhang   Middle Tennessee State University, USA
Pengfei Wang   Dalian University of Technology, China
Philip Brown   University of Colorado Colorado Springs, USA
Roshan Ayyalasomayajula   University of California San Diego, USA
Shigeng Zhang   Central South University, China
Shuo Yu   Dalian University of Technology, China
Shuxin Zhong   Rutgers University, USA
Xiaoyang Xie   Meta, USA
Yi Ding   Massachusetts Institute of Technology, USA
Yin Zhang   University of Electronic Science and Technology
                                  of China, China
Yukun Yuan   University of Tennessee at Chattanooga, USA
Zhengxiong Li   University of Colorado Denver, USA
Zhihan Fang   Meta, USA
Zhou Qin   Rutgers University, USA
Zonghua Gu   Umeå University, Sweden
Geng Sun   Jilin University, China

Track 9: Performance Modeling and Evaluation

Zhibin Yu (Chair)   Shenzhen Institute of Advanced Technology,
                                  Chinese Academy of Sciences, China
Chao Li   Shanghai Jiao Tong University, China
Chuntao Jiang   Foshan University, China
Haozhe Wang   University of Exeter, UK
Laurence Muller   University of Greenwich, UK

Lei Liu	Beihang University, China
Lei Liu	Institute of Computing Technology, Chinese Academy of Sciences, China
Jingwen Leng	Shanghai Jiao Tong University, China
Jordan Samhi	University of Luxembourg, Luxembourg
Sa Wang	Institute of Computing Technology, Chinese Academy of Sciences, China
Shoaib Akram	Australian National University, Australia
Shuang Chen	Huawei, China
Tianyi Liu	Huawei, China
Vladimir Voevodin	Lomonosov Moscow State University, Russia
Xueqin Liang	Xidian University, China

Reviewers

Dezun Dong
Chao Wang
Chentao Wu
Chi Lin
Deze Zeng
En Shao
Fei Lei
Haikun Liu
Hailong Yang
Junlong Zhou
Kejiang Ye
Lei Wang
Massimo Cafaro
Massimo Torquati
Mengying Zhao
Roman Wyrzykowski
Rui Wang
Sheng Ma
Songwen Pei
Susumu Matsumae
Weihua Zhang
Weixing Ji
Xiaoli Gong
Youyou Lu
Yu Zhang
Zichen Xu
Patrick P. C. Lee
Erci Xu

Xiaolu Li
Shujie Han
Mi Zhang
Jing Gong
Radu Prodan
Wei Wang
Himansu Das
Rong Gu
Yongkun Li
Ladjel Bellatreche
Meng Shen
Ruidong Li
Bin Wu
Chao Li
Chaokun Zhang
Chuan Zhang
Chunpeng Ge
Fuliang Li
Fuyuan Song
Gaopeng Gou
Guangwu Hu
Guo Chen
Guozhu Meng
Han Zhao
Hai Xue
Haiping Huang
Hongwei Zhang
Ioanna Kantzavelou

Jiawen Kang

Jie Li

Jingwei Li

Jinwen Xi

Jun Liu

Kaiping Xue

Laurent Lefevre

Lanju Kong

Lei Zhang

Li Duan

Lin He

Lingling Wang

Lingjun Pu

Liu Yuling

Meng Li

Minghui Xu

Minyu Feng

Ning Hu

Pengfei Liu

Qi Li

Qian Wang

Raymond Yep

Shaojing Fu

Shenglin Zhang

Shu Yang

Shuai Gao

Su Yao

Tao Yin

Tingwen Liu

Tong Wu

Wei Quan

Weihao Cui

Xiang Zhang

Xiangyu Kong

Xiangyun Tang

Xiaobo Ma

Xiaofeng Hou

Xiaoyong Tang

Xuezhou Ye

Yaoling Ding

Yi Zhao

Yifei Zhu

Yilei Xiao

Yiran Zhang

Yizhi Zhou

Yongqian Sun

Yuchao Zhang

Zhaoteng Yan

Zhaoyan Shen

Zhen Ling

Zhiquan Liu

Zijun Li

Li Chen

Alfredo Cuzzocrea

Heng Qi

Marc Frincu

Mingwu Zhang

Qianhong Wu

Qiong Huang

Rongxing Lu

Shuo Yu

Weizhi Meng

Wenbin Pei

Xiaoyi Tao

Xin Xie

Yong Yu

Yuan Cao

Zhiyang Li

Wei Bao

Jun Li

Dong Yuan

Francesco Palmieri

George Bosilca

Humayun Kabir

Jaya Prakash Champati

Peter Kropf

Pedro Soto

Wenjuan Li

Xiaojie Zhang

Chuang Hu

Hang Qiu

Ang Li

Daniel Andresen

Di Wu

Fawad Ahmad

Haonan Lu

Silvio Barra

Weitian Tong

Xu Zhang

Yitao Hu

Zhixin Zhao
Wei Yang
Dezhi Ran
Hanlin Chen
Jun Shao
Jinguang Han
Mirazul Haque
Simin Chen
Wenyu Wang
Yitao Hu
Yueming Wu
Zhengkai Wu
Zhiqiang Li
Zhixin Zhao
Ze Zhang
Ravishka Rathnasuriya
Yu Yang
Qun Song
Chenhan Xu
Mahbubur Rahman
Guang Wang
Houcine Hassan
Hua Huang
Junlong Zhou
Letian Zhang
Pengfei Wang
Philip Brown
Roshan Ayyalasomayajula

Shigeng Zhang
Shuo Yu
Shuxin Zhong
Xiaoyang Xie
Yi Ding
Yin Zhang
Yukun Yuan
Zhengxiong Li
Zhihan Fang
Zhou Qin
Zonghua Gu
Geng Sun
Zhibin Yu
Chao Li
Chuntao Jiang
Haozhe Wang
Laurence Muller
Lei Liu
Lei Liu
Jingwen Leng
Jordan Samhi
Sa Wang
Shoaib Akram
Shuang Chen
Tianyi Liu
Vladimir Voevodin
Xueqin Liang

Contents – Part II

LearnedSync: A Learning-Based Sync Optimization for Cloud Storage

Yuxuan Zhou[1], Suzhen Wu[1,2], Shengzhe Wang[1], Chunfeng Du[1], Jiayang Guo[3], Yijie Pan[4,5], Naian Xiao[6], and Bo Mao[1(✉)]

[1] Department of Informatics, Xiamen University, Xiamen, China
maobo@xmu.edu.cn
[2] Wuhan National Laboratory for Optoelectronics, Wuhan, China
[3] Department of Hematology, School of Medicine, Xiamen University, Xiamen, China
[4] Department of Computer Science and Technology, Tsinghua University, Beijing, China
[5] Eastern Institute for Advanced Study, Eastern Institute of Technology, Ningbo, China
[6] Department of Neurology, The Third Hospital of Xiamen, Xiamen, China

Abstract. Cloud sync refers to the synchronization (sync) between devices for files that live on cloud storage. Its efficiency is critical to delivering on the promise of anywhere and anytime access for individuals, groups, or enterprises for cloud storage. However, existing cloud sync optimizations can be characterized as either full or delta sync with human-driven configurations. This paper proposes a machine learning-based cloud sync optimization, LearnedSync, that utilizes machine learning to optimize the cloud sync process. LearnedSync combines three sync methods with different characteristics based on workload characteristics and environmental conditions. It can learn from actual sync scenes and achieve the learning effect of offline training. The key idea of LearnedSync is to (1) record the sync information during each sync and verify whether the sync method is optimal, (2) train the verified records by using the multilayer perceptron (MLP) network to select for appropriate sync method, and (3) regularly update the network to improve the accuracy of decision-making continuously. Our experimental results show that the efficiency of LearnedSync is higher than existing full sync, FSC-based delta sync, and CDC-based delta sync. Moreover, LearnedSync increases the cloud sync speed by at least 41.4% when compared to PandaSync, the state-of-the-art sync scheme, and sync traffic is reduced by 9.6%.

Keywords: Cloud Storage · Sync Optimization · Machine Learning

1 Introduction

The cloud is no longer optional, with over 48% companies planning to move most of their data to the cloud, and the global cloud computing market has reached $623.3 billion by the end of 2022. By 2025, cloud-native platforms will host 95% of all new workloads, making cloud usage essential for businesses to remain competitive [1]. As a result, cloud vendors such as Amazon, Microsoft, Google,

Z. Tari et al. (Eds.): ICA3PP 2023, LNCS 14488, pp. 1–21, 2024.
https://doi.org/10.1007/978-981-97-0801-7_1

and Alibaba have invested in cutting-edge work to address the performance of distributed cloud infrastructures on a global scale [2]. Cloud synchronization (sync) technology is desirable and urgent, with about a third of an average company's IT expenses going toward cloud services.

For companies to stay competitive, embracing the cloud and modernizing IT is essential. To achieve cloud migration for enterprises and users, an efficient and cost-effective data sync technology is required. This technology, known as cloud sync, synchronizes different devices for real-time files stored on the cloud. Changes or updates to these files and any new uploads are reflected across devices within a short time. Maximum cloud sync performance is not only desirable from an end-user's perspective, but also brings advantages to the cloud storage provider. Specifically, fast sync latency increases system throughput, ultimately improving data center performance and reducing costs. As such, it has an enormous impact on the user experience, and so sync performance must be managed carefully to ensure accuracy between clients/devices and the cloud data center [3,4].

Existing cloud sync approaches can be divided into two categories: full sync and delta sync [5–9]. Full sync involves sending the entire file to the cloud and replacing the old version with it. Delta sync, on the other hand, uploads only the changed parts of the file, verifying the existence of the older file in the cloud and then calculating the hash fingerprints of the new file. In order to determine which pieces of the file have been modified, it divides the file into chunks and computes the respective fingerprints to compare them. Delta sync can be further demarcated into Fixed-Size Chunking (FSC)-based delta sync and Content Defined Chunking (CDC)-based delta sync. FSC-based delta sync is more straightforward but could possibly miss some redundant data, whereas CDC-based delta sync while being more computationally intensive, is able to discern more data redundancies.

Further studies of enterprise and cloud environments have found that 80% of user operations involve small files [10,11]. As these small files form a substantial portion of the employed working files, the access performance of these small files directly affects enterprise efficiency and user experience [12]. In the past, solutions such as QuickSync [5], DeltaCFS [6], WebRSync [7], and Dsync [8] have all relied on delta sync with limited efficiency in segmenting small files. However, PandaSync [13] presents a hybrid sync scheme which takes into account essential environmental factors of file size and network RTT and is thus more adaptive and responsive to changing conditions. Our empirical evaluations and analysis in Sect. 4 have revealed that PandaSync may not be as efficient in networks with fluctuating bandwidth.

Recent progress and achievements in machine learning (ML) have demonstrated the feasibility and potential of enhancing system and cloud efficiency [14,15]. Inspired by this research, we propose LearnedSync, a learning-based cloud sync scheme. LearnedSync can dynamically switch between full sync, FSC-based delta sync, and CDC-based delta sync to adapt to workload characteristics and environmental conditions, including file size, redundancy rate, bandwidth, RTTs, and the number of running threads. The guiding principle behind LearnedSync is the multilayer perceptron (MLP), which uses historical sync records to train on environmental information, sync method, and sync time and dynamically selects between the three sync methods based on the environment. Specifically, this paper makes the following contributions:

1. Our experiments demonstrate that the sync time of full sync, FSC-based delta sync, and CDC-based delta sync varies significantly in different environments, and each method can only achieve the minimum sync time under specific conditions, indicating that no single method is universally applicable.
2. We propose a learning-based sync scheme, LearnedSync, which trains the verified records using the MLP network to select appropriate sync methods according to environmental conditions.
3. We implement a prototype of LearnedSync and conduct extensive experiments. Our performance results show that LearnedSync achieves the lowest sync time compared to other cloud sync schemes and improves cloud sync speed by at least 41.4% when compared to PandaSync, the state-of-the-art sync scheme, and sync traffic is reduced by 9.6%.

The remainder of this paper is organized as follows: Sect. 2 presents the background and motivation, Sect. 3 describes the design of LearnedSync, Sect. 4 presents the performance evaluation, Sect. 5 reviews related work, and Sect. 6 concludes the paper.

2   Background and Motivation

This section provides background information on cloud sync and some essential observations from our preliminary experiments and analysis. We introduce machine learning methods as a motivation for the LearnedSync study (Fig. 1).

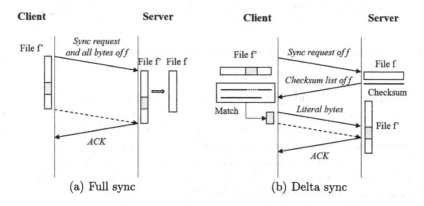

(a) Full sync   (b) Delta sync

Fig. 1. Workflow and procedure for full sync and delta sync.

2.1 Cloud Sync

Cloud storage is a service provider that allows users to purchase storage space to store file data [1]. The cloud service provider is responsible for ensuring the data is available anytime, anywhere via the Internet. Cloud sync ensures that cloud data is up-to-date and consistent across all user devices. There are two cloud sync methods: full sync and delta sync. Full sync involves transferring the entire new file to the cloud and overwriting the old file, whereas delta sync matches the changed contents between two files and transfers only the changed parts to the cloud.

In addition, Content-Defined Chunking (CDC) and Fixed-Size Chunking (FSC) divide delta sync into two forms. CDC-based delta sync calculates the hash value of the sliding window during the chunking, while FSC-based delta sync divides the chunks quickly but leads to many undetected redundant chunks. Figure 2 illustrates the working mechanism of these two methods for a 10KiB file with a 4KiB chunk size and 2KiB of new content inserted. FSC employs a fixed chunk length, so the chunk changes after the insert and redundant data chunks go undetected. In contrast, CDC generates boundaries only at specific bytes, so the insertion does not easily affect the previous chunk location [5]. However, the byte-by-byte detection used by CDC adds time-consuming overhead [8].

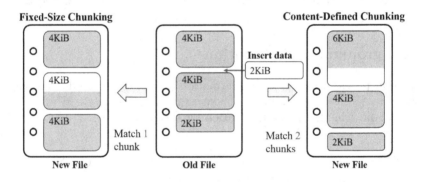

Fig. 2. Mechanisms for Fixed-Size Chunking and Content-Defined Chunking.

Our performance results for synchronizing a single file using three different sync methods in various environments are presented in Fig. 3. Specifically, Fig. 3(a) illustrates the outcome obtained in environments with 96Mbps bandwidth, 60ms network round-trip time and 40% file redundancy rate; Fig. 3(b) and (c) reflected the results obtained in environments with 16Mbps bandwidth, 10ms network round-trip time and 80% file redundancy rate. Additionally, we have additionally introduced several unrelated threads in the Fig. 3(c). In Fig. 3(a), full sync is shown to have the shortest sync time for small files, high bandwidth, and high network RTT. In Fig. 3(b), for large files with high redundancy, low bandwidth, and low network RTT, CDC-based delta sync (CDCSync) is the most efficient method as it can discover the most redundancy. In Fig. 3(c), with further increased file size and the addition of unrelated threads, FSC-based delta sync (FSCSync) outperforms CDCSync due to its low computational overhead.

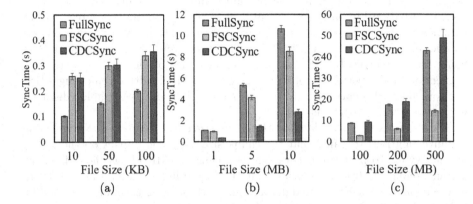

Fig. 3. The sync time results of the three sync methods in different environments. The respective bandwidth, RTT, and redundant rate in (a) 96 Mbps, 60 ms, and 40%; (b) and (c) 16 Mbps, 10 ms, and 80%. Note that some additional threads are added in (c).

The performance results show that each method performs optimally under specific conditions. Full sync is more efficient than delta sync for small files, as the contribution of small files to reducing sync time is relatively insignificant [13]. CDC-based delta sync performs well in large files and low bandwidth environments, while FSC-based delta sync performs well in compute-intensive environments. Table 1 summarizes the differences among three sync methods.

Table 1. Features of the three sync methods.

Sync Method	Calculation Overhead	Detect Redundancy	Network Rond-Trip Frequency
Full sync	Low	None	Low
FSC-based delta sync	Medium	Medium	High
CDC-based delta sync	High	High	High

2.2 Workload Characteristics and Environmental Conditions

Efficient upper-layer applications rely on a well-designed storage system that considers workload characteristics and environmental conditions. Our evaluations highlight several factors that are critical to sync performance. For instance, recent research on cloud tracking has shown that most files (77%) are small, less than 100 KB in size, yet receive the majority of file accesses (over 80%) [10]. This emphasizes the need for storage systems that optimize performance for small files

[13]. Additionally, the redundancy rate of files affects delta sync performance, as higher redundancy rate between new and old files result in fewer delta chunks needing to be transferred, leading to faster sync speed.

CPU performance is also critical, particularly in CDC-based delta sync, where CPU utilization changes significantly impact sync performance. Moreover, the relationship between CPU utilization and computing resources is often nonlinear [16], making the study of sync time complex. Network conditions are another critical factor frequently fluctuating, with bandwidth and network RTT being crucial metrics [17]. Since the cumulative size of files sent over a device connection in any period is not fixed, the device's bandwidth often changes [18]. Network RTT also varies due to various factors, including geographic distance, wireless network impacts, etc. [19].

We also took into consideration the chunk size of the delta sync. Nevertheless, our investigations revealed that the chunk size has little impact on the timing of delta sync. To perform a delta sync, we chose a 3MB file as the old version file and replaced 10% of its content at the beginning, middle, and end as the new version file. For FSC-based delta sync, we chose various chunk sizes, and for CDC-based delta sync, we chose various chunk maximum and minimum ranges. Figure 4 depicts the results of our measurements of the sync time under different circumstances. It can be observed that the sync time stays nearly constant regardless of the chunk size. Though larger chunks could accelerate the chunking process, they can lead to inefficient redundancy detection and reduce sync efficiency.

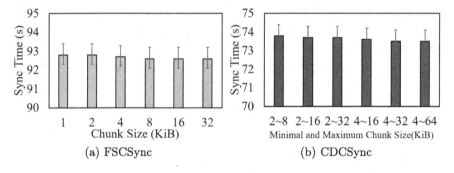

Fig. 4. Performance impact of chunk size on FSCSync and CDCSync when synchronizing 3MB files.

2.3 Motivation

Due to various environmental factors, no single sync method can achieve the minimum sync time in multiple environments. Moreover, the workload characteristics and the network conditions directly affect the sync efficiency. Their effect on the sync efficiency is complicated and interference. Meanwhile, it is impractical to analyze the effect of each factor on the sync time, and various challenges will be encountered. The previous human-driving configurations are not comprehensive and ineffective from our preliminary experiments.

On the other hand, the recent achievements in ML show that ML-based methods are very effective in computer systems and cloud computing [14,15]. These studies inspire us to propose LearnedSync, which effectively exploits the workload characteristics and environmental conditions to select the sync methods. LearnedSync is committed to dynamically selecting sync methods by sensing the workload and network to obtain the minimum sync time.

3 The Design of LearnedSync

This section first introduces the system overview of LearnedSync and then illustrates in detail the three modules of LearnedSync, including State Monitor, Sync Method Selector and Sync Record Directory.

3.1 System Overview

LearnedSync is a learning-based sync scheme that aims to select the most efficient sync method based on workload characteristics and environmental conditions. It dynamically switches among three methods: full sync, FSC-based delta sync, and CDC-based delta sync to achieve the minimum sync time.

The guiding principle behind LearnedSync is the multilayer perceptron (MLP), one of the most well-known models of artificial neural networks [20]. MLP is well-suited to learning from large datasets, making it ideal for scenarios where a continuous stream of samples can improve prediction accuracy. The cloud sync scenario is particularly well-suited to MLP because: (1) different sync methods are appropriate for different scenarios, so the optimal application scenario of a single sync method can be easily aggregated; (2) files are continuously synchronized from the client to the cloud server, and all kinds of synchronized information constitute a continuous sample flow; and (3) if the sample feature is environmental information and the label is the optimal method, then existing samples can be trained to predict the optimal method of new samples to complete the decision, which is also the responsibility of MLP. We have considered other learning methods but chose MLP for its fast decision-making and lightweight computing. This shows that LearnedSync is not rigidly bound to a particular learning method but is a framework for efficient and low-cost cloud sync.

Figure 5 illustrates the system architecture of LearnedSync on the client side. It comprises three modules: State Monitor, Sync Method Selector, and Sync Record Directory. These modules are integral to the sync process of LearnedSync, which works as follows: the State Monitor first gathers environmental data and conveys it to the Sync Method Selector; Sync Method Selector then utilizes the decision model to identify the proper sync method and carry out the sync; following this, the environmental information and sync time are recorded as training samples in the Sync Record Directory. While the mechanism of LearnedSync is simple to comprehend, an essential consideration for its efficient functioning is that several questions must be answered. We need to investigate how the Sync Method Selector makes informed decisions and predicts the most suitable sync method, as well as how prediction errors may influence the synchronizing of samples. Likewise, as the number of synchronizations grows, the quantity of training samples may impact sync performance. In the rest of the chapter, we will discuss in detail the mechanism of LearnedSync and answer these inquiries.

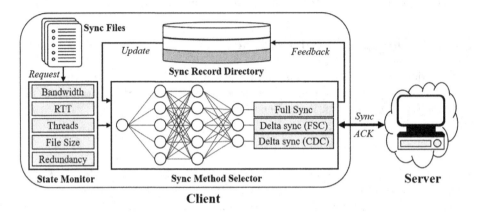

Fig. 5. System architecture of LearnedSync.

It is worth mentioning that the sample format of LearnedSync is octets, which contain seven characteristic values and a label. The sample first obtains five environmental parameters from the State Monitor, then obtains the sync method and sync time from the Sync Method Selector. The label is "unreliable" by default and becomes "reliable" when the sample meets the conditions in the Sync Record Directory.

3.2   State Monitor

Figure 6 shows the structure of the State Monitor. It obtains five environmental parameters: bandwidth, RTTs, the running threads, file size, and redundancy rate between the new and old files. It is necessary to detect the file size and redundancy rate before synchronizing for different files. The file size is easy to obtain, while the redundancy rate must be obtained by comparing the server's old files with the client's new files. To do this, the client should first receive the weak hash value of the old file from the server and then complete the file chunk matching to get the redundancy before sync. Additionally, the device bandwidth and network RTT will be changing dynamically, the former being calculated from the file size ratio to transmission time, the latter being determined by the propagation time of the file. Lastly, it isn't easy to accurately measure CPU utilization as it fluctuates. To better estimate the CPU's working capacity, we measured the number of running threads, as CPU utilization depends on this number.

Specifically, the current number of running threads was considered for single-core devices. For multi-core devices, the ratio of the current number of running threads to the number of CPUs was supposed to reflect the CPU utilization level.

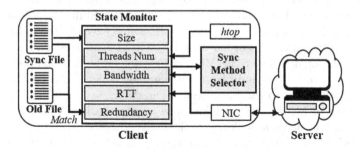

Fig. 6. Structure of State Monitor

The five parameters discussed above are independent of each other. File transfer volume depends on file size and redundancy rate, while network performance depends on bandwidth and RTT. File size indicates the maximum transmission capacity and redundancy rate means the amount of transfer that can be saved. Combining the size and redundancy between an old and a new file is arbitrary and unrelated. Moreover, the redundancy detection and transmission reduction of small files is small compared to the overall transmission time, leading most deduplication schemes to ignore them. Bandwidth is the maximum data transfer rate on a given path, and RTT is the time taken for sending a signal and confirming its receipt. Therefore, there is no correlation between the two. Finally, CPU utilization measures thread execution efficiency and has nothing to do with the other parameters. In conclusion, these five environmental factors are independent and have no relation.

The State Monitor acquires these parameters in real time and transmits them to the Sync Method Selector. It is worth noting that LearnedSync is not restricted to specific environmental parameters; it can even leverage other quantifiable environmental data. In the future, if we find that chunk size, method, or other factors negatively or positively impact sync performance, we can parametrize and factor this into the model. This article, however, focuses on analyzing and evaluating five key parameters that have the most significant effect on sync performance.

3.3   Sync Method Selector

Sync Method Selector periodically learns the reliable samples in the Sync Record Directory (the reliable samples are described in Sect. 3.4) and uses the training model for prediction. It uses MLP to build a four-layer neural network, with five neurons in the input layer, which is used to obtain five environmental parameters; The output layer neuron is three, which is used for the output sync method.

There are also two hidden layers, each with 64 neurons. The activation function is the sigmoid activation function, which trains the sample 2000 times each time. However, since no samples are trained prior to its execution, the prediction model cannot be generated at the outset. Consequently, the Sync Method Selector outputs the sync method randomly until enough synchronizations are completed, and only then is a prediction model generated. Once the State Monitor sends the five-dimensional parameters, the model identifies and executes the optimal sync method. The sync time is recorded once the sync is completed and is essential for training the model. The Sync Method Selector adds the sync method and sync time based on the five-dimensional parameters and submits this seven-dimensional parameter to the Sync Record Directory.

In addition to MLP, we have taken into account various learning methods when exploring the decision-making problem of sync methods, such as K-Nearest Neighbor (KNN) and Support Vector Machines (SVM). KNN operates by classifying objects based on pre-classified data and proximity to the object [21], while LearnedSync's training data is derived from the sync process, making it problematic to provide quick and accurate sync decisions. SVM exhibits efficiency with linear and nonlinear image classification, yet can be problematic when presented with large quantities of training data [22]. In contrast, MLP, which is a neural network with detailed mesh structure features and capable of processing huge datasets, is a favorable choice for LearnedSync. It is important to note, however, that LearnedSync is not confined to certain learning methods and will select and compare more suitable learning methods with the advancement of machine learning to ensure the learning process of LearnedSync is upheld.

3.4   Sync Record Directory

Sync Record Directory accepts and stores samples from the Sync Method Selector. However, these samples cannot be used for training because (1) LearnedSync is not to predict the sync time of each method but to predict the optimal sync method and (2) Sync Method Selector is not always reliable because there may

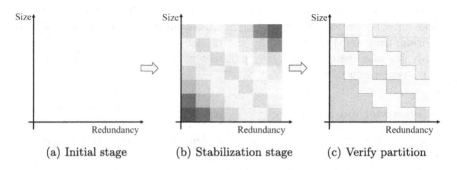

(a) Initial stage   (b) Stabilization stage   (c) Verify partition

Fig. 7. Sync Record Directory continuously collects samples and selects reliable samples from a two-dimensional perspective.

be deviations in the prediction. Therefore, the label of these samples is "unreliable" by default.

Figure 7 illustrates the process of generating reliable samples in a two-dimensional environment. Sync Record Directory first partitions the possible range of environmental parameters. In Fig. 7(a), all partitions are blank since no initial samples exist. In Fig. 7(b), the status of each partition is shown after a certain number of samples have been accumulated. The best methods in the partition are represented by the colors blue, red, and yellow, which stand for full sync, CDC-based delta sync, and FSC-based delta sync, respectively. The performance of the ideal approach is more obvious the darker the partition color is. Thus, it can be found that the optimal method for different partitions has been different, and the partition with the same optimal method has a clustering effect. In Fig. 7(c), the Sync Record Directory calculates the average time of the three sync methods for each partition separately and sets a maximum value α. When the ratio of the average value of the optimal method of a partition to the average value of the suboptimal method (the second least time-consuming method among the three methods) is lower than α, this partition is called the "firm partition" of method M (colored partition); otherwise, it is called the "swing partition" (gray partition). The samples with method M in the firm partition of method M will transform their label into "reliable," while the samples without method M or in the swing partition still need to be verified over time.

The Sync Record Directory regularly calculates the average time of the three methods for all non-blank partitions to increase firm partitions and obtain more samples. Subsequent samples appearing in the firm partition will adopt method M, while the result of model prediction will be used in the swing partition. Over time, the Sync Method Selector learns more and more samples to predict the swing partition more accurately, making the performance of the optimal method gradually generated in the swing partition more apparent, thus converting part of the swing partition into a firm partition. Although it may be challenging to identify the optimal method for certain partitions, utilizing multiple methods for these partitions can result in comparable sync times. Because the optimal method is selected in almost all samples, the objective of LearnedSync to reduce sync time is achieved, as verified in the performance evaluation (Sect. 4).

3.5   The Training Process for LearnedSync

LearnedSync's predictive ability is trained in a random environment. In this environment, we generated files of random sizes and changed the redundancy rate between new and old versions, while bandwidth, network RTT, and the number of additional threads were randomly changed at each sync completion. Their variation ranges were 10 KB–100 MB, 0–100%, 8–96 Mbps, 20–100 ms, and 0–20, respectively. LearnedSync kept synchronizing, accumulating, training samples, and establishing network prediction models.

We have set up 3000 partitions for the Sync Record Directory, each of which records its coverage range, the number of samples in the partition, the average time of the three sync methods, and the sample information of the ten recent synchronizations for each method. This limited sample information is retained as too many samples can hinder training speed and adversely affect the sync method decision-making. A partition with less than 30 samples is considered a blank partition and needs to be synchronized later to ensure each method is chosen ten times. In contrast, the partition is deemed "firm" if the optimal method is 10% lower than the sub-optimal method (i.e., $\alpha = 0.9$) when the sample size reaches 30. For every 1000 files synchronized, the Sync Record Directory retrieves the firm partition, using the samples from the optimal method to complete the training of the prediction model. This operation is conducted in tandem in a separate thread without disrupting the file sync process. In an ideal scenario, the Sync Record Directory preserves the recent 90000 samples and 30000 reliable samples to perform quality training and keep the training period smaller than the sync period for LearnedSync to operate harmoniously.

We assessed the performance of LearnedSync by testing and comparing the time required to synchronize files using the "MechSync" and "Optimal" methods. MechSync is a mechanism-based sync method developed through extensive testing across a total of 32,000 different simulated environments. This testing incorporated eight different file sizes, five redundancy rates, eight bandwidths, ten network RTTs, and ten thread counts, which are included in the LearnedSync training environment previously mentioned. Results were used to establish a linear regression to chart the relationship between sync time and environmental variables. All data was used to develop formulas to determine the appropriate sync method for each environment. We organized the MechSync time prediction formula to choose three sync methods:

$$Time_{FullSync} = 1.179 * Size/Band. + 1.304 * RTT \qquad (1)$$

$$Time_{DeltaSync} = 1.179 * (1 - Rdd.) * Size/Band. + 2.871 * RTT \qquad (2)$$

To produce $T_{FullSync}$ and $T_{DeltaSync}$, the environmental parameters file size, bandwidth, network RTT, and redundancy rate are added. If $T_{FullSync}$ is less than $T_{DeltaSync}$, FullSync is adopted since the predicted sync time is shorter. If $T_{DeltaSync}$ is smaller, the number of running threads is determined. If the number is less than 4.4, the FSCSync should be used; otherwise, the CDCSync is selected. And the "Optimal" sync method endeavors to select the most efficient sync method for each file. To achieve this, we compare the results of FullSync, FSCSync, and CDCSync to identify the method which requires the least amount of time. Thus, Optimal can be seen as a God's perspective method, as it consistently opts for the most efficient solution. The experimental results are shown in Fig. 8.

Firstly, LearnedSync was not as good as MechSync in the initial stage. Figure 8(a) shows that the slope of MechSync and Optimal is almost fixed, and their sync time tend to the expected time of random changes in the environment. However, LearnedSync cannot train samples for prediction at the initial stage and can only randomly select the sync method, thus having the highest initial slope. This was further reflected in the increasing number of "firm" partitions and accuracy from 0 and 33%, respectively, as shown in Fig. 8(b). Secondly,

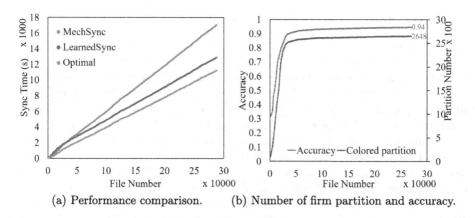

(a) Performance comparison.   (b) Number of firm partition and accuracy.

Fig. 8. Training effect experiment of LearnedSync.

after synchronizing a certain number of files, LearnedSync exceeded the performance of MechSync. LearnedSync accumulated some samples with the label "reliable" and could start to train the model to make decisions. At this point, LearnedSync outperformed MechSync in performance and time consumption with 13,000 and 32,000 files, respectively. This is because the number of "firm partitions" has expanded quickly, generating a rapid improvement in accuracy. Finally, LearnedSync maintained high accuracy after stabilization, and its performance was close to Optimal. The number of "firm" partitions and accuracy eventually reached 2648 and 94% in Fig. 8(b), respectively, with an almost fixed slope similar to that of Optimal in Fig. 8(a), showcasing that LearnedSync has achieved satisfactory training results, which can be maintained continuously in subsequent sync processes.

4   Performance Evaluations

This section describes the prototype implementation of LearnedSync, followed by the experimental setup and methodology. Finally, we present trace-driven performance results for different sync methods and sensitivity analysis.

4.1   Prototype Implementation

The LearnedSync prototype is implemented based on FastCDC [23] and hashmap. We use Python code to link the C codes of the three sync methods and establish an MLP training model. After the network connection is established, the SmokePing [24] module embedded in the network-aware dynamic threshold module starts to monitor the network RTT, the *nload* command monitors the network bandwidth and *htop* command monitors the running threads.

4.2 Experimental Setup and Methodology

The evaluations are performed on two cloud servers with the same specifications: Intel Xeon (Ice Lake) Platinum 8369B 2.7GHz quad-core processor and 16GB memory. Their operating system is Ubuntu 20.04 (Linux kernel version 5.15). By default, the client bandwidth is 98.1Mbps, and the network RTT is 20ms, which is dynamically changed. We use the *tc* command to adjust the network bandwidth and RTT between the client and server and the *stress* command to add additional worker threads to obtain various environments. We use scripts to adjust the network bandwidth and RTT periodically to change the network environments dynamically. We compare LearnedSync with full sync (we created in C language), rsync-3.2.3 (FSC-based delta sync), Seafile (CDC-based delta sync), and the state-of-the-art PandaSync. For rsync, We set the chunk size to 4KiB. For Seafile, we use variable chunk sizes ranging from 2KiB to 32KiB, with an average of 8KiB.

4.3 Performance Results and Analysis

Trace-Driven Experiments: We collect the sync files from publicly available websites, such as Linux kernel [25] and GitHub [26]. Table 2 shows the characteristics of these trace-driven experiments in terms of initial and updated version and volume. We conducted experiments to measure the sync time of six tracking drivers using multiple cloud sync methods under fluctuating bandwidth, network RTT, and the number of additional threads ranging from 8-96Mbps, 20–100ms, and 0–20, respectively. Each experiment was run at least three times, and we report the average cloud sync time in Fig. 9. We separated the sync time and distinguished it with different colors since some of the experimental methods are single sync methods or mixed sync methods. In the bar graph, blue, yellow, and red show the sync time used with full sync, FSC-based delta sync, or CDC-based delta sync.

Table 2. The workload characteristics of the six traces.

Trace Name	Initial Version	Initial Volume	Update Version	Update Volume
OpenCV	4.6.0	193.3 MB	4.7.0	198.7 MB
Seafile	9.0.9	235.4 MB	9.0.10	239.7 MB
Tensorflow	2.10.0	260.6 MB	2.10.1	260.7 MB
Linux-4.9	4.9.2	666.2 MB	4.9.336	669.4 MB
Linux-5.8	5.8.13	947.8 MB	5.8.14	947.8 MB
Linux-6.1	6.1.1	1316.6 MB	6.1.3	1316.8 MB

Our results show that LearnedSync consistently achieves the minimum sync time compared to multiple methods, with a speed of at least 41.4% faster than

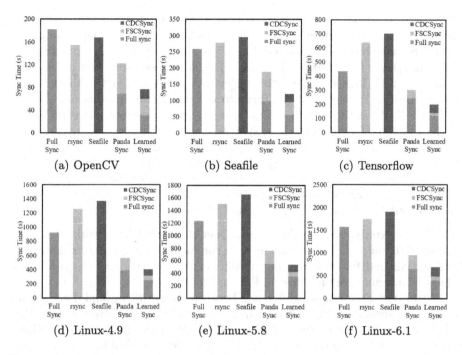

Fig. 9. The cloud sync times under the six traces for the different cloud sync methods.

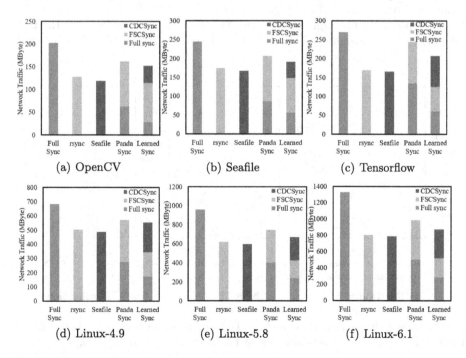

Fig. 10. The total data volume transferred over the network driven by the six traces for the different cloud sync methods.

state-of-the-art PandaSync. FullSync performed well in the last four experiments due to the large proportion of small files, which makes delta sync more time-consuming. We have also discovered that when selecting full sync, LearnedSync's sync time is lower than that of PandaSync. This is because PandaSync is designed to operate with higher bandwidth and cannot adapt to bandwidth changes, whereas LearnedSync effectively adjusts to varying network conditions.

Figure 10 shows the total amount of data transferred over the network driven by the six traces for different cloud sync schemes. We observed that rsync and Seafile significantly reduce sync traffic by 33.9% and 36.8%, respectively, primarily due to the optimization of large files, as small files often exceed their traffic after delta sync. In contrast, the sync traffic of LearnedSync was 9.6% lower than that of PandaSync, since LearnedSync chooses less full sync, and the sync traffic of delta sync is also lower than that of PandaSync.

Sensitivity Analysis: To evaluate the impact of bandwidth, network RTT, and number of threads on sync performance, we conducted sensitivity experiments on LearnedSync and PandaSync using six different traces with varying parameter values. The results are presented in Fig. 11.

Firstly, the performance difference between LearnedSync and PandaSync remains consistent as the bandwidth increases. This demonstrates that the unique CDC-based delta sync of LearnedSync significantly reduces the time expense. Secondly, with the increase in RTT, the performance difference between LearnedSync and PandaSync gradually increases. PandaSync assumes that the number of delta sync will increase with the rise of RTT, causing more overhead when the RTT is high [13]. Finally, the performance gap between the two schemes reduces gradually with an increase of unrelated threads. This is because

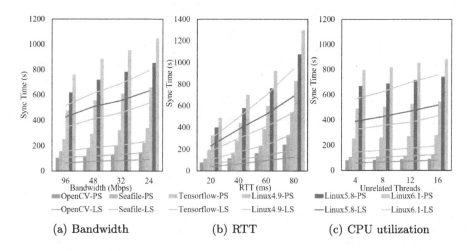

(a) Bandwidth   (b) RTT   (c) CPU utilization

Fig. 11. Sensitivity analysis of bandwidth, RTT, and CPU utilization. *-PS* and *-LS* in the legend indicate that the PandaSync or LearnedSync is used for testing.

the CDC-based delta sync in LearnedSync is computationally expensive, making it more vulnerable to changes in CPU utilization than PandaSync.

5 Related Work

Synchronization is a vital and indispensable technique for cloud storage services, enabling clients to keep their local files synchronized with those stored in the data centers in the cloud. Existing cloud sync schemes can be classified as full sync or delta sync, with delta sync being the most widely studied [13]. Delta sync only transmits the altered parts of the file instead of synchronizing the entire file, thereby decreasing sync time and conserving bandwidth during the sync process.

There is already a range of delta sync algorithms, such as Vcdiff [27] and rsync [28], which were among the first and provided an essential foundation for their subsequent development. More and more cloud storage providers have adopted delta sync, including Dropbox, which was the first to add this for PC-based file sync, and Seafile [29], which uses an approach known as Content-Defined Chunking (CDC). Furthermore, various studies have enabled improvements to rsync, including DeltaCFS [6], WebRsync [7], and PandaSync [13] - DeltaCFS presents a new framework for cloud storage services that minimize client and server overhead while still maintaining network efficiency, while WebRsync was the initial attempt at delta sync on the web browser and was further improved by WebR2sync+ as a viable solution for cloud storage services; PandaSync balances the advantages of full and delta sync depending on file size and round-trip-time.

Other studies based on delta synchronizations, such as QuickSync [5], Dsync [8], and FeatureSync [30], leverage the advantage of CDC to reduce sync times. QuickSync is the first model to analyze the sync efficiency of mobile cloud storage services [5]. It optimizes bandwidth utilization by employing an adaptive partitioning and deduplication strategy based on real-time network conditions. Dsync introduces a new weak hash, FastFp, to streamline the chunk-matching process and revamps the client-server communication protocol to minimize computational overhead and network traffic [8]. FeatureSync, a feature-based encryption sync system, is characterized by its lightweight matching algorithm. All the investigations and studies of various cloud storage services reaffirm that no one cloud sync method suffices for all users [30].

Some studies have also extensively explored cloud storage services, examining various sync services' effectiveness and potential optimizations. Li et al. developed the CloudCmp system comparator to evaluate the performance and cost of cloud service providers; they found that many of the major providers had variations in implementation [31]. Drago et al. then designed a method to monitor the Dropbox cloud storage traffic and proposed countermeasures to enhance its performance [32]. Additionally, they researched and compared five cloud storage services, uncovering their various system architectures and functions and pointing out potential areas for improvement [33]. Li et al. then proposed the UDS update batch delayed sync mechanism [34] and the TUE (traffic usage efficiency) metric [10], which is applied for research and characterization of 6 popular cloud storage services to reduce overhead and boost file sync speed.

Recent advances in machine learning show that learning-based approaches are much more viable and effective than human-driven ones. Zhang et al., with the AutoSync end-to-end pipeline, provided automated optimization of sync strategies given the model architecture and resource specifications, achieving significantly better results than manual optimization strategies [14]. Li et al. used reinforcement learning in an IoT device scanning system, demonstrating that the system was able to detect more IP-device changes than random and sequential scanning methods [35]. Laskaridis et al. proposed SPINN, a distributed inference system that deploys devicecloud computing and progressive inference, enabling rapid and robust Convolutional Neural Network (CNN) inference in various settings [36]. Notably, SPINN features a scheduler that co-optimizes the early-exit policy and CNN division during runtime, allowing adaptation to fluctuating conditions and meeting user-defined service-level requirements.

In conclusion, existing studies show that the success of a sync scheme is contingent on both the workload characteristics and the network conditions. To this end, we propose LearnedSync, a learning-based sync scheme that selects the most advantageous sync method using machine learning models. This approach lets us consider dynamic workload characteristics and current network conditions, significantly reducing sync latency. To our knowledge, this is the first study to apply machine learning to cloud sync procedures for improved efficiency.

6   Conclusion

To ensure data remains accurately synchronized between the device and cloud storage, we must strive for optimum sync performance. This paper introduces LearnedSync, an ML-based sync scheme that can dynamically select one of three cloud sync methods - full sync, FSC-based delta sync, and CDC-based delta sync - based on environmental conditions by building an MLP network model with sync records. This helps to improve the cloud sync efficiency and shorten the sync time. Moreover, LearnedSync records each environment parameter, the method used, and the sync time after the completion of every file sync so that it continues to learn how best to synchronize data between the device and the cloud storage. Experiments on the lightweight prototype LearnedSync demonstrate that it boosts the sync performance better than other cloud sync methods.

LearnedSync is an ongoing research project with two main future directions: (1) studying the impact of other factors on cloud sync efficiencies, such as file modification and wireless networks, which can have a significant impact on sync performance; and (2) exploring other ML methods, such as deep reinforcement learning, to improve the learning process and accuracy.

Acknowledges. This work was supported in part by the National Natural Science Foundation of China under Grants U22A2027 and 61972325, in part by the Open Project Program of Wuhan National Laboratory for Optoelectronics under Grant

2021WNLOKF011, and in part by the Research Project of Zhejiang Lab under Grant 2021DA0AM01/002, Key Research and Development (Digital Twin) Program of Ningbo City under Grant No. 2023Z219, and Young Tech Innovation Leading Talent Program of Ningbo City under Grant No.2023QL008.

References

1. Six Cloud Computing Trends for 2022 (and Beyond) (2022). https://phoenixnap.com/blog/cloud-computing-trends

2. Pan, T., et al.: Sailfish: accelerating cloud-scale multi-tenant multi-service gateways with programmable switches. In: Proceedings of the ACM SIGCOMM 2021 Conference (2021)

3. Abebe, M., Daudjee, K., Glasbergen, B., Tian, Y.: EC-store: bridging the gap between storage and latency in distributed erasure coded systems. In: Proceedings of the 38th IEEE International Conference on Distributed Computing Systems (2018)

4. Singh, A.K., Cui, X., Cassell, B., Wong, B., Daudjee, K.: MicroFuge: a middleware approach to providing performance isolation in cloud storage systems. In: Proceedings of the IEEE 34th International Conference on Distributed Computing Systems (2014)

5. Cui, Y., Lai, Z., Wang, X., Dai, N., Miao, C.: QuickSync: improving synchronization efficiency for mobile cloud storage services. In: Proceedings of the 21st Annual International Conference on Mobile Computing and Networking (2015)

6. Zhang, Q., et al.: DeltaCFS: boosting delta sync for cloud storage services by learning from NFS. In: Proceedings of the 37th IEEE International Conference on Distributed Computing Systems (2017)

7. Xiao, H., et al.: Towards web-based delta synchronization for cloud storage services. In: Proceedings of the 16th USENIX Conference on File and Storage Technologies (2018)

8. He, Y., et al.: Dsync: a lightweight delta synchronization approach for cloud storage services. In: Proceedings of the 36th Symposium on Mass Storage Systems and Technologies (2020)

9. Wu, S., et al.: FASTSync: a FAST delta sync scheme for encrypted cloud storage in high-bandwidth network environments. ACM Trans. Storage (2023)

10. Li, Z., et al.: Towards network-level efficiency for cloud storage services. In: Proceedings of the 14th Internet Measurement Conference (2014)

11. Zhang, S., Catanese, H., Wang, A.: The composite-file file system: decoupling the one-to-one mapping of files and metadata for better performance. In: Proceedings of the 14th USENIX Conference on File and Storage Technologies (2016)

12. Meyer, D.T., Bolosky, W.J.: A study of practical deduplication. In: Proceedings of the 9th USENIX Conference on File and Storage Technologies (2011)

13. Wu, S., Liu, L., Jiang, H., Che, H., Mao, B.: PandaSync: network and workload aware hybrid cloud sync optimization. In: Proceedings of the 39th IEEE International Conference on Distributed Computing Systems (2019)

14. Zhang, H., Li, Y., Deng, Z., Liang, X., Carin, L., Xing, E.P.: AutoSync: learning to synchronize for data-parallel distributed deep learning. In: Proceedings of the 34th Annual Conference on Neural Information Processing Systems (2020)
15. Tang, Y., Lu, H., Li, X., Chen, L., Yuan, M., Zeng, J.: Learning-aided heuristics design for storage system. In: Proceedings of the International Conference on Management of Data (2021)
16. Wang, Z., et al.: DeepScaling: microservices AutoScaling for stable CPU utilization in large scale cloud systems. In: Proceedings of the 13th Symposium on Cloud Computing (2022)
17. Miyazawa, K., Yamaguchi, S., Kobayashi, A.: Mechanism of cyclic performance fluctuation of TCP BBR and CUBIC TCP communications. In: Proceedings of the 44th IEEE Annual Computers, Software, and Applications Conference (2020)
18. Sackl, A., Casas, P., Schatz, R., Janowski, L., Irmer, R.: Quantifying the impact of network bandwidth fluctuations and outages on Web QoE. In: Proceedings of the 7th International Workshop on Quality of Multimedia Experience (2015)
19. Dang, T., Mohan, N., Corneo, L., Zavodovski, A., Ott, J., Kangasharju, J.: Cloudy with a chance of short RTTs: analyzing cloud connectivity in the Internet. In: Proceedings of the 21st Internet Measurement Conference (2021)
20. Meyer, B.H., Zola, W.M.N.: Towards a GPU accelerated selective sparsity multi-layer perceptron algorithm using K-nearest neighbors search. In: Workshop Proceedings of the 51st International Conference on Parallel Processing (2022)
21. Chern, F., Hechtman, B., Davis, A., Guo, R., Majnemer, D., Kumar, S.: TPU-KNN: K nearest neighbor search at peak FLOP/s. In: Advances in Neural Information Processing Systems (2022)
22. Lv, S., Wang, J., Liu, J., Liu, Y.: Improved learning rates of a functional lasso-type SVM with sparse multi-Kernel representation. In: Advances in Neural Information Processing Systems (2021)
23. Xia, W., et al.: FastCDC: a fast and efficient content-defined chunking approach for data deduplication. In: Proceedings of the 13th USENIX Annual Technical Conference (2016)
24. SmokePing (2018). https://oss.oetiker.ch/smokeping/
25. Linux Kernel Archive (2022). https://www.kernel.org/
26. Github (2022). https://github.com/
27. Korn, D.G., Vo, K.: Engineering a differencing and compression data format. In: Proceedings of the 2002 USENIX Annual Technical Conference (2002)
28. RSYNC Open Source Utility (2022). https://rsync.samba.org/
29. Seafile (2022). https://www.seafile.com/en/home
30. Wu, S., Tu, Z., Wang, Z., Shen, Z., Mao, B.: When delta sync meets message-locked encryption: a feature-based delta sync scheme for encrypted cloud storage. In: Proceedings of the 41st IEEE International Conference on Distributed Computing Systems (2021)
31. Li, A., Yang, X., Kandula, S., Zhang, M.: CloudCmp: comparing public cloud providers. In: Proceedings of the 10th ACM SIGCOMM Internet Measurement Conference (2010)
32. Drago, I., Mellia, M., Munafò, M.M., Sperotto, A., Sadre, R., Pras, A.: Inside dropbox: understanding personal cloud storage services. In: Proceedings of the 12th ACM SIGCOMM Internet Measurement Conference (2012)
33. Drago, I., Bocchi, E., Mellia, M., Slatman, H., Pras, A.: Benchmarking personal cloud storage. In: Proceedings of the 13th Internet Measurement Conference (2013)

34. Li, Z., et al.: Efficient batched synchronization in dropbox-like cloud storage services. In: Proceedings of the ACM/IFIP/USENIX 14th International Middleware Conference (2013)
35. Qu, J., et al.: Landing reinforcement learning onto smart scanning of the Internet of Things. In: Proceedings of the IEEE Conference on Computer Communications (2022)
36. Laskaridis, S., Venieris, S.I., Almeida, M., Leontiadis, I., Lane, N.: SPINN: synergistic progressive inference of neural networks over device and cloud. In: Proceedings of the The 26th Annual International Conference on Mobile Computing and Networking (2020)

Optimizing CSR-Based SpMV on a New MIMD Architecture Pezy-SC3s

Jihu Guo[1,2], Jie Liu[1,2(✉)], Qinglin Wang[1,2], and Xiaoxiong Zhu[1,2]

[1] Laboratory of Digitizing Software for Frontier Equipment,
National University of Defense Technology, Changsha 410073, China
[2] National Key Laboratory of Parallel and Distributed Computing,
National University of Defense Technology, Changsha 410073, China
{guojihu,liujie}@nudt.edu.cn

Abstract. Sparse matrix-vector multiplication (SpMV) is extensively used in scientific computing and often accounts for a significant portion of the overall computational overhead. Therefore, improving the performance of SpMV is crucial. However, sparse matrices exhibit a sporadic and irregular distribution of non-zero elements, resulting in workload imbalance among threads and challenges in vectorization. To address these issues, numerous efforts have focused on optimizing SpMV based on the hardware characteristics of computing platforms. In this paper, we present an optimization on CSR-Based SpMV, since the CSR format is the most widely used and supported by various high-performance sparse computing libraries, on a novel MIMD computing platform Pezy-SC3s. Based on the hardware characteristics of Pezy-SC3s, we tackle poor data locality, workload imbalance, and vectorization challenges in CSR-Based SpMV by employing matrix chunking, applying Atomic Cache for workload scheduling, and utilizing SIMD instructions during performing SpMV. As the first study to investigate SpMV optimization on Pezy-SC3s, we evaluate the performance of our work by comparing it with the CSR-Based SpMV and SpMV provided by Nvidia's CuSparse. Through experiments conducted on 2092 matrices obtained from SuiteSparse, we demonstrate that our optimization achieves a maximum speedup ratio of x17.63 and an average of x1.56 over CSR-Based SpMV and an average bandwidth utilization of 35.22% for large-scale matrices ($nnz \geq 10^6$) compared with 36.17% obtained using CuSparse. These results demonstrate that our optimization effectively harnesses the hardware resources of Pezy-SC3s, leading to improved performance of CSR-Based SpMV.

Keywords: SpMV · Optimization · CSR-Based SpMV · Pezy-SC3s

1 Introduction

Sparse matrix-vector multiplication (SpMV) is a fundamental operation denoted by $y = Ax$, where A is a sparse matrix and x and y are dense vectors. SpMV finds widespread application in high-performance computing domains such as graph

Z. Tari et al. (Eds.): ICA3PP 2023, LNCS 14488, pp. 22–39, 2024.
https://doi.org/10.1007/978-981-97-0801-7_2

analysis, machine learning, deep learning, and more. Its significance extends to solving sparse linear systems, eigenvalue systems, Krylov subspace methods, and similar problems [5,9–11,21,23,26,27,31]. Given the impact of SpMV performance on these domains, improving the efficiency of sparse matrix-vector multiplication becomes crucial, as SpMV often dominates the computational overhead of related tasks.

However, optimizing SpMV poses significant challenges. Sparse matrices exhibit a sparse and irregular distribution of non-zero elements, with different matrices having distinct sparse patterns. Consequently, unbalanced workload distribution among threads and difficulty in vectorization. Numerous studies have introduced well-designed sparse matrix storage formats, including ELL [4], DIA, and BCSR [8], among others, tailored to different sparse patterns. Workload balancing approaches such as CSR5 [17], Merge-based CSR [19], yaSpMV [35], and HCC [16] have also been proposed to address workload imbalance between threads. Additionally, methods like CVR [33], SpV8 [14], ELL-R [28], ELLR-T [29], BiELL [36], have tackled the issue of limited vectorization. These endeavors have predominantly focused on CPU and GPU platforms. However, this paper presents SpMV optimization on a novel MIMD platform called Pezy-SC3s, as illustrated in Fig. 1. Pezy-SC3s comprises two Prefectures, each equipped with an Atomic Cache and an HBM2 module. The Atomic Cache facilitates atomic operations on internal data, and Table 1 presents the specific parameters associated with the Atomic Cache. Max Operations/Clock column refers to the maximum 32 operations per clock cycle that can be maintained when the number of operations performed by the atomic cache in each clock cycle is not greater than this value. Otherwise, the performance will drop dramatically.

Each Prefecture encompasses 16 cities, and each city contains 4 villages. Within each city, there exists an L2 Cache that is shared by 128 threads.

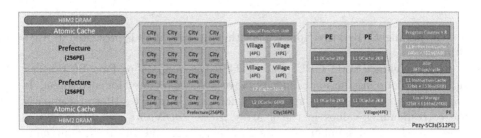

Fig. 1. Pezy-SC3s Block Diagram

Table 1. Parameters of Atomic Cache

Size (Chip Total)	WAY	Line size	Max Operations/Clock
1 KB (16 KB)	4	256 B	32

Additionally, a Special Function Unit is present in each city to handle division, modulo, and square root operations. Within each Village, there are 4 Processing Elements (PEs), each accompanied by an L1 Cache. The PEs house a Local Storage of up to 24 KB, and the size of the Local Storage can be adjusted by modifying the stack space size of threads. Furthermore, Pezy-SC3s supports 128-bit SIMD instructions.

We have selected CSR-based SpMV as the fundamental basis for our optimization approach. There are several reasons for this choice. Firstly, no SpMVs have been specifically designed for Pezy-SC3s thus far. Secondly, CSR-based SpMV has been extensively utilized and established as benchmarks in numerous previous studies [14,17,19,33,35]. Lastly, CSR-based SpMV is also the supported SpMV in various high-performance scientific sparse computing libraries such as MKL [30] and CuSparse [3]. Please refer to Fig. 2 for an illustration of the CSR format. *rowDelimiters* stores the indexes range of non-zero elements of rows, and *colIdx* and *vals* store the corresponding column indexes and values of the non-zero elements.

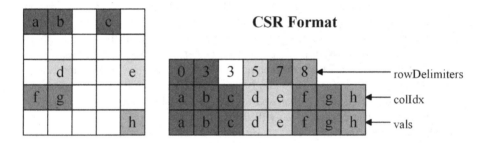

Fig. 2. CSR Format

The parallel CSR-Based SpMV algorithm on Pezy-SC3s is presented in Algorithm 1.

Unlike Nvidia's SIMT GPUs, parallel CSR-Based SpMV on Pezy-SC3s necessitates a for-loop instruction to specify the operations required for each thread. In Fig. 1, each thread handles computations for a single row at a time and updates the corresponding values in the y array. Here, *rowNums* and *threadNums* represent the total number of rows in the matrix and the maximum number of threads supported by the hardware platform, respectively. However, this straightforward parallelization approach can result in workload imbalance among threads and limited data reuse in the cache, especially when the number of threads (*threadNums*) becomes excessively large, leading to row-based workload imbalance [20]. Furthermore, the parallel CSR-Based SpMV primarily focuses on inter-row parallelism, with all scalar operations performed within each row.

To address these issues, we optimize the CSR-Based SpMV based on hardware characteristics of Pezy-SC3s. Firstly, we introduce matrix chunking to

Algorithm 1. Parallel CSR-Based SpMV on Pezy-SC3s

Input: x, rowDelimiters, colIdx, vals;
Output: y;
1: gid = getgid(); // Get thread id.
2: **for** i = gid; i < rowNums; i += threadNums **do**
3:     sum = 0.0;
4:     **for** j = rowDelimiters[i]; j < rowDelimiters[i+1]; j ++ **do**
5:         sum += x[colIdx[j]] × vals[j];
6:     y[i] = sum;
7: flush_L2(); // Write back y by flush_L2.
8: **return** y;

---

enhance data locality. The number of matrix chunks is determined through careful testing of various chunking parameters. Additionally, we utilize the Atomic Cache to dynamically guide threads in fetching the next row to be computed, thereby achieving intra-block workload balance. Finally, we incorporate 128-bit SIMD instructions within each row to improve vectorization. By fully leveraging the hardware capabilities of Pezy-SC3s, we effectively mitigate issues related to poor data locality, workload imbalance, and difficult vectorization in CSR-Based SpMV.

Through extensive experiments conducted on 2092 matrices obtained from SuiteSparse [6], our optimization demonstrates remarkable results. We achieve a maximum speedup ratio of x17.63 compared to the CSR-Based SpMV. For small-scale matrices ($nnz < 10^6$), we achieve a maximum speedup ratio of x10.84 and an average speedup ratio of x1.24. Similarly, for large-scale matrices ($nnz \geq 10^6$), we achieve a maximum speedup ratio of x17.63 and an average speedup ratio of x2.97 (average speedup on the total 2092 matrices is x1.57). Moreover, our average bandwidth utilization on large-scale matrices reaches 35.22%, which can further increase to 62.54% when the matrix size becomes larger ($nnz \geq 10^7$). These figures demonstrate a satisfactory performance compared with the corresponding figures of 36.17% and 45.56% obtained using the Nvidia CuSparse library, highlighting the effective utilization of Pezy-SC3s hardware resources in our approach.

2   Related Work

Sparse matrices typically contain a small percentage (usually only 1%) of non-zero elements compared to the total number of elements in the matrix. To conserve space and enhance operation efficiency, we commonly utilize sparse matrix formats to store and perform operations solely on the non-zero elements. However, the distribution of these non-zero elements often exhibits irregular patterns. As a result, significant variations in the number of non-zero elements between rows arise, leading to issues such as unbalanced load across operating units and challenging vectorization during SpMV computations.

Various types of sparse matrices exhibit distinct sparse patterns, prompting numerous studies to propose novel storage formats tailored to different patterns. For instance, the ELL format compresses all elements to the left and employs two-dimensional arrays to store the values and column coordinates of the non-zero elements. The row coordinate in the array corresponds to the y coordinate for writing, and the length of each row depends on the maximum number of non-zero elements in the matrix. The DIA format arranges non-zero elements in a diagonal structure, using an $OFFSET$ array to record the offset of each column relative to the main diagonal. BCSR, on the other hand, adopts a blocked CSR format, storing non-zero elements in matrix blocks. It employs two arrays to track row offsets and column coordinates of these blocks, utilizing dense matrix operations within the blocks.

Efforts to address workload imbalance between threads have also been explored. CSR5 [17], for example, chunks the matrix into tiles based on the CSR format to ensure that each block contains an equal number of non-zero elements. These blocks are then evenly distributed among the threads. Merge-based CSR [20] not only prioritizes balancing the number of non-zero elements but also considers the load associated with writing back to the y array. It accomplishes this by merging the $rowDelimiters$ array of CSR into the $colIdx$ array, which is subsequently divided equally among threads to achieve workload balance. Another approach employed in yaSpMV [35] involves chunking matrices in a row- and column-insensitive manner, resulting in significant compression ratios. HCC [16] utilizes fixed-distance chunking in the column direction and further chunks based on the number of non-zero elements in the row direction. This strategy ensures both data locality of x and workload balance.

To tackle the challenge of insufficient vectorization, various techniques have been proposed. CVR [33] schedules tasks to each thread based on row granularity, enabling full utilization of SIMD instructions. Other works [15,34] also aim at the full utilization of SIMD components. SpV8 [14] separates vectorization and scalarization, maximizing vectorization operations for aligned non-zero elements and resorting to scalarization for non-aligned elements, thereby achieving enhanced vectorization. BiELL [36], an extension of the ELL format, further compresses ELL by employing folding techniques to avoid excessive padding resulting from excessively long lines with numerous zero elements.

These advancements have predominantly focused on CPU or GPU architectures. However, in this paper, we aim to optimize the widely used CSR-Based SpMV specifically for the Pezy-SC3s, a MIMD computing platform. Our goal is to address challenges associated with poor data localization, workload imbalance, and insufficient vectorization encountered during SpMV computations in the CSR-Based SpMV on Pezy-SC3s and fully utilize the MIMD computing resources.

Algorithm 2. Optimized Parallel CSR-Based SpMV on Pezy-SC3s

Input: x, rowDelimiters, colIdx, vals, blockDelimiters;
Output: y;
1: GLOBAL nextRowIdx[4096];
2: blockSize = 64;
3: gid = getgid(); // Get thread id.
4: beginRow = blockDelimiters[gid / blockSize], endRow = blockDelimiters[gid / blockSize + 1];
5: i = beginRow + gid % blockSize;
6: pz_atomic_store(&nextRowIdx[gid / blockSize × 64], beginRow + blockSize);
7: **for** i **do** = 0; i < endRow; i = pz_atomic_inc(&nextRowIdx[gid / blockSize]
8:     sum2 = {0.0, 0.0};
9:     eleBegin = rowDelimiters[i], eleEnd = rowDelimiters[i + 1];
10:    span = (eleEnd - eleBegin) & ~ 1 ;
11:    tail = (eleBegin $^\wedge$ eleEnd) & 1;
12:    **for** j **do** = eleBegin; j < eleEnd; j ++
13:        vv = {vals[j], vals[j+1]};
14:        vx = {x[colIdx[j]], x[colIdx[j+1]]};
15:        sum2 += vv × vx;
16:    y[i] = sum2.x + sum2.y + tail × (x[colIdx[eleEnd-1]] × vals[eleEnd-1]);
17: flush_L2(); // Write back y by flush_L2.
18: **return** y;

---

3   Methodology

SpMV is an operation that is memory-intensive and constrained by limited bandwidth. The inherent sparsity of the matrices introduces workload imbalance, limiting the efficient utilization of computing resources. Furthermore, the irregular distribution of non-zero elements in sparse matrices presents an additional challenge in achieving optimal hardware performance. The irregularity hinders the full vectorization of the operation [1,2,7,12,18,22]. To address these challenges, we propose optimizations for CSR-Based SpMV for the Pezy-SC3s platform. Our approach involves chunking the matrix based on row granularity to enhance data locality, utilizing the Atomic Cache for efficient workload balance, and leveraging SIMD instructions to improve intra-row computational parallelism. The optimized algorithm is presented in Algorithm 2, and detailed explanations of the algorithm will be provided in the subsequent subsections.

3.1   Row Granularity Matrix Partition

Based on the hardware architecture of Pezy-SC3s, it is observed that there is no Last Level Cache (LLC). Instead, data is delivered directly to the L2 Cache through HBM2 when required by the PEs. As a result, the L2 Cache stores almost all the reusable data among threads. To improve data locality in the L2 Cache, we partition the matrix into chunks to make threads that share the same

L2 Data Cache to compute the adjacent rows, enabling threads to collaborate effectively.

The performance is tested with various matrix chunking numbers, and we select af_1_k101 and $F1$ matrices which have the sufficient number of non-zero elements to test the performance of SpMV to plot the results in Fig. 3. The findings demonstrate that SpMV performance improves with an increase in the number of matrix chunkings but sharply drops when the number becomes excessively high. The optimal performance is achieved with 64 chunkings, with 64 threads assigned to each chunking, respectively. These threads share the L2 Cache. Since the best performance is observed with 64 chunkings, it is chosen as the chunking number (Algorithm 2, line 2).

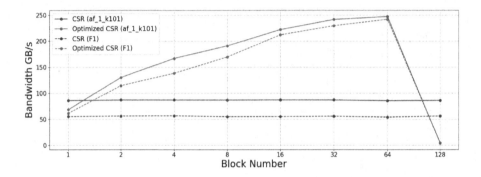

Fig. 3. Bandwidth obtained in different number of matrix chunks

During chunking, the number of non-zero elements in each row is determined using the $rowDelimiters$ array. The matrix is then divided into 64 blocks using row granularity division, aiming to ensure an approximately equal total number of non-zero elements in each block. Based on the $rowDelimiters$, the total number of non-zero elements in the range is calculated as $rowDelimiters[numRows] - rowDelimiters[0]$. The division point, denoted as m, corresponds to the maximum coordinate to make $rowDelimiters[m] - rowDelimiters[0]$ less than or equal to $(rowDelimiters[numRows] - rowDelimiters[0])/2$. Using m, the original range is divided into $[0, m)$ and $[m, numRows)$. The chunk with a higher number of non-zero elements, assuming it to be $[m, numRows)$, is further partitioned, and this step is repeated until 64 segments are obtained. Subsequently, each segment is assigned to the corresponding thread block in order. The chunking process is illustrated in Fig. 4. Finally, an array of $blockDelimiters$ with a length of 65 is obtained, where $blockDelimiters[i]$ represents the starting row of the matrix block assigned to the ith thread group, and $blockDelimiters[i + 1]$ represents the ending row of the matrix block assigned to the ith thread group (line 3 of Algorithm 2).

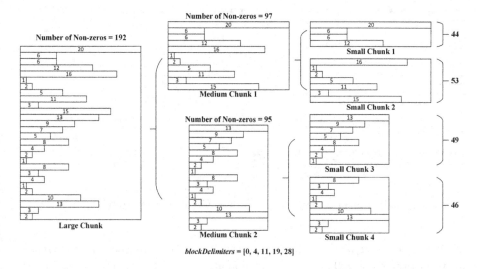

Fig. 4. In this matrix partition process, we first divide the matrix (Large Chunk) with 28 rows into 2 Medium Chunks. Then, divide Medium Chunk 1 since Medium Chunk 1 has more non-zero elements ($97 \geq 95$). Finally, divide Medium Chunk 2.

3.2   Workload Balance Within Matrix Chunks

During the chunking phase, the matrix is divided based on row granularity, with each chunk consisting of complete rows. When a thread is assigned a single row, it can directly write back the result once its computation is complete. However, when multiple threads are responsible for a row, they must wait for each other to finish their computations before performing the write-back. In this case, only one thread needs to write back, but multiple threads have to wait for this write-back operation. On Pezy-SC3s with 4096 threads, using multiple threads to compute a row introduces significant delays as threads wait for each other. Conversely, assigning a single thread to a row causes load imbalance within the Chunk. To address this issue, we utilize the atomic cache hardware in Pezy-SC3s, which has high performance and supports atomic operations, for single thread computing single row workload balance.

In Algorithm 1, *threadNums* instructs the next row to be computed by a thread. For example, with 4096 threads, thread i would compute lines i, $i+4096$, $i+4096 \times 2$, and so on. This approach results in workload imbalance, as depicted in Fig. 5(a). The red thread is assigned most of the total tasks, while the blue and yellow threads are assigned only a small percentage. Consequently, the yellow, blue, and green threads spend considerable time waiting for the red thread, leading to the underutilization of computing resources. By leveraging Atomic Cache, we can guide the computation order for each group of threads, ensuring a roughly equal distribution of tasks among threads and reducing the waiting time between them, thereby improving workload balance.

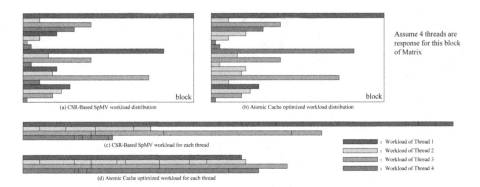

Fig. 5. Workload Assignment by Atomic Cache. (Color figure online)

The array $nextRowIdx$ (line 1 of Algorithm 2) is an int ($sizeof(int)$ = 4Byte) array of size 4096. It stores a value at every 64 locations, corresponding to the cache line size of the atomic cache (256Byte). This arrangement ensures that the values used by each thread group are stored in a single cache line, and the cache line is exclusively accessed and modified by the threads in the same group, enhancing the performance of atomic operations. The value stored in $nextRowIdx$ guides the threads in a thread group to the next line they should operate on. For instance, if the group size is 64, there will be 4096/64=64 threads in each group. The value in $nextRowIdx$ is loaded into the atomic cache using the pz_atomic_store instruction (line 6 of Algorithm 2), and each thread increments the value when it fetches the value. The fetch-increment operation is achieved through $pz\_atomic\_inc(\&nextRowIdx[gid/blockSize])$, which returns the value stored in $nextRowIdx[gid/blockSize]$ and adds 1 to it in the atomic cache (Algorithm 2, line 6).

The atomic add operation on Pezy-SC3s' atomic cache differs from that on Nvidia GPUs. Pezy-SC3s' Atomic Cache delivers exceptionally high performance, capable of executing up to 32 operations in a single clock (refers to Table 1 for details). Each operation retrieves the $nextRowIdx$ for one thread, enabling the thread to identify the corresponding row for computation. Moreover, since the distribution of elements within each row of the sparse matrix is typically uneven, the time required to complete a row varies among most threads in the group. Consequently, it is uncommon for all threads to wait together for the atomic cache to compute $nextRowIdx$. This dynamic scheduling of thread tasks effectively achieves workload balance among threads with row-granularity partitioning, as illustrated in Fig. 5(d).

Moreover, it is also essential to minimize the $Interval$ between threads, which refers to the number of rows between any two threads, since increasing the $Interval$ leads to reduced data reuse between threads, as depicted in Fig. 6.

In this figure, assuming there are 4 threads processing a matrix chunk and the workload of each row equals to x2 L2 Cache Line Size. When the $Interval$ between threads is less than or equal to 3, data can be well reused. However,

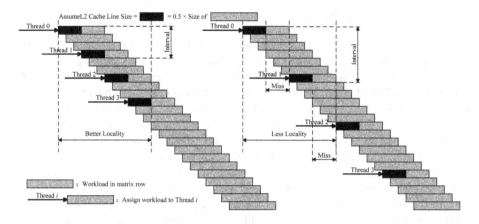

Fig. 6. Minimizing the Interval between threads to reduce cache miss.

if the *Interval* exceeds 3, it results in Cache Miss, which increases with the workload of each row. By applying Atomic Cache, we ensure that every thread will compute the adjacent row (*Interval* = 1), which also dramatically improves the data locality in the shared cache level [33].

3.3 Vectorization

Two basic vectorization strategies in SpMV are cross-row and in-row vectorization [14]. Pezy-SC3s owns 4096 threads, enables simultaneous processing of 4096 rows, resulting in inherent cross-row parallelism. To leverage this parallelism, we have partitioned the matrix, ensuring that each block contains complete rows. Within each block, 64 threads work in parallel to compute the result for each row, without any dependencies among the threads. Additionally, Pezy-SC3s provides 128-bit SIMD instructions, allowing us to increase the in-row computation parallelism by processing two numbers per thread at a time. This further enhances the SIMD parallelism of SpMV (see Algorithm 2, lines 8, 13–16).

4 Experimental Results and Evaluation

We conducted SpMV performance testing on 2092 matrices collected from the SuiteSparse [6] sparse matrix database, comparing our optimized SpMV implementations to the CSR-Based SpMV and CuSparse SpMV. Each SpMV execution consisted of 1000 warm-up executions and 3000 actual executions. On Pezy-SC3s, we evaluated the optimized SpMV based on double-precision floating-point performance, bandwidth utilization, and the speedup ratio achieved compared to the CSR-Based SpMV [13]. On Nvidia 3090ti GPU with CuSparse, due to the different hardware configurations, we mainly evaluate the bandwidth utilization achieved by our optimized CSR-Based SpMV and CuSparse SpMV. We used the CSR-Based SpMV code provided by CuSparse for its double floating-point performance and bandwidth utilization test.

4.1   Preprocessing Overhead

In our method, the only preprocessing step involved is the chunking process. As mentioned in Sect. 3.1, we select 64 as the chunking number, resulting in a fixed total of 63 chunking operations and a size of $65 \times sizeof(int)$ space overhead. Since the number of chunks remains constant regardless of matrix size, the preprocessing time for chunking is negligible compared to the time required for executing a single SpMV. We measured the preprocessing time and the average time of a CSR-Based SpMV for each of the 2092 matrices. The results are illustrated in Fig. 7, demonstrating that the preprocessing time overhead is minimal and consistently lower than the time required for two CSR-Based SpMV. Regarding space overhead, the additional space required is a size of $65 \times sizeof(int)$ due to the fixed number of chunks. This means that the number of bytes added is fixed to $65 \times sizeof(int)$ for any matrix, which is negligible compared to the overall memory usage in real applications and the space occupied by sparse matrices.

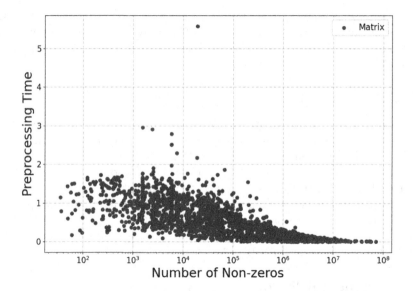

Fig. 7. Preprocessing Time (Normalize to average single CSR-Based SpMV time)

4.2   Floating-Point Performance

To obtain the floating-point performance, we considered one multiplication and one addition operation for each non-zero element. Assuming a total of nnz non-zero elements and a SpMV execution time of $time$, $GFlops$ is denoted as follows:

$$GFlops = \frac{2 \times nnz}{time} \tag{1}$$

We performed tests on the 2092 matrices by running CSR-Based SpMV, our optimized SpMV on Pezy-SC3s, and the CSR-Based SpMV provided by CuSparse on 3090ti. We measured the corresponding *GFlops* achieved by executing SpMV for each matrix, and the results are shown in Figs. 8 and 9.

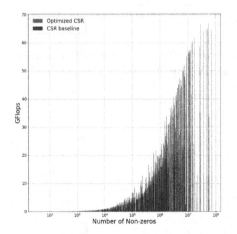
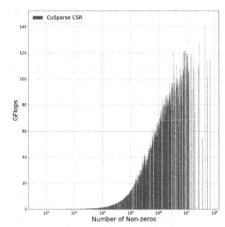

Fig. 8. Pezy-SC3s SpMV GFlops   **Fig. 9.** 3090ti SpMV GFlops

The results indicate that the computing power of the hardware platform becomes more evident as the number of non-zero elements increases. When the number of non-zero elements is small $(nnz < 10^6)$, all three SpMVs achieve relatively low GFlops, which gradually increase as the number of non-zero elements in the matrix grows. When the number of non-zero elements becomes sufficiently large $(nnz \geq 10^6)$, our optimization of CSR-Based SpMV on Pezy-SC3s demonstrates a more significant improvement. CSR-Based SpMV also performs better when the data volume is very small $(nnz < 10^4)$ due to shorter waiting times between threads and no need to wait for Atomic Cache workload balance. However, for practical applications, larger matrices are more commonly encountered and tend to have a more significant time overhead. The optimization effect becomes noticeable when the number of non-zero elements reaches around 10^6 and continues to improve with increasing non-zero elements.

4.3   Bandwidth Utilization

Bandwidth utilization is a crucial performance metric for SpMV implementations [13]. Due to the bandwidth-constrained and access-intensive nature of SpMV, achieving high bandwidth utilization is a critical optimization goal in computing platform with many cores [24,32]. In our experiments, we focused on the number of bytes occupied by non-zero elements as the key factor in bandwidth calculations, which is a common approach in SpMV optimization efforts [17,20,35]. Assuming the floating-point type is *double*, a matrix with nnz non-zero elements, and an average time of *time* required for a single SpMV operation, the bandwidth calculation formula is:

$$Bandwidth = \frac{nnz \times sizeof(double)}{time} \qquad (2)$$

Figures 10 and 11 present the achieved bandwidth values for 2092 matrices on Pezy-SC3s and 3090ti using the three SpMV implementations, respectively.

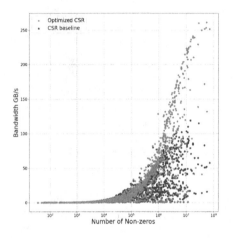

Fig. 10. Pezy-SC3s SpMV BandWidth

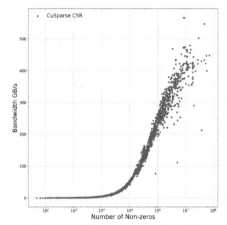

Fig. 11. 3090ti SpMV BandWidth

The results demonstrate that the bandwidth values for all three methods increase as the number of non-zero elements increases. This trend aligns with the measured *GFlops* in Figs. 8 and 9 and holds consistent in general. On Pezy-SC3s, our method exhibits more significant improvements in bandwidth utilization as the number of non-zero elements increases compared to CSR-Based SpMV. The effect of our optimization may not be apparent when the number of non-zero elements is less than 2^6, but as the number of non-zero elements in the matrix grows, the importance of workload balance between threads becomes more pronounced. The workload imbalance between different threads increases as the number of non-zero elements in the matrix rises, making our optimization increasingly effective in addressing the workload imbalance issue. Notably, for matrices with the number of non-zero elements larger than 2^6, our optimization demonstrates considerable advantages and even achieves multiplicative performance gains compared to CSR-Based SpMV on certain matrices.

Furthermore, we conducted a comparison of SpMV bandwidth utilization on Nvidia GPUs. We evaluated the bandwidth of Pezy-SC3s and 3090ti using the open-source bandwidth test code STREAM benchmark [25]. The maximum value from the STREAM benchmark test results is denoted as R_{max} (307.62 GB/s on Pezy-SC3s and 929.88 GB/s on 3090ti), and the average bandwidth measured for each matrix during SpMV execution on both platforms is represented as R. The bandwidth utilization formula is as follows:

$$BandwidthUtilization = \frac{R}{R_{max}} \times 100\% \qquad (3)$$

We tested SpMV on the 3090ti using Nvidia's highly optimized sparse library CuSparse. For each matrix, we performed 1000 warm-up SpMV operations followed by 3000 SpMV operations, averaging the total elapsed time to obtain the single SpMV elapsed time. The final comparison results are presented in Fig. 12. Our optimization method achieves an average bandwidth utilization of 35.22% on large-scale matrices ($nnz \geq 10^6$), reaching 62.54% on very large-scale matrices ($nnz \geq 10^7$). This result surpasses the bandwidth utilization achieved by Nvidia CuSparse libraries on very large-scale matrices ($nnz \geq 10^7$), which is 36.17% and 45.56%, respectively. The analysis reveals that our method achieves better bandwidth utilization compared to CuSparse when the matrix contains a sufficient number of non-zero elements. However, our bandwidth utilization slightly lags behind CuSparse for matrices with fewer non-zero elements ($nnz \leq 10^7$). As shown in Fig. 12, the performance of our method is similar to CuSparse on average, but the performance variance is higher (some points are close to the X-axis in Fig. 12). This can be attributed to our workload balance approach, which involves assigning at least one entire row at a time. As a result, we can

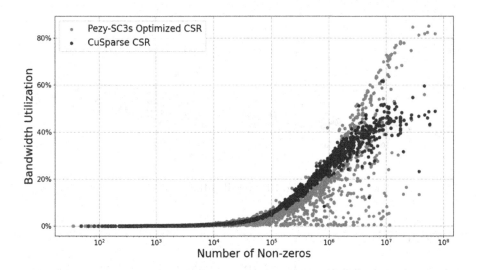

Fig. 12. Bandwidth Utilization On Pezy-SC3s and Nvidia 3090ti GPU

not obtain a complete workload balance, making it unsuitable for matrices with specific sparse patterns. Consequently, this leads to relatively significant performance variance across different matrices.

4.4 SpeedUp

We also calculated the speedup ratio of the optimized SpMV over the CSR-Based SpMV for the 2092 matrices on Pezy-SC3s. Assuming the average time of the optimized SpMV is denoted as $time_{opt}$, and the average time of the CSR-Based SpMV is denoted as $time_{baseline}$, the speedup ratio ($SpeedUp$) can be expressed as:

$$SpeedUp = \frac{time_{baseline}}{time_{opt}} \tag{4}$$

As shown in Fig. 13, our optimized SpMV achieves a maximum speedup ratio of x17.63 compared to the CSR-Based SpMV. CSR-Based SpMV also performs better when the data volume is very small ($nnz \leq 10^4$). However, for practical applications, larger matrices that impose significant time overhead are more common. Our optimized SpMV demonstrates increasingly noticeable benefits as the number of non-zero elements reaches approximately 10^6, and the performance improvement continues to grow with a higher number of non-zero elements.

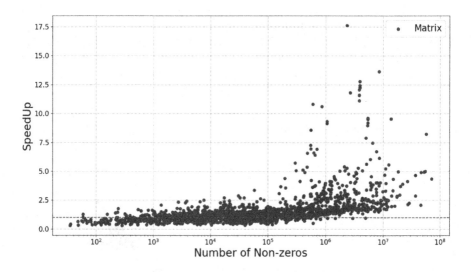

Fig. 13. SpeedUp Rate

5 Conclusion

In summary, we have optimized the CSR-Based SpMV on a new MIMD platform Pezy-SC3s. Our optimization approach involves matrix chunking at the cache level, with the best performance achieved when the number of chunks is set to 2^6. Fast matrix partitioning with less than two CSR-Based SpMV average time in most cases. Additionally, we utilize the Atomic Cache, a high-performance component provided by Pezy-SC3s, to schedule thread workload distribution in CSR-Based SpMV. This workload scheduling ensures that as many threads as possible are actively working simultaneously, avoiding thread waiting and computing resource waste, thereby addressing the workload imbalance problem. Finally, we apply 128-bit SIMD instructions to enhance parallelism. We test our method on 2092 matrices from the SuiteSparse collection, with the final results showing a maximum speedup ratio of x17.63 compared to the CSR-Based SpMV. Besides, we significantly improve bandwidth utilization. A comparison with Nvidia's highly optimized sparse library CuSparse on the 3090ti GPU indicates that our optimization method achieves an average bandwidth utilization of 35.22% on large-scale matrices ($nnz \geq 10^6$), reaching 62.54% on very large-scale matrices ($nnz \geq 10^7$). This result surpasses the bandwidth utilization achieved by Nvidia CuSparse libraries on very large-scale matrices ($nnz \geq 10^7$), which is 36.17% and 45.56%, respectively. These results highlight that our optimization method can effectively utilize the bandwidth and computing resources of Pezy-SC3s.

Acknowledgments. This research was funded by the R&D project 2023YFA1011704, and we would like to thank the ICA3PP 2023 reviewers for their valuable revision comments. We will continue our research in the future to explore more efficient SpMV implementations on Pezy-SC3s and other platforms.

References

1. Ashari, A., Sedaghati, N., Eisenlohr, J., Parthasarathy, S., Sadayappan, P.: Fast Sparse Matrix-Vector Multiplication on GPUS for Graph Applications. IEEE
2. Ashari, A., Sedaghati, N., Eisenlohr, J., Sadayappan, P.: An efficient two-dimensional blocking strategy for sparse matrix-vector multiplication on GPUs (2014)
3. Bell, N., Garland, M.: Implementing sparse matrix-vector multiplication on throughput-oriented processors. In: Conference on High Performance Computing Networking (2009)
4. Bell, N., Garland, M.: Implementing sparse matrix-vector multiplication on throughput-oriented processors. In: Conference on High Performance Computing Networking (2009)
5. Bolz, J., Farmer, I., Grinspun, E., Schrder, P.: Sparse Matrix Solvers on the GPU: Conjugate Gradients and Multigrid. ACM (2003)
6. Davis, T.A., Hu, Y.: The university of florida sparse matrix collection. ACM Trans. Math. Softw. **38**(1), 1–25 (2011)
7. Heller, M., Oberhuber, T.: Adaptive row-grouped CSR format for storing of sparse matrices on GPU. Comput. Sci. (2012)
8. Karakasis, V., Goumas, G.I., Koziris, N.: Perfomance models for blocked sparse matrix-vector multiplication kernels. In: International Conference on Parallel Processing (ICPP 2009), Vienna, 22–25 September 2009 (2009)

9. Kepner, J., Bade, D., Buluc, A., Gilbert, J., Mattson, T., Meyerhenke, H.: Graphs, matrices, and the graphblas: seven good reasons. arXiv preprints (2015)
10. Kepner, J., Gilbert, J.: Graph algorithms in the language of linear algebra. In: Opencoursesfree Org, pp. 315–337 (2011). https://doi.org/10.1137/1. 9780898719918
11. Khairoutdinov, M.F., Randall, D.A.: A cloud resolving model as a cloud parameterization in the ncar community climate system model: preliminary results. Geophys. Res. Lett. **28**(18), 3617–3620 (2001)
12. Krotkiewski, M., Dabrowski, M.: Parallel symmetric sparse matrix-vector product on scalar multi-core cpus. Parall. Comput. **364**, 181–198 (2010)
13. Langr, D., Tvrdik, P.: Evaluation criteria for sparse matrix storage formats. IEEE Trans. Parallel Distrib. Syst. **27**(2), 428–440 (2015)
14. Li, C., Xia, T., Zhao, W., Zheng, N., Ren, P.: Spv8: pursuing optimal vectorization and regular computation pattern in SPMV. In: 2021 58th ACM/IEEE Design Automation Conference (DAC), pp. 661–666. IEEE (2021)
15. Li, Y., et al.: VBSF: a new storage format for SIMD sparse matrix-vector multiplication on modern processors. J. Supercomput. **76**, 2063–2081 (2020)
16. Liang, Y., Tang, W.T., Zhao, R., Lu, M., Huynh, H.P., Goh, R.S.M.: Scale-free sparse matrix-vector multiplication on many-core architectures. IEEE Trans. Comput. Aided Des. Integr. Circuits Syst. **36**(12), 2106–2119 (2017). https://doi.org/ 10.1109/TCAD.2017.2681072
17. Liu, W., Vinter, B.: Csr5: an efficient storage format for cross-platform sparse matrix-vector multiplication. In: Proceedings of the 29th ACM on International Conference on Supercomputing, pp. 339–350 (2015)
18. Maggioni, M., Bergerwolf, T.Y.: Adell: an adaptive warp-balancing ell format for efficient sparse matrix-vector multiplication on gpus. IEEE Comput. Soc. 11–20 (2013)
19. Merrill, D., Garland, M.: Merge-based parallel sparse matrix-vector multiplication. In: Proceedings of the International Conference for High Performance Computing, Networking, Storage and Analysis (SC 2016) (2017)
20. Merrill, D., Garland, M.: Merge-based sparse matrix-vector multiplication (SPMV) using the CSR storage format. ACM Sigplan Notices **51**(8), 1–2 (2016)
21. Mohri, M.: Semiring frameworks and algorithms for shortest-distance problems. J. Automata. Lang. Combinator. (2002)
22. Mu, S., et al.: GPU accelerated sparse matrix-vector multiplication and sparse matrix-transpose vector multiplication. Concurr. Comput. Pract. Exp. (2015)
23. Ravishankar, M., et al.: Distributed memory code generation for mixed irregular/regular computations. In: Proceedings of the 20th ACM SIGPLAN Symposium on Principles and Practice of Parallel Programming, pp. 65–75 (2015)
24. Shalf, J., Dosanjh, S.S., Morrison, J.: Exascale computing technology challenges. In: High Performance Computing for Computational Science - 9th International conference, Berkeley, 22–25 June 2010 (VECPAR 2010), Revised Selected Papers (2010)
25. STREAM: Sustainable memory bandwidth in high performance computers. http:// www.cs.virginia.edu/stream/
26. Sundaram, N., et al.: Graphmat: high performance graph analytics made productive. arXiv preprints (2015)
27. Venkat, A., Hall, M., Strout, M.: Loop and data transformations for sparse matrix code. ACM SIGPLAN Notices **50**(6), 521–532 (2015)
28. Vázquez, F., Fernández, J., Garzón, E.: A new approach for sparse matrix vector product on nvidia gpus. Concurr. Comput. Pract. Exp. **23**(8), 815–826 (2011)

29. Vázquez, F., Ortega, G., Fernández, J., Garzón, E.: Improving the performance of the sparse matrix vector product with gpus. In: IEEE International Conference on Computer and Information Technology (2010)

30. Wang, E., et al.: Intel math kernel library. In: Wang, E., et al. (eds.) High-Performance Computing on the Intel® Xeon Phi^{TM}: How to Fully Exploit MIC Architectures, pp. 167–188. Springer, Cham (2014). https://doi.org/10.1007/978-3-319-06486-4_7

31. Wang, Y., et al.: Gunrock: GPU Graph Analytics (2017)

32. Wright, M.: The Opportunities and Challenges of Exascale Computing (2010)

33. Xie, B., et al.: CVR: efficient vectorization of SPMV on x86 processors. In: Proceedings of the 2018 International Symposium on Code Generation and Optimization, pp. 149–162 (2018)

34. Yan, J., Chen, X., Liu, J.: CSR&RV: an efficient value compression format for sparse matrix-vector multiplication. In: Liu, S., Wei, X. (eds.) Network and Parallel Computing: 19th IFIP WG 10.3 International Conference, NPC 2022, Jinan, 24–25 September 2022, Proceedings, pp. 54–60. Springer, Cham (2022). https://doi.org/10.1007/978-3-031-21395-3_5

35. Yan, S., et al.: YASPMV: yet another SPMV framework on GPUS. ACM SIGPLAN Notices (2014)

36. Zheng, C., Gu, S., Gu, T.X., Yang, B., Liu, X.P.: Biell: a bisection ellpack-based storage format for optimizing SPMV on GPUS. J. Parall. Distrib. Comput. **74**(7), 2639–2647 (2014)

Intrusion Detection Method for Networked Vehicles Based on Data-Enhanced DBN

Yali Duan[1], Jianming Cui[1], Yungang Jia[2], and Ming Liu[2(✉)]

[1] School of Information Engineering, Chang'an University, ShaanXi 710064, China
cjianming@chd.edu.cn
[2] National Computer Network Emergency Response Technical Team/Coordination Center of China, Beijing 100029, China
liuming@cert.org.cn

Abstract. At present, cyber attacks on vehicle network have are proliferating, one of the most significant difficulties in the current detection methods is that the malicious flows are small and discrete in the whole link. In view of the above issues, this paper proposed a detection model based on the integration of Generative Adversarial Networks (GANs) and Deep Belief Networks (DBN). In this model, GANs is first used to enhance the few malicious flow samples, and then an improved DBN is used to evaluate the effect of data generation, so as to improve the uneven distribution of samples in the data set. In the testing section, open data set CIC-IDS2017 was selected for data enhancement and evaluated the performance of the proposed model. The experimental results show that the proposed model has significantly improved the detection performance of few cyber attacks samples compared with traditional detection algorithms. In addition, compared with the method of merge-generate data set approach, the accuracy rate, recall rate, F1 value and other evaluation indexes of the proposed model for the few samples detection have been greatly improved. Therefore, it can be considered that the proposed model is effective than current methods in dealing with the uneven distribution of data sets in traditional cyber attack detection.

Keywords: Generative Adversarial Networks · Networked vehicles · Intrusion detection · Sample distribution

1 Introduction

Traditional vehicular networks (VANETs) [1] are gradually evolving into intelligent vehicular networks. While achieving network communication, vehicles are vulnerable to malicious network flow and may lead to privacy leakage due to the lack of security mechanisms such as firewalls and gateways in some of the devices [1–3]. Improving the active defense capability and security of vehicular

This work is financially supported by the National Natural Science Foundation of China under Grant 62106060.

networks is an important and popular research direction [3,4]. Traditional intrusion detection techniques can detect ongoing and existing malicious attacks in a timely manner. However, in uneven distribution massive network flow, malicious cyber attacks often hide in a large amount of normal data, making traditional intrusion detection methods difficult to deal with evolving malicious attacks and network threats [1,5].

Currently, the main methods for handling uneven distribution data [6] include resampling methods [7], cost-sensitive algorithms, ensemble methods, feature representation and classification decoupling, etc. These methods attempt to rebalance the class weight norms in the machine learning model by increasing the number of samples of minority attacks. However, these traditional algorithms still have some problems. Generative adversarial networks (GANs) [8,9] can learn the distribution of given data and generate new sample data. Currently, GANs are mostly used in natural images [10,11], and have achieved significant results. Inspired by its success in these fields, scholars are gradually starting to use GANs to generate adversarial network flow for intrusion detection.

It can be argued that the current intrusion detection approach for vehicular networks has the following two drawbacks: (1) The network flow data is uneven distribution and the number of samples in the minority class is too small. The commonly used data augmentation algorithm, SMOTE algorithm [12,13] does not consider noise data and boundary issues, which may cause overlap between different categories, leading to decreased accuracy and overfitting problems. (2) Some current intrusion detection models for vehicular networks perform a lower detection rate and weak classification ability. On this basis, a new intrusion detection model is proposed, which combines GANs with DBN [14], and uses a GAN-based data augmentation method [15] to generate adversarial attack samples for the minority class, in order to expand the dataset CIC-IDS2017, an improved DBN classifier is designed to evaluate the effectiveness of this method [14,16]. How to achieve better classification effect, higher classification accuracy and higher precision, which is the main innovation and challenge of this paper.

2   Intrusion Detection Model Based on Data Augmentation

2.1   Data Processing and Augmentation

Dataset Analysis and Preprocessing. In this paper, since the vehicle data set may lead to user privacy leakage, the general vehicle data set is not open to the public, so the open source network intrusion detection data set is adopted. In the existing open source datasets, CIC-DDOS2019 dataset proposes an attack classification method for DDOS, and the CSE-CIC-IDS2018 is mostly used for anomaly detection. However, this paper studies data enhancement and intrusion detection for a few categories in the vehicle network. So, the dataset used is the CIC-IDS2017 dataset [17] provided by the Canadian Institute for Cybersecurity. It contains 5 d of normal and attack flow data collected by the institute, with each record having 78 network features. The dataset includes the latest cyber attacks and meets all standards of real-world attacks, and can fully simulate the attack to vehicle network [18].

After merging the 8 csv files of the CIC-IDS2017 dataset, missing instances were removed from the dataset along with NaN values to avoid redundancy and exploding gradients when training the model. The dataset was then transformed into numerical values and normalized, with the most frequently occurring class being labeled as 'Normal' and all other classes labeled as 'Attack' to meet the conditions of inputting the dataset into the GAN network.

GANs-Based Data Augmentation Methods. To address the problem of highly uneven distribution data in vehicle network datasets, a data augmentation method based on Generative Adversarial Networks (GANs) is used to generate adversarial attack samples for the minority class, in order to expand the dataset CIC-IDS2017 [17] used in the study.

Algorithm 1. Data Augmentation Algorithm based on GANs

Input: $s = (r, y)$ r is the eigenvector y is the category label
Output: $S_G = [G(z, y'), y']$
1: **while** $0 <= p_{\text{data}} <= \frac{1}{2}$ **do** /* train GAN*/
2:      **for** k steps **do** /* train Discriminator*/
3:          Sample random noise $\left\{ z^{(1)}, z^{(2)}, \ldots z^{(m)} \right\}$ from $p_z(z)$
4:          Sample real data $\left\{ x^{(1)}, x^{(2)}, \ldots x^{(m)} \right\}$ from $p_{\text{data}}(x)$
5:          $\eta_{\theta_D} \leftarrow \nabla \theta_D \frac{1}{m} \sum_1^m \left[\log D\left(x^{(i)} \right) + \log \left(1 - D\left(G\left(z^{(i)} \right) \right) \right) \right]$
6:          /* update the weight parameters and gradients */
7:          $\theta_D \leftarrow \theta_D + \alpha_D \cdot \text{Adam}(\theta_D, \eta_{\theta_D})$
8:      **end for**
9:      Sample random noise $\left\{ z^{(1)}, z^{(2)}, \ldots z^{(m)} \right\}$ from $p_z(z)$/* train Generator*/
10:      $\eta_{\theta_D} \leftarrow \nabla \theta_G \frac{1}{m} \sum_1^m \left[\log \left(1 - D\left(G\left(z^{(i)} \right) \right) \right) \right]$ /* update the weight parameters and gradients */
11:      $\theta_D \leftarrow \theta_G - \alpha_G \cdot \text{Adam}(\theta_G, \eta_{\theta_G})$
12: **end**
13: **return** data /* Generate data */

Algorithm 1 is the data augmentation training process of GANs. Where Θ_G, η_{θ_G}, θ_D, η_{θ_D} are the generated weight parameters, gradients, and discriminator weight parameters and gradients, respectively. Its main steps are:

- (1) The category label y' of the preprocessed minority data is input to the generator G with the random noise vector z for training and the data sample S_G generated;
- (2) Fix the generator G, train the discriminator D, and gradually update the weight parameter θ_D of the discriminator;

- (3) Fixed discriminator D, train generator G, and gradually update the weight parameter θ_G of the generator;
- (4) Loop (1)–(3) until $p_g = p_{data} = 1/2$, the discriminator cannot distinguish between the two distributions, so that the generated sample keeps approaching the real data sample.

Figure 1 shows the occurrence of classes in the original dataset and the minority class after data augmentation, it can be observed that data augmentation is effective in increasing the number of samples in classes with fewer than 5000 samples in the dataset, particularly for extremely rare classes such as Heartbleed and Infiltration, which are increased from 11 and 36 samples, respectively, to 5632 and 8704 samples.

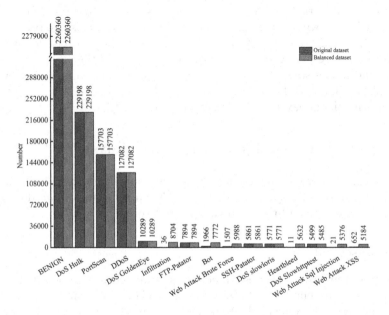

Fig. 1. Comparison of Original Data and Quantity after Data Augmentation

2.2 Improved DBN Model

Structurally, Deep Belief Networks [19] is a probabilistic generative model composed of multiple layers of unsupervised Restricted Boltzmann Machines (RBM) and a supervised Back-Propagation (BP) network, DBN is composed of multiple stacked RBMs, each consisting of a hidden layer and a visible layer.

The training of DBN [20] consists of the layer-wise pre-training stage and the back-propagation fine-tuning stage. In the pre-training stage, it uses the Contrastive Divergence (CD) algorithm proposed by Hinton to quickly train RBM to obtain an approximate representation of the input vector v. In the back-propagation fine-tuning stage, the BP algorithm and stochastic gradient descent are used to optimize the connection weights in the DBN to obtain the optimal model parameters. For the pre-training stage, the Mean Square Error (MSE) and Pseudo-Likelihood (PL) loss functions were used to evaluate the accuracy of RBM training. The calculation method for MSE is Eq. (1):

$$MSE = \frac{1}{m} \sum_{i=1}^{m} (x_i - \bar{x})^2 \tag{1}$$

where m is the number of samples, $x_i (i = 1, 2, 3 \ldots m)$ is the sample, $\bar{x}$ is the average of m samples. For the reverse tuning stage, in order to determine the learning rate of the model, the loss and accuracy of the training set and the validation set are used to evaluate its performance.

Algorithm 2 is the specific algorithm trained by DBN. Where W, W^k are the weight matrix with the training stage and the fine-tuning stage respectively, $a_i (i = 1, 2, 3 \ldots M)$, $b_j (j = 1, 2, 3 \ldots N)$ are the biases of the visible layer and the hidden layer respectively, ϵ, ϵ_{ft} are the learning rate of the pre-training stage and the fine-tuning stage, $V = v_1, v_2 \ldots, v_m$ is the training sample of RBM, l is the number of layers of RBM, a^k, b^k $(k = 1, 2, 3 \ldots l)$ are the bias of the visible layer and the hidden layer at the kth layer, respectively.

3   Experimental Design and Result Analysis

The experiments were conducted in a Win10 environment, with a 64-bit Intel(R) Xeon(R) Silver 4100 CPU and 32 GB RAM. The implementation was done using Python 3.8 language and the Pytorch 1.9 framework.

3.1   Dataset Labels

The proposed intrusion detection model was evaluated using the CIC-IDS2017 dataset. After data enhancement, similar attack classes with similar characteristics and behaviors were merged into a new class, and the dataset was re-labeled. The final standard dataset was divided into 9 classes. Table 1 shows the number of labels in the standardized dataset after re-labeling.

3.2   Experiments and Analysis

To evaluate the detection performance of the proposed intrusion detection model, the following experiments were designed.

Experiments on Training GANs-DBN Model. GANs were used to generate samples for the 8 minority classes in the dataset. The GANs training parameters are shown in Table 2. After training, the dataset was re-labeled to generate the standard dataset, which was then divided into training set, testing set, and validation set in a 60%, 20%, 20% ratio. Finally, the data was input into

---

Algorithm 2. DBN Training Algorithm

Initialize $W = W^k = a_i = b_j = 0$

1: **the first phase: Train RBM**
2:     **for** s steps **do**/*Set the number of iterations s*/
3:         **for** v_i **do** $i = 1, 2 \ldots m$
4:             **for** k **do** /*Gibbs sampling*/
5:                 **for** i **do** $i = 1, 2 \ldots M$
6:                     $h_j^{(k)} \leftarrow p\left(h_j \mid v^{(k)}\right)$
7:                 **end for**
8:                 **for** j **do** $j = 1, 2 \ldots N$
9:                     $v_i^{(k+1)} \leftarrow p\left(v_i \mid h^{(k)}\right)$
10:                 **end for**
11:                 **for** i, j **do**
12:                     $i = 1, 2 \ldots M j = 1, 2 \ldots N$ /* Update weights and biases */
13:                     $W_{ij}^{(k)} \leftarrow W_{ij}^{(k+1)} + \varepsilon\left(p\left(h_j \mid v^{(0)}\right)\right)$
14:                     $a_i^{(k+1)} \leftarrow a_i^{(s)} + \varepsilon\left(v_i^{(0)} - v_i^{(k)}\right)$
15:                     $b_i^{(k+1)} \leftarrow b_i^{(s)} + \varepsilon\left(p\left(h_j \mid v^{(s)}\right)\right) - p\left(h_j \mid v^{(k)}\right)$
16:                 **end for**
17:             **end for**
18:         **end for**
19: **end for**
20: **the second phase: Fine Tune DBN**
21: % Forward propagation
22: **for** l **do**
23:     Initialization $\dot{W}^k = a^k = b^k = 0,\quad \varepsilon = \varepsilon_0$
24:     **Train RBM**   /* Traverse each layer RBM*/
25: **end for**
26: **for** i **do**
27:     $i = 1, 2 \ldots m$
28:     Compute $o_i\left(x_i\right)$
29: **end for**
30: % Backpropagation
31: **for** k **do** $k = l, l - 1 \ldots 1$
32:     **if** $k = l$
33:         $\delta_k \leftarrow o_k\left(1 - o_k\right)\left(t_k - o_k\right)$
34:     **else**
35:         $\delta_h \leftarrow o_h\left(1 - o_k\right)\sum_{k \in \text{ outputs}} \varepsilon_{kh}\delta_k$
36:         $\theta_{ji} \leftarrow \theta_{ji} + \Delta\theta_{ji},\quad \theta_{ji} = \varepsilon_{ft}\delta_j x_j$
37:     **end if**
38: **end**

Table 1. Number of Labels After Relabeling

New labels	Origin labels	Numbers
Benign	BENIGN	2260360
Brute Force	SSH-Patator	13755
	FTP-Patator	
DoS	DoS-Hulk	250743
	DoS-GoldenEye	
	DoS-slowloris	
	DoS-Slowhttptest	
Heartbleed	Heartbleed	5632
Infiltration	Infiltration	8704
Web Attack	Web Attack-Brute Force	16548
	Web Attack-XSS	
	Web Attack-Sql Injection	
DDoS	DDoS	127082
PortScan	PortScan	157703
Bot	Botnet ARES	7772

Table 2. GANs Training Parameters

	Parameter
Activation function	Leaky ReLU
Learning rate	0.0002
Optimister	Adam
Loss function	Cross-entropy
Batch_size	5

Table 3. DBN Network Parameters

Parameter	Pre-training phase	Fine-tuning phase
Epochs	10	5
Learning rate	0.015	0.005
Batch size	64	128
Optimister	SGD	Adam
Gibbs steps	1	

the DBN classifier for model evaluation. The parameter settings for the DBN classifier are shown in Table 3.

During the training of the DBN classifier, in the pre-training stage, the learning rate of the RBM were determined by changing the learning rate within an approximate range of [0.001, 0.1]. As shown in Fig. 3, when the learning rate lr = 0.015, MSE = 0.538, PL = -0.818. Compared with other learning rates, the training accuracy of RBM is optimal at this learning rate. Therefore, this method

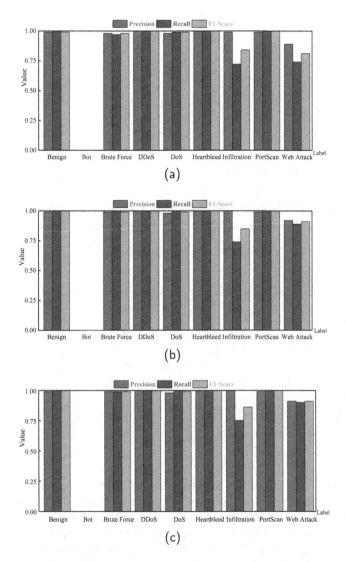

Fig. 2. Dataset Classification Results:(a) Training Set Classification Results. (b)Testing Set Classification Results. (c) Verification Set Classification Results.

selected the learning rate lr = 0.015 as the learning rate for RBM training. In the back-propagation fine-tuning stage, the performance of the model was evaluated using the loss and accuracy of the training set and the validation set to determine the optimal learning rate for this stage. From Fig. 4, it can be clearly seen that when the learning rate is lr = 0.005, the loss reaches its minimum value with train-loss = 0.453 and val-loss = 0.419, and the accuracy of the training set and validation set reaches its maximum value with train-acc = 0.993 and val-acc = 0.990. However, when the learning rate is too high, such as lr = 0.1, the model's

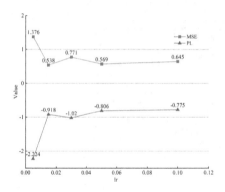 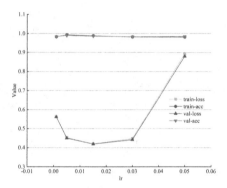

Fig. 3. Comparison of RBM Training Performance at different learning rates.

Fig. 4. Comparison of Accuracy and Loss at different learning rates

Table 4. Testing Set Confusion Matrix

True Label	Predicted Label									Recall
	Benign	Bot	Brute Force	DDoS	DoS	Heartbleed	Infiltration	PortScan	Web Attack	
Benign	451120	0	30	33	988	3	4	188	203	0.994
Bot	0	0	0	0	0	0	0	0	0	0
Brute Force	12	0	2733	0	1	0	0	0	0	0.989
DDoS	51	0	0	25138	3	0	0	0	0	0.998
DoS	167	0	0	0	49741	0	0	0	46	0.978
Heartbleed	0	0	0	0	0	1100	0	0	0	0.997
Infiltration	455	0	0	0	0	0	1270	0	0	0.997
PortScan	24	0	0	6	17	0	0	31440	2	0.994
Web Attack	276	0	0	0	92	0	0	0	2976	0.922
Precision	0.997	0	0.995	0.998	0.996	1.000	0.736	0.998	0.890	

loss reaches 0.98 and the model fails to converge. Therefore, based on the above results, this method selected the learning rate lr=0.005 as the training learning rate for the back-propagation fine-tuning stage of the model.

The classification results of the proposed model for the training set, test set, and validation set are shown in Fig. 2, and the confusion matrix for the predicted classes in the test set is presented in Table 4. It show that the proposed model can correctly classify most of the network flow, with high precision, recall, and F1 score. The precision, recall, and F1 score for the minority classes such as Brute Force, DDoS and PortScan are close to 1. Additionally, the precision, recall, and F1 score for extremely rare classes like Heartbleed and Infiltration are also above 70%, with a recall rate of 99.7% and a precision rate of 100% for Heartbleed. Therefore, the proposed model has strong detection performance for attacks on minority classes in the vehicular networks while maintaining high performance in detecting other attacks.

Performance Comparison Experiments of Different Data Augmentation Methods. To verify the effectiveness of the proposed data augmentation method, this study compared different data augmentation methods, including the SMOTE algorithm, class weight strategy, combination of SMOTE algorithm and class weight strategy, and GANs and combines it with DBN classifier for model evaluation. The parameters of the DBN network were kept consistent for each method, and the specific parameter settings are shown in Table 3.

Table 5. Comparison of GANs with Other Data Augmentation Methods

Model	Accuracy	F1-score	AUC
Class Weights+DBN	96	84	97
SMOTE+DBN	98	82	96
SMOTE+Class Weights+DBN	95	80	91
GANs+DBN	99	86	99

Table 5 shows that compared to the other three commonly used data augmentation methods, the proposed model improves accuracy, F1 score, and AUC by at least 1%, 2%, and 2%, respectively. Figure 5 compares the offline AUCs for different classes using various data augmentation methods. It can be concluded that the proposed intrusion detection method based on GANs-DBN outperforms other classification algorithms in overall performance, although it may not perform as well as some other methods for certain classes. Overall, this method greatly improves the accuracy of intrusion detection for each class.

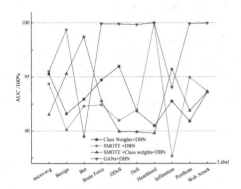

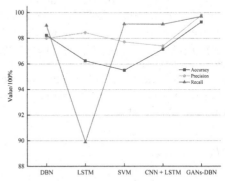

Fig. 5. Comparison of AUC for Different Data Augmentation Methods

Fig. 6. Performance Comparison of Different Models

Performance Comparison Experiment of Different Models. To verify the intrusion detection performance of the proposed model, the performance of GANs-DBN was compared with several existing intrusion detection models using the CIC-IDS2017 dataset. Othmane Belarbi [21] To verify the intrusion detection performance of the proposed model, the performance of GANs-DBN was compared with several existing intrusion detection models using the CIC-IDS2017 dataset. Monika Roopak et al. [22] proposed deep learning models including LSTM, CNN + LSTM, and SVM, and evaluated DDoS attack detection using the CIC-IDS2017 dataset. For the LSTM model, the final accuracy reached 86.34%; for the CNN + LSTM model, the final accuracy reached 97.16%; for the SVM model, the accuracy reached 95.5%. By comparing the performance data of the above reference papers with the GANs-DBN model used in this paper, the detection performance of various intrusion detection models was evaluated.

From Fig. 6, it can be seen that the proposed model outperformed other models in all three indicators, reaching 99.27%, 99.80%, and 99.70% respectively, which represents at least a 1.03%, 1.36%, and 0.58% improvement, respectively. Thus, the proposed model significantly improved the detection performance for multi-class intrusion detection compared to other models.

4   Conclusion

This paper presents an integrated network intrusion detection model, GANs-DBN, designed to address the issue of low detection performance for small quantities of malicious flow in vehicle networks due to the discrete distribution of network attacks. The performance of the model is evaluated using the CIC-IDS2017 dataset. Specifically, GANs are employed for data augmentation, expanding the dataset and enriching its distribution, while an improved DBN classifier is utilized to assess the model's classification capability. Experimental results demonstrate that the proposed model outperforms alternative methods in overall detection performance, effectively enhancing the detection rate for specific classes of attacks and thereby improving overall accuracy. However, it is worth noting that the current research only partially simulates the real network conditions, and future efforts should focus on identifying and defending against the complex traffic characteristics encountered in actual vehicle networks, particularly APT attacks.

References

1. Cui, J., Ma, L., Wang, R., Liu, M.: Research and optimization of GPSR routing protocol for vehicular ad-hoc network. China Commun. **19**(10), 194–206 (2022)
2. Zhang, Y., Cui, J., Liu, M.: Research on adversarial patch attack defense method for traffic sign detection. In: Lu, W., Zhang, Y., Wen, W., Yan, H., Li, C. (eds.) Cyber Security: 19th China Annual Conference, CNCERT 2022, Beijing, China, August 16–17, 2022, Revised Selected Papers, pp. 199–210. Springer, Singapore (2022). https://doi.org/10.1007/978-981-19-8285-9_15

3. Liu, M., et al.: Modeling and analysis of the decentralized interactive cyber defense approach. China Commun. **19**(10), 116–128 (2022)

4. Kang, M.J., Kang, J.W.: Intrusion detection system using deep neural network for in-vehicle network security. PLoS ONE **11**(6), e0155781 (2016)

5. Qu, F., Wu, Z., Wang, F.Y., Cho, W.: A security and privacy review of vanets. IEEE Trans. Intell. Transp. Syst. **16**(6), 2985–2996 (2015)

6. Zhang, Y., Li, X., Gao, L., Wang, L., Wen, L.: Imbalanced data fault diagnosis of rotating machinery using synthetic oversampling and feature learning. J. Manuf. Syst. **48**, 34–50 (2018)

7. He, H., Yang, B., Garcia, E., Li, S.A.: Adaptive synthetic sampling approach for imbalanced learning. In: Proceedings of the 2008 IEEE International Joint Conference on Neural Networks (IEEE World Congress on Computational Intelligence), Hong Kong (2008)

8. Creswell, A., White, T., Dumoulin, V., Arulkumaran, K., Sengupta, B., Bharath, A.A.: Generative adversarial networks: an overview. IEEE Signal Process. Mag. **35**(1), 53–65 (2018)

9. Goodfellow, I., et al.: Generative adversarial networks. Commun. ACM **63**(11), 139–144 (2020)

10. Wang, Z., She, Q., Ward, T.E.: Generative adversarial networks in computer vision: a survey and taxonomy. ACM Comput. Surv. **54**(2), 1–38 (2021)

11. Yu, X., Cui, J., Liu, M.: An embedding carrier-free steganography method based on Wasserstein GAN. In: Lai, Y., Wang, T., Jiang, M., Xu, G., Liang, W., Castiglione, A. (eds.) Algorithms and Architectures for Parallel Processing (ICA3PP 2021). LNCS, vol. 13156. Springer, Cham (2022). https://doi.org/10.1007/978-3-030-95388-1_35

12. She, X., Sekiya, Y.: A convolutional autoencoder based method with smote for cyber intrusion detection. In: 2021 IEEE International Conference on Big Data (Big Data), pp. 2565–2573. IEEE (2021)

13. Soltanzadeh, P., Hashemzadeh, M.: Rcsmote: range-controlled synthetic minority over-sampling technique for handling the class imbalance problem. Inf. Sci. **542**, 92–111 (2021)

14. Zhang, Y., Li, P., Wang, X.: Intrusion detection for iot based on improved genetic algorithm and deep belief network. IEEE Access **7**, 31711–31722 (2019)

15. Tanaka, F.H.K.d.S., Aranha, C.: Data augmentation using gans. arXiv preprint arXiv:1904.09135 (2019)

16. Liu, J., Wu, N., Qiao, Y., Li, Z.: Short-term traffic flow forecasting using ensemble approach based on deep belief networks. IEEE Trans. Intell. Transp. Syst. **23**(1), 404–417 (2020)

17. Sharafaldin, I., Lashkari, A.H., Ghorbani, A.A.: Toward generating a new intrusion detection dataset and intrusion traffic characterization. ICISSp **1**, 108–116 (2018)

18. Frid-Adar, M., Klang, E., Amitai, M., Goldberger, J., Greenspan, H.: Synthetic data augmentation using gan for improved liver lesion classification. In: 2018 IEEE 15th International Symposium on Biomedical Imaging (ISBI 2018), pp. 289–293. IEEE (2018)

19. Sohn, I.: Deep belief network based intrusion detection techniques: a survey. Expert Syst. Appl. **167**, 114170 (2021)

20. Gao, N., Gao, L., Gao, Q., Wang, H.: An intrusion detection model based on deep belief networks. In: 2014 Second International Conference on Advanced Cloud and Big Data, pp. 247–252. IEEE (2014)

21. Belarbi, O., Khan, A., Carnelli, P., Spyridopoulos, T.: An intrusion detection system based on deep belief networks. In: Su, C., Sakurai, K., Liu, F. (eds.) Science of Cyber Security: 4th International Conference, SciSec 2022, Matsue, 10–12 August 2022, Revised Selected Papers, pp. 377–392. Springer, Cham (2022). https://doi.org/10.1007/978-3-031-17551-0_25

22. Roopak, M., Tian, G.Y., Chambers, J.: Deep learning models for cyber security in IoT networks. In: 2019 IEEE 9th Annual Computing and Communication Workshop and Conference (CCWC), pp. 0452–0457. IEEE (2019)

A Multi-source Domain Adaption Approach to Minority Disk Failure Prediction

Wang Wang[1,2], Xuehai Tang[1(✉)], Biyu Zhou[1], Yangchen Dong[1],
Yuanhang Feng[1], Jizhong Han[1], and Songlin Hu[1]

[1] Institute of Information Engineering, Chinese Academy of Sciences, Beijing, China
{wangwang,tangxuehai,zhoubiyu,dongyangchen,fengyuanhang,
hanjizhong,husonglin}@iie.ac.cn
[2] School of Cyber Security, University of Chinese Academy of Sciences,
Beijing, China

Abstract. Frequent happening of disk failures affects the reliability of the storage system, which can cause jittering of performance or even data loss of services and thus seriously threaten the quality of service. Although a host of machine (deep) learning-based disk failure prediction approaches have been proposed to prevent system breakdown due to unexpected disk failure, they are able to achieve high performance based on the assumption that the disk model has plenty of samples (especially failure samples). However, new disk models continuously appear in data centers with the evolution of disk manufacturing technology and the expansion of storage system capacity. Limited by the deploying time, these disk models have few failure samples and are called minority disks. The minority disks are widespread in large-scale data centers and contain amounts of disks while existing approaches cannot reach satisfying performance on such disks due to the lack of failure samples. What's worse, failure prediction models trained on other disk models cannot be directly applied to these minority disks either due to the commonly existing distribution shift among disk models. In this work, we propose DiskDA, a novel multi-source domain adaption-based solution that can fully utilize knowledge from other disk models to predict failures for minority disks having no failure samples. Our experimental results on real-world datasets show the superiority of DiskDA against previous approaches on minority disks with a few failure samples. What's more, DiskDA also shows its good adaptivity on minority disks having no failure samples, whereas previous works are unusable.

Keywords: Fault tolerance · Disk failure prediction · Domain adaption · Cloud computing

1 Introduction

Disk failures are common in modern large-scale data centers, accounting for more than 70% of hardware replacement events [5,13,16]. Frequent happening

Z. Tari et al. (Eds.): ICA3PP 2023, LNCS 14488, pp. 53–72, 2024.
https://doi.org/10.1007/978-981-97-0801-7_4

of disk failure can lead to service performance jitter or even data loss which severely affects the availability and reliability of cloud applications [7,17]. To ensure the availability and reliability of cloud applications from unexpected disk failures, operators should proactively predict the upcoming disk failure events before they actually happen, so as to take preventive measures in time, such as virtual machine migration.

The Self-Monitoring, Analysis, and Reporting Technology (SMART) has been widely implemented by hard disk drive (HDD) and solid-state drive (SSD) manufacturers to monitor the status of individual disk drives. The values of SMART attributes related to disk health status are helpful to disk health tendency assessment.

Recently, with the development of machine learning, a host of supervised learning-based approaches has been proposed to predict disk failures with the SMART values [10,20,22,23]. With sufficient samples (both healthy and failure samples) provided, these methods are able to train binary classifiers and classify newly coming disk samples collected periodically from data centers to predict failures for each disk with high accuracy.

Table 1. Statistics of Disk Population

Data Center	Disk Type	Type Number	Type Percentage	Disk Number
Backblaze	Majority	12	11.65%	114,570
	Minority	**91**	**88.35%**	**34,978**
Tencent	Majority	8	13.33%	52,235
	Minority	**52**	**86.67%**	**18,996**

However, the condition of sufficient failure samples can hardly be satisfied by all disk models. With the evolution of disk manufacturing technology and the expansion of storage system capacity, disks from different models are continuously added to data centers. Limited by the deploying scale and time, the newly coming disk models usually have only a few or even no failure samples for most cases and are named minority disks [11,24]. According to the studies [9,24] in large-scale data centers (i.e. Backblaze, Tencent, and AliCloud), minority disks are generally existing in modern data centers. As shown in Table 1 [24], the minority disks dominate the disk models (over 85%) and contain a great number of disks (tens of thousands of disks). Unfortunately, the traditional supervised ML method cannot be applied to predict failures for minority disks, otherwise, it will suffer from over-fitting or cold-start issues [3,6,24,25]. What's worse, the prediction model trained on other disk models cannot be applied to minority disks either and we illustrate this in our extensive experiments. Because the commonly existing distribution shifts across disk models break the assumption of independent identical distribution holding between training and test set.

More recently, several transfer learning (TL)-based methods [2,9,15,19,24] and semi-supervised learning approaches [3,6,25] are proposed. Based on the fact that the failure modes are common for different disk models (e.g. all disks will fail due to too many bad sectors) [5], the TL-based methods try to adapt failure prediction knowledge extracted from other disk models to minority disk models by directly selecting samples similar to the minority disks as training set or transforming the SMART attribute distribution of minority disks to that of other disk models via a heuristic statistical model. However, the existing TL-based approaches can only transfer partial knowledge from a single source domain (other disk models) since they have to drop many useful samples due to their dissimilarity to minority disks or abandon critical features that hard to be transformed. Although using multiple source domains is more likely to introduce more failure modes, the large number of samples contained in it also means the complex distribution of source domains, which will lead to negative migration problems in the existing TL-based methods. In addition, all existing TL-based approaches need a certain number of failure samples of minority disks in their transfer procedure, which suggests they can only handle very limited cases. As for the semi-supervised learning approaches, they though can train their model with only healthy samples, large quantities of minority disk failure samples are needed to set appropriate classification thresholds.

In this work, we are exploring extracting the transferable failure modes from multiple source domains and aligning their semantics across source and target domains so that the full knowledge can be leveraged to enhance the failure prediction of minority disks. To this end, we model our problem as an unsupervised domain adaption problem and propose DiskDA, a multi-source domain adaption-based failure prediction approach for minority disks. It is able to extract failure modes from samples of multiple disk models and utilize them in minority disk (even with no failure samples) failure prediction with high performance. The goal is achieved because of two key designs. Firstly, although minority disks do not have complete class distribution, we find that the particularity of the distribution of disk samples can be leveraged to ensure the execution of domain adaption. Based on this, we use a representor to extract failure modes from source domain samples and align their semantics across two domains using only healthy samples in the target domain. And a Wasserstein distance measurement is adopted to guarantee the effectiveness of the domain adaption even when the distribution of two domains is distant. We also prove the rationality of this strategy by analyzing the generalization error bound in this case. Secondly, DiskDA adopts a confidence-based sample selection to filter out irrelevant samples in the source domains, so as to eliminate the negative transfer issue. By running the two processes alternatively, DiskDA can successfully extract transferable failure modes from multiple source domain disk models and utilize them in minority disk failure prediction with high accuracy.

The main contributions are summarized as:

1) We explore the problem of failure prediction for minority disk without failure samples so that it can be adaptive to all minority disk models.

2) To the best of our knowledge, we are the first to propose a Wasserstein distance-based domain adaption solution for the minority disk failure prediction problem and first analyze the generalization error bound theoretically under this condition.

3) Guided by the generalization error bound, we design a novel unsupervised domain adaption framework, DiskDA, to minimize the generalization error of the failure predictor in the target domain.

4) We conduct evaluations to demonstrate the superiority of DiskDA on 9 disk models from 3 vendors, collected from 2 large-scale data centers. The evaluation results reveal that DiskDA can improve the F1-score by an average of 20.02% compared with the best competitor when less than a dozen failure samples are provided. **More importantly, DiskDA can still obtain a satisfactory F1-Score of about 0.93 when no failure samples are provided (most minority disks face), while all existing TL-based approaches will fail.**

2   Related Work

- **Supervised Learning-Based Failure Prediction Approaches.** Li et al. [10] propose a Classification And Regression Trees (CART) based model which can give disks a health assessment. Xu et al. [20] present a Recurrent Neural Networks (RNN [12]) method to leverage sequential information in hard disk failure prediction. Yang et al. [22] design a disk failure prediction model by using L1-regularized logistic regression. Zhang et al. [23] adopt the Siamese network [4] to improve the applicability and adaptivity of the disk failure prediction model. All these supervised learning-based approaches can achieve high performance based on the assumption that large quantities of failure samples are provided. However, this is harsh for minority disks.

- **Semi-supervised Learning-Based Failure Prediction Approaches.** The main idea of the semi-supervised learning-based approach is to model the distribution of healthy samples and predict failure samples based on their reconstruction errors. Once the reconstruction errors surpass a predefined threshold, the disk samples are classified as failure samples. Jiang et al. [6] propose a GAN (Generative Adversarial Network)-based anomaly prediction approach that adopts an encoder-decoder-encoder architecture. They define the reconstruction error as the difference between two encoders' outputs and predict failures by comparing the error with a threshold. Zhou et al. [25] and Chakraborttii et al. [3] predict failures for SSDs with similar approaches. The performance of such approaches relies on the manually set thresholds and the operators are able to find appropriate thresholds only when a certain number of reconstruction errors of failure samples are provided. Since the failure samples of minority disks are limited, it is hard for such methods to reach satisfying performance in minority disk failure prediction.

- **Transfer Learning-Based Failure Prediction Approaches.** The target of the TL-based approach is to adapt a failure prediction model trained from existing disk models (source domain) to the minority disk (target domain). MirelaMadalina Botezatu et al. [2] propose an instance-based de-bias approach. They select samples from the source domain disk model based on their similarity degree given by a domain classifier and a Regularized Greedy Forests (RGF [8]) trained on the augmented minority disk dataset is adopted as the failure prediction model. Xie et al. [19] select the source domain based on the performance similarity of the failure prediction model on the minority disk and each candidate source domain disk model. Then the minority disk failure prediction model is trained on the union set of source and target domains. Zhang et al. [24] propose to utilize the Kullback-Leibler divergence (KLD) of the specific SMART attribute to select the source domain disk model and adopt the Tradaboost algorithm trained on both domains as the minority disk failure prediction model. Sun et al. [15] take another approach and propose to use a statistic-based feature transformation to align cumulative SMART attribute (e.g. SMART_5 represents reallocated sector count) distribution. They find the same cumulative SMART attribute of disks from different vendors/models have similar distributions and align their distributions based on the ratio of failed to healthy devices so as to adapt the failure prediction model trained from one disk model to others. Lan et al. [9] also try to transfer knowledge from the source domain by utilizing a domain classifier to learn the domain invariant representation of source and target domain samples. It is worth noting that the domain invariant representation learning guided by the domain classifier will fail (gradient vanishing) if the distribution of the source and target domain is distant [18]. And this has been shown in their experiment where 50% of the transfer process (domain adaption) failed due to the large distribution divergence of source and target domain samples. In a word, existing TL-based can only utilize limited information from the source domain due to the drop of samples and critical attributes. In addition, they can only work when a certain number of failure samples from minority disks are provided, while this can be harsh for minority disks.

To sum up, DiskDA differs from previous approaches in three aspects:

- Compared to supervised learning-based approaches, DiskDA extracts failure prediction knowledge from large amounts of samples from other disk models. And this strategy protects DiskDA from overfitting caused by the limited failure samples of minority disks.
- DiskDA adopts a binary classifier built on labeled samples to automatically discriminate the healthy and failed samples rather than manually setting the classification threshold as semi-supervised learning approaches.
- Compared to existing TL-based approaches, DiskDA does not choose to drop samples or critical attributes but tries to fuse the distribution of source and target domain samples, so as to fully utilize the failure prediction knowledge from source domain disk models. DiskDA avoids the gradient vanishing problem by adopting the Wasserstein distance to guide the domain invariant representation learning process. Because the Wasserstein distance can always provide stable gradients no matter how distant the distributions are [18].

3   Motivation

3.1   Problem Statement

In the problem of minority disk failure prediction based on domain adaption (MDFP-DA), we suppose a labeled dataset $X^s = \{(x_i^s, y_i^s)_{i=1}^{n_s}\}$ including n_s samples from **multiple** disk models of the data center, which are sufficient to train a high precision prediction model. Furthermore, we assume a dataset $X^t = \{(x_m^t, y_m^t)_m\}$ from the minority disk, where x_i^t refers to the sample collected online in future and y_i^t is the corresponding label. The samples from X^s and X^t ①share the same feature space (this can be ensured by keeping their common SMART attributes), but ②follow different marginal distributions, $\mathbb{P}_s$ and $\mathbb{P}_t$. Although X^t is unreachable in reality, we can collect quantities of healthy samples $X_H^t = \{(x_j^t, 0)_{j=1}^{n_t}\}$ from the minority disk through short-term deployment, which is always held in disk failure prediction [3,6,25]. And we denote the marginal distribution of the healthy samples (from both X^t and X_H^t) as $\mathbb{P}_{t_H}$. Here, we regard MDFP-DA as a binary classification problem and label the healthy samples as '0', and the failure samples as '1'. Now we give the definition of the MDFP-DA problem:

Definition 1. *The MDFP-DA problem is to learn a transferable classification model $h(s)$ to minimize the risk $\epsilon_t(h) = Pr_{(x,y)\sim X^t}[h(x) \neq y]$ using X^s and X_H^t.*

3.2   Generalization Error Bound Analysis

We analyze the generalization error bound by introducing the unsupervised domain adaption problem. The unsupervised domain adaption studies the problem of adapting a classifier trained in the source domain to target domain ①sharing the same feature space while ②having different data distribution.

Obviously, in the case of given X^t, the MDFP-DA problem can be converted to an unsupervised domain adaption problem. Although the failure samples of the minority disk are lost in the MDFP-DA problem, we have amounts of its healthy samples X_H^t. Let $\epsilon_t(h)$ denote the generalization error bound of a classification function h in target domain t. Let $W_1(P, Q)$ denote the Wasserstein distance between P and Q. In this case, the following Theorem holds.

Theorem 1. *For any classification function h to the MDFP-DA problem satisfying K-Lipschitz, the following holds:*

$$\epsilon_t(h) \leq \epsilon_s(h) + 2KW_1(\mathbb{P}_s, \mathbb{P}_{t_H}) + \lambda + C \tag{1}$$

where C is the Wasserstein distance of distribution of target domain $\mathbb{P}_t$ and its healthy samples $\mathbb{P}_{t_H}$.

Proof. See Appendix for details.

Considering the fact that the healthy samples dominate the whole disk samples (with an average ratio of 9997:10000 [2]), the distribution of samples of a disk model is actually similar to its healthy samples, which suggests that C is a small constant. We have verified this by randomly selecting 4 disk models and calculating C and $W_1(\mathbb{P}_{t_H}, \mathbb{P}_s)$. And the results show that C is small in scale of 10^{-3} and $W_1(\mathbb{P}_{t_H}, \mathbb{P}_s)$ are hundreds of times of C, so it can be ignored in practice.

Remark. Theorem 3.1 implies that the generalization error of a prediction model in the target domain (i.e., $\epsilon_t(h)$) is smaller than the sum of the generalization error of the prediction model in the source domain (i.e., $\epsilon_s(h)$), the Wasserstein distance of source domain samples and minority disk healthy samples (i.e., $W_1(\mathbb{P}_s, \mathbb{P}_{t_H})$), and a constant (i.e., $\lambda + C$) much smaller than the former two. In other words, the generalization error of the prediction model in the minority disk ($\epsilon_t(h)$) can be optimized if we are able to reduce $\epsilon_s(h)$ and $W_1(\mathbb{P}_s, \mathbb{P}_{t_H})$. Once $\epsilon_t(h)$ is optimized, the performance of the failure prediction model in the minority disk can be improved. To sum up, it not only proves the generalization error bound of the MDFP-DA problem but also indicates the optimization direction in the absence of failure samples.

4 Method

4.1 Overview of DiskDA

Figure 1 illustrates the framework of DiskDA. As seen, the DiskDA consists of two processes, ① the domain invariant representation learning and ② the confidence-based sample selection.

① The first process mainly involves three modules:

- **Representor**: a deep neural network that projects the samples from source domain disk models (X^s) and minority disk (X_H^t) into a unified latent space (representation vector in Fig. 1).
- **Distance Estimator**: measures the Wasserstein distance of representation vectors from X^s and X_H^t.
- **Failure Predictor**: classifies whether a sample from X^s is a failure sample based on their representation vectors.

The representor is able to reduce the Wasserstein distance of the source domain and minority disk samples (i.e. reducing $W_1(\mathbb{P}_s, \mathbb{P}_{t_H})$) under the guidance of distance estimator. In the meantime, it helps the failure predictor to reach high performance in the source domain (i.e., reducing $\epsilon_s(h)$) via extracting discriminant information. In this way, the performance of the failure predictor in the minority disk can be optimized according to Theorem 3.1.

② The second process mainly includes the sample selector module. Its main purpose is to avoid negative transfer which may occur in the first process. The principle is to eliminate the samples in X^s that hinder the further narrowing of the distance between X^s and X_H^t, which also corresponds to reducing $W_1(\mathbb{P}_s, \mathbb{P}_{t_H})$.

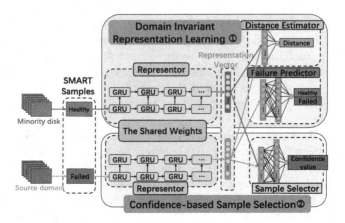

Fig. 1. The architecture of DiskDA.

In the training stage, DiskDA iterates the two processes alternatively. For example, the process ① runs every N (e.p., 100) iterations, and then the process ② runs M (e.p.,1) iterations. In this way, DiskDA can fully utilize the failure prediction knowledge of the source domain and lose the least information. The alternating iteration can be stopped until the parameters are converged or the iteration times reach a threshold.

In the online prediction, the SMART instances of a minority disk are collected daily to a sample pool. And these instances are combined to form samples as in the training stage. All these samples will be input into the representor to generate corresponding representation vectors. Then the failure predictor will predict whether a sample is a failure sample based on its representation. Once a sample of a disk (minority disk) is predicted as a failure sample, it suggests that the disk will fail soon and the alarm system will inform the operator to repair/exchange the disk in time.

4.2   Domain Invariant Representation Learning

To effectively adapt the failure predictor trained in the source domain disks to the minority disk, we need the representor to learn domain invariant representations of samples from both domains. That is, the distributions of representations from both domains projected by the representor should have a small divergence. Besides, the representations should retain key information that can be used to classify failure samples.

Firstly, all samples are projected to a d-dimensional space by the representor. The representations of X^s and X_H^t are denoted as $f_r(X^s)$ and $f_r(X_H^t)$, where f_r is the mapping function of the representor. The distance estimator is then used to measure the distribution divergence of representations from source domain $\mathbb{R}^s$ and minority disk $\mathbb{R}_H^t$. Here, we introduce Wasserstein distance as the measurement metric because it can measure the divergence between two arbitrary distributions even if they are distant.

Based on Kantorovich Rubinstein theorem, the dual representation of the first Wasserstein distance of two Borel probability measures $\mathbb{P}$ and $\mathbb{Q}$ can be formalized as

$$W_1(\mathbb{P}, \mathbb{Q}) = \sup_{\|f\|_L \leq 1} \mathbb{E}_{x \sim \mathbb{P}}[f(x)] - \mathbb{E}_{x \sim \mathbb{Q}}[f(x)] \tag{2}$$

where L-Lipschitz condition is defined as $\|f\|_L = \frac{\sup|f(x) - f(y)|}{\rho(x,y)} \leq L$. Accordingly, the Wasserstein distance of source domain and minority disk $W_1(\mathbb{R}^s, \mathbb{R}^t_H)$ in the latent space can be calculated as:

$$W_1(\mathbb{R}^s, \mathbb{R}^t_H) = \sup_{\|f_d\|_L \leq 1} \mathbb{E}_{x \sim \mathbb{P}_s}[f_d(f_r(x))] - \mathbb{E}_{x \sim \mathbb{P}_{t_H}}[f_d(f_r(x))] \tag{3}$$

where f_d is the function learned by the distance estimator with its parameters θ_d to map representations h to real numbers. Then, we can approximate the empirical Wasserstein distance of representation distribution of source and target domain via maximizing domain critic loss $\mathcal{L}_{wd}$ with respect to θ_d:

$$\mathcal{L}_{wd} = \frac{1}{n_s} \sum_{x^s \in X^s} f_d(f_r(x^s)) - \frac{1}{n_t} \sum_{x^t \in X^t_H} f_d(f_r(x^t)) \tag{4}$$

Note that the f_d should satisfy the 1-Lipschitz condition when calculating the first Wassertein distance. Therefore, a gradient penalty term $\mathcal{L}_{grad} = \mathbb{E}_{h \sim [\mathbb{R}^s, \mathbb{R}^t_H]}[(\|\nabla_h f_d(h)\|_2 - 1)^2]$ is added to $\mathcal{L}_{wd}$. And the final objective function of distance estimator ($\mathcal{L}_{dist}$) can be written as:

$$\max_{\theta_d}\{\mathcal{L}_{wd} + \lambda \mathcal{L}_{grad}\} \tag{5}$$

where the λ is used to balance the $\mathcal{L}_{wd}$ and $\mathcal{L}_{grad}$.

The failure predictor is used to predict failures for the source disk models. Its inputs are representations from source domain samples and the labels of representations are consistent with their corresponding source domain samples. The objective function of failure predictor ($\mathcal{L}_C$) can be formalized as:

$$\min_{\theta_c} \frac{1}{N} \sum_{h \in \mathbb{R}^s} -[y_i \cdot log(f_c(h_i)) + (1 - y_i) \cdot log(1 - f_c(h_i))] \tag{6}$$

The task of the representor is to ① reduce the representation divergence of the minority disk and source domain disk models, and ② extract the discriminative information from samples in representation learning. Therefore, the objective of representor ($\mathcal{L}_R$) can be formalized as:

$$\min_{\theta_r}\{\mathcal{L}_C + \gamma \mathcal{L}_{wd}\} \tag{7}$$

where γ is the coefficient to balance between discriminative and transferable feature learning.

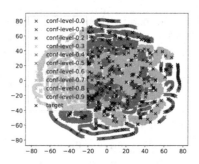

Fig. 2. Visualization of the representations from the target domain and the source domain at different confidence thresholds (i.e., conf-level-*)

4.3   Confidence-Based Sample Selection

While introducing more disk models into the source domain helps bring in more failure modes, it also complicates the distribution of the source domain, leading to negative transfer in representation learning. Therefore, it is necessary to filter out irrelevant samples from the source domain to avoid the probable negative transfer.

Specifically, DiskDA achieves this via a confidence-based sample selection process. The confidence of source domain samples is measured based on the similarities of their representations to those from the minority disk. The confidence is given by the domain classifier which is a supervised learning-based binary classifier. The inputs are representations of samples from both domains and the labels are 0/1. The representations from the source domain are labeled as '0' and those from the target domain are labeled as '1'. To train a domain classifier, we randomly select a small proportion of representations of samples from the source domain and minority disk. And the objective of the domain classifier can be formulated as:

$$\min_{\theta_D} \frac{1}{N} \sum_i -[y_i \cdot log(f_c(h_i)) + (1 - y_i) \cdot log(1 - f_c(h_i))] \tag{8}$$

where $x \in [\mathbb{R}^s, \mathbb{R}_H^t], y \in \{0, 1\}$.

When the parameters converge, the domain classifier is applied to the adjusted source domain as a filter. Since the domain classifier is realized by a deep neural network and the results are given by a sigmoid function in the last layer, the output of the domain classifier is the probability that a representation belongs to the minority disk. So we use this probability to measure the confidence of a sample. DiskDA discards the samples in the source domain whose confidence is lower than the pre-defined threshold and uses the remaining samples in the source domain with higher confidence for further representation learning. By filtering out the 'low quality' samples, the distribution divergence between the source domain and the minority disk is reduced.

In Fig. 2, we visualize the distribution of sample representations for the minority disk and the source domain under different confidence thresholds via t-SNE, where the colored points denoted as "conf-level-*" represent the representations of source domain samples with different confidence values and the black point represent the representations of minority disk samples. t-SNE is short for t-Distributed Stochastic Neighbor Embedding which allows us to project high-dimensional embedding spaces into 2D spaces for visualization while keeping their relative distance. In other words, the points close in the figure have a small distance in the original space. Since the dimension of representations has been compressed, the x and y axes of points have no specific meaning. As seen, the representations of minority disk samples are located closer to that of the source domain samples when a higher confidence threshold is selected, which indicates the rationality of our sample selection.

5   Experiment

In this section, we conduct experiments to evaluate the performance of DiskDA. We first describe the methodology and then show the experimental comparison results among DiskDA and 7 state-of-the-art solutions. Finally, we show the results of sensitivity analysis to explore how the critical hyper-parameters affect the failure prediction performance of DiskDA.

5.1   Methodology

Datasets. The disk models used in our experiments are from two real-world datasets. We select ST4000DM000 (Disk_1), ST6000DX000 (Disk_2), ST3000DM001 (Disk_3), Hitachi HDS5 C4040ALE630 (Disk_4), Hitachi HDS722020ALA330 (Disk_5), Hitachi HDS723030ALA640 (Disk_6), HGST HMS5C4040BLE640 (Disk_7), HGST HMS5C4040ALE640 (Disk_8), HGST HUH728080ALE600 (Disk_9) in Backblaze[1] with a period from 2015-01-01 to 2019-12-31. We select MC1 (SSD_1), MC2 (SSD_2) and MA1 (SSD_3) from Alibaba Cloud [21] with a period from 2019-01-01 to 2019-12-31. All disk models are selected randomly. Each record in both datasets is labeled as healthy or failed on a daily basis.

Attribute Selection. Not all SMRAT attributes are useful for disk failure prediction, we select SMART 1,4,7,12,190,192,193,194,196,197,199 for HDD failure prediction and SMART 1,5,9,12,171,172,174,175,183,190,232,233 for SSD failure prediction via correlation coefficient analysis. Min-max normalization (i.e. $x_{norm} = \frac{x - x_{min}}{x_{max} - x_{min}}$, where x is the raw value of the SMART attribute, x_{max} and x_{min} are the maximum and minimum values of the SMART attribute in the training set) is used to normalize the values of different SMART attributes.

---

[1] https://www.backblaze.com/b2/hard-drive-test-data.html.

Experiment Setup. Regarding records, close to actual failure, will disturb the failure prediction model, a commonly used approach is to label k continuous healthy samples before the actual failure as failure records too [23,24]. And k is determined via change-point detection and set to be 3 in our experiments. The representation vector length is set to 128 for both representation ability and cost saving (detailed in Sect. 5.2). The coefficient of gradient penalty term λ is set to 10, which is consistent with the setting commonly used in ML models based on Wasserstein distance [1,14]. The parameter γ used to balance the weight of discriminative and domain invariant representation learning is set to 1e-2, which is determined through grid search. In each domain invariant representation learning iteration, the distance estimator runs 10 steps then the parameters of the representor and the failure predictor update once. The sample selection process runs one time every 500 representation learning iterations. And the confidence threshold is set to 0.2, which delivers the optimal transfer learning performance (detailed in Sect. 5.2).

Evaluation Metrics. The failure prediction rate (FDR, also called recall), false alarm rate (FAR, also called false positive rate), and F1-Score are adopted as the metrics to measure the disk failure prediction performance. A good disk failure prediction method should reach a high FDR with a low FAR. And the F1-Score is the balance between the FDR and FAR, thus is the most important metric to measure the performance of the prediction model.

Benchmarks. We test three types of benchmarks.

- **Supervised Learning-Based:** We first measure the performance of three supervised learning-based failure prediction methods (i.e. GBRT [10], HDDse [23] and RGF [2]) with only minority disk samples.
- **Semi-supervised Learning-Based:** We explore the performance of the semi-supervised learning model (VAE-LSTM [25]) which models the healthy samples and classifies the failure samples by comparing the reconstruction error with a pre-defined threshold.
- **TL-Based:** We evaluate 3 state-of-the-art TL-based failure prediction models (i.e. TLDFP [24], SSDB [2] and FLBT [15]). Note that TLDFP and SSDB are instance-based TL approaches, and FLBT is a feature-based TL approach. Note that we also test ADA-CBAN [9] in our experiments while its performance is not stable. 3 (i.e., Exp_1, Exp_2, and Exp_4) of the 5 experiments failed due to the large distribution divergence of the source and target domain. In addition, the results of the ADA-CBAN in the last two experiments are worse than those of DiskDA, so we omit to show its results in our comparison. Since all the TL-based methods require a base failure prediction model, a bidirectional gated recurrent unit network (Bi-GRU) is adopted as the base failure prediction model in this paper. Note that the base model can also be replaced by any neural network-based failure prediction model. In fact, we choose Bi-GRU because it shows simplicity, robustness, and accuracy in experiments.

Table 2. The Details about Dataset Used in Experiments.

No.	Training Set	Testing Set
Exp.1	Source: Disk_1, Disk_2 Target: Disk_3 (110, 6)	Disk_3 (1058, 100)
Exp.2	Source: Disk_4, Disk_5 Target: Disk_6 (100, 4)	Disk_6 (918, 40)
Exp.3	Source: Disk_7, Disk_8 Target: Disk_9 (100, 3)	Disk_9 (985, 15)
Exp.4	Source: Disk_8, Disk_9 Target: Disk_4 (300, 4)	Disk_4 (2360, 38)
Exp.5	Source: SSD_1, SSD_2 Target: SSD_3 (399, 13)	SSD_3 (39563, 1357)

Considering that the three comparative TL-based methods are all single source-based TL models, we traverse all source domain disk models and select the one with the best performance as the source domain.

The detailed setup of training and testing datasets is shown in Table 2. Note that Disk_a (x, y) denotes that the number of healthy disks is x and the number of the failed disk is y. Generally, the number of healthy samples in a real dataset is much larger than that of failure samples. In order to avoid model bias caused by data imbalance, we only use the randomly selected healthy samples with the same number of failure samples. Cross-validation is done for each method, and we average their performance as the final result.

Table 3. Performance Comparison of Various Disk Failure Prediction Models

Method	Exp.1			Exp.2			Exp.3			Exp.4			Exp.5		
	FDR	FAR	F1	FDR	FAR	F1	FDR	FAR	F1	FDR	FAR	F1	FDR	FAR	F1
RGF	0.382	0.021	0.536	0.698	0.056	0.725	0.621	0.017	0.726	0.618	0.06	0.557	0.747	0.034	0.793
GBRT	0.605	0.106	0.666	0.674	0.035	0.744	0.5	0.014	0.637	0.532	0.017	0.624	0.684	0.013	0.801
HDDse	0.634	0.057	0.702	0.784	0.009	0.891	0.715	0.013	0.757	0.573	0.023	0.649	0.624	0.003	0.799
VAE-LSTM	0.756	0.103	0.807	0.821	0.095	0.886	0.796	0.031	0.848	0.697	0.046	0.723	0.785	0.067	0.815
Bi-GRU	0.491	0.098	0.582	0.684	0.011	0.792	0.68	0.021	0.761	0.355	0.04	0.405	0.51	0.004	0.672
DiskDA	0.917	0.027	0.976	0.991	0.002	0.998	0.913	0.003	0.974	0.907	0.001	0.969	0.973	0.043	0.993

5.2 Experimental Results

Performance Comparison with Supervised and Semi-supervised Learning Methods. We first show the performance comparison results of

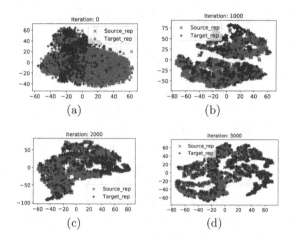

Fig. 3. Visualization of representation from samples of minority disk and source domain disk models

DiskDA with supervised learning methods (RGF, GBRT, and HDDse) and semi-supervised learning methods (VAE-LSTM). The training set used is the dataset indicated as "Target" in Table 2.

It can be seen from Table 3 that

- The supervised learning-based methods perform poorly due to insufficient training data. Although HDDse adopts a metric learning method that actually increases the size of the training set by taking pairs of samples as input, its prediction performance is only slightly better than the other two.
- Compared with the above supervised-learning methods, the FDR value of VAE-LSTM is significantly improved. However, because the threshold-based method adopted by VAE-LSTM cannot well classify the failure samples from the healthy, the FAR is also higher.
- DiskDA has the best failure prediction performance, with an F1-Score 20.86% higher than VAE-LSTM on average. In addition, we also find that the F1-Score of DiskDA is 61.75% higher than that of the base model, which also proves the necessity and effectiveness of domain adaption. DiskDA can reach the best performance because ① it can extract failure prediction knowledge from large amounts of source domain disk samples rather than the limited samples provided by the minority disk; ② it adopts a supervised learning-based approach that can automatically discriminate the healthy and failure samples without setting classification threshold manually.

Performance Comparison with TL-Based Methods. Next, we show the performance comparison results of DiskDA with TL-based methods (SSDB, TLDFP, and FLBT). Since all these TL-based methods can only work if there are failure samples in the target domain, DiskDA also uses failure samples in the target domain as other methods for fair comparison (see Table 2 for details of the training set). However, the DiskDA can work without failure samples, which is superior to other solutions. To highlight that, we further implement DiskDA in a more restrictive case where no failure sample of minority disks is provided and the results are indicated as "DiskDA*". To facilitate analysis, we also test the performance of the base model trained in the source domain on the testing set as the baseline (indicated as "Src"). From Table 4 we can see that

- The F1-Score of Src is only 0.592 on average. The root cause is the distribution shift of SMART attributes among disks, so simply reusing the prediction model trained upon other disks will fail in practice.
- Among the three TL-based methods, the instance-based TL approach TLDFP has the best performance. By continuously enhancing the weight of misclassified samples, the ability of the failure predictor can be improved to a certain extent. Another instance-based TL approach, SSDS, uses a sample selection strategy based on similarity to adjust the source domain samples. However, its performance is even worse than the baseline in some cases. The performance of FLBT is worse than that of TLDFT. All the TL-based approaches can just reach sub-optimal performance as they drop useful samples and attributes and can only utilize partial information from the source domain.

Table 4. Performance Evaluation of Transfer Learning Based Failure Prediction Models

Method	Exp.1			Exp.2			Exp.3			Exp.4			Exp.5		
	FDR	FAR	F1	FDR	FAR	F1	FDR	FAR	F1	FDR	FAR	F1	FDR	FAR	F1
Src	0.486	0.09	0.584	0.979	0.605	0.618	0.916	0.762	0.537	0.97	0.919	0.505	0.977	0.472	0.715
SSDB	0.34	0.467	0.302	0.911	0.384	0.801	0.991	0.84	0.707	0.95	0.68	0.685	0.71	0.672	0.597
TLDFP	0.785	0.094	0.836	0.751	0.001	0.858	0.811	0.422	0.773	0.648	0.001	0.786	0.804	0.117	0.843
FLBT	0.685	0.194	0.636	0.747	0.131	0.834	0.611	0.134	0.683	0.588	0.112	0.716	0.722	0.093	0.837
DiskDA*	0.84	0.048	0.889	0.987	0.002	0.994	0.852	0.005	0.92	0.825	0.002	0.904	0.962	0.092	0.947
DiskDA	0.917	0.027	0.976	0.991	0.002	0.998	0.913	0.003	0.974	0.907	0.001	0.969	0.973	0.043	0.993

- DiskDA performs the best, with the F1-Score 20.02% higher than the best competitor. This is because ① DiskDA can extract failure prediction knowledge from multiple source domain disk models while existing instance-based TL approaches can only benefit from a single source domain (just one disk model in the source domain) to prevent negative transfer. ② DiskDA tries to fully utilize the source domain disk samples by fusing the distribution of source and target domain samples in the latent space, rather than directly dropping samples or attributes as existing TL-based approaches. We visualize the fusion of representations from the source and target domain via t-SNE in Fig. 3, where the red points represent the representations from the source domain and the green ones represent the representations from the target domain. As seen, the green and red points are fused constantly as the

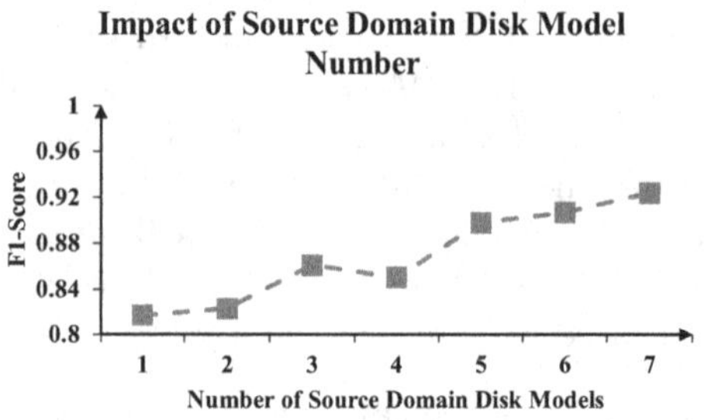

**Impact of Source Domain Disk Model
Number**

Fig. 4. Performance of introducing more disk models to source domain

increment of iterations, which suggests that the domain invariant representation learning process can effectively fuse the representations from the source and target domain and thus the failure prediction knowledge extracted from the source domain can adapt to the target domain(minority disk).

- In the absence of failure samples from the minority disk, the F1-Score of DiskDA can still reach a satisfactory 0.93, 66.75% higher than that of the baseline. The results prove the effectiveness of our theorem that DiskDA can still adapt the failure prediction knowledge extracted from source domain disk models to minority disks with no failure sample. And this shows that DiskDA has much higher adaptivity compared to existing TL-based approaches. We further explore whether DiskDA can continuously benefit from the increment of source domain failure modes. And we investigate this problem by measuring the performance of DiskDA as the increase of source domain disk models because more disk models potentially contain more failure modes. As shown in Fig. 4, the performance of DiskDA can be continuously improved by adding more disk models to the source domain. Thanks to the sample selection process, DiskDA can timely filter out the source domain samples deteriorating the domain invariant representation learning process and effectively transfer failure prediction knowledge from the source domain to the minority disk. The results motivate us to add more disk models to the source domain to reach high performance without worrying about the negative transfer problem.

5.3 Sensitivity Study

Impact of Hidden Size. In domain invariant representation learning, the representor projects samples from both domains to fixed-length vectors as their representations. And we evaluate how the size of length affects the performance of DiskDA. As seen in Fig. 5.a, the performance of DiskDA is improved as the increase of hidden size and then steady until the hidden size reaches 128. The results indicate that a small size will limit the representation ability and seriously affect the performance of DiskDA. And we set the hidden size as 128 to reach a balance between the performance and computation cost.

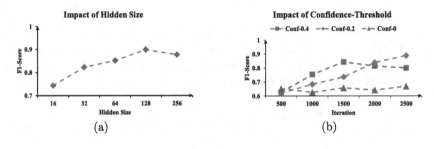

Fig. 5. Sensitivity Study

Impact of Confidence-Threshold. The confidence threshold determines which samples to select in representation learning afterward. And we explore how different confidence thresholds affect the performance of DiskDA. In the experiment, we experiment with different confidence threshold values and summarize the results as 3 representative curves corresponding to thresholds of 0.4 (red curve), 0.2 (green curve), 0 (blue curve), shown in Fig. 5.b. As seen, a low (i.e., ranging from 0 to 0.1) or high threshold (i.e., ranging from 0.4 to 1) can both deteriorate the performance of DiskDA since the negative transfer caused by irrelevant samples remaining in the source domain or the loss of relevant samples filtered out in confidence-based sample selection process. And we set the threshold as 0.2 to achieve a balance between the performance of domain adaption and the loss of source domain samples.

6 Conclusion

In this work, we investigate the problem of minority disk failure prediction in data centers. Based on the fact that the failure modes are common for different disk models, our basic idea is to utilize full of the failure prediction knowledge learned from other disk models to the minority disk. We model this as an unsupervised domain adaption problem and analyzed the generalization error bound of the prediction model in the target domain (minority disk) theoretically. Guided by the generalization error bound, we design a framework which can effectively optimize the error bound by elaborately combining the domain invariant representation and confidence-based sample selection processes. Our experiments on real-world datasets show the effectiveness of our approach. Moreover, our approach can still reach a satisfying F1-score of 0.93 on average for minority disks even with no failure samples, which suggests that our approach can fit for more broad cases compared to existing approaches.

7   Appendix

Proof. The discrepancy between the source and target domain is measured using the Wasserstein distance in DiskDA. Specifically, the p-th Wasserstein distance between two Borel probability measures $\mathbb{P}$ and $\mathbb{Q}$ is defined as:

$$W_p(\mathbb{P}, \mathbb{Q}) = (\inf_{\mu \in \Gamma(\mathbb{P}, \mathbb{Q})} \int \rho(x, y)^p d\mu(x, y))^{1/p} \tag{9}$$

where the $\Gamma(\mathbb{P}, \mathbb{Q})$ is the set of all joint distributions $\mu(x, y)$ whose marginal distribution are $\mathbb{P}$ and $\mathbb{Q}$. The $\mu(x, y)$ can be viewed as a policy for transporting a unit quantity of material from x to y and the $\rho(x, y)$ is the corresponding cost. And the Wasserstein distance between $\mathbb{P}$ and $\mathbb{Q}$ represents the minimum expected transport cost. As Wasserstein distance satisfies the triangle inequality, the following equation holds

$$W_p(\mathbb{P}_s, \mathbb{P}_t) \leq W_p(\mathbb{P}_s, \mathbb{P}_{t_H}) + W_p(\mathbb{P}_{t_H}, \mathbb{P}_t) \tag{10}$$

Shen et al. [14] prove the generalization error bound of a classification function h in the target domain for unsupervised domain adaption based on Wasserstein distance as

$$\epsilon_t(h) \leq \epsilon_s(h) + 2K W_1(\mathbb{P}_s, \mathbb{P}_t) + \lambda \tag{11}$$

where the K means that all hypotheses h are K-Lipschitz continous, λ is the combined error of the optimal hypothesis $h*$ which minimizes the combined error $\epsilon_s(h) + \epsilon_t(h)$, $\mathbb{P}_s$ and $\mathbb{P}_t$ are distributions of source and target domain, respectively. Let C denote $2K W_1(\mathbb{P}_{t_H}, \mathbb{P}_t)$. By substituting inequality (11) for (10), Theorem 3.1 is derived.

References

1. Arjovsky, M., Chintala, S., Bottou, L.: Wasserstein generative adversarial networks. In: International Conference on Machine Learning, pp. 214–223. PMLR (2017)
2. Botezatu, M.M., Giurgiu, I., Bogojeska, J., Wiesmann, D.: Predicting disk replacement towards reliable data centers. In: Proceedings of the 22nd ACM SIGKDD International Conference on Knowledge Discovery and Data Mining, pp. 39–48 (2016)
3. Chakraborttii, C., Litz, H.: Improving the accuracy, adaptability, and interpretability of SSD failure prediction models. In: Proceedings of the 11th ACM Symposium on Cloud Computing, pp. 120–133 (2020)
4. Chopra, S., Hadsell, R., LeCun, Y.: Learning a similarity metric discriminatively, with application to face verification. In: 2005 IEEE Computer Society Conference on Computer Vision and Pattern Recognition (CVPR 2005), vol. 1, pp. 539–546. IEEE (2005)
5. Ghemawat, S., Gobioff, H., Leung, S.T.: The google file system. In: Proceedings of the Nineteenth ACM Symposium on Operating Systems Principles, pp. 29–43 (2003)

6. Jiang, T., Zeng, J., Zhou, K., Huang, P., Yang, T.: Lifelong disk failure prediction via gan-based anomaly detection. In: 2019 IEEE 37th International Conference on Computer Design (ICCD), pp. 199–207. IEEE (2019)
7. Jiang, W., Hu, C., Zhou, Y., Kanevsky, A.: Are disks the dominant contributor for storage failures? A comprehensive study of storage subsystem failure characteristics. ACM Trans. Storage (TOS) 4(3), 1–25 (2008)
8. Johnson, R., Zhang, T.: Learning nonlinear functions using regularized greedy forest. IEEE Trans. Pattern Anal. Mach. Intell. 36(5), 942–954 (2013)
9. Lan, X., et al.: Adversarial domain adaptation with correlation-based association networks for longitudinal disk fault prediction. In: 2021 International Joint Conference on Neural Networks (IJCNN), pp. 1–8. IEEE (2021)
10. Li, J., et al.: Hard drive failure prediction using classification and regression trees. In: 2014 44th Annual IEEE/IFIP International Conference on Dependable Systems and Networks, pp. 383–394. IEEE (2014)
11. Lu, S., Luo, B., Patel, T., Yao, Y., Tiwari, D., Shi, W.: Making disk failure predictions smarter! In: FAST, pp. 151–167 (2020)
12. Mikolov, T., Kombrink, S., Burget, L., Černockỳ, J., Khudanpur, S.: Extensions of recurrent neural network language model. In: 2011 IEEE International Conference on Acoustics, Speech and Signal Processing (ICASSP), pp. 5528–5531. IEEE (2011)
13. Schroeder, B., Gibson, G.A.: Understanding disk failure rates: what does an MTTF of 1,000,000 hours mean to you? ACM Trans. Storage (TOS) 3(3), 8-es (2007)
14. Shen, J., Qu, Y., Zhang, W., Yu, Y.: Wasserstein distance guided representation learning for domain adaptation. In: Proceedings of the AAAI Conference on Artificial Intelligence, vol. 32 (2018)
15. Sun, X., et al.: System-level hardware failure prediction using deep learning. In: Proceedings of the 56th Annual Design Automation Conference 2019, pp. 1–6 (2019)
16. Vishwanath, K.V., Nagappan, N.: Characterizing cloud computing hardware reliability. In: Proceedings of the 1st ACM Symposium on Cloud Computing, pp. 193–204 (2010)
17. Wang, Y., Miao, Q., Ma, E.W., Tsui, K.L., Pecht, M.G.: Online anomaly detection for hard disk drives based on mahalanobis distance. IEEE Trans. Reliab. 62(1), 136–145 (2013)
18. Wilson, G., Cook, D.J.: A survey of unsupervised deep domain adaptation. ACM Trans. Intell. Syst. Technol. 11(5), 1–46 (2020)
19. Xie, Y., Feng, D., Wang, F., Zhang, X., Han, J., Tang, X.: OME: an optimized modeling engine for disk failure prediction in heterogeneous datacenter. In: 2018 IEEE 36th International Conference on Computer Design (ICCD), pp. 561–564. IEEE (2018)
20. Xu, C., Wang, G., Liu, X., Guo, D., Liu, T.Y.: Health status assessment and failure prediction for hard drives with recurrent neural networks. IEEE Trans. Comput. 65(11), 3502–3508 (2016)
21. Xu, F., Han, S., Lee, P.P., Liu, Y., He, C., Liu, J.: General feature selection for failure prediction in large-scale SSD deployment. In: 2021 51st Annual IEEE/IFIP International Conference on Dependable Systems and Networks (DSN), pp. 263–270. IEEE (2021)
22. Yang, W., Hu, D., Liu, Y., Wang, S., Jiang, T.: Hard drive failure prediction using big data. In: 2015 IEEE 34th Symposium on Reliable Distributed Systems Workshop (SRDSW), pp. 13–18. IEEE (2015)

23. Zhang, J., Huang, P., Zhou, K., Xie, M., Schelter, S.: HDDSE: enabling high-dimensional disk state embedding for generic failure detection system of heterogeneous disks in large data centers. In: Proceedings of the 2020 USENIX Conference on Usenix Annual Technical Conference, pp. 111–126 (2020)
24. Zhang, J., et al.: Minority disk failure prediction based on transfer learning in large data centers of heterogeneous disk systems. IEEE Trans. Parallel Distrib. Syst. **31**(9), 2155–2169 (2020)
25. Zhou, H., et al.: A proactive failure tolerant mechanism for SSDS storage systems based on unsupervised learning. In: 2021 IEEE/ACM 29th International Symposium on Quality of Service (IWQOS), pp. 1–10. IEEE (2021)

Sequenced Quantization RNN Offloading for Dependency Task in Mobile Edge Computing

Tan Deng, Shixue Li, Xiaoyong Tang$^{(\boxtimes)}$, Wenzheng Liu, Ronghui Cao, Yanping Wang, and Wenbiao Cao

School of Computer and Communications Engineering,
Changsha University of Science and Technology, Changsha 410114, China
tangxy@csust.edu.cn

Abstract. Mobile edge computing (MEC) allows terminals to send tasks to adjacent edge servers for calculation to reduce the burden on terminals and task completion time. With the widespread use of wireless devices (WDs) and the increasing complexity of applications, how to partially offload tasks to minimize task completion time has become a huge challenge. We propose a sequenced quantization based on recurrent neural network (SQ-RNN) algorithm that makes reasonable partial offload decisions for subtasks with dependencies. Specifically, the SQ-RNN algorithm first inputs the environment information into the RNN, and uses the RNN to generate a task offloading strategy. Then the algorithm quantifies the offloading strategy generated by the RNN into multiple binary offloading actions according to a certain method, and selects the action with the lowest computational delay from the multiple binary offloading actions as the offloading decision of the task. In addition, the algorithm also configs RNN with a fixed-size memory space to store the latest unloading strategy generated by RNN for further training of RNN. Experiments have proved that the SQ-RNN offloading algorithm described in our study generates better offloading decisions than those made by conventional offloading techniques.

Keywords: Dependent task · Task offloading · Neural networks

1 Introduction

There has been a continued increase in the number of smart mobile devices (SMDs) connected to the Internet, with the rapid development of the internet of things (IoT), thus resulting in large-scale data. This has caused problems such as bandwidth load, slow response, poor security, and poor privacy in traditional cloud computing models [1]. MEC proposes to move computing to the edge of the network closer to the user. In MEC, tasks can be offloaded from SMDs to edge servers, such as small base stations (SBSs) with computing and storage resources, thereby reducing data transmission, network latency and load on the cloud resources. However, offloading all the computing tasks to an edge server

Z. Tari et al. (Eds.): ICA3PP 2023, LNCS 14488, pp. 73–91, 2024.
https://doi.org/10.1007/978-981-97-0801-7_5

leads to long processing delays and high energy consumption for computing tasks [2]. Thus, a key challenge in MEC research is specifying which tasks should be offloaded and which ones should be executed locally to maximize resources' utilization and reduce latency and energy consumption [3].

The sustainable development of IoT technology is inseparable from computation offloading, which mainly involves two issues: offloading decision-making and resource allocation. Task offloading includes two categories: coarse-grained offloading (also known as binary offloading) and fine-grained offloading (also known as partial offloading). In partial unloading, a task is divided into multiple subtasks to be unloaded separately. Existing studies on partial offloading almost ignore the complex dependencies generated when a task is divided into multiple subtasks. The dependencies of subtasks play a decisive role in the execution sequence and waiting time of subtasks, which further affect the unloading decision and completion time of subtasks, which cannot be ignored.

Partial offloading can achieve lower latency and higher energy savings compared with binary offloading [4]. The goal of this paper is to make reasonable unloading decisions on subtasks to minimize task completion time. In addition, we take into account the general dependencies between subtasks, i.e., a subtask within a task may depend on one or more previous subtask results. We examines the edge computing involving an edge server, multiple WDs and multiple tasks, and each WD can choose either to offload its subtasks to the edge server or execute them locally. We proposes a SQ-RNN algorithm to make task offloading decisions given the uncertainty of the time-variant MEC, allowing for flexible task scheduling between the edge layer and local devices. The main contributions of this paper include the following:

- We propose a SQ-RNN offloading algorithm based on the edge environment of multi-user and multi-task, which uses the RNN to generate offload strategy. The offloading algorithm considers the dependencies between subtasks and can make task offloading decisions for mobile users arriving randomly for each task within the coverage of the edge server to minimize the task completion time.
- We design a sequence quantization method to generate task offloading actions, which quantizes the offloading policies output by the RNN into K binary offloading actions. This method can efficiently generate offloading actions. Compared with traditional methods, it can generate more candidate actions and increase the chance of finding better offloading decisions.
- We config the RNN with an experience pool with a fixed memory size, which can store the latest M offloading decisions generated by RNN itself. Every once in a while, the RNN randomly selects a batch of data from the experience pool for training, and updates its parameters to make better unloading decisions.
- We conduct simulation experiments to evaluate the performance of the SQ-RNN algorithm. Experimental results show that, compared with other basic offloading algorithms, SQ-RNN can effectively reduce the average completion time of tasks.

Section 2 of this paper reviews related work; Section 3 introduces the system model and problem formulation; Section 4 presents the proposed offloading algorithm; Simulation experiments are conducted, and the results are analyzed in Sect. 5, before the last Sect. 6 concludes the paper.

2 Related Work

MEC computation offloading is a hot research area for scholars both domestically and internationally. Most studies have converted the computation offloading and resource allocation problem into mixed-integer nonlinear programming (MINLP) problems, which feature mixed variables and nonlinear constraints. Because of this, using traditional mathematical optimization algorithms to obtain a guaranteed optimal solution in a reasonable time is challenging. There are many methods for computing offloading, among which the use of intelligent swarm algorithms and machine learning (ML) to address MEC computation offloading problems has received widespread attention in recent years.

Swarm intelligence algorithms, a branch of biologically inspired algorithms, have been extensively used in solving MINLP problems. Some popular swarm intelligence algorithms include genetic algorithms (GA), the ant colony optimization (ACO), and particle swarm optimization (PSO). [5] uses a set of chromosomes to represent possible offloading decisions for a set of subtasks, and generates new chromosomes through GA selection, crossover, and mutation, before identifying the best offloading decision based on fitness values and feasibility indicators of chromosomes. [6] proposes a two-layer optimization method where the upper layer uses sorting-enabled ACO to find the optimal offloading decision, and the lower layer uses monotonic optimization to achieve the optimal resource allocation for each offloading decision. A PSO based on genetic simulated annealing has been proposed in [7], which integrates the genetic operation of GA and the Metropolis rule of simulated annealing (SA). In so doing, particles can update their velocity and position based on their learning experience and the current situation, thus finding the optimal unloading strategy. Swarm intelligence algorithms have been frequently employed to solve MEC computing offloading problems, however, their drawbacks have gradually emerged. Firstly, swarm intelligence algorithms may get stuck in local optima during the optimization process, decreasing the search precision; secondly, swarm intelligence algorithms may not be able to effectively handle constraints, as a result of which, the optimal solution may be infeasible; thirdly, the performance of swarm intelligence algorithms is significantly impacted in high-dimensional scenarios, resulting in a decline in their search capability.

ML has a wide range of applications in many fields [8–10]. In recent years, a new research field called Edge Intelligence [11, 12] has emerged, which applies ML to edge computing offloading. Some commonly used ML methods for MEC computing offloading include deep learning (DL), reinforcement learning (RL) and deep reinforcement learning (DRL).

With robust learning and reasoning, DL can extract useful information from massive real-time data generated by SMDs and accordingly make appropriate offloading decisions with low power consumption and low latency. The primary advantage of DL lies in high accuracy in decision making and high computational speed for training models [13]. [14] proposed a DL-based offloading technique that considers partial task offloading and task heterogeneity, and uses DNN to compute and make task offloading decisions. [15] developed a distributed DL-based computation offloading and resource allocation algorithm that uses multiple parallel DNNs to generate the optimal offloading decisions and resource scheduling. [16] used the MAPE-K loop to simulate the offloading problem in various contexts, and DNN to select the optimal location for task offloading, and subsequently employed the hidden Markov model (HMM) to choose the most suitable upstream transmission media formats.

Compared with DL, RL does not require labeled data for training. Alternatively, RL interacts with the MEC network directly to identify the optimal strategy. A Q-learning-based task computing and resource allocation method was proposed in [17]. It defines the Q function values of the actions of the edge in each state, then selects an action for next state, updates the immediate cost caused by the transition with the Q function, and finally selects the optimal strategy based on the Q function values. Nevertheless, conventional RL cannot scale effectively with the increase in the number of agents, as traditional tabular methods tend to be infeasible in solving the state space explosion problem [18].

Fusing RL with DNN, DRL learns effective strategies through interacting with the environment, achieving flexible and adaptive task offloading in cases of no expert knowledge. This is particularly effective for solving complex decision-making problems in high-dimensional state and action spaces. A new framework that integrates Lyapunov optimization and DRL was presented in the study [19]. The Lyapunov optimization was utilized to decouple multi-stage MINLP into deterministic per-frame MINP subproblems, before a model-free actor-critic architecture was adopted to solve per-frame MINLP problems. [20] designed a counter-factual multi-agent DRL algorithm based on a Markov chain model, which introduces a centralized critic and counter-factual baselines. To be specific, a centralized critic takes the joint state of each user as its input and calculates a baseline that represents the contribution of its action to the global reward based on the centralized critic for each agent. The paper [21] adopted the deep deterministic policy gradient (DDPG), which learn the offloading strategies of each user on the continuous action spaces. Without any prior knowledge, each mobile user can independently develop a dynamic offloading strategy based on its local observation of the system. [22] designed a deep meta RL-based offloading algorithm that combines muliple parallel DNNs with Q-learning to make optimal unloading decisions.

Combining ML techniques with MEC systems to obtain unloading strategies is effective for computation offloading optimization in MEC. The aforementioned research is not without flaws, though. Some papers have overlooked the fact that the number of SMDs is rapidly increasing. When there are excessive devices within the coverage area of edge servers, it is not realistic to execute all tasks locally or entirely offload to the edge. Considering the partial offloading of tasks is more in line with the current network environment, achieving lower latency and higher energy saving. In addition, although some studies have suggested partial offloading of tasks, they tend to ignore the dependencies between tasks. Subtask dependencies determine their execution sequences and delay time, which have a substantial impact on the offloading and computing process and must be taken into consideration. When the number of subtasks increases, the dependencies between subtasks are more complex, so we propose to use the powerful learning ability of DL to extract useful information between subtasks and make low-latency offloading decisions on them.

3   System Model and Problem Formula

3.1   System Model

This study takes into account an edge server with many WDs and multiple tasks, as depicted in Fig. 1. The edge server can be a 4G/5G macrocellular base station or a SBS, or a wireless access point (AP), which can provide services to all mobile devices within its signal range. The main notations of the paper are shown in the Table 1, and the following content will explain these notations in detail. Assume that users can move within the range of the MEC server, $N = \{1, 2, \cdots, n\}$ represents the set of WDs in the edge server range. Within a slot, a user's location and system resource status remain constant, while its computing needs and task requests fluctuate dynamically across slots. Each user has the option of using local computing or edge offloading for task processing. The former means that the user chooses to process the task locally on the terminal, whilst the latter means that the user chooses to offload the task to the edge server.

Task Model. Each WD needs to compute the randomly arriving tasks in each slot, where each task consists of multiple subtasks with a general dependency relation. Assume that the computation task that arrives at each WD has Q subtasks, the task arriving at WD n in slot τ can be denoted as $S_{n,\tau} = \{S_{n,\tau}^1, S_{n,\tau}^2, \cdots, S_{n,\tau}^Q, \}$, and the i-th subtask of the task is denoted as $S_{n,\tau}^i = \{D_{n,\tau}^i, C_{n,\tau}^i\}$, where $D_{n,\tau}^i$ and $C_{n,\tau}^i$ respectively represent the data size and computation density of the i-th subtask. Given the general dependencies between the subtasks, matrix $A_{n,\tau[Q*Q]}$ is used to represent the dependency relationships between the subtasks in task $S_{n,\tau}$ arriving at WD n in slot τ. $A_{n,\tau}[i][j] = 1$ indicates that subtask j depends on subtask i, where $i < j$ and $i, j \leq Q$.

Communication Model. The WDs use orthogonal frequency-division multiple access (OFDMA) to communicate with AP, with all users sharing the same frequency band. When a user offloads the task to the edge, the task is wirelessly

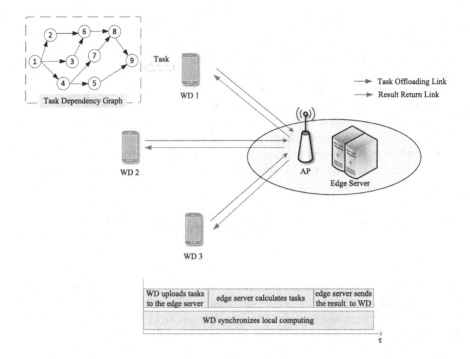

Fig. 1. System Model of an MEC network with multiple WDs and multiple dependent tasks

Table 1. Main notations

Notation	Description
N	the set of WDs
T	index set of the time slots
$S_{n,\tau}^i$	i-th subtask from WD n at slot τ
$D_{n,\tau}^i$, $C_{n,\tau}^i$	data size and computation density of subtask $S_{n,\tau}^i$, respectively
$A_{n,\tau[Q*Q]}$	dependency relationships between the subtasks in task $S_{n,\tau}$
$T_{n,\tau}^{i,start}$, $T_{n,\tau}^{i,end}$	start time and end time of subtask $S_{n,\tau}^i$, respectively
B	bandwidth
P_{max}	maximum value of transmission power
$f_{n,\tau}^l$	computing capacity of WD n at slot τ
$f_{n,\tau}^e$	computing capacity of the edge server at slot τ
$a_{n,\tau}^i$	offloading decision of subtask $S_{n,\tau}^i$
$r_{n,\tau}$, $p_{n,\tau}$	transmission rate and transmission efficiency of WD n at slot τ, respectively
$h_{n,\tau}$	channel gain of WD n at slot τ
h^0	channel gain when the reference distance is 1m
$d_{n,\tau}$	straight-line distance between WD n and the edge server at slot τ
v_n, s	moving speed and walking distance of WD n, respectively
σ, β	noise power of the channel and path loss exponent, respectively
z	initial distance between WD n and the base station
e, x	vertical height and coverage range of the base station, respectively

delivered to the relevant AP and computed by the edge server. The user then synthesizes the results after receiving the feedback from the AP when the calculation is finished. Since all WDs in the system use orthogonal access, there is no interference among mobile devices. The transmission rate is determined as follows:

$$r_{n,\tau} = B \cdot \log_2 \left(1 + \frac{p_{n,\tau} \cdot h_{n,\tau}}{\sigma} \right) \tag{1}$$

where B represents the bandwidth of WD n at slot τ, σ represents the noise power of the channel, $p_{n,\tau}$ and $h_{n,\tau}$ represent the transmission efficiency and channel gain of WD n at slot τ, respectively.

The channel gain $h_{n,\tau}$ can be expressed as:

$$h_{n,\tau} = h^0 \left(d_{n,\tau} \right)^{-\beta} \tag{2}$$

where h^0 represents the channel gain when the reference distance is 1m, β is the path loss exponent, and $d_{n,\tau}$ represents the straight-line distance between WD n and the edge server at slot τ. The distance between WDs and the edge server changes over time due to user movement. If the moving speed of WD n is v_n, then the straight-line distance between WD n and the edge server changes as follows:

$$d_{n,\tau} = z + \left(x - \sqrt{e^2 + \left(\frac{s}{2} - v_n\tau \right)^2} \right) \tag{3}$$

The walking distance of WD n within the coverage range of the base station is denoted as $s = 2\sqrt{x^2 - e^2}$. e and x represent the vertical height and coverage range of the base station, respectively, while z represents the initial distance between WD n and the base station.

Computational Model. Each WD has the option of executing computing tasks locally or offloading them to the edge server. The offloading decision of task $S_{n,\tau}$ arriving at WD n in slot τ can be represented as $a_{n,\tau} = \{a_{n,\tau}^1, a_{n,\tau}^2, \cdots, a_{n,\tau}^Q\}$, where $a_{n,\tau}^i$ takes the value 0 or 1, with $1 \le i \le Q$. $a_{n,\tau}^i = 0$ indicates that the i-th subtask of the task $S_{n,\tau}$ is executed locally, while $a_{n,\tau}^i = 1$ indicates that it is offloaded to the edge server for execution. $T_{n,\tau}^{i,start}$ and $T_{n,\tau}^{i,end}$ are used to represent the start and end times of the i-th subtask of the task $S_{n,\tau}$.

When the subtask $S_{n,\tau}^i$ is executed locally on WD n, $T_{n,\tau}^{i,start}$ is defined as follows:

$$T_{n,\tau}^{i,start} = \begin{cases} 0, & i = 1 \\ 0, & A_{n,\tau}[j][i] = 0 \&\& a_{n,\tau}^j = 1 \quad for\forall j < i \\ \max\left(T_{n,\tau}^{j,end}\right), & A_{n,\tau}[j][i] = 1 || a_{n,\tau}^j = 0 \quad for\exists j < i \end{cases} \tag{4}$$

And $T_{n,\tau}^{i,end}$ is defined as follows:

$$T_{n,\tau}^{i,end} = T_{n,\tau}^{i,start} + \frac{C_{n,\tau}^i \cdot D_{n,\tau}^i}{f_{n,\tau}^l} \tag{5}$$

where $f_{n,\tau}^l$ represents the computing capacity of WD n in slot τ.

When the subtask $S_{n,\tau}^i$ is offloaded to the edge for computation, $T_{n,\tau}^{i,start}$ is defined as follow:

$$T_{n,\tau}^{i,start} = \begin{cases} 0, & i = 1 || A_{n,\tau}[j][i] = 0 \quad for \forall j < i \\ \max\left(T_{n,\tau}^{i,end}\right), & A_{n,\tau}[j][i] = 1 \quad for \exists j < i \end{cases} \quad (6)$$

The time it takes to transmit the computed result back to the terminal device can be ignored because the computed result is quite small in number compared with the uploaded computation task and wireless transmission rate. $T_{n,\tau}^{i,end}$ is defined as follows:

$$T_{n,\tau}^{i,end} = T_{n,\tau}^{i,start} + T_{n,\tau}^{i,trans} + \frac{C_{n,\tau}^i \cdot D_{n,\tau}^i}{f_{n,\tau}^e} \quad (7)$$

where $f_{n,\tau}^e$ represents the computing capacity of the edge server in slot τ. $T_{n,\tau}^{i,trans}$ represents the transmission time needed to offload subtask $S_{n,\tau}^i$ to the edge:

$$T_{n,\tau}^{i,trans} = \frac{D_{n,\tau}^i}{r_{n,\tau}} \quad (8)$$

3.2   Problem Formula

Our study seeks to reduce the task completion time by making wise offloading decisions for computational tasks that randomly arrive at WDs in T sequential time frames. The time required for WD n to complete the task $S_{n,\tau}$ that arrives in slot τ is denoted as:

$$T_{n,\tau} = \max_{i \in [1,Q]}\left(T_{n,\tau}^{i,end}\right) \quad (9)$$

The optimization goal is:

$$\min\left(\sum_{\tau=1}^{T} T_{n,\tau}\right) \quad (10a)$$

$$subject \quad to \quad T_{n,\tau} \leq \tau \quad (10b)$$

$$a_{n,\tau}^i \in \{0,1\} \quad (10c)$$

$$0 \leq p_{n,\tau} \leq P_{max} \quad (10d)$$

where P_{max} represents the maximum value of transmission power.

4   Sequenced Quantization Based on RNN Offloading Algorithm

The SQ-RNN algorithm proposed in the paper consists of three main stages, namely, generating unload actions, sequenced quantization offloading actions, and experience pool recycling. The details of the algorithm are shown in Algorithm 1. Our study aims to minimize the computation completion time by making optimal unload decisions for the tasks that are delivered at consecutive time intervals T from the WDs that are within the edge server's coverage area. In order to make optimal task unload decisions, we need to design a function to generate unload actions for different users at different time intervals. As shown in Fig. 2, in SQ-RNN algorithm, the generation of task unload actions relies on RNN, so we use the parameterized function of RNN to generate task unload actions. In slot τ, there are multiple WDs within the coverage of the edge server, and each WD has multiple computing tasks to arrive. The AP first collects the environmental information in the current slot as the input of RNN, and the RNN will output the unload actions of these tasks based on the current offloading strategy. Then, the order-preserving quantization method is used to quantify the unload actions of each task into K binary unload actions, and the one with the minimum delay is chosen as the task unload decision. The latest M unload decisions generated by RNN are stored in the experience pool. When there exist M' new unload decisions, a batch of data is randomly selected from the experience pool to train and update the parameter θ of the RNN. The latest parameter function is adopted to create offload decisions for the task in the subsequent slot. The following subsection will detail the three stages.

Algorithm 1. SQ-RNN Algorithm

Input: $\theta, f_{n,\tau}^l, f_{n,\tau}^e, d_{n,\tau}, D_{n,\tau}, A_{n,\tau[Q*Q]}$
Output: $a_{n,\tau}$
1: set T, M, M', N and K;
2: initialize the RNN with parameter θ;
3: initialize an empty experience pool of size M;
4: set SUM $\leftarrow 0$;
5: **for** $\tau \leftarrow 1, 2, \cdots, T$ **do**
6:     **for** $n \leftarrow 1, 2, \cdots, N$ **do**
7:         AP collect $In_n \leftarrow \{f_{n,\tau}^l, d_{n,\tau}, D_{n,\tau}, A_{n,\tau[Q*Q]}\}$;
8:     **end for**
9:     AP collect $f_{n,\tau}^e$;
10:     take $\{f_{n,\tau}^e, In_1, In_2, \cdots, In_N\}$ as input to the RNN(θ) ;
11:     RNN(θ) generates $\tilde{a}_{1,\tau}, \tilde{a}_{2,\tau}, \cdots, \tilde{a}_{N,\tau}$;
12:     **for** $n \leftarrow 1, 2, \cdots, N$ **do**
13:         $a' \leftarrow argminT(L(\tilde{a}_{n,\tau}))$;
14:         store $\{In_n, a'\}$ to the experience pool;
15:         SUM $\leftarrow$ SUM $+ 1$;
16:         **if** SUM $\% M' == 0$ **then**
17:             randomly select a batch of data from the experience pool;
18:             train the RNN and update θ;
19:         **end if**
20:     **end for**
21: **end for**

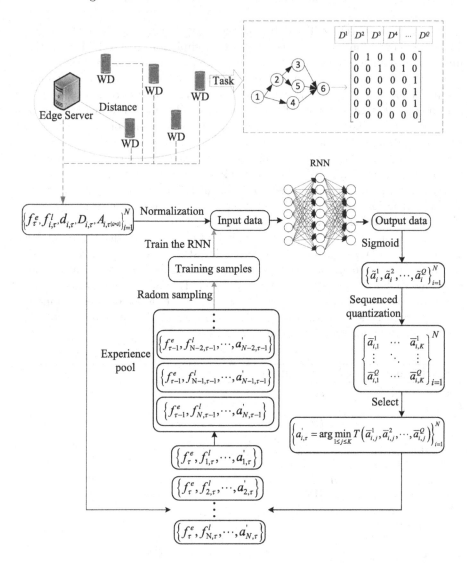

Fig. 2. The SQ-RNN Algorithm

4.1   Generation of Unload Actions

Each WD must complete the computation of randomly assigned tasks in each slot. Each task contains multiple subtasks with general dependencies, which significantly affects the execution order and wait time for the subtasks. The complexity of subtask dependencies increases as the number of subtasks increases. More factors need to be considered when making unloading decisions for subtasks with complex dependencies. The decision of a subtask may affect multiple subtasks, thereby affecting the entire task. Tasks randomly arrive at the WDs in sequential time frames, each WD's decision about task offloading for the previous time slot affects the network environment in the subsequent time slot. This, in turn, has an impact on the task offloading decision for the subsequent time slot. Considering the above two points, we employ a RNN to generate the task offloading action. First of all, RNN is an algorithm belonging to DL, which has strong learning ability and reasoning ability, and can make reasonable unloading decisions for tasks in complex task dependencies. Second, RNN is a type of neural networks with short-term memory. In an RNN, neurons can exchange information with other neurons and combine their own information to build a network loop. The information flow in a regular neural network is unidirectional, which makes the network easier to learn but somewhat reduces the effectiveness of the neural network model. The input of RNN at different time steps is related to its previous time state, while its output is the sum of the input at that time and all prior inputs.

According to the well-known approximation theorem, a hidden layer with sufficient hidden neurons can approximate any continuous mapping if appropriate activation functions, such as sigmoid, ReLu, and tanh functions, are applied to the neurons [23]. We use the sigmoid activation function in the output layer, and the task offloading actions for task arrived at WD n in slot τ can be output as $\tilde{a}_{n,\tau} = \left(\tilde{a}_{n,\tau}^1, \tilde{a}_{n,\tau}^2, \cdots, \tilde{a}_{n,\tau}^Q \right)$, where $\tilde{a}_{n,\tau}^i \in (0,1)$, $1 \leq i \leq Q$. Using the sigmoid function can avoid the loss of mean square error and the decline in learning rate during the gradient descent process.

4.2 Sequenced Quantization Offloading Actions

In MEC, the edge server has an AP for collecting information about servers, users and tasks. At the beginning of each slot, the AP collects the previously mentioned information for that slot and sends it to the RNN. The RNN takes the environmental and task parameters as its input, and then outputs the unload action to be taken by the task arriving at WDs in that slot after calculation.

Most existing methods convert offloading actions $\tilde{a}_{n,\tau}$ with values between 0 and 1 to binary offloading actions $a_{n,\tau}$ using the following approach.

$$a_{n,\tau}^i = \begin{cases} 0, & \tilde{a}_{n,\tau}^i < 0.5 \\ 1, & \tilde{a}_{n,\tau}^i \geq 0.5 \end{cases} \tag{11}$$

for $i = 1, 2, \cdots, Q$. The traditional action generation method cannot effectively convert the neural network output into corresponding binary actions. The SQ-RNN algorithm proposed in the paper uses a order-preserving quantization method to quantize the output of the RNN to obtain K binary unload actions. The quantization function L is defined as follows:

$$L : \tilde{a}_{n,\tau} \mapsto \left\{ \bar{a}_{n,\tau,k} \mid \bar{a}_{n,\tau,k} \in \{0,1\}^Q, k = 1, 2, \cdots, K \right\} \tag{12}$$

where K is a set parameter, $K \in \left[1, 2^Q \right]$, and the larger the value of K, the better the solution obtained, and the higher the computational complexity. The basic idea behind the order-preserving quantization method used in the SQ-RNN algorithm is to maintain order during the quantization process. It means that the order of the subtasks in the generated quantized actions $\bar{a}_{n,\tau,k}$ is the same as that of subtasks in the unload action $\tilde{a}_{n,\tau}$. Compared with traditional methods, the order-preserving quantization method causes greater distances between the generated offloading actions, enriches the diversity of the candidate set, and increases the chance of finding local maxima. For a given parameter K, the method of generating the quantized action set$\{\bar{a}_{n,\tau,k}\}$ is shown in Fig. 3, and the details are as follows.

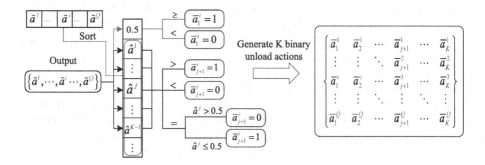

Fig. 3. Sequenced quantization offloading actions

The method for generating the first binary actions $\bar{a}_{n,\tau,1}$ from the unload action $\tilde{a}_{n,\tau}$ is as follows:

$$\bar{a}_{n,\tau,1}^i = \begin{cases} 0, & \tilde{a}_{n,\tau}^i < 0.5 \\ 1, & \tilde{a}_{n,\tau}^i \geq 0.5 \end{cases} \tag{13}$$

for $i = 1, 2, \cdots, Q$.

To generate the remaining $K - 1$ binary actions, the entries of the unload are first sorted in ascending order according to their distances to 0.5, and the sorted unload action is represented as $\hat{a}_{n,\tau} = \{\hat{a}_{n,\tau}^1, \hat{a}_{n,\tau}^2, \cdots, \hat{a}_{n,\tau}^Q\}$, where $|\hat{a}_{n,\tau}^1 - 0.5| \leq |\hat{a}_{n,\tau}^2 - 0.5| \leq \cdots \leq |\hat{a}_{n,\tau}^Q - 0.5|$. The index $i(1 < i \leq Q)$ in $\hat{a}_{n,\tau}$ no longer represents the order of the subtasks. The method for generating the k-th$(k = \{2, 3, \cdots, K\})$ unload action is as follows:

$$\bar{a}_{n,\tau,k}^i = \begin{cases} 0, & \tilde{a}_{n,\tau}^i < \hat{a}_{n,\tau}^{k-1} \\ 0, & \tilde{a}_{n,\tau}^i = \hat{a}_{n,\tau}^{k-1} \&\& \hat{a}_{n,\tau}^{k-1} > 0.5 \\ 1, & \tilde{a}_{n,\tau}^i = \hat{a}_{n,\tau}^{k-1} \&\& \hat{a}_{n,\tau}^{k-1} \leq 0.5 \\ 1, & \tilde{a}_{n,\tau}^i > \hat{a}_{n,\tau}^{k-1} \end{cases} \tag{14}$$

for $i = 1, 2, \cdots, Q$. After generating K binary actions, the optimal offloading decision for the task is selected in line with (10a).

4.3   Experience Pool Recycling

The SQ-RNN algorithm has an experience pool that stores the latest M offloading decisions generated by the RNN. After generating unload decisions for the tasks in the current slot, the algorithm will save the latest unload decisions to the experience pool with limited capacity. Once the memory is full, newly generated data will replace the previous data. Each time M' new unload decisions are stored in the experience pool, a batch of data will be randomly selected for the self-learning of the RNN. To lower the average cross-entropy loss, the paper updates the RNN's parameters using the Adam algorithm. The experience pool provides the following benefits: first, batch data updates are less complex than updates utilizing the entire dataset; second, random sampling helps quicken convergence by reducing the correlation between training samples; and third, real-time data can be used to enhance the RNN parameters.

5   Simulation Experiments and Result Analysis

5.1   Experimental Setup

This paper examines quasi-static scenarios of a single server with multiple WDs, where the positions of the WDs and the state of the system resources remain unchanged within a slot. Our proposed SQ-RNN algorithm makes offloading decisions for multiple computation tasks generated by five WDs in 200 continuous slots, where each computing task contains dependent subtasks. Other experiment parameters are shown in Table 2. We implement the SQ-RNN algorithm in Python with Pytorch. The batch_task data in the cluster-trace-v2018 dataset is used for simulation experiments. The task_name, plan_mem, and plan_cpu attributes in the batch_task data are required for our experiment, where task_name contains the DAG information of the task, and plan_mem and plan_cpu represent the task size and the number of CPU cycles required. To better train the neural network parameters, 80% of the dataset is used for training purposes and the remaining 20% is used to test the model. The activation functions for the hidden layer and the output layer of the neural network are set to 'ReLu' and 'Sigmoid', respectively. The Loss function used is BCELoss, and Adam is selected as the optimizer.

5.2   Algorithm Parameter Selection

To generate the most suitable offloading strategy using the SQ-RNN algorithm, some key parameters have to be properly configured, including the learning rate of the RNN, M', and K. The SQ-RNN algorithm is used to compute the offloading decisions for each task that arrives in T consecutive time slots after the RNN has been trained with various parameter values. The average time taken to complete all tasks for each 25 slot is then calculated as the SQ-RNN algorithm performance metric.

Table 2. Parameter settings

Parameter	Value	Parameter	Value
M	2048	h^0	$-30dB$
$D_{n,\tau}^i$	$(0,2]\,Mb$	β	4
$C_{n,\tau}^i$	$[50,100]\,Cycles/bit$	v_n	$[0,1]\,m/s$
$f_{n,\tau}^l$	$[4,6]\times 10^7\,Cycles/s$	e	35 m
$f_{n,\tau}^e$	$5\times 10^9\,Cycles/s$	x	300 m
B	$[10,20]\,MHz$	σ	-104 dBm
$p_{n,\tau}$	$[30,40]$ dBm		

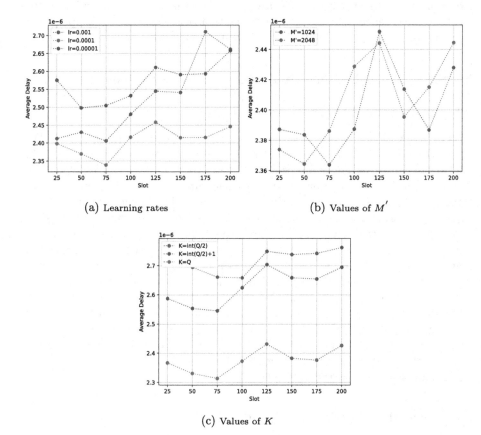

(a) Learning rates

(b) Values of M'

(c) Values of K

Fig. 4. Performance of the SQ-RNN algorithm with different neural network parameters

The RNN cannot converge to an optimum if the learning rate is either too high or too low. Figure 4(a) depicts the simulation based on various learning rates. When the learning rates are 0.001 and 0.00001, the average delay of the unload strategies generated by the SQ-RNN algorithm in each slot is usually always greater than the average delay of the unload decisions generated by the algorithm with a learning rate of 0.0001. Therefore, the optimal learning rate value for the algorithm is 0.0001.

Figure 4(b) shows how the performance of the SQ-RNN algorithm was affected by various values of M'. A memory experience pool was installed in the MEC network under consideration. Each time when M' fresh offloading decisions are stored in the experience pool, a batch of data is then randomly chosen to update the RNN's parameters. M' was examined at two different values, namely, 1024 and 2048. It can be seen from Fig. 4(b) that both $M' = 1024$ and $M' = 2048$ provide offloading decisions with significant fluctuations in average latency. The average delay of the decisions made by $M' = 1024$ is slightly smaller than that of $M' = 2048$, nevertheless. Therefore, in subsequent experiments, a batch of data can be randomly selected from the experience pool to train the RNN when the experience pool stores every 1024 new offloading decisions.

The paper uses the order-preserving quantization method to quantize the output of the RNN into K binary offloading actions, increasing the diversity of candidate actions and the likelihood of finding local maxima. Figure 4(c) shows the impact of different K values on the performance of the SQ-RNN algorithm. The larger the value of K, the lower the average latency in offloading decisions in each slot. Therefore, the optimal value of K is Q.

5.3   Result Analysis

In order to analyze the performance of the SQ-RNN algorithm trained with optimal parameters in more detail, in this part, we take the average computation time required by all tasks arriving at each slot as a comparison index. Figure 5(a) shows a performance comparison of the SQ-RNN model trained with optimal parameters and the greedy algorithm. We list all possible offloading decision combinations for the greedy algorithm and choose the one with the minimum latency. It should be noted that the greedy approach is very time-consuming, especially when when there are many users and tasks. As shown in the Figure, in each time slot, the average latency in offloading decisions by the SQ-RNN algorithm differs from that by the greedy algorithm by only $0.296e^{-6}$ seconds. Therefore, the generated offloading decisions for our proposed SQ-RNN algorithm are almost optimal if relevant parameters can be appropriately defined.

To further evaluate the performance of the proposed SQ-RNN algorithm in the article, we compare it with the following scheme:

- All local computing (ALC): All task arriving at WDs in each time slot is computed locally.
- Random offloading (RO): Each task arriving at WD in each time slot is randomly offloaded to the edge server or computed locally.
- GA based offloading (GA): GA represents the potential offloading decision of each task in each slot as a set of genetic parameters, and generates new chromosomes through operations such as selection, crossover, and mutation.

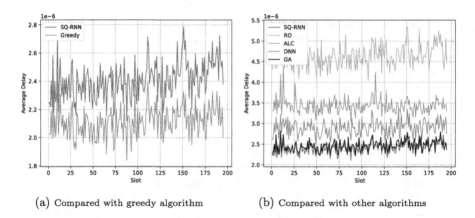

(a) Compared with greedy algorithm   (b) Compared with other algorithms

Fig. 5. Comparison of the performance between the SQ-RNN algorithm and other algorithms

The optimal unloading decision is then made based on chromosome fitness and feasibility criteria.

- DNN based offloading (DNN): We train a simple DNN with two hidden layers to make unloading decisions for tasks considering the edge environment and task size, but do not consider the dependencies between subtasks.

In terms of GA parameters, the population iteration number and the chromosome population size are both set to 10. Figure 5(b) displays the average delay of the unload decisions for all tasks arriving at different slots, generated by the five unload schemes. As seen from the Figure, random offloading has the worst performance, followed by all local computing. The SQ-RNN algorithm, GA and DNN deliver better results. The calculation shows that the average delay of offloading decisions generated by the SQ-RNN algorithm in each time slot is typically $0.491e^{-6}$ seconds and $0.056e^{-6}$ seconds lower than that generated by the DNN and the GA. It can be seen from the Fig. 5(b) that when the $Q = 6$, the performance of the GA is basically the same as that of the SQ-RNN. However, the time required for GA to make an offloading decision increases with the number of subtasks, while the time required for SQ-RNN to generate an offloading decision is not affected by the number of subtasks. In summary, the SQ-RNN algorithm can better adapt to partial offloading scenarios involving multiple MDs and multiple tasks.

5.4   The Impact of the Number of Subtasks on the SQ-RNN Algorithm

The aforementioned experiment only takes the scenario of $Q = 6$ into consideration. If Q increases or decreases, the complexity of the dependencies between subtasks changes, which accordingly affects the algorithm's performance. As shown in Fig. 6, we experimented with different values of Q, keeping all other conditions and parameters constant, calculated the average completion time of all tasks arriving within 200 consecutive time slots, and used it as the algorithm performance index. Figure 6 shows that for $Q = 3$, the unload decisions generated by the DNN and GA are better than the unload decision generated by the SQ-RNN algorithm. When $Q = 5, 7$, and 9, the SQ-RNN algorithm makes better unload decisions than the DNN and GA. From the above experimental results, it can be seen that the SQ-RNN algorithm proposed in this paper performs better on more complex task dependencies.

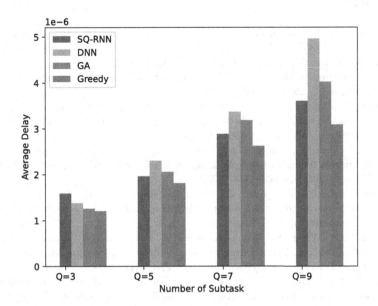

Fig. 6. The SQ-RNN algorithm performance under different number of subtasks

6   Conclusion

This paper has examined the problem of subtask offloading with general dependencies in quasi-static scenarios of a single server with multiple WDs and multiple tasks. The SQ-RNN algorithm has been proposed to solve the subtask offloading problem to minimize the delay in completing tasks. The SQ-RNN algorithm first generates unloading decisions with RNN, and then quantifies them into K binary unloading actions using the order-preserving quantization method. The SQ-RNN algorithm also uses the latest unloading decisions to train and update the parameters of the RNN so that it can make a better unloading decision. The proposed SQ-RNN algorithm has been implemented in Python and tested extensively using the cluster-trace-v2018 dataset. The experimental results demonstrate that when the number of subtasks is 6, the superiority of the SQ-RNN algorithm over DNN, GA, random offloading, and all local calculations. And the average delay of the unload decision generated by the SQ-RNN algorithm in each slot is $0.296e^{-6}$ seconds longer compared with the greedy algorithm. In addition, the proposed SQ-RNN algorithm performs better as the subtask dependency complexity increases.

Acknowledgements. This work was supported in part by the National Natural Science Foundation of China (Grant Nos. 61972146, 62002032) and Postgraduate Scientific Research Innovation Project of Hunan Province (CX20220942).

References

1. Cao, K., Liu, Y., Meng, G., Sun, Q.: An overview on edge computing research. IEEE access **8**, 85714–85728 (2020)
2. Raeisi-Varzaneh, M., Dakkak, O., Habbal, A., Kim, B.S.: Resource scheduling in edge computing: architecture, taxonomy, open issues and future research directions. IEEE Access **11**, 25329–25350 (2023)
3. Yeganeh, S., Sangar, A.B., Azizi, S.: A novel q-learning-based hybrid algorithm for the optimal offloading and scheduling in mobile edge computing environments. J. Netw. Comput. Appl. **214**, 103617 (2023)
4. Saleem, U., Liu, Y., Jangsher, S., Tao, X., Li, Y.: Latency minimization for d2d-enabled partial computation offloading in mobile edge computing. IEEE Trans. Veh. Technol. **69**(4), 4472–4486 (2020)
5. Al-Habob, A.A., Dobre, O.A., Armada, A.G., Muhaidat, S.: Task scheduling for mobile edge computing using genetic algorithm and conflict graphs. IEEE Trans. Veh. Technol. **69**(8), 8805–8819 (2020)
6. Huang, P.Q., Wang, Y., Wang, K., Liu, Z.Z.: A bilevel optimization approach for joint offloading decision and resource allocation in cooperative mobile edge computing. IEEE Trans. Cybern. **50**(10), 4228–4241 (2019)
7. Yuan, H., Bi, J., Duanmu, S.: Cost optimization for partial computation offloading and resource allocation in heterogeneous mobile edge computing. In: 2021 IEEE International Conference on Systems, Man, and Cybernetics (SMC), pp. 3089–3094. IEEE (2021)
8. Li, Y., Li, K., Chen, C., Zhou, X., Zeng, Z., Li, K.: Modeling temporal patterns with dilated convolutions for time-series forecasting. ACM Trans. Knowl. Disc. Data (TKDD) **16**(1), 1–22 (2021)
9. Chen, C., Li, K., Zhongyao, C., Piccialli, F., Hoi, S.C., Zeng, Z.: A hybrid deep learning based framework for component defect detection of moving trains. IEEE Trans. Intell. Transp. Syst. **23**(4), 3268–3280 (2020)
10. Zou, X., Zhou, L., Li, K., Ouyang, A., Chen, C.: Multi-task cascade deep convolutional neural networks for large-scale commodity recognition. Neural Comput. Appl. **32**(10), 5633–5647 (2020)
11. Xu, D., et al.: Edge intelligence: Architectures, challenges, and applications. arXiv preprint arXiv:2003.12172 (2020)
12. Deng, S., Zhao, H., Fang, W., Yin, J., Dustdar, S., Zomaya, A.Y.: Edge intelligence: the confluence of edge computing and artificial intelligence. IEEE Internet Things J. **7**(8), 7457–7469 (2020)
13. Ali, Z., Jiao, L., Baker, T., Abbas, G., Abbas, Z.H., Khaf, S.: A deep learning approach for energy efficient computational offloading in mobile edge computing. IEEE Access **7**, 149623–149633 (2019)

14. Ali, Z., Abbas, Z.H., Abbas, G., Numani, A., Bilal, M.: Smart computational offloading for mobile edge computing in next-generation internet of things networks. Comput. Netw. **198**, 108356 (2021)
15. Wang, Z., Lv, T., Chang, Z.: Computation offloading and resource allocation based on distributed deep learning and software defined mobile edge computing. Comput. Netw. **205**, 108732 (2022)
16. Shakarami, A., Shahidinejad, A., Ghobaei-Arani, M.: An autonomous computation offloading strategy in mobile edge computing: a deep learning-based hybrid approach. J. Netw. Comput. Appl. **178**, 102974 (2021)
17. Dab, B., Aitsaadi, N., Langar, R.: Q-learning algorithm for joint computation offloading and resource allocation in edge cloud. In: 2019 IFIP/IEEE Symposium on Integrated Network and Service Management (IM), pp. 45–52. IEEE (2019)
18. Sutton, R.S., Barto, A.G.: Reinforcement learning: An introduction. MIT press (2018)
19. Bi, S., Huang, L., Wang, H., Zhang, Y.J.A.: Lyapunov-guided deep reinforcement learning for stable online computation offloading in mobile-edge computing networks. IEEE Trans. Wireless Commun. **20**(11), 7519–7537 (2021)
20. Liu, C., Tang, F., Hu, Y., Li, K., Tang, Z., Li, K.: Distributed task migration optimization in mec by extending multi-agent deep reinforcement learning approach. IEEE Trans. Parallel Distrib. Syst. **32**(7), 1603–1614 (2020)
21. Chen, Z., Wang, X.: Decentralized computation offloading for multi-user mobile edge computing: a deep reinforcement learning approach. EURASIP J. Wirel. Commun. Netw. **2020**(1), 1–21 (2020)
22. Qu, G., Wu, H., Li, R., Jiao, P.: Dmro: a deep meta reinforcement learning-based task offloading framework for edge-cloud computing. IEEE Trans. Netw. Serv. Manage. **18**(3), 3448–3459 (2021)
23. Goyal, M., Goyal, R., Venkatappa Reddy, P., Lall, B.: Activation functions. Deep learning: Algorithms and applications, pp. 1–30 (2020)

KylinArm: An Arm Gesture Recognition System for Mobile Devices

Shikun Zhao[1], Jingxuan Hong[1,2], Zixuan Zhang[4], Xuqiang Wang[4], Jin Zhang[1,2], and Xiaoli Gong[1,2,3(✉)]

[1] College of Computer Science, Nankai University, Tianjin, China
gongxiaoli@nankai.edu.cn
[2] Tianjin Key Laboratory of Brain Science and Intelligent Rehabilitation, Tianjin, China
[3] State Key Laboratory of High-End Server and Storage Technology, Jinan, China
[4] State Grid Tianjin Information and Communication Company, Tianjin, China

Abstract. Gesture-based Human-Computer Interaction (HCI) has become a primary means of device control due to its naturalness and humanized characteristics, making it applicable for tasks such as drone control and gaming. Gesture recognition using an inertial measurement unit (IMU) has emerged as a major trend in this field. However, due to the intricate nature of the arm structure and the diversity of gestures, relying on a single IMU system for gesture recognition results in limited accuracy. Modern mobile devices, such as smartphones and smartwatches, are equipped with IMUs that allow for convenient data acquisition methods and offer computing resources for deep learning model inference. In this paper, we propose a real-time arm gesture recognition method, called *KylinArm*, which achieves high-precision gesture recognition by coordinating 2 IMUs. The *KylinArm* method is optimized for mobile devices and based on a dual-branch 1D-CNN classifier. It supports the classification of 12 arm gestures with an optimized strategy for mobile devices that have limited computation resources and power supply. Additionally, we adopt an optimization method based on CORrelation ALignment (CORAL) to address the decreasing accuracy that occurs when new users are introduced. Finally, we evaluate *KylinArm* and test it in real scenes, achieving a recognition accuracy of over 98%.

Keywords: Arm Gesture Recognition · Inertial Measurement Unit · Human-Computer Interaction · Mobile Devices · Deep Learning

1 Introduction

As artificial intelligence develops rapidly, Human Activity Recognition (HAR) [22,29] has become an important technology in many fields, including health

This work is supported in part by Natural Science Foundation of China (62172239), Key Technologies R&D Program of Guangdong Province, China (2021B0101310002), and Shandong Provincial Natural Science Foundation, China (ZR2022LZH009).

care and human-computer interaction (HCI) [11]. There has been a growing interest in HCI as a research area that caters to the ever-changing needs of human progress with the integration of smart devices into daily lives. Within HCI, gesture recognition, a subfield of HAR, presents a flexible and practical method of transmitting information in complex and dynamic environments.

There are two primary types of gesture recognition: finger gestures and arm gestures. Finger gestures are more complex and may be affected by hand occupation or injury. Arm gestures can free up the hands, improve efficiency, and be used by more individuals to control machines.

Advancements in microelectromechanical system have led to the development of smaller and lighter sensors, making it easier to produce wearable devices based on inertial measurement units (IMU). Customized devices such as wristbands and rings can be used in gesture recognition tasks. However, smartwatches and mobile phones equipped with IMUs offer a cost-effective and convenient way to acquire data for gesture recognition, while also providing an inference environment. This simplifies the hardware composition and makes the HCI process much more natural and user-friendly. The success of projects like EatingTrak [33], MoRSE [14] and the work of Wei et al. [28] demonstrates the feasibility of arm gesture recognition based on IMU in smart mobile devices. However, due to the intricate nature of the arm structure and the diversity of gestures, relying on a single IMU system for gesture recognition results in limited accuracy.

To successfully recognize arm gestures and create a HCI process, obtaining data that comprehensively captures the characteristics of arm gestures is of utmost importance. In this paper, we propose *KylinArm*, a solution for recognizing arm gestures based on dual IMU data collection and optimized for mobile devices to address the aforementioned issues. By leveraging two IMUs located at the wrist and elbow, more data can be obtained to distinguish between arm gestures with similar wrist motion trends.

In *KylinArm*, a dual-branch convolutional neural network (CNN) model is designed that relies on one-dimensional convolution and accounts for data collected by two IMUs to classify 12 arm gestures for HCI with high precision. Additionally, *KylinArm* is optimized for the limited computing resources and energy supply of mobile devices by integrating a Wakeup-Detection-Classification module.

To enhance the generalization of arm gesture recognition, a real-time data alignment strategy based on CORrelation ALignment (CORAL) [25] is implemented. This is necessary as different users and the variance of sensor positions may change the features of data collected during gesture recognition.

The main contributions of this paper are as follows:

- We propose a new arm gesture recognition framework called *KylinArm*. This framework supports 12 arm gestures using 2 IMUs for data collection and a dual-branch convolutional neural network (CNN) model for classification.
- We optimize *KylinArm* for ubiquitous mobile devices with a Wakeup-Detection-Classification mechanism and real-time data alignment strategy, which ensures efficient utilization of limited computing resources and energy supply.

- We implement *KylinArm* on Android-based mobile devices and evaluate its effectiveness by testing with collected datasets and real-world scenarios. The results demonstrate the high precision and robustness of *KylinArm* in recognizing arm gestures.

The rest of the paper is organized as follows. Section 2 presents related works on IMU-based arm gesture recognition and algorithms for sensor-based gesture recognition. Section 3 provides details on the design and implementation of *KylinArm*. Section 4 evaluates the proposed arm gesture recognition method. Finally, Sect. 5 concludes the paper.

2   Related Work

2.1   IMU-Based Gesture Recognition

IMU typically includes three sensors: an accelerometer, a gyroscope, and a magnetometer, each of which measures data along the XYZ axes. The magnetometer is an optional component, and the IMU without it is commonly referred to as a 6-axis IMU. The accelerometer measures linear acceleration, while the gyroscope measures angular velocity.

IMUs are capable of accurately capturing the inertial data generated during the execution of gestures. Bianco et al. [2] used an IMU wristband to classify 6 letters, 6 numbers, and 12 simple forearm gestures. Cui et al. [6] used three 6-axis IMUs to recognize three arm activities. Kim et al. [15] identified handwritten digits with a handheld IMU.

With the advancement of mobile devices, smartphones and smartwatches are often equipped with built-in IMUs and possess stronger computing capabilities, which can support deep learning model inference. Kurz et al. [17] utilized a self-assembled 6-axis IMU and a smartwatch to collect data and successfully recognize 12 different gestures on a smartphone. Kang et al. [12] used a smartwatch to recognize 7 gestures while walking. Guo et al. [7] used a smartwatch and a camera to recognize 12 hand gestures. Kasnesis et al. [14] recognized 5 arm gestures representing different emergency signals using the 6-axis IMU in a smartwatch.

2.2   Algorithms for Sensor-Based Gesture Recognition

The design of the classifier plays a crucial role in achieving high-accuracy gesture recognition. Classifiers based on time-series data collected from IMUs can be categorized into two groups: traditional machine learning and deep learning. Common machine learning methods include KNN [8], Random Forest [10] and SVM [4,23,27], while deep learning methods can be further divided into three types: CNN [1,5,14,26], RNN [2,7], and their combinations [21,31]. Ulysse et al. [5] proposed a CNN model based on continuous wavelet transform to classify 7 gestures. Bianco et al. [2] built an arm gesture recognition system based on GRU and LSTM. Yuan et al. [31] utilized the CNN+LSTM method to recognize sign language.

Moreover, the generalization of classifiers is also an important issue. Both [2] and [30] discovered that deep learning models can differentiate the same gesture done by different people. [4] and [14] noted that the accuracy of arm gesture recognition is affected when using the model with people whose motion data are not included in training, but they only described the phenomenon without proposing improvements. In [7], gesture recognition was performed on smartwatch. To alleviate the impact of new users on accuracy, data was re-collected and trained on the server. Although effective, this process can be cumbersome.

3 Design and Implementation of KylinArm

3.1 Framework Design of KylinArm

We develop an arm gesture recognition system, named *KylinArm*, that is designed to achieve high-precision recognition and perform real-time inference on mobile devices. The overall structure of *KylinArm* is illustrated in Fig. 1.

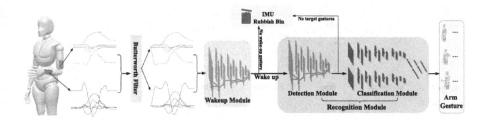

Fig. 1. Framework of KylinArm

To ensure accurate capture of data that represents the differ characteristics of arm gestures, we adopt a dual-IMU data acquisition mechanism. Specifically, the user wears two 6-axis IMUs, one on the wrist and the other on the upper arm near the elbow. The IMU data is collected at a sampling frequency of 30 Hz. The classifier in *KylinArm*, which is referred to as the Classification Module in Fig. 1. From the acquisition of signals to input into the classifier, the continuous IMU data is filtered using a low-pass Butterworth filter to remove noise caused by the environment and equipment. In addition, a sliding window with a size of 3 s and a step of 200 ms is used to segment the continuous data.

KylinArm is set to sleep mode by default to conserve computing resources and prevent unintended operations. Wakeup Module is responsible for determining when the recognition procedure should be activated. The preprocessed data windows are initially directed to Wakeup Module to determine if they constitute a wake-up gesture. However, if a wake-up gesture is detected, Wakeup Module stops working and the recognition procedure starts. Once the recognition procedure is woken up, windows are fed directly into Recognition Module. The windows are then processed by Detection Module to prepare input for Classification Module, with only one window being input to Classification Module for each arm gesture. Finally, the recognized arm gesture type is output as a control command.

3.2 Gesture-Controlled Command Set

Gestures are fundamental to HCI, and the design of arm gestures should be tailored to the actual operational requirements of the devices. It is essential to consider the distinction between control gestures and daily arm gestures as well. We are inspired by Zhang et al.'s study [32] on the naturalness of upper limb movements and design twelve arm gestures that involve movement of the entire arm, which are illustrated in Fig. 2. The correspondence between these gestures and their respective potential control commands is also shown in the same figure.

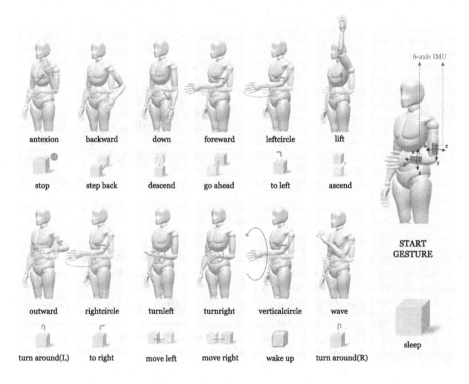

Fig. 2. Designed arm gestures and corresponding control commands

3.3 Classification Module: Dual-Branch 1D-CNN Classifier

For the 12-class arm gesture recognition task, we design a classification model named dual-branch 1D-CNN (*Dual-CNN*). The model is based on one-dimensional (1D) convolution and uses two convolutional branches to extract features from the time-series data obtained from each IMU to enhance classification accuracy. The network structure is shown in Fig. 3 and more detail can be found in Table 1. This model is applied in Classification Module to classify arm gestures. During training, only windows containing the 12 arm gestures are used.

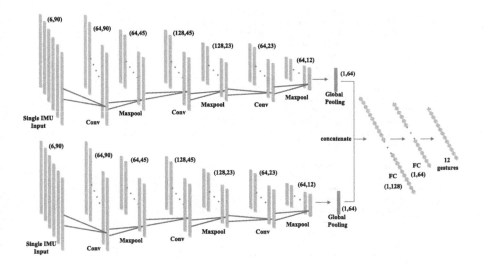

Fig. 3. Structure of Dual-CNN

Dual-CNN is designed to run on a mobile device and is composed of two convolutional branches, each of which includes three 1D convolutional layers. The input data of each branch has a shape of (6,90). After convolution, a MaxPooling [20] layer, a Batch Normalization (BN) [9] layer and a Dropout [24] layer are applied. The MaxPooling and Dropout layers serve to prevent model overfitting, while the BN layer accelerates the model's convergence speed. Global Average Pooling (GAP) [19] is then applied to reduce the number of parameters in *Dual-CNN*. The concatenated features are fed into two fully connected (FC) layers, which map extracted features to classification labels as the final output.

3.4   Arm Gesture Recognition

To adapt limited computing resources of mobile devices, we propose a Wakeup-Detection-Classification process. This process utilizes *Dual-CNN* discussed in Sect. 3.3 for arm gesture classification, as well as two much lighter models trained separately for the Wakeup Module and Detection Module.

Wakeup Module. utilizes a binary classification model, called CNN_{wake}, which is lightweight and efficient. CNN_{wake} consists of three 1D convolutional

Table 1. Details of Dual-CNN

Layer	Layer Type	Input Size	Output Size	Kernel Size	Activation
Layer 1-1	1DConvolution	(6,90)	(64,90)	8	ReLU
	Batch Normalization				
	Dropout				
Layer 1-2	1DMaxpooling	(64,90)	(64,45)	3	
Layer 1-3	1DConvolution	(64,45)	(128,45)	5	ReLU
	Batch Normalization				
	Dropout				
Layer 1-4	1DMaxpooling	(128,45)	(128,23)	3	
Layer 1-5	1DConvolution	(128,23)	(64,23)	3	ReLU
	Batch Normalization				
	Dropout				
Layer 1-6	1DMaxpooling	(64,23)	(64,12)	3	
	Global Average Pooling	(64,12)	(1,64)		
Layer 2-1	1DConvolution	(6,90)	(64,90)	8	ReLU
	Batch Normalization				
	Dropout				
Layer 2-2	1DMaxpooling	(64,90)	(64,45)	3	
Layer 2-3	1DConvolution	(64,45)	(128,45)	5	ReLU
	Batch Normalization				
	Dropout				
Layer 2-4	1DMaxpooling	(128,45)	(128,23)	3	
Layer 2-5	1DConvolution	(128,23)	(64,23)	3	ReLU
	Batch Normalization				
	Dropout				
Layer 2-6	1DMaxpooling	(64,23)	(64,12)	3	
	Global Average Pooling	(64,12)	(1,64)		
Layer 3	Fully-Connected	(1,128)	(1,64)		
	Dropout				
Output Layer	Fully-Connected	(1,64)	12		Softmax

layers and two fully connected layers. Each convolutional layer is followed by a Maxpooling layer, a BN layer, and a Dropout layer. With only 4994 parameters, CNN_{wake} is better suited for frequent use on mobile devices compared to *Dual-CNN*.

The gesture *verticalcircle* (shown in Fig. 2) is chosen as the wake-up gesture. This particular gesture is relatively specific and less commonly used in daily life, which makes it a suitable choice for ensuring the reliability of Wakeup Module. CNN_{wake} takes IMU data as input by the sliding window with a size of 3 s and a step of 200 ms. To improve the accuracy of capturing the wake-up gesture, a window will be classified as positive if it contained data of a gesture for more than 80%, and negative otherwise. During the *verticalcircle* gesture, CNN_{wake} outputs consecutive 1 s. If 0 appears, it indicates that the gesture has not occurred or has been completed. Therefore, we define the occurrence of *verticlecircle* as CNN_{wake} outputting 3 or more consecutive 1.

Detection Module. is responsible for capturing windows that contain the complete gesture as much as possible and ensuring that the gesture classifier only runs when necessary. This helps to reduce the computational cost of the Recognition Module and improve the accuracy of arm gesture recognition. Detection Module uses a binary classification model on the IMU data processed by sliding window mechanism, determining whether it contains data of one of the 12 arm

---

Algorithm 1. Main Window Selection Mechanism

Input: The results of Detection(0 or 1), Classifier Dual-CNN
Output: The inference result of Recognition
1: **if** (0 appear and len(pool)$\neq$0) or len(pool)$\geq$7 **then**
2:     **if** len(pool) is odd **then**
3:         center_idx = $\lfloor$len(pool) / 2$\rfloor$
4:     **else**
5:         center_idx = len(pool) / 2 - 1
6:     **end if**
7:     standard_window = pool[center_idx] $\triangleright$ Select the main window
8:     result = Dual-CNN(standard_window) $\triangleright$ Classify the selected window
9:     clear pool
10: **else if** 1 appear **then**
11:     append current window to pool $\triangleright$ Accumulate windows
12: **end if**

---

gestures. It is similar to the role of CNN_{wake}. Therefore, Detection Module utilizes the same model structure as CNN_{wake}, which called CNN_{detect}. When the user performs one of the 12 gestures, CNN_{detect} first outputs 0, then outputs a continuous segment of 1, and finally returns to outputting 0. The Main Window Selection (MWS) mechanism is employed to provide the window that contains as much gesture data as possible for Classification Module.

As illustrated in Algorithm 1, when an arm gesture is performed, CNN_{detect} outputs consecutive 1 s and the corresponding windows are added to the window pool with a maximum size of 7. Window accumulating process continues until either a zero appears or the pool reaches its maximum size. Subsequently, a method similar to finding the median is applied to select the main window, which is then passed to Classification Module. The pool is cleared and MWS mechanism awaits the next sequence of 1 s.

3.5 CORAL-Based Generalization Optimization

Arm gesture recognition is a user-sensitive task. There are distributional differences even in the performance of the same gesture by different users. Arm gesture classifier may not perform optimally on data from new users that have not been previously trained on. In the case of data with different distributions, Domain Adaption [16] typically involves mapping data from both the source and target domains to a common distribution space through feature changes. And then the migrated data is used to train the model improving its performance on the target domain data. However, implementing this process solely on mobile devices is often infeasible. Consequently, we propose a new real-time generalization strategy based on CORAL [25], named as $CORAL_{REVERSE}$, which shifts the focus from a model-to-data fit to a data-to-model fit approach.

$CORAL_{REVERSE}$ migrates the new data (target domain) to the original data (source domain) in real time. As outlined in Algorithm 2, the new users' data is represented by D_t, while D_s represents the original data used to train the model. The process of data alignment involves two steps: first, the new data D_t is whitened, and then, the whitened data is re-colored using the statistical characteristics of original data D_s.

Algorithm 2. $CORAL_{REVERSE}$

Input: Source Data D_s, New Data D_t
Output: Transferred New Data D_t^*
1: $C_s = cov(D_s) + eye(size(D_s, 2))$
2: $C_t = cov(D_t) + eye(size(D_t, 2))$
3: $D_t = D_t * C_t^{-\frac{1}{2}}$ ▷ Whitening New Data D_t
4: $D_t^* = D_t * C_s^{\frac{1}{2}}$ ▷ Re-coloring New Data D_t

---

IMU data is continuously input to the inference system. Each new data needs to choose a source domain to perform whitening and re-coloring, which produce a lot of repetitive work. Since $CORAL_{REVERSE}$ implements data migration based on two matrix multiplication operations, the migration matrix $coral_t$, which represents the data gap between the target and source domains, is calculated in advance.

$$coral_t = C_t^{-\frac{1}{2}} \cdot C_s^{\frac{1}{2}} \tag{1}$$

where C_t is the feature matrix of gesture data for new users, and C_s is the feature matrix of original data. During the real-time inference process, $coral_t$ is dot-producted with the continuously IMU windows to align the new data with the original data in real-time. It improves the overall recognition accuracy of Recognition Module.

To effectively employ transfer learning, it is crucial to identify a source domain that is similar to the target domain. Therefore, for new users with different characteristics, we utilize different data transfer strategies for data migration that incorporated the migration matrix calculation method. The method of using $CORAL_{REVERSE}$ in $KylinArm$ is shown in Algorithm 3.

The $coral_t$ matrix is calculated prior to performing real-time data alignment. Therefore, a gesture sample set containing 12 gestures, denoted as D_t', is captured in advance to implement Algorithm 3. D_t' contains one item for each gesture. Then, the source domain data is selected based on the distribution of the pre-collected gestures set D_t'. The classification accuracy of D_t' is used as the criterion for selecting the data transfer strategy. These two data transfer strategies are referred to as *Select_Person* and *Select_Overall* in Algorithm 3. If the accuracy exceeds threshold A, the former strategy is chosen; otherwise, the latter is chosen.

D_O contains all gesture data for all individuals which is the training dataset of the gesture classifier. In *Select_Overall* strategy, the global source domain data D_s is fixed and a K-means [18] method is used to select representative data from the training dataset for Classification Module. D_s contains the original data of each class of gestures for each person. In *Select_Person* strategy, D_s

Algorithm 3. New Data Transfer Process

Require: Overall Original Data D_O, Sample Target Data D_t', Classifier *Recognition*
Input: Real-time IMU Windows D_{IMU}
Output: Transferred Real-time IMU Windows D_{IMU}^*
 1: $label_{predict} = Recognition(D_t')$
 2: $accuracy = accuracy\_score(lable_{sample}, label_{predict})$
 3: **if** $accuracy > A$ **then**
 4:    Source Data $D_s = Select\_Person(D_O)$
 5: **else**
 6:    Source Data $D_s = Select\_Overall(D_O)$
 7: **end if**
 8: **for** act_{t_i} in D_t' **do**
 9:    Select action data act_{s_i} with the same label as act_{t_i} from D_s
10:    $coral_{t_i} = CORAL_{REVERSE}(act_{t_i}, act_{s_i})$
11: **end for**
12: $CORAL_t = Average\{coral_{t_0}, coral_{t_1}, ...\}, coral_{t_{11}}$
13: $D_{IMU}^* = D_{IMU} \cdot CORAL_t$

---

includes data from the person who is most similar to D_t'. We adapt a classification accuracy-based method to assess the similarity. Specifically, original data samples from each person are used as training sets in turn to train corresponding classifiers. D_t' of new user is then put into every classifier to obtain the accuracy. The data from the person whose classifier achieves the highest accuracy is selected as the source domain D_s.

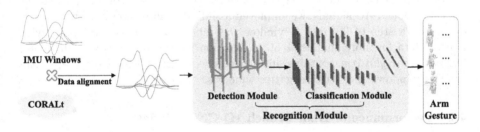

Fig. 4. Usage of the transformation migration matrix $CORAL_t$

After selecting the source domain data, $CORAL_{REVERSE}$ is used to calculate migration matrix $CORAL_t$. In the process of calculating $CORAL_t$, the source and target domain data with the same label are calculated according to Algorithm 1 to obtain $coral_{t_i}$. $CORAL_t$ is computed as the average of multiple $coral_{t_i}$ matrices. Once the feature linear transformation migration matrix $CORAL_t$ is obtained, $CORAL_{REVERSE}$ can be used to perform whitening and re-coloring operations via dot product between $CORAL_t$ and D_{IMU}, resulting in the aligned data D_{IMU}^*. And then D_{IMU}^* is inputted into Recognition Module for gesture recognition, as shown in Fig. 4.

4   Evaluation

To fully validate the effectiveness of *KylinArm*, we evaluate the performance of *Dual-CNN* and each module in *KylinArm* using our datasets. At last, we implement it on smartphones and test the system in a real scene.

4.1   DataSet Description

We used two IMUs placed on the wrist and upper arm to collect data from 15 volunteers (9 males and 6 females) with ages ranging from 22 to 27 years, heights ranging from 160–185 cm, and weights ranging from 50–80 kg. Data was collected at a frequency of 30 Hz, with each gesture performed for 3 s. Both Dataset for *Dual-CNN* and Dataset for *KylinArm* are derived from the raw IMU data collected, and they are processed separately and labeled in different ways to accommodate different task needs. The former dataset is only used to train and test the designed classification model. The latter dataset simulates the data generated in real-world scenarios and contains windows of continuous time, which is used to train and test each module in the system inference phase.

Dataset for *Dual-CNN* only contains the data of 12 gestures. Each of the 15 volunteers repeated each action 10 times according to the rule of performing for 3 s and resting for 3 s. The final dataset contains 1800 items, where each item has the shape of (12,90).

Dataset for *KylinArm* includes data collected from 15 volunteers, each of whom performed each gesture 10 times within a 10-minute time frame. The dataset contains a total of 1800 gestures and spans a duration of 3 h. To segment the continuous IMU signals, we employed a sliding window approach with a size of 3 s and a step of 200 ms. A window was labeled as 1 if it overlapped with gesture data for more than 80% of its duration, and as 0 otherwise. Therefore, Dataset for *KylinArm* contains 46230 windows data, of which there are 7605 positive windows and 38625 negative windows.

4.2   Performance of Dual-Branch 1D-CNN Model

5-fold cross-validation is performed on five baseline models and *Dual-CNN* using Dataset for *Dual-CNN* to evaluate classification accuracy. Adam optimizer and cross-entropy loss function are used, batch size is 8, and 50 epochs are trained. The single-branch feature extraction variant of *Dual-CNN* is 1D-CNN, which takes the data from two IMUs and concatenates them as input. CNN-LSTM [34] and EMGHandNet [13] perform well in HAR tasks.

To provide a comprehensive assessment of the effectiveness of each model, we utilized four commonly-used metrics: Accuracy, Precision, Recall, and F1 score. The results of the evaluation, which include the *Dual-CNN* model and five baseline models, are presented in Table 2. The *Dual-CNN* model outperforms all other models across all metrics. These results highlight the significant difference in classification ability between machine learning and deep learning methods.

Table 2. The global 5-fold cross-validation results for the different models

Model	Accuracy (%)	Precision (%)	Recall (%)	F1 (%)
SVM	91.61 ± 2.04	92.58 ± 2.44	91.61 ± 2.04	91.61 ± 2.03
Decision Tree	91.94 ± 1.65	92.19 ± 1.61	91.94 ± 1.65	91.94 ± 1.67
1D-CNN	99.56 ± 0.33	99.55 ± 0.35	99.53 ± 0.31	99.53 ± 0.34
CNN-LSTM [34]	99.67 ± 0.22	99.67 ± 0.20	99.65 ± 0.22	99.65 ± 0.21
EMGHandNet [13]	99.50 ± 0.21	99.52 ± 0.20	99.50 ± 0.21	99.50 ± 0.21
Dual-CNN	**99.78 ± 0.27**	**99.78 ± 0.26**	**99.78 ± 0.27**	**99.78 ± 0.27**

While the accuracy of all tested deep learning models exceeded 99%, our proposed *Dual-CNN* model stands out as the most efficient solution. It has the least number of parameters and fewer FLOPs as shown in Table 3. We then evaluate the inference time cost of *Dual-CNN* on the Honor 30 smartphone, which runs on HarmonyOS 2.0, has a memory size is 8.0 GB and is powered by Kirin 985 processor. *Dual-CNN* takes an average of less than 5ms to perform an inference on Honor 30. The efficiency of the *Dual-CNN* model makes it more mobile device-friendly and enables real-time gesture recognition with greater accuracy.

Table 3. FLOPs and parameters of four deep learning model

Model	FLOPs	Parameters
1D-CNN	3087616	172428
CNN-LSTM [34]	10816928	504748
EMGHandNet [13]	3572199680	2634572
Dual-CNN	**5482752**	**147788**

To evaluate the generalization, we performed 15-fold cross-validation by selecting one individual's data as the test set and the data of the other 14 individuals as the training set. The model settings, training settings are the same as 5-fold cross-validation. The results are shown in Table 4 and Fig. 5. *Dual-CNN* exhibits superior generalization performance when facing untrained new data. Compared to the results of the 5-fold cross-validation, *Dual-CNN* has the least decrease in accuracy when classifying unfamiliar data. While the results of 1D-CNN is comparable to that of *Dual-CNN*, merging feature extraction appears to weaken it's generalization ability across individuals.

Table 4. Results of the 15-fold cross-validation for different models

Model	Accuracy (%)	Precision (%)	Recall (%)	F1 (%)
SVM	80.83	84.27	80.83	77.88
Decision Tree	75.00	74.79	75.00	71.65
1D CNN	94.61	95.58	94.61	93.68
CNN-LSTM [34]	93.50	94.01	93.50	92.49
EMGHandNet [13]	93.94	94.65	93.94	93.66
Dual-CNN	**96.00**	**96.91**	**96.00**	**95.33**

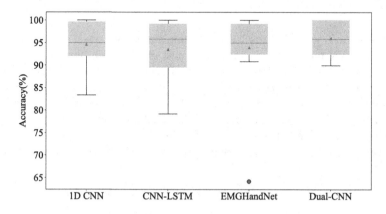

Fig. 5. Accuracy comparison of four deep learning models in 15-fold cross-validation (triangles represent means and dotted lines represent medians)

4.3　Performance of Inference Modules

Wakeup Module. The CNN_{wake} is trained based on Dataset for *KylinArm*. However, due to the fact that the positive examples of Wakeup Module are only related to windows of *verticalcircle*, the number of positive examples is greatly reduced. This results in severe imbalances between positive and negative labels, which significantly impacts the model's discrimination ability. To address this issue, we use SMOTE [3] to increase the number of positive examples. Wakeup Module achieves an accuracy of 100% for *verticalcircle* in a 5-fold cross-validation experiment conducted on Dataset for *KylinArm*. When using all data from a person as the test set and using the remaining data as the training set, the recognition accuracy of the module is 98.46%. The complexity of *verticalcircle* makes it relatively easy to distinguish. Additionally, SMOTE can increase the diversity of data, which has a positive effect on compensating for differences between individuals' gestures.

Recognition Module. The CNN_{detect} is trained using Dataset for *KylinArm*. To fully validate the effectiveness and generalization of Recognition Module, we designed the same 5-fold and 15-fold cross-validation experiments as in Sect. 4.2.

As shown in Table 5, continuous windows containing 360 gestures are inputed into Recognition Module for test. Detection Module achieves an accuracy of 98.67%, while Classification Module achieves an accuracy of 99.72%. As a result, Recognition Module exhibited an overall accuracy of 98.39% for recognizing gestures. Table 6 shows the results of 15-fold cross-validation. We can find that the accuracy of Detection Module is 96.72%, which is 1.95% lower than the 5-fold cross-validation result. The accuracy of Classification Module remains almost unchanged at 99.54%. The overall accuracy of Recognition Module decreases by 2.11% to 96.28%. While Detection Module alleviates the performance degradation caused by the sensitivity of gesture classifier to unfamiliar data, data not included in the training dataset still caused Detection Module to miss a certain number of windows, limiting the high-precision classification ability of Recognition Module.

Table 5. Accuracy (%) of global 5-fold cross-validation

Fold #	$Detect.$	$Classif.$	$Recogn.$
0	100	99.44	99.44
1	99.44	99.16	98.61
2	97.78	100	97.78
3	96.94	100	96.94
4	99.17	100	99.17
avg.	98.67	99.72	98.39

Table 6. Accuracy (%) of 15-fold cross-validation

Person	$Detect.$	$Classif.$	$Recogn.$
P_0	96.67	100	96.67
P_1	100	100	100
P_2	100	100	100
P_3	99.17	99.16	98.33
P_4	100	100	100
P_5	94.17	100	94.17
P_6	100	99.17	99.17
P_7	100	100	100
P_8	100	100	100
P_9	94.17	99.12	93.33
P_{10}	79.17	98.95	78.33
P_{11}	98.33	100	98.33
P_{12}	89.17	100	89.17
P_{13}	100	100	100
P_{14}	100	96.67	96.67
avg.	96.72	99.54	96.28

4.4   Performance of $CORAL_{REVERSE}$

The experiment primarily focuses on verifying the effect of adding $CORAL_{REVERSE}$ to Recognition Module on the recognition of gestures. As described in Sect. 3.5, we firstly pre-collect 12 gestures as samples, which are then verified using the Classification model for initial distribution verification. If the accuracy is close to 100%, samples are aligned with 12 gestures of one individual. Otherwise, they are aligned with the global source gesture library. The global library contains data for each arm gesture performed by each volunteer, resulting in a total dataset size of 12×15.

By adding $CORAL_{REVERSE}$ to Recognition Module, the rate of gestures captured by Detection Module increased from 96.72% to 97.89%, while the overall gesture recognition accuracy increased from 96.28% to 97.56%. The detailed results of each test group are presented in Fig. 6. The results indicate that 7 groups exhibited varying degrees of improvement, while 6 groups remained unchanged. Among them, P9 and P10 have an accuracy increase of 6.67% and 7.50%, respectively. However, P6 and P8 experienced a decrease in the number of recognized gestures by 2. The reduction in accuracy observed in the P6 and P8 groups can be attributed to the transformation of accurately classifiable gesture data during data alignment based on $CORAL_{REVERSE}$. This process has the potential to convert the data into other similar gestures when the arm gesture is not standard enough.

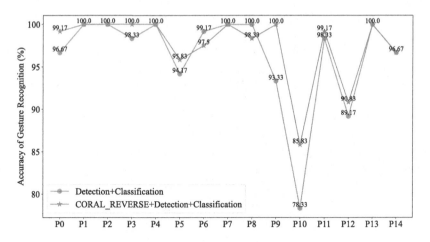

Fig. 6. Comparison of recognition accuracy before and after generalization optimization

4.5   Real-World Evaluation of KylinArm

To assess the real-world performance of *KylinArm*, we deploy it on two Android mobile phones. Figure 7 shows the arm gesture recognition system, which consists of two IMUs provided by respective mobile phones. The system utilizes

MQTT for communication with the broker running on a laptop. The mobile phone located on the upper arm is responsible for collecting IMU data, while that located on the wrist handles preprocessing of IMU data, as well as the recognition of arm gestures. The recognition result is then transmitted to the target device. The mobile application running on the mobile phones plays a crucial role in the system. It is responsible for preprocessing the IMU data and running *KylinArm* for arm gesture recognition. The main GUI of mobile application is presented in Fig. 8(a) whereas Fig. 8(b) shows the GUI where new users can perform gesture calibration to achieve data alignment before using the system.

First of all, five volunteers are recruited to test the arm gesture recognition system. After familiarizing themselves with the actions, each volunteer performed 12 actions according to their own ideas, ensuring that each gesture was completed 10 times. Table 7 presents the statistical results, which indicate that the arm gesture recognition accuracy of *KylinArm* is 98.5%.

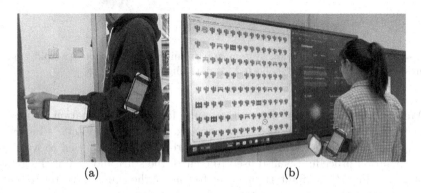

(a)   (b)

Fig. 7. System test (As an illustration, we implement KylinArm on two smartphones. However, the use of two smartphones is not necessarily required, one of them can be replaced by a smartwatch or other IMU-equipped device.): (a) IMUs Position (b) Interact with the game through arm gestures

Table 7. Accuracy of arm gesture recognition system

User	Num of correct	Accuracy (%)
1	120	100
2	118	98.33
3	118	98.33
4	116	96.67
5	119	99.17
avg		98.50

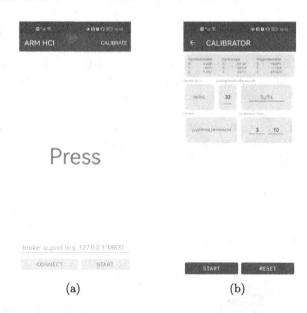

Fig. 8. Arm gesture recognition system app running KylinArm on Android smartphones: (a) Main GUI (b) Pre-collected for data alignment

We also developed a PC interactive game for further evaluation. All components are under the same local area network. The game character's behavior corresponds to 11 arm gestures, and the gesture *verticalcircle* used to wake up the recognition procedure is also used to control the start and end of the game. As shown in Fig. 7(b), volunteers are able to control the game character to reach the end point without any prompts. A video which demonstrating the operation process of each arm gesture during the game, as well as the game instructions, can be found in the GitHub repository[1]. KylinArm running on smartphones continued to provide normal services throughout this process, fully demonstrating its stability and usability.

5   Summary

We propose a new framework called *KylinArm*, which is designed to recognize arm gestures and is optimized for mobile devices. This framework employs two IMUs for data collection and a dual-branch CNN model called *Dual-CNN* for classification, which enables the high-accuracy classification of 12 arm gestures. In order to achieve real-time arm gesture recognition and conserve resources, we have optimized *KylinArm* for ubiquitous mobile devices by implementing a Wakeup-Detection-Classification mechanism and a real-time data alignment strategy. As a result, the *KylinArm* system can operate effectively on mobile devices that have constrained computing resources and energy supply. Finally,

---

[1] https://github.com/NKU-EmbeddedSystem/KylinArm.

we have implemented *KylinArm* on Android-based mobile devices and evaluated its effectiveness by testing it with collected datasets and real-world scenarios. The results demonstrate the high precision and robustness of *KylinArm* in recognizing arm gestures.

References

1. Bhattacharya, D., Sharma, D., Kim, W., Ijaz, M.F., Singh, P.K.: Ensem-HAR: an ensemble deep learning model for smartphone sensor-based human activity recognition for measurement of elderly health monitoring. Biosensors **12**(6), 393 (2022)
2. Bianco, S., Napoletano, P., Raimondi, A., Rima, M.: U-wear: User recognition on wearable devices through arm gesture. IEEE Transactions on Human-Machine Systems **52**(4), 713–724 (2022)
3. Chawla, N.V., Bowyer, K.W., Hall, L.O., Kegelmeyer, W.P.: SMOTE: synthetic minority over-sampling technique. J. Artif. Intell. Res. **16**, 321–357 (2002)
4. Colli Alfaro, J.G., Trejos, A.L.: User-independent hand gesture recognition classification models using sensor fusion. Sensors **22**(4), 1321 (2022)
5. Côté-Allard, U., et al.: Deep learning for electromyographic hand gesture signal classification using transfer learning. IEEE Trans. Neural Syst. Rehabil. Eng. **27**(4), 760–771 (2019)
6. Cui, J.W., Li, Z.G., Du, H., Yan, B.Y., Lu, P.D.: Recognition of upper limb action intention based on IMU. Sensors **22**(5), 1954 (2022)
7. Guo, K., Zhou, H., Tian, Y., Zhou, W., Ji, Y., Li, X.Y.: Mudra: a multi-modal smartwatch interactive system with hand gesture recognition and user identification. In: IEEE INFOCOM 2022-IEEE Conference on Computer Communications, pp. 100–109. IEEE (2022)
8. Hellara, H., et al.: Classification of dynamic hand gestures using multi sensors combinations. In: 2022 IEEE 9th International Conference on Computational Intelligence and Virtual Environments for Measurement Systems and Applications (CIVEMSA), pp. 1–5. IEEE (2022)
9. Ioffe, S., Szegedy, C.: Batch normalization: accelerating deep network training by reducing internal covariate shift. In: International Conference on Machine Learning, pp. 448–456. PMLR (2015)
10. Ji, L., Liu, J., Shimamoto, S.: Recognition of Japanese sign language by sensor-based data glove employing machine learning. In: 2022 IEEE 4th Global Conference on Life Sciences and Technologies (LifeTech), pp. 256–258. IEEE (2022)
11. Jindal, S., Sachdeva, M., Kushwaha, A.K.S.: Deep learning for video based human activity recognition: review and recent developments. In: Bansal, R.C., Zemmari, A., Sharma, K.G., Gajrani, J. (eds.) Proceedings of International Conference on Computational Intelligence and Emerging Power System. AIS, pp. 71–83. Springer, Singapore (2022). https://doi.org/10.1007/978-981-16-4103-9_7
12. Kang, P., Li, J., Fan, B., Jiang, S., Shull, P.B.: Wrist-worn hand gesture recognition while walking via transfer learning. IEEE J. Biomed. Health Inform. **26**(3), 952–961 (2021)
13. Karnam, N.K., Dubey, S.R., Turlapaty, A.C., Gokaraju, B.: EMGHandNet: a hybrid CNN and Bi-LSTM architecture for hand activity classification using surface EMG signals. Biocybern. Biomed. Eng. **42**(1), 325–340 (2022)
14. Kasnesis, P., Chatzigeorgiou, C., Kogias, D.G., Patrikakis, C.Z., Georgiou, H.V., Tzeletopoulou, A.: MoRSE: deep learning-based arm gesture recognition for search and rescue operations. arXiv preprint arXiv:2210.08307 (2022)

15. Kim, M., Cho, J., Lee, S., Jung, Y.: IMU sensor-based hand gesture recognition for human-machine interfaces. Sensors **19**(18), 3827 (2019)
16. Kouw, W.M., Loog, M.: An introduction to domain adaptation and transfer learning. arXiv preprint arXiv:1812.11806 (2018)
17. Kurz, M., Gstoettner, R., Sonnleitner, E.: Smart rings vs. Smartwatches: utilizing motion sensors for gesture recognition. Appl. Sci. **11**(5), 2015 (2021)
18. Likas, A., Vlassis, N., Verbeek, J.J.: The global k-means clustering algorithm. Pattern Recogn. **36**(2), 451–461 (2003)
19. Lin, M., Chen, Q., Yan, S.: Network in network. arXiv preprint arXiv:1312.4400 (2013)
20. Nagi, J., et al.: Max-pooling convolutional neural networks for vision-based hand gesture recognition. In: 2011 IEEE International Conference on Signal and Image Processing Applications (ICSIPA), pp. 342–347. IEEE (2011)
21. Nan, Y., Lovell, N.H., Redmond, S.J., Wang, K., Delbaere, K., van Schooten, K.S.: Deep learning for activity recognition in older people using a pocket-worn smartphone. Sensors **20**(24), 7195 (2020)
22. Punithavathi, D., Janakiraman, R., Santhoshkumar, S., Srikanth, R.: Human activity recognition using deep learning techniques: a review. J. Ambient. Intell. Humaniz. Comput. **12**(6), 5669–5695 (2021)
23. Shahzad, W., Ayaz, Y., Khan, M.J., Naseer, N., Khan, M.: Enhanced performance for multi-forearm movement decoding using hybrid IMU-SEMG interface. Front. Neurorobot. **13**, 43 (2019)
24. Srivastava, N., Hinton, G., Krizhevsky, A., Sutskever, I., Salakhutdinov, R.: Dropout: a simple way to prevent neural networks from overfitting. J. Mach. Learn. Res. **15**(1), 1929–1958 (2014)
25. Sun, B., Feng, J., Saenko, K.: Return of frustratingly easy domain adaptation. In: Proceedings of the AAAI Conference on Artificial Intelligence, vol. 30 (2016)
26. Tam, S., Boukadoum, M., Campeau-Lecours, A., Gosselin, B.: A fully embedded adaptive real-time hand gesture classifier leveraging HD-SEMG and deep learning. IEEE Trans. Biomed. Circuits Syst. **14**(2), 232–243 (2019)
27. Wahid, M.F., Tafreshi, R., Al-Sowaidi, M., Langari, R.: Subject-independent hand gesture recognition using normalization and machine learning algorithms. J. Comput. Sci. **27**, 69–76 (2018)
28. Wei, W., Kurita, K., Kuang, J., Gao, A.: Real-time 3D arm motion tracking using the 6-axis IMU sensor of a smartwatch. In: 2021 IEEE 17th International Conference on Wearable and Implantable Body Sensor Networks (BSN), pp. 1–4. IEEE (2021)
29. Wu, H., Zhang, C., Zhang, W., Wang, J.: Monocular 3D human pose estimation by predicting the 2D pose and depth map simultaneously. In: Proceedings of the IEEE Conference on Computer Vision and Pattern Recognition, pp. 4500–4509 (2019)
30. Yu, Y., Chen, X., Cao, S., Zhang, X., Chen, X.: Exploration of Chinese sign language recognition using wearable sensors based on deep belief net. IEEE J. Biomed. Health Inform. **24**(5), 1310–1320 (2019)
31. Yuan, G., Liu, X., Yan, Q., Qiao, S., Wang, Z., Yuan, L.: Hand gesture recognition using deep feature fusion network based on wearable sensors. IEEE Sens. J. **21**(1), 539–547 (2020)
32. Zhang, D., et al.: Fine-grained and real-time gesture recognition by using IMU sensors. IEEE Trans. Mob. Comput. **22**(4), 2177–2189 (2023). https://doi.org/10.1109/TMC.2021.3120475

33. Zhang, R., et al.: EatingTrak: detecting fine-grained eating moments in the wild using a wrist-mounted IMU. Proc. ACM Human-Comput. Interact. **6**(MHCI), 1–22 (2022)
34. Zhang, X.: Application of human motion recognition utilizing deep learning and smart wearable device in sports. Int. J. Syst. Assur. Eng. Manage. **12**(4), 835–843 (2021)

FCSO: Source Code Summarization by Fusing Multiple Code Features and Ensuring Self-consistency Output

Donghua Zhang[1], Gang Lei[2], Jianmao Xiao[2,4(✉)], Zhipeng Xu[1], Guodong Fan[3], Shizhan Chen[3], and Yuanlong Cao[2]

[1] School of Digital Industry, Jiangxi Normal University, Shangrao 334000, China
{Dong_hua,zp_xu}@jxnu.edu.cn

[2] School of Software, Jiangxi Normal University, Nanchang 330022, China
{leigang,jm_xiao,ylcao}@jxnu.edu.cn

[3] College of Intelligence and Computing, Tianjin University, Tianjin 300350, China
{Guodongfan,shizhan}@tju.edu.cn

[4] Jiangxi Provincial Engineering Research Center of Blockchain Data Security and Governance, Nanchang 330022, China

Abstract. Source code summarization is the process of generating a concise and generalized natural language summary from a given source code, which can facilitate software developers to comprehend and use the code better. Currently, most research on source code summarization generation focuses on either converting the source code into abstract syntax tree (AST) sequences or directly converting it into code segments and then feeding these representations into deep learning models. However, these single representation approaches ignore the semantic features of source code and destroy the structure of the abstract syntax tree, which affects the quality of the generated source code summarization. In this paper, we propose a novel source code summarization approach that fuses multiple code features into self-consistency output (FCSO). Our approach is based on a graph neural network encoder and a Code-BERT encoder with a self-attention mechanism. It extracts the sentence feature attention vector and the AST feature attention vector of the source code for feature fusion. Then, it inputs them into the Transformer decoder. Furthermore, to generate more accurate source code summaries, we adopt a new decoding strategy called self-consistency. It samples different inference paths, uses a penalty mechanism to calculate their similarity scores, and ultimately selects the most consistent answer. Our experimental results demonstrate that our proposed approach outperforms standard baseline approaches. On the Python dataset, the BLEU score, METEOR score, and ROUGE_L score increase by 11.13%, 9.12%, and 7.88%, respectively. These results show that our approach provides a promising direction for future research on source code summarization.

Keywords: Source code summarization · Code feature Fusion · Self-consistency · Transformer

© The Author(s), under exclusive license to Springer Nature Singapore Pte Ltd. 2024
Z. Tari et al. (Eds.): ICA3PP 2023, LNCS 14488, pp. 112–129, 2024.
https://doi.org/10.1007/978-981-97-0801-7_7

1   Introduction

In the current era, the Internet is growing rapidly, expanding the size of software systems for companies. Unfortunately, every software development and maintenance operation requires developers to re-familiarize themselves with the source code, ultimately reducing operational efficiency [1]. To address this concern, high-quality source code summarization is essential as it enables programmers to swiftly comprehend and use the source code, thereby enhancing their work efficiency. High-quality code summaries improve software development and maintenance efficiency by providing accurate information on the function of the module's code [2], promoting rapid industry growth.

Research work in the area of Source Code Summarization generally falls into three categories: artificial templates, information retrieval, and deep learning models. The first approach, based on artificial templates for generating source code summaries, is the most traditional approach. Sridhara et al. [3] utilized the Software Word Usage Model (SWUM) to produce descriptive summaries of Java approaches, while Merono et al. [4] employed heuristics and natural language processing to generate Java code summarization. The second approach, based on information retrieval, involves extracting code semantic information by marking code feature information and applying information retrieval techniques to generate code summaries. Wong E et al. [5] proposed a probabilistic and statistical AutoComment model based on a large dataset and fed the mapping relationship of a vast amount of data into the AutoComment model to generate a code summary. However, the first two approaches have limitations due to their poor reusability and low accuracy of the generated code summaries. Researchers have gradually shifted to new models to carry out their work.

The most promising approach in current research on Source Code Summarization is the third category - deep learning-based models. In earlier studies, deep learning networks were commonly used for code summarization tasks. Iyer et al. [6] employed LSTM (long short-term memory network), a widely-used deep learning model, to build CODE-NN, an automatic code summary generation model capable of creating code summaries for SQL query statements. Hu et al. [7] proposed TL-CodeSum, which utilizes API information to enhance the quality of code summary generation. This approach uses two encoders to process API information and source code vocabulary information separately to improve the accuracy of generated summaries. To capture the code's semantic information more comprehensively, researchers have started focusing on generating code summaries by improving the abstract syntax tree of the code. Hu et al. [8] converted the source code into AST using an attention mechanism and presented the DeepCom approach, which inputs the AST sequence into the encoder for encoding. Wang et al. [9] fine-tuned the Transformer model and introduced the TranS approach, which leverages the Actor-Critic network to encode code vocabulary and indentation structure. The results indicate that this technique generates better summaries corresponding to source code fragments.

Although the research mentioned above has achieved the goal of code summary generation, there are still some limitations. First of all, the single use of code sequence or AST path in the task ignores the structural characteristics of AST, which will lose part of the code information. The second problem is that the previous decoder output uses the traditional Beam Search [10] strategy for path reasoning and finally selects the sequence with the highest score from all candidate sequences in the termination state as the output. However, the code summary candidate sequence obtained in this way only considers the candidate with the highest local score and cannot guarantee the global optimal solution, so there may be some repetitions or unreasonable situations in the output sequence.

To solve the above problems, we propose a source code summarization approach (FCSO) that fuses code features into self-consistent output to solve it. We found that CodeBERT [11], as a Transformer-based pre-training model, learned the semantic representation of code. It provides a robust feature extraction function, which can better capture the semantic information of the code. We have also seen that the graph neural network (GCN) can aggregate the information of AST neighbor nodes to help the model learn code structure information and context dependencies. Therefore, as an inspiration, we take whether they can integrate the semantic information and structural information of the code as a challenge to get the answer to the problem. For the first question, we use CodeBERT encoder and GCN encoder to extract sequence and AST features and then perform feature fusion so that the fused code feature self-attention vector can be input into the Transformer decoder to preserve the source code to the greatest extent-semantic and syntactic information to improve the accuracy of code summary generation. For the second question, since the current code summary generation task requires higher and higher accuracy and consistency, we abandoned the previous greedy random output strategy. To achieve this goal, we introduce a new decoding approach, Self-consistency in the Transformer decoder. After Beam search calculates the probability distribution of the final time step, it randomly samples the output inference path. Then, it obtains the most consistent answer by judging the similarity score.

In the following experiments, we used Java code and Python database as the corpus to train the model. After comparing the standard baseline method, we found that the BLEU and METEOR indicators have been improved accordingly, and the ablation experiment proved the feasibility of the FCSO approach. The main contributions of this paper are as follows:

- A source code summarization approach that fuses code features into self-consistency output (FCSO) is proposed. This approach can extract and fuse code sequence and AST features, improving code summarization generation quality.
- We break the traditional decoding strategy and add a new decoding strategy, Self-consistency. By defining a penalty mechanism, calculating the similarity score of multiple output sequences ensures the consistent output of the code summary and improves the generation accuracy.
- We compare the standard baseline approach in the experiment, and the BLEU score, METEOR score, and ROUGE_L score on the Python dataset increase by 11.13%, 9.12%, and 7.88%, respectively. It proves that our approach is effective and provides a good idea for future research.

2   Related Work

At present, the field of code summarization is mainly based on deep learning research, which is generated by improving AST traversal approaches, GCN embedding approaches, commonly used LSTM networks, encoder-decoder architectures, and Transformer models. Huo et al. [12] used the LSTM and CNN networks to learn a control flow graph (CFG) representation so that valuable information can be focused on a graphical representation. LeClair et al. [13] feed the AST as a sequence into the encoder to generate Java code annotations. Shi et al. [14] built a neural network encoder to recursively decompose the subtree of AST and then encode the processed data to generate a code summary. Hu et al. [8] proposed a structure-based traversal (SBT) approach by improving AST into a flattened sequence to solve the problem of the traditional AST sequence losing the global information of the code. LeClair et al. [15] used the SBT approach to conduct experiments and found that decoupling the code structure and code tags can better generate code summaries.

The above research mostly starts with improving code structure and AST structure. In recent years, researchers have gradually begun to use deep learning networks to solve the problem of code summarization generation. Due to the inability of traditional RNN models [16] or LSTM networks with attention mechanisms [17] to capture long-term dependency relationships, researchers have found that the Transformer model [18] utilizes self-attention mechanisms to solve this problem. Ahmad et al. [19] used a relatively encoded Transformer model to ensure the dependency of code information, and experiments found that using only an unimproved model resulted in much higher performance in code summarization generation than common deep learning networks such as RNN, indicating that using an encoder-decoder is a good approach.

In addition, some researchers have begun to start with code feature fusion, providing a new idea for code summary generation. The online learning of social performance (DeepWalk) proposed by Bryan et al. [20] is applied to CFG in a learning manner. Then it connects nodes through a convolutional neural network to achieve the goal of error positioning. Wang et al. [21] applied Self-consistency to the language model, allowing a complex reasoning problem to allow many different ways of thinking, and finally selected the only correct answer, which improved the reasoning ability of the thinking chain. Cheng et al. [22] proposed the GN-Transformer approach, which combines sequence and graph learning representations to improve the quality of code summarization generation. Wang et al. [23] constructed two encoders to fuse code-informed attention weights by learning mixture representations of codes. Similarly, Gao et al. [24] innovatively proposed a multi-modal and multi-scale approach to fuse the feature information of the code and input the code feature into the modified Transformer model decoder to improve the code summary generation performance. The above research mainly focuses on how to decompose each feature of the code. Their research more or less ignores the attention weight of the code feature or all randomly generates summary results in the traditional Beam Search method. The accuracy of the code generated is insufficient, so further research is needed.

3   The Architecture of Approach

The FCSO approach proposed in our paper consists of four main parts: data preprocessing, feature extraction, feature fusion, and self-consistency output. Firstly, the input source code is preprocessed and parsed into an abstract syntax tree and a token sequence. Next, these two parts of data are embedded into the GCN encoder and CodeBERT encoder we set up for feature extraction. The token feature vector and AST feature vector generated by the encoder are then input into the Transformer decoder for source code summarization. Finally, the self-consistency penalty mechanism is utilized in the decoder to calculate the similarity score, and the sequence with the highest similarity is retained by comparison to derive a consistent summary answer. The overall framework of the FCSO is illustrated in Fig. 1.

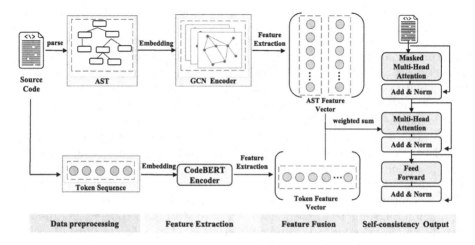

Fig. 1. The overall framework of our approach.

3.1   Data Preprocessing

In order to preserve the information of the source code more completely, we divide the code of the Java dataset and the Python dataset into two parts for data preprocessing. Part of it is processed as code sequence information, and part is converted into AST to save code structure information. In the former, we use the word segmentation toolkit to convert each piece of input code into a sequence and save it as an original suffix file as the input for the next stage. When the latter is converted into an abstract syntax tree since our experimental training is Java code and Python code, the javalang3 toolkit [23] is used to parse the Java code into AST, and the asttokens toolkit [24] is used to parse the Python code into AST.

At the same time, for the consistency of the data sets during training, we set the same length limit for the two data sets. This includes setting code tags, the average length of natural language, maximum node tree, maximum depth, and other parameters. For tag sequences that are less than or greater than the maximum length in the dataset, we pad and truncate them, respectively. In addition, we also divided the data set. Java data set and Python data set are set into the training set, validation set, and test set, which are divided into 6:2:2 and 8:1:1, respectively, to ensure that it has the same segmentation ratio as the standard baseline [4].

As shown in Fig. 2, it is an example of implementing the Java code of the decrement function and converting it to AST. In Fig. 2 (b), the gray color refers to nodes and branches and does not represent a complete tree structure. As can be seen from the figure, the AST branches from the MethodDeclaration node, and the type attribute indicates the return type of the method, which is void as a leaf node. The name attribute represents the method's name, which is also decreasingForLoop. The Param attribute represents the method's parameter list, which contains two FormalParameter nodes, representing the startValue and endValue parameters, respectively.

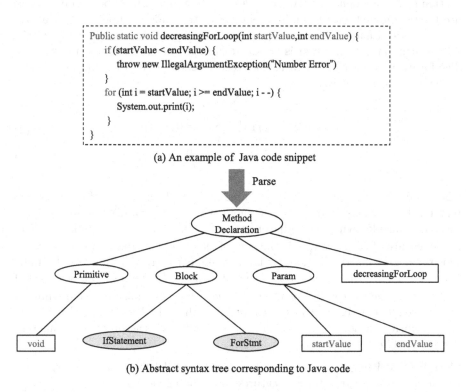

(a) An example of Java code snippet

(b) Abstract syntax tree corresponding to Java code

Fig. 2. Example of converting Java code to AST.

3.2   Feature Extraction

After the data preprocessing in the previous step, to obtain the code's local and global information, we need to continue processing the obtained sequence and AST. Based on the Transformer model, we retained the original self-attention mechanism, improved the embedding method of the encoder, and constructed two encoders: GCN Encoder and CodeBERT Encoder.

GCN Encoder: In this encoder, we treat each node in the AST as a node in the graph data. Each node is a code block, and each edge connects two adjacent code blocks, and the normalized adjacency matrix is constructed using the connection relationship between them [25]. Initially, each node has an eigenvector and then uses the weighted average of the eigenvectors of neighboring nodes to update the eigenvector of the node. In each layer, GCN uses this method to update the feature vector of each node until the desired number of layers or feature convergence is reached [26]. Specifically, graph convolution is applied to each node, and two-layer graph convolution is used to capture AST structure information and dependencies between nodes. The computed neighbor node features are then combined with the initial features of the nodes to update the features of each node. Finally, this embedding vector is input into the encoder, and the AST feature attention vector is obtained through the self-attention mechanism. The formula for GCN propagation between layers and GCN calculation of node eigenvectors is as follows:

$$\widetilde{D}_{ii} = \sum j \widetilde{A}_{ij} \tag{1}$$

$$H^{(l+1)} = \sigma(\widetilde{D}^{-\frac{1}{2}} \widetilde{A} \widetilde{D}^{-\frac{1}{2}} H^{(l)} W^{(l)}) \tag{2}$$

$$h_i^{(l+1)} = \sigma\Big(\sum_{j \in N(i)} \frac{1}{c_{ij}} W^{(l)} h_j^{(l)} + \theta^{(l)} h_i^{(l)} \Big) \tag{3}$$

where the H is the characteristics of these nodes to form an $N \times D$ dimensional matrix, the relationship between each node will also form an $N \times N$ dimensional matrix A, also known as the adjacency matrix. In (1) and (2) formulas, each H is the feature of each layer, $\widetilde{A}$ is the sum of the A matrix and I identity matrix, $\widetilde{D}$ is the degree matrix of $\widetilde{A}$, and σ is the nonlinear activation function. Formula (3) can calculate the eigenvector of the GCN node. In the neighbor set $N(i)$ of node i, the eigenvector $h_i^{(l)}$ of node i in the l layer is multiplied and summed by the weight matrix $W^{(l)}$ in the l layer, and then the bias item $\theta^{(l)}$ is added. c_{ij} is a normalization factor, which is used to alleviate the problem of different degrees of different nodes.

CodeBERT Encoder: There are two parts: CodeBERT and Encoder. CodeBERT is used to extract code features, and the encoder continues to process feature vectors to generate self-attention vectors. We first tokenize the code and divide each word or symbol into tokens. For the CodeBERT model to distinguish different types of text sequences, it is also necessary to add the "[CLS]" tag at the beginning of the code and the "[SEP]" tag at the end of the sentence. Appropriate tags also need to be added at the start and end of the abstract. Then concatenate the code fragment and its abstract to form an input sequence and map the word-segmented sequence to BERT's vocabulary to obtain the digital ID representation of each token. Finally, input the preprocessed code dataset, load the trained BERT model, and extract the corresponding features. Among them, the feature vector of each code fragment can be obtained through the output of the last layer of BERT. Specifically, the vector corresponding to the last layer "[CLS]" tag can be used as the feature vector of the code fragment; similarly, the vector corresponding to the last layer "[SEP]" tag can be used as the feature vector of the summary [27]. The formula for extracting code features by the CodeBERT model is as follows:

$$h_{code} = [BERT([p_1, p_2, p_3, ..., p_n])] \tag{4}$$

where $[p_1, p_2, p_3, ..., p_n]$ is the token sequence obtained after the code sequence is mapped through the vocabulary, and the $BERT$ function converts it into the corresponding feature vector matrix.

Continuing to process the feature vectors generated by CodeBERT in the encoder can further improve the efficiency of the model. First of all, the massive text data used in CodeBERT pre-training has carried out unsupervised learning on the model so that it has a stronger semantic understanding ability; secondly, when extracting code features, CodeBERT can effectively capture the critical information in code fragments to improve the representation ability and expression efficiency of the encoder. Therefore, combining CodeBERT features with Transformer encoders can lead to better code generation results [28]. At the model structure level, both CodeBERT and Transformer use the same attention mechanism and multi-layer perceptron structure, and there are similar network structures and parameter settings between them, so they are compatible with each other. Therefore, we use the CodeBERT pre-trained model to take the output of the last layer as the input of the Transformer encoder and proceed to the next step. The sequence representation output in this encoder is the extracted sequence feature vector. The formula for the encoder using the self-attention mechanism is as follows:

$$Q_i = XW_i^q, K_i = XW_i^k, V_i = XW_i^v \tag{5}$$

$$Attention(Q, K, V) = softmax(\frac{QK^T}{\sqrt{d_k}})V \tag{6}$$

where Q_i(Query), K_i(Key), and V_i(Value) is obtained by multiplying the weights W_i^q, W_i^k, and W_i^v, respectively, and finally, the $softmax$ activation function is calculated.

3.3 Feature Fusion

After obtaining the sequence feature attention vector and AST feature attention vector of the source code in the previous two steps, we need to fuse these two vectors before entering the multi-head attention mechanism in the Transformer decoder. The specific way of fusion is described below. First, the similarity between two vectors is calculated by the dot product operation, and a scalar value can be obtained. Then, we use the Softmax function to normalize the similarity and convert it into an attention weight to indicate the importance of the two features in the fusion process. Finally, the attention weight is weighted and summed with the corresponding feature vector to obtain the fusion of the subsequent feature representation. This method determines the importance of features by calculating the similarity and normalization weights and then weights and sums the features according to the attention weights. This enables the adaptive fusion of different features to better capture the correlation and importance between features. At the same time, in the fusion process, we minimize the cross-entropy loss function and use the gradient descent algorithm and dropout to optimize the parameters in the model. The negative log-likelihood loss function is as follows:

$$L = -\frac{1}{N} \sum_{i=1}^{N} \sum_{t=1}^{Z} \log[p(y_t^i)] \tag{7}$$

where X_i is a source code segment given in the formula, $Y_i = [y_1^i, y_2^i, y_3^i, ..., y_N^i]$ is the target digest segment prepared, N is the number of data points in the training decoder, and Z is the maximum length of the target digest given to training.

3.4   Self-consistency Output

In the inference process on the decoder side, we use parameter weights after training and fusing multiple code features to load. Then Beam Search calculates the probability distribution of each time step and stores the updated candidate path ranking in a heap for subsequent summary output. For the self-consistency output, we use first samples these candidate inference paths, use the cosine similarity method to calculate the similarity score, compare the scores through the penalty mechanism of the Self-consistency scoring function, and finally retain the sequence with the highest similarity to obtain more accurate and consistent answers [21]. The calculation formula for cosine similarity and penalty mechanism is as follows:

$$cos(x, y) = \frac{x \cdot y}{||x|| \cdot ||y||} = \frac{\sum_{i=1}^{n} x_i y_i}{\sqrt{\sum_{i=1}^{n} x_i^2} \sqrt{\sum_{i=1}^{n} y_i^2}} \tag{8}$$

$$f(s) = g(s) - w \sum_{j=1}^{K} D_{i,j} \tag{9}$$

where D is the penalty score, K is the number of candidate sequences obtained by the Beam Search algorithm, and $D_{i,j}$ represents the penalty score between the i and j candidate sequences. Formula (8) uses the cosine similarity method to measure the similarity between n dimensional vectors x and y. At the same time, using this similarity formula, we define a new scoring function $f(s)$ for scoring each sequence and analyzing the optimal solution. Where $g(s)$ represents the initial score of the sequence s, and w is a non-negative weight coefficient used to control the degree of similarity penalty. The penalty mechanism is to subtract the penalty degree between the sequence s and other candidate sequences based on the original score $g(s)$ to save the sequences with higher similarity and filter out the sequences with lower similarity.

Figure 3 below is a Python code example of a function to find the longest common prefix of a string, showing the principle of self-consistency output. First, the source code is decoded by the trained Transformer model, and then the similarity calculation is performed on the candidate sequences in Beam Search. Because the value set by our beam size is 5, after processing the Self-consistency scoring function among the five random output sequences, sequence 4, with the highest similarity score, is reserved for output.

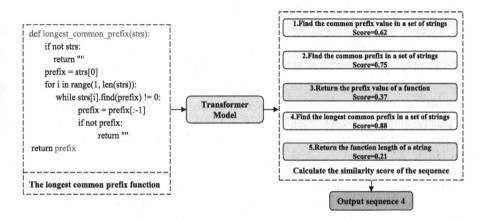

Fig. 3. Self-consistency output example.

4   Experiment

In the experiment, we first set up the database, evaluation indicators, model parameters, benchmark methods, etc., required for the experiment. Then, we performed model training to compare the evaluation indicators with the benchmark method. To demonstrate the effectiveness of our approach, we conduct qualitative ablation experiments. Finally, compare the summary generated by the benchmark method and the summary generated by our approach as an example.

4.1   Experiment Settings

Experimental Datasets: Our experiments are based on two databases well-recognized in source code summarization research, the Java database [7] and the Python database [29]. There are as many as 80,000 training sets, verification sets, test sets, and millions of Tokens which can fully train the parameters of the model. At the same time, we only use words with high frequency, and other words will be replaced by marks. For words beyond the maximum length, we perform a data truncation. Table 1 shows the number of training sets, AST nodes, and total number of tokens for Java and Python datasets.

Table 1. Statistical analysis of the Java and Python dataset.

Dataset	Java	Python
Train	69708	55538
Validation	8714	18505
Test	69708	18502
Unique nodes in ASTs	57478	101283
Unique tokens in code	66650	307596
Unique tokens in summary	46895	56189
Avg.node in AST	131.72	104.11
Avg.tokens in code	120.16	47.98
Avg.tokens in summary	17.73	9.48

Evaluation Metrics: To qualitatively compare the effect of the experimental generation, this paper adopts three indicators recognized in the field of machine translation and code summarization, BLEU [30], METEOR [31], and ROUGE_L [32].

The BLEU (Bilingual Evaluation Understudy) indicator is a standard machine translation indicator that measures the quality of translation by comparing the n-gram coincidence between the translation result and the reference translation.

The METEOR (Metric for Evaluation of Translation with Explicit ORdering) indicator combines various linguistics and machine learning techniques to obtain a comprehensive score by comparing the similarity in vocabulary, grammar, and semantics between the translation output and the reference translation.

The ROUGE_L (Recall-Oriented Understudy for Gisting Evaluation) indicator is often used to evaluate tasks such as text summarization and machine translation. It is more comprehensive, using the longest common subsequence (LCS) algorithm to measure the matching between the translation output and the reference translation. This enables a more comprehensive translation of the resulting code summary score.

Baselines: In order to prove that our research work is effective, the representative baseline models and methods in recent years are selected below. Metrics are generated by summarizing code replicating these methods and compared to our scores.

- CODE-NN [4]. The basic architecture of the model is LSTM with attention, which is an entirely driven generative model. The code is embedded through the encoder and decoder framework, resulting in the final C and SQL code digest.

- Code2Seq [33]. By inputting the processed AST path into the LSTM model as a sequence and outputting a fixed-length vector, the code summary is finally output under the calculation of the attention mechanism.
- Tree2Seq [34]. The model is based on end-to-end syntax, which extends from tree structure to sequence structure for input. At the same time, the model decoder also has a code summary corresponding to the output of the attention mechanism.
- DeepCom [6]. This method proposes a reversible AST traversal sequence method (SBT) for code comment generation, which provides a suitable method for future research under the same conditions as a reference.

Hyper-Parameters Setting: To better train the encoder and decoder architecture, we set appropriate parameters, as shown in Table 2. We set the size d_e in the embedding layer to 768 and the number of embeddings l_g in the GCN encoder to 300. Note that the number of $head_g$ is 8, and the number of L_g layers is 4. Similarly, the number of self-attention layers L_b in the CodeBERT encoder is 12, the d_k and d_v values are 64, and finally, the output size of the feed-forward network d_{ff} is 2048. At the same time, the length l_s in training in the decoder will be truncated to 100. During training, we set the embedding layer dropout to 0.2, the learning rate to 0.0001, and the batch size to 32.

Table 2. Hyper-Parameters Setting.

	Hyper-Parameters	Value
Embedding	d_e	768
GCN-Encoder	l_g	300
	h_g	768
	$head_g$	8
	L_g	4
CodeBERT-Encoder	l_b	400
	L_b	12
	$head_b$	12
	d_{model}	768
	d_k, d_v	64
	d_{ff}	2048
Decoder	l_s	100
	l_d	6
Training	dropout	0.2
	optimizer	Adam
	learning rate	0.0001
	batch size	32
Testing	beam size	5

4.2   Comparison Experiment and Ablation Experiment

After preparing the experimental setup, we conduct our experiments. First, we look for more excellent methods and models in recent years as the baseline and obtain the corresponding evaluation index data by reproducing their methods. Then, on the two data sets, Java code and Python code, we trained our model for 56 h before and after and compared the calculated evaluation index data with the baseline. Finally, we performed ablation experiments by removing each module to demonstrate that our various code feature attention vectors can help improve code summarization performance.

For comparative experiments, we compare the obtained baseline evaluation index data with the FCSO data, as shown in Table 3 below. The approach column in the table is each method, and the BLEU column, METEOR column, and ROUGE_L column are the corresponding method indicator percentages. It can be seen from the table that the standard baseline method is DeepCom [6], which has a BLEU score of 39.25%, a METEOR score, and a ROUGE_L score of 23.06% and 52.67% on the Java dataset. And our approach to FCSO is in bold in the table. It can be observed that with the baseline method DeepCom, the BLEU score, METEOR score, and ROUGE_L score on the Java dataset are increased by 5.45%, 3.80%, and 1.84%, respectively. On the Python dataset, the BLEU score, METEOR score, and ROUGE_L score increased by 11.13%, 9.12%, and 7.88%, respectively. From these data, it can be seen that our approach has a corresponding improvement compared with the baseline method on the Java dataset and the Python dataset, and it also shows that the FCSO approach we proposed is beneficial to the research work of source code summarization. At the same time, it can also be found that the evaluation index data on the Python dataset is lower than that of the corresponding Java dataset. This is because the Python code itself is lower than the semantic information contained in the Java code, so when converting the sequence to a semantic map such as AST, When it contains less information, there is a gap between the final generated summary and the reference value matching. Future research seems to improve the accuracy of Python dataset generation by modifying the nodes of the AST semantic tree to increase the semantic information on the Python dataset.

For the ablation experiments, we split the FCSO approach and removed each module. Then, increase quantitatively one by one and observe the change of the corresponding evaluation index data for each module added. The specific data is shown in Table 3 below. The first is to add a separate CodeBERT encoder and GCN encoder module and put the generated code feature attention vector into the Transformer decoder for scoring output. Their respective evaluation index data in the Java dataset are as expected, with BLEU scores of 32.68% and 38.55%. This data shows that a single code sequence feature and AST feature are not enough to reflect the semantic information of the code. The second is our multiple code feature attention fusion, using GCN-Encoder and CodeBERT-Encoder on the Java dataset, to score BLEU, METEOR score and ROUGE_L score 42.64%, 26.13%, and 53.12% have corresponding improvements. Finally, the Self-consistency module is added based on the above experiment. Compared with the previous experiment, the BLEU score, METEOR score and ROUGE_L score on the Java dataset increased by 1.31%, 1.74%, and 2.20%, respectively. The ablation experiments show that each module in our approach is indispensable, and the fusion of multiple features and Self-consistency output can improve the evaluation index of code summarization. For the experimental device, the experiments in this paper are conducted on an Ubuntu GPU server, with two GPUs of Tesla P40 and a graphics memory of 24 GB.

Table 3. Comparison of our proposed approach with the baseline approaches.

Approach	Java			Python		
	BLEU	METEOR	ROUGE_L	BLEU	METEOR	ROUGE_L
CODE-NN	27.51%	12.59%	40.30%	17.28%	9.16%	37.68%
Code2Seq	37.12%	20.14%	51.37%	19.88%	10.33%	37.80%
Tree2Seq	37.75%	21.95%	51.49%	20.07%	8.96%	35.64%
DeepCom	39.25%	23.06%	52.67%	20.78%	9.98%	37.35%
FCSO	**44.70%**	**26.86%**	**54.51%**	**31.91%**	**19.10%**	**45.23%**
Ablation Study						
CodeBERT-Encoder	32.68%	18.71%	43.59%	19.50%	8.64%	36.57%
GCN-Encoder	38.55%	21.43%	49.74%	20.86%	8.92%	37.31%
(GCN+BERT) Encoder	42.62%	25.13%	52.12%	30.29%	17.98%	43.67%
Self-consistency	43.93%	26.87%	54.32%	31.41%	18.38%	44.58%

4.3   Code Summarization Examples

At the end of the experiment, we compared the summary generated by the previous benchmark model with the summary generated by our approach, as shown in Fig. 4 and Fig. 5. In the examples of Java and Python codes, we use a section of the MYSQL database connection method code to observe the differences in abstracts generated by various models and methods. The blue color in the figure indicates the key information of the code, while the red color indicates the information that our approach has not lost compared to the benchmark method. In Java functions, the first few methods, such as CODE-NN and Code2Seq, are only summaries and do not involve the Class. forName method used to connect to the database. In Python functions, the same benchmark method is a single grab header function, and our approach diagram is highlighted in blue, where we found the connect method and the judgment statement.

Specifically, it can be found from the figure that in the CODE-NN method, because the LSTM structure is used, it is not easy to process the semantic information of long sequences, and the key information is lost in the summary. At the same time, the Code2Seq method adds self-attention weights to the LSTM framework, and the summary output can briefly explain the connection method of the database. The SBT method implemented by DeepCom also processes AST sequences in a single way, and the summary cannot retain important judgment statements. The approach FCSO we proposed can retain the connect and if statements, and express the summary more completely, which is beneficial to the programmer's code development.

```
public static Connection getConnection() {
    try {
        Class.forName("com.mysql.jdbc.Driver");
        conn = DriverManager.getConnection(url, user, password);
    } catch (SQLException e) {
        System.out.println("Error: Failed to connect to database!");
        e.printStackTrace();
    } return conn;
}
```

CODE-NN: Connect to database and return connection data

Code2Seq: Connect to MySQL database and return the error information

Tree2Seq: Connect to database with a conn number

DeepCom: Connect to MySQL database and return connection data

FCSO: Connect to MySQL database using Class. forName driver to return
             connection data

Fig. 4. Comparing Summary Results with Baseline on Java Code Methods. (Color figure online)

```
def create_connection():
    conn = None
    try: conn = mysql.connector.connect(
            host="localhost",
            user="username",
            password="password",
            database="mydatabase")
        if conn.is_connected():
            print("Connected to MySQL database")
    except mysql.connector.Error as e:
            print(f"Error connecting to MySQL database: {e}")
    return conn
```

CODE-NN: Use the connect method to establish a database connection

Code2Seq: Connect to MySQL database and return connection data

Tree2Seq: Use the connect method to establish a database connection

DeepCom: Use the connect method to establish a database connection

FCSO: Determining whether the connection is successful, use the connect method to
             establish a database connection

Fig. 5. Comparing Summary Results with Baseline on Python Code Methods. (Color figure online)

5   Conclusion

In this paper, we propose a source code summarization approach that fuses code features into self-consistency output (FCSO). This approach first constructs a GCN encoder and a CodeBERT encoder. It outputs the AST feature attention vector and sequence feature attention vector of the source code, respectively. Then, input these two feature vectors into the Transformer model decoder for feature fusion to ensure subsequent model training. Finally, we use the self-consistency decoding strategy to calculate the similarity score of the sequence output by beam search and output a consistent code summary answer, which improves the accuracy of the generated summary. In particular, we conducted comparative benchmark experiments and ablation experiments on two data sets, Java and Python, and the indicators of the experimental results have been significantly improved. It proves that our approach can learn and express the global and local features of the code more accurately and retain the semantic information of the code more completely to help programmers better understand and use the code and improve the development efficiency of software code.

In the future, our research will pay more attention to modifying the structure in the decoder and think about how to realize the fusion of the front encoder information and the back decoder information to improve the performance of source code summarization. Or it may also study code summary generation in other fields, such as helping the summary generation of the current popular blockchain smart contract code, which can promote the rapid development of this field.

Acknowledgements. This work is supported by Jiangxi Provincial Natural Science Foundation under Grant No. 20224BAB212015, the Foundation of Jiangxi Educational Committee Under Grant No. GJJ210338, the National Natural Science Foundation of China (NSFC) under Grant No. 62363015, 61962026 and the National Natural Science Key Foundation of China Grant No. 61832014.

References

1. Ko, A.J., Myers, B.A., Aung, H.H.: Six learning barriers in end-user programming systems. In: 2004 IEEE Symposium on Visual Languages-Human Centric Computing, pp. 199–206. IEEE, September 2004
2. Eddy, B.P., Robinson, J.A., Kraft, N.A., Carver, J.C.: Evaluating source code summarization techniques: replication and expansion. In: 2013 21st International Conference on Program Comprehension (ICPC), pp. 13–22. IEEE, May 2013
3. Sridhara, G., Hill, E., Muppaneni, D., Pollock, L., Vijay-Shanker, K.: Towards automatically generating summary comments for Java methods. In: Proceedings of the IEEE/ACM International Conference on Automated Software Engineering, pp. 43–52, September 2010
4. Moreno, L., Aponte, J., Sridhara, G., Marcus, A., Pollock, L., Vijay-Shanker, K.: Automatic generation of natural language summaries for Java classes. In: 2013 21st International Conference on Program Comprehension (ICPC), pp. 23–32. IEEE, May 2013

5. Wong, E., Yang, J., Tan, L.: AutoComment: mining question and answer sites for automatic comment generation. In: 2013 28th IEEE/ACM International Conference on Automated Software Engineering (ASE), pp. 562–567. IEEE, November 2013

6. Iyer, S., Konstas, I., Cheung, A., Zettlemoyer, L.: Summarizing source code using a neural attention model. In: Proceedings of the 54th Annual Meeting of the Association for Computational Linguistics, vol. 1: Long Papers, pp. 2073–2083, August 2016

7. Hu, X., Li, G., Xia, X., Lo, D., Lu, S., Jin, Z.: Summarizing source code with transferred API knowledge (2018)

8. Hu, X., Li, G., Xia, X., Lo, D., Jin, Z.: Deep code comment generation. In: Proceedings of the 26th Conference on Program Comprehension, pp. 200–210, May 2018

9. Wang, W., Zhang, Y., Zeng, Z., Xu, G.: TranS3: a transformer-based framework for unifying code summarization and code search. arXiv preprint arXiv:2003.03238 (2020)

10. Freitag, M., Al-Onaizan, Y.: Beam search strategies for neural machine translation. arXiv preprint arXiv:1702.01806 (2017)

11. Feng, Z., et al.: CodeBERT: a pre-trained model for programming and natural languages. arXiv preprint arXiv:2002.08155 (2020)

12. Huo, X., Li, M., Zhou, Z.H.: Control flow graph embedding based on multi-instance decomposition for bug localization. In: Proceedings of the AAAI Conference on Artificial Intelligence, vol. 34, no. 04, pp. 4223–4230, April 2020

13. LeClair, A., Haque, S., Wu, L., McMillan, C.: Improved code summarization via a graph neural network. In: Proceedings of the 28th International Conference on Program Comprehension, pp. 184–195, July 2020

14. Shi, E., et al.: Cast: enhancing code summarization with hierarchical splitting and reconstruction of abstract syntax trees. arXiv preprint arXiv:2108.12987 (2021)

15. LeClair, A., Jiang, S., McMillan, C.: A neural model for generating natural language summaries of program subroutines. In: 2019 IEEE/ACM 41st International Conference on Software Engineering (ICSE), pp. 795–806. IEEE, May 2019

16. Sutskever, I., Vinyals, O., Le, Q.V.: Sequence to sequence learning with neural networks. In: Advances in Neural Information Processing Systems, vol. 27 (2014)

17. Luong, M.T., Pham, H., Manning, C.D.: Effective approaches to attention-based neural machine translation. arXiv preprint arXiv:1508.04025 (2015)

18. Vaswani, A., et al.: Attention is all you need. In: Advances in Neural Information Processing Systems, vol. 30 (2017)

19. Ahmad, W.U., Chakraborty, S., Ray, B., Chang, K.W.: A transformer-based approach for source code summarization. arXiv preprint arXiv:2005.00653 (2020)

20. Perozzi, B., Al-Rfou, R., Skiena, S.: DeepWalk: online learning of social representations. In: Proceedings of the 20th ACM SIGKDD International Conference on Knowledge Discovery and Data Mining, pp. 701–710, August 2014

21. Wang, X., Wei, J., Schuurmans, D., Le, Q., Chi, E., Zhou, D.: Self-consistency improves chain of thought reasoning in language models. arXiv preprint arXiv:2203.11171 (2022)

22. Cheng, J., Fostiropoulos, I., Boehm, B.: GN-Transformer: fusing sequence and graph representation for improved code summarization. arXiv preprint arXiv:2111.08874 (2021)

23. Wang, Y., Dong, Y., Lu, X., Zhou, A.: GypSum: learning hybrid representations for code summarization. In: Proceedings of the 30th IEEE/ACM International Conference on Program Comprehension, pp. 12–23, May 2022

24. Gao, Y., Lyu, C.: M2TS: multi-scale multi-modal approach based on transformer for source code summarization. In: Proceedings of the 30th IEEE/ACM International Conference on Program Comprehension, pp. 24–35, May 2022

25. Veličković, P., Cucurull, G., Casanova, A., Romero, A., Lio, P., Bengio, Y.: Graph attention networks. arXiv preprint arXiv:1710.10903 (2017)

26. Kipf, T.N., Welling, M.: Semi-supervised classification with graph convolutional networks. arXiv preprint arXiv:1609.02907 (2016)

27. Devlin, J., Chang, M.W., Lee, K., Toutanova, K.: BERT: pre-training of deep bidirectional transformers for language understanding. arXiv preprint arXiv:1810.04805 (2018)

28. Sun, Z., Zhu, Q., Xiong, Y., Sun, Y., Mou, L., Zhang, L.: TreeGen: a tree-based transformer architecture for code generation. In: Proceedings of the AAAI Conference on Artificial Intelligence, vol. 34, no. 05, pp. 8984–8991, April 2020

29. Barone, A.V.M., Sennrich, R.: A parallel corpus of Python functions and documentation strings for automated code documentation and code generation. arXiv preprint arXiv:1707.02275 (2017)

30. Papineni, K., Roukos, S., Ward, T., Zhu, W.J.: BLEU: a method for automatic evaluation of machine translation. In: Proceedings of the 40th Annual Meeting of the Association for Computational Linguistics, pp. 311–318, July 2002

31. Banerjee, S., Lavie, A.: METEOR: an automatic metric for MT evaluation with improved correlation with human judgments. In: Proceedings of the ACL Workshop on Intrinsic and Extrinsic Evaluation Measures for Machine Translation and/or Summarization, pp. 65–72, June 2005

32. Lin, C.Y.: ROUGE: a package for automatic evaluation of summaries. In: Text Summarization Branches Out, pp. 74–81, July 2004

33. Alon, U., Brody, S., Levy, O., Yahav, E.: code2seq: generating sequences from structured representations of code. arXiv preprint arXiv:1808.01400 (2018)

34. Tai, K.S., Socher, R., Manning, C.D.: Improved semantic representations from tree-structured long short-term memory networks. arXiv preprint arXiv:1503.00075 (2015)

Graph Structure Learning-Based Compression Method for Convolutional Neural Networks

Tao Wang, Xiangwei Zheng$^{(\boxtimes)}$, Lifeng Zhang, and Yuang Zhang

School of Information Science and Engineering, Shandong Normal University, Jinan 250300, Shandong, China
xwzhengcn@163.com, zhangyuang@sdnu.edu.cn

Abstract. Convolutional neural networks (CNNs) have achieved remarkable performance in diverse applications. Nevertheless, the substantial scale and computational intricacy limit the practical implementation of CNNs, particularly on resource- starved devices. This paper presents a compression technique based on graph structure learning (GSL) for CNNs. This method aims to capture the correlations among parameters in each neural network layer and compress the model by leveraging the strength of these correlations. Firstly, the neural network parameters are modeled as a graph structure by adopting a graph construction methodology. Subsequently, the graph is fed into a dual-branch GSL module. This module introduces a constraint that optimize and refine the original graph topology and maximize the difference in feature information obtained between the two channels. Through the process of graph learning, the existing correlations among the parameters of the CNNs are demonstrated. Finally, based on the correlations between the parameters of the CNNs, the parameters with lower relative importance are selected and the neural network parameters are compressed. The proposed method significantly reduces the parameter and floating-point computation complexity of CNNs, thereby diminishing the model's intricacy. Furthermore, the operational efficiency of the network model is improved without compromising its prediction accuracy. The effectiveness of the proposed method is validated on the VGG-16 and ResNet-101 models. The compressed models' accuracy, efficiency, and memory consumption are then compared with the original models to demonstrate the effectiveness of the method.

Keywords: Graph · Graph structure learning · Model compression · Convolution neural networks

1 Introduction

Deep learning, a branch of machine learning, relies on artificial neural networks to extract complex patterns and make predictions. Convolutional Neural Networks (CNNs) are Deep Neural Networks (DNNs) with multi-layer network

Z. Tari et al. (Eds.): ICA3PP 2023, LNCS 14488, pp. 130–146, 2024.
https://doi.org/10.1007/978-981-97-0801-7_8

structures. CNNs have exhibited significant advantages in areas like natural language processing and computer vision. However, their parameter count is often prohibitively high, reaching billions, which hinders practical implementation. The deployment of large-scale DNNs on devices that are resource-constrained, with limited memory and computational capabilities, poses a formidable task [1]. As an example, the VGG16 [2] comprises over 130 million parameters, resulting in significant storage and computational overheads. Consequently, the application of deep networks on edge devices becomes severely limited. Many practical situations require prompt decision making with DNNs, such as object detection in autonomous vehicles [3]. These tasks require real-time computation of multiple objects, such as pedestrians, vehicles, and animals, as well as distances, to achieve effective autonomous driving. Hence, it is imperative to reduce the parameter redundancy and compact the model to enable its execution on resource-constrained devices.

In previous studies, LeCun et al. [4] substantiated that not all parameters in neural networks carry equivalent significance, while Denil et al. [5] argued that the network is over-parameterized due to the redundancy of numerous parameters that contribute minimally to error reduction and overall generalization. In order to address this issue, some researchers introduced various innovative algorithms to strike a balance between the accuracy and computational complexity of models. Generally, most compression algorithms endeavored to transform a large and intricate model into a more compact form. In addition, some researchers investigated the design of simplified network architectures and the training of smaller models from scratch. Based on the extent of modifications applied to the network structure during the compression process, model compression can be broadly classified as frontend compression and backend compression. Frontend compression comprises approaches that reduce complexity while retaining the original network structure, such as knowledge distillation [6], compact model structure design [7], and filter-level pruning [8]. Frontend compression techniques primarily focus on reducing the number of layers or filters while preserving the network structure. In contrast, backend compression aims to minimize the model size, typically involving substantial transformations to the original network structure, such as low-rank approximation [9], unconstrained pruning [10], parameter quantization [11], and binary networks [12]. Backend compression techniques strive for the utmost compression ratio, necessitating significant modifications to the original network structure, not only in terms of volume but also in terms of runtime. The core objective of backend compression is to decrease resource consumption, achieved through parameter quantization and similar methods. Parameter quantization is a popular approach wherein similar weights are grouped together and condensed into a unified value, resulting in significant storage reduction as well as improvement in overall computational efficiency of the network model.

Current approaches to compress CNNs predominantly concentrate on examining the interaction between layers or filters and the unique features of a specific task while also assessing their influence on prediction accuracy for that particular task. However, within this paper, the existence of intrinsic correlations or dependencies are supposed among the parameters within CNNs. During training, the network acquires knowledge of these dependencies, which significantly influences its performance. Consequently, our research centers on the analysis and remediation of parameter relationships within CNNs and the formulation of compression strategies to enhance network performance. Hence, it is imperative to consider modeling parameter relationships in CNNs.

The contributions of the paper are as follows:

(1) A compression method based on Graph Structure Learning (GSL) for CNNs is proposed. It utilizes the graph learning to mine the correlation information between the parameters of filters within the network layer, and then the CNNs is pruned based on this correlation information. Ultimately, CNNs are compressed with negligible loss in accuracy while significantly improving its operational efficiency.
(2) The connections among neurons in neural networks can be viewed as a graph structure. In this paper, graphs are employed to depict the parameter relationships in CNNs and assess the relative significance of these parameters within the network structure from an alternative perspective.
(3) A graph structure learner is proposed, which is essentially a dual-branch graph neural network that maps certain data with non-Euclidean properties into the feature space of the graph, and then their potential features through the aggregation transformation of GNNs are obtained. With GSL, into the deeper correlations between parameters in CNNs can be delved.

The following sections of this paper are structured as follows: Sect. 2 introduces the key research on deep network model compression, along with the development and advantages of graph learning. Section 3 provides a comprehensive explanation of the deep model compression method we proposed based on GSL. Section 4 demonstrates the effectiveness of the proposed method through experimental results. Section 4 presents the experimental results obtained on CNNs, which confirms the effectiveness of the proposed theoretical method. Finally, we conclude with a discussion on future work.

2   Related Work

2.1   Compression of CNNs

In the realm of CNNs, a significant challenge revolves around reducing the computational burden and storage requirements of training models, thereby enabling their deployment on devices with limited resources [13]. Li et al. [14] have emphasized the benefits of deploying deep learning models in the cloud, such as ample computational power and storage availability. However, they have also highlighted drawbacks associated with poor throughput and extended response times. Nevertheless, there is a growing demand to shift the inference process from the cloud to edge devices in real-time applications, such as object detection and video segmentation. The current performance limitations of edge devices pose a constraint on the real-time inference of CNNs. Furthermore, due to the high cost of network transmission, data transfer over the network consumes more energy compared to local data processing.

Pruning serves as a widely employed compression technique aimed at reducing the storage requirements of CNNs, rendering them more storage-efficient. Through the removal of parameters or filters from convolutional layers, pruning also reduces the computational workload, leading to faster inference. Srivastava et al. [15] demonstrated that pruning parameters from dense layers can effectively reduce the model size, further suggesting that such pruning aids in mitigating overfitting in neural networks. Han et al. [16] proposed the deletion, or zeroing, of parameter connections that fall below a predefined threshold or are deemed redundant during unimportant parameter pruning. In the context of filter pruning, Li et al. [8] ranked filters based on their importance, typically calculated using L1/L2 norms or alternative methods, and subsequently removed the least important (lowest-ranked) filters from the network. Srinivas et al. [17] proposed the deletion of redundant individual neurons, entailing the removal of all input and output connections associated with the respective neuron. Similarly, Chen et al. [18] suggested that certain layers can be pruned within a deep network, contributing to overall model compression. Within the realm of CNNs, parameters are commonly stored as 32-bit floating-point numbers. However, through the reduction of the bit width used to represent weights and activations, it becomes feasible to significantly reduce the number of Multiply-Accumulate (MAC) operations needed and diminish the size of trained DNNs. Fiesler et al. [19] and Balzer et al. [20] first introduced the concept of quantized neural networks, which aimed to render neural networks more compatible with hardware implementations.

2.2 Graph Structure Learning

Graphs offer unique advantages over other data representations for depicting objects and capturing their intricate interactions. Graph Neural Networks (GNNs), serving as powerful instruments for structured data learning, have discovered extensive applications in a variety of analytical tasks spanning different domains. The successes of GNNs can be credited to their ability to exploit the abundant and inherent information embedded within graph structures and attributes. However, the graphs provided for analysis often suffer from incompleteness and noise, presenting formidable challenges when employing GNNs in real-world scenarios. GNNs generate node embeddings through a process of iterative aggregation of information from neighboring nodes, and this iterative mechanism, as astutely highlighted by Gilmer et al. [21], exhibits a cascading effect wherein even minute noise can propagate to the surrounding neighborhood, resulting in a deterioration of the quality of numerous representations.

For example, a social network graph where nodes represent users and edges signify friendship relationships. In such cases, the recursive aggregation scheme employed by GNNs can inadvertently create spurious links between fraudulent and genuine accounts. This, in turn, facilitates the dissemination of misinformation throughout the network, thereby posing challenges in accurately estimating the credibility of user accounts. Furthermore, studies conducted by Dai et al. [22] and Zhu et al. [23] demonstrate that even subtle and imperceptible alterations in the graph structure, known as adversarial attacks, can easily lead to incorrect predictions by most GNNs. Acknowledging these observations, Luo et al. [24] argue that GNNs typically necessitate high-quality graph structures to acquire informative representations. In response, Newman et al. [25] propose methods to refine the graph structure by fostering intra-cluster connections based on the network homophily assumption, wherein edges have a tendency to link similar nodes. This approach results in a more concise structural representation. Additionally, Li et al. [26] introduce AGCN, a structure learning model rooted in metric learning. AGCN computes the generalized Mahalanobis distance between every pair of node features and subsequently optimizes the topology using a Gaussian kernel of size K, guided by the calculated distances.

Continuing the trajectory established by the previously mentioned studies, our approach begins by utilizing graph generation methods to initially capture the desired topology from raw data. Subsequently, we meticulously eliminate the noise that arises during the generation process and improve the graph structure using graph learning techniques. This facilitates the revelation of the underlying topological insights inherent in the original data, culminating in a more condensed and concise structural representation. Finally, by leveraging the knowledge obtained from the interconnections among parameters discovered through GSL, we adeptly adjust the parameters within the network, thereby accomplishing network compression.

3   The Proposed Method

3.1   Overview

The proposed method comprises 5 essential steps, as illustrated in Fig. 1, which are Training and Parameter Collection, Graph Construction, Graph Structure

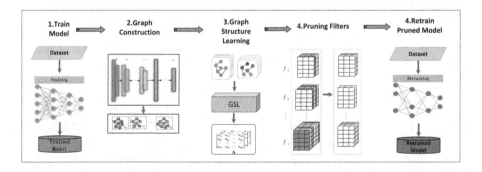

Fig. 1. A Compression Method for CNNs Based on GSL.

Learner, Pruning, and Model Retraining, respectively. Firstly, the filter parameters of the original model are trained to establish the baseline of the network. Then, the Graph Construction module takes these parameters as input and transforms the parameter connections into the node relationships within the graph. This generated graph functions as the training dataset for the GSL phase, which captures intricate dependencies among filter parameters. These dependencies are represented through a collection of matrices that denote their relative significance.

Leveraging the established relationships among the filters, we execute pruning on the network's layers, thereby generating a new model. This process entails selectively removing redundant parameters while preserving the essential ones. The pruned and unpruned layers are then merged into the new model. The new model then proceeds to retraining phase to enhance its performance. Compared to the original model, it boasts significantly fewer parameters and lower computational demands, while maintaining near-identical predictive accuracy. Thus, the new model consumes less storage space, requires lower memory usage during runtime, and executes more efficiently, all contributing to enhanced overall efficiency.

3.2   Graph Construction

To examine the relationships among CNN parameters using graph learning techniques, it's crucial to map these parameter relationships onto the connections between nodes in graphs. Each convolutional layer of CNNs is transformed into a corresponding graph, which acts as training data for graph learning. This section introduces two methods for constructing these graphs, where the neural network parameters are integrated as node features, leading to the creation of graphs with two distinct feature types. These two types of graphs are employed as training data and input into a graph structure learner.

This section introduces two methods for constructing graphs, where the neural network parameters are embedded as node features, resulting in graphs with two different types of features. These two types of graphs are used as training samples and are input into a graph structure learner.

A concise depiction of the graph construction process is presented in Fig. 2. Throughout the neural network training process, all filter parameters collected within the same training epoch are gathered. These parameters are subsequently structured into a matrix format, signifying the nodes and their attributes in the graph model. Through the utilization of the graph construction technique, nodes and edges are generated to establish a dedicated graph model tailored to the CNN parameters. Each parameter set L_n from every layer in the CNNs corresponds to a distinct graph G_n within that specific training epoch.

K-Nearest Neighbor Graphs. The k-nearest neighbor (kNN) algorithm is one of the simpler machine learning algorithms commonly used for classification problems. In this work, we utilize the idea of kNN to construct a kNN graph [27].

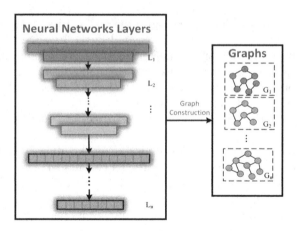

Fig. 2. Simple schematic diagram of graph construction.

During the graph construction process, the parameters of the filters are embed in each channel within the layers of CNNs as node features X in the graph. To grasp the inherent structure of nodes within the feature space, a kNN graph $G_t = (A_t, X)$ is constructed based on the node feature matrix X, where A_t is the adjacency matrix of the kNN graph. Specifically, we first compute the cosine similarity between n nodes to obtain the similarity matrix $S \in R^{n \times n}$. The cosine similarity measures the similarity between two vectors based on the cosine of the angle between them, and it is calculated using the following formula:

$$S_{ij} = \frac{x_i \cdot x_j}{|x_i| \, |x_j|}, \tag{1}$$

where x_i and x_j are the feature vectors of nodes i and j, respectively. We then select the k most similar nodes to each node and connect them to obtain the adjacency matrix A_t. This method generates a graph with strong structural characteristics.

Gaussian Kernel Graphs. The Gaussian kernel function, alternatively known as the radial basis function (RBF), is a scalar function characterized by radial symmetry. It serves the purpose of mapping finite-dimensional data into a high-dimensional space and is commonly used for similarity calculations. The formula for the Gaussian kernel function is expressed as follows:

$$k(x, y) = e^{-\frac{\|x-y\|^2}{2\sigma^2}}, \tag{2}$$

where k(x,y) represents the Gaussian kernel function, and σ is a hyperparameter that controls the range of the Gaussian kernel function. The graph construction method based on the Gaussian kernel calculates the potential relationships between nodes in a DNN by measuring the high-dimensional manifold distance.

It constructs a graph based on the similarity between filters in the CNNs. Firstly, the filter parameters of each channel within the layers of the network are flattened into one-dimensional vectors. The similarity between filters in each channel is calculated. If the dimensions of two filter parameters are different, the lower-dimensional vector is padded with zeros to match the dimensions, facilitating the similarity calculation between vectors. This step yields a similarity matrix for filters within each channel.

Next, a threshold ξ is set, and based on this threshold, the relationship between two filters in a channel is determined. If the similarity between two filters exceeds the threshold ξ, it is considered that there exists a certain degree of similarity between these two filters, and a connection is established between them. Conversely, if the similarity falls below the threshold, no connection is established.

Finally, the filter parameters of each channel within the layers of CNNs are embedded as the feature vectors X of the graph. The relationships established between filters in the previous step are mapped as connections between nodes in the graph, resulting in the adjacency matrix A_f of the graph. Thus, the graph $G_f = (A_f, X)$ is obtained. Let RBF_{mn} denote the similarity between filter m and filter n in a channel, the weight e_{mn} of the edge between node m and node n is given by:

$$e_{mn} = \begin{cases} 0, & \text{if } RBF_{mn} > \xi \\ RBF_{mn}, & \text{else} \end{cases} \tag{3}$$

3.3   Graph Structure Learner with Dual-Branch

The task of the GSL module is to explore the dependencies between parameters in the original neural network through the process of graph learning. Specifically, we train a dual-branch graph convolutional neural network (GCN) by inputting graphs with different feature information. During the graph learning process, the information learned from the two branches is fused to filter out noise in the original graph data and obtain a new adjacency matrix that better represents the relationships between nodes in the graph.

The graph structure learning module consists mainly of a dual-branch GCN, as shown in Fig. 3. It takes two parts of the graph as inputs. The upper branch is the structural convolutional network, which aims to learn the structural information in the training samples. It processes the parameters of CNNs using the K-nearest neighbor method, during which the parameters are mapped to nodes. For each node, the method calculates the similarity between it and all other nodes, resulting in a similarity matrix. From this matrix, the top K most similar nodes are selected and connected to establish the adjacency matrix. This approach generates a graph with strong structural features. By employing GCN, the exploration and learning of information within the underlying structural space of the graph can be delved. The output of structural space is as follows:

$$\mathbf{Z}_t^{(l)} = \text{ReLU}\left(\widetilde{\mathbf{D}}_t^{-\frac{1}{2}} \widetilde{\mathbf{A}}_t \widetilde{\mathbf{D}}_t^{-\frac{1}{2}} \mathbf{Z}_t^{(l-1)} \mathbf{W}_t^{(l)} \right), \tag{4}$$

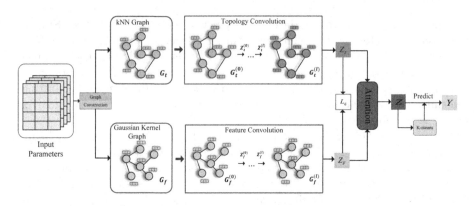

Fig. 3. The task of the GSL module is to mine the dependencies between parameters in CNNs through the graph learning process.

where $\boldsymbol{W}_t^{(l)}$ is the weight matrix of the lth layer in the GCN, ReLU is the activation function, and the initial $\boldsymbol{Z}_t^{(0)}$ set as $\boldsymbol{X}$. $\widetilde{\boldsymbol{A}}_t = \boldsymbol{A}_t + \boldsymbol{I}_t$, and $\widetilde{\boldsymbol{D}}_t$ is the diagonal degree matrix of $\widetilde{\boldsymbol{A}}_t$. The embedded representation that captures specific information outputted by the last layer of this branch is denoted as $\boldsymbol{Z}_T$. The lower branch is the feature convolutional network, which aims to learn the node feature information in the training samples, specifically the features of the original neural network parameters themselves. Therefore, When constructing the graph, the parameters were embedded as node features, with an effort to retain the original features to the greatest extent possible. We use the Gaussian kernel method to obtain the adjacency matrix of the graph. By using the feature convolutional network, we capture information from the graph feature space and obtain the output as:

$$\mathbf{Z}_f^{(l)} = \mathrm{ReLU}\left(\widetilde{\mathbf{D}}_f^{-\frac{1}{2}}\widetilde{\mathbf{A}}_f\widetilde{\mathbf{D}}_f^{-\frac{1}{2}}\mathbf{Z}_f^{(l-1)}\mathbf{W}_f^{(l)}\right). \tag{5}$$

The embedded representation that captures specific information, obtained from the last layer of the lower branch, is denoted as $\boldsymbol{Z}_F$. In the final step, an attention mechanism is applied to assign weights to $\boldsymbol{Z}_T$ and $\boldsymbol{Z}_F$, resulting in the final output $\boldsymbol{Z}$. The process is illustrated as follows:

$$\boldsymbol{Z} = \boldsymbol{\alpha}_t \cdot \boldsymbol{Z}_T + \boldsymbol{\alpha}_f \cdot \boldsymbol{Z}_F, \tag{6}$$

where $\alpha_t, \alpha_f \in \boldsymbol{R}^{n \times 1}$ represent the attention values of n nodes with embeddings $\boldsymbol{Z}_T$ and $\boldsymbol{Z}_F$, respectively. These attention values are learned through the attention mechanism $att(\boldsymbol{Z}_T, \boldsymbol{Z}_F)$, as shown below:

$$(\boldsymbol{\alpha}_t, \boldsymbol{\alpha}_f) = \mathrm{att}\left(\boldsymbol{Z}_T, \boldsymbol{Z}_F\right). \tag{7}$$

In the upper branch, taking node n as an example, its embedding in $\boldsymbol{Z}_T$, is represented by $\mathbf{z}_T^n \in \boldsymbol{R}^{1 \times h}$. Firstly, a non-linear transformation is applied to

$\mathbf{z}_T^n$, and then an attention value ω_T^n is obtained using a shared attention vector $a \in \mathbf{R}^{1 \times h'}$, as follows:

$$\omega_T^n = \mathbf{a}^T \cdot \tanh\left(\mathbf{W}_T \cdot (\mathbf{z}_T^n)^T + \mathbf{b}\right), \tag{8}$$

$\mathbf{W}_T \in \mathbf{R}^{h' \times h}$ is the weight matrix, and $\mathbf{b}$ is the bias vector. Similarly, we can obtain the attention value ω_F^n for node n in the embedding matrix $\mathbf{Z}_F$. Then, we normalize the attention values ω_F^n and ω_T^n using the softmax function as follows:

$$\alpha_t^n = \text{softmax}\left(\omega_T^n\right) = \frac{\exp\left(\omega_T^n\right)}{\exp\left(\omega_T^n\right) + \exp\left(\omega_F^n\right)} \tag{9}$$

The final weights for $\mathbf{Z}_T$ are denoted as $\alpha_t = [\alpha_t^n]$, and similarly, we can obtain the final weights for $\alpha_f = \left[\alpha_f^n\right] = [\text{softmax}\left(\omega_F^n\right)]$.

To ensure the independence of the two branches and to capture feature information from different spaces, we introduce a diversity constraint L_d, using the Hilbert-Schmidt Independence Criterion (HSIC) [28]. HSIC is a simple yet effective measure of independence used to enhance the dissimilarity between these two embeddings:

$$L_d = \text{HSIC}\left(\mathbf{Z}_T, \mathbf{Z}_F\right) = (n-1)^{-2} \operatorname{tr}\left(\mathbf{R}\mathbf{K}_T\mathbf{R}\mathbf{K}_F\right), \tag{10}$$

where K_T and K_F are Gram matrices with elements $k_{T,ij} = k_T\left(\mathbf{z}_T^i, \mathbf{z}_T^j\right)$ and $k_{F,ij} = k_F\left(\mathbf{z}_F^i, \mathbf{z}_F^j\right)$. $\mathbf{R} = \mathbf{I} - \frac{1}{n}\mathbf{e}\mathbf{e}^T$, where $\mathbf{I}$ is an identity matrix and $\mathbf{e}$ is a column vector. In the loss calculation, the cross-entropy loss function is utilized:

$$L_c = \frac{1}{N}\sum_i L_i = -\frac{1}{N}\sum_i \sum_{c=1}^C y_{ic}\log\left(p_{ic}\right), \tag{11}$$

where C is the total number of classes in the data. If the true class of sample i is equal to c, y_{ic} is 1; otherwise, it is 0. p_{ic} is the probability of sample i belonging to class c. Considering the differential constraint, the complete objective function is formally defined as follows:

$$L = L_c + L_d. \tag{12}$$

Due to the lack of explicit labels for the graph or its nodes, as the graph data is derived from the parameters of a CNN, we utilize self-supervised learning in the final prediction stage. We perform clustering on the output $\mathbf{Z}$ of the graph neural network, and the resulting clusters are used as labels for prediction and loss calculation. At the beginning of the training process, the matrix $\mathbf{A}$ randomly is initialized. During the learning process, the information from the two branches is continuously used to update the matrix $\mathbf{A}$. Finally, the matrix $\mathbf{A}^i$ is obtained, where i represents the network layer. The importance matrix $\mathbf{A}^i$ serves as a measurement criterion for subsequent pruning of the filters of CNNs.

3.4　Pruning Filters

The main objective of model pruning is to remove less important filters or parameters from a trained model in order to improve computational efficiency while minimizing the loss in accuracy. After obtaining the filter importance matrix A^i through the GSL module described in Sect. 3.3, each row of the matrix corresponds to the connection weights between nodes in the newly constructed graph. More precisely, it signifies the relevance scores between filters in a particular channel of the lth layer and filters in other channels. We assess the relative significance of filters within their respective layers by computing the node degrees within the graph and subsequently trim filters with comparatively lower importance.

Let n_i represent the number of input channels of the ith convolutional layer, and h_i/w_i denote the height/width of the input feature map $x_i \in R^{n_i \times h_i \times w_i}$. The convolutional layer transforms the input features into output features $x_{i+1} \in R^{n_{i+1} \times h_{i+1} \times w_{i+1}}$, which serve as the input features for the next convolutional layer. This process is achieved by applying n_{i+1} 3D filters $f_{i,j} \in R^{n_{i+1} \times k \times k}$ (j=1,2,...,n_{i+1}) on the n_i input channels, where each filter generates one feature map. Each filter consists of n_i 2D convolution kernels $K \in R^{k \times k}$. All filters together form the convolutional kernel matrix $f_i \in R^{n_i \times n_{i+1} \times k \times k}$. The number of operations for each convolutional layer is $n_{i+1} \times n_i \times k^2 \times h_{i+1} \times w_{i+1}$. As shown in Fig. 4, when a filter $f_{i,j}$ is pruned, its corresponding feature map $x_{i+1,j}$ is also removed, reducing the number of operations by $n_i \times k^2 \times h_{i+1} \times w_{i+1}$. Moreover, the kernels applied to the pruned feature maps from the next convolutional layer are also removed, saving additional $n_{i+2} \times k^2 \times h_{i+2} \times w_{i+2}$ operations. Therefore, pruning m filters in layer i will reduce the computation cost of layers i and $i+1$ by m/n_{i+1}.

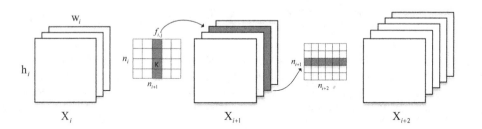

Fig. 4. Pruning a filter causes its corresponding feature map and associated convolutional kernel to be deleted in the next layer.

The process of pruning the m filters of the convolution kernel matrix f_i of the ith layer in the network is as follows:

(1) For the jth filter $f_{i,j}$ ($j \leq m < n_{i+1}$) of the ith layer in the network, by calculating the degree of the corresponding node of the filter $f_{i,j}$ in the relative importance matrix A^i, the relative importance $P_j = \sum_{c=1}^{n_{i+1}} A^i_{j,c}$ in its layer is obtained;

(2) Sort the filters according to the value of P_j;
(3) Prune the m filters with the smallest P_j value, and the kernel in the next convolutional layer corresponding to the number of pruned channels is also removed;
(4) Create a new kernel matrix for the ith layer and the $i+1$th layer, and copy the remaining convolution kernel parameters to the new model.

4 Experiments

In the experimental section, we selected the VGG-16 and ResNet-101 models for compression and trained them using the CIFAR-10 dataset. VGG-16 and ResNet-101 are widely used and well-established architectures of CNNs in the field of computer vision, often serving as building blocks for many other deep learning models. To obtain the baseline accuracy of each network, the original networks are trained. The CIFAR-10 dataset is used for this process, where pre-trained models are loaded, and the hyperparameter settings remained unchanged. After pruning the filters, a new model with fewer filters is created, and the modified network layers along with the remaining parameters of unaffected layers are copied to the new model. Additionally, if pruning is applied to convolutional layers, the weights of subsequent batch normalization layers are also removed. Subsequently, the original and new models are retrained for 50 epochs on the CIFAR-10 dataset, maintaining a constant learning rate of 0.001. Table 1 shows the experimental results on both models.

Table 1. Overall results. Reported the validation accuracy during the training process (presented as error rate). GSL-pruned-retrain is the data obtained by retraining the new model obtained by tailoring the trained model. GSL-pruned-train is the data obtained from training the new model from scratch after pruning the original model.

Method	Model	Error %	FLOPs	Pruned %	Parameters	Pruned %
	VGG-16	6.16%	4.2E+08		1.2E+08	
GSL-pruned- retrain	VGG-16	**6.18%**	1.3E+08	58%	5.1E+06	**66%**
GSL-pruned-train	VGG-16	8.84%	1.5E+08	58%	5.1E+06	66%
L1-norm(Li et al. [8])	VGG-16	6.60%	2.06e+08	34.2%	5.4e+06	64%
	ResNet-101	5.22%	5.3E+08		3.6E+07	
GSL-pruned-retrain	ResNet-101	**5.88%**	3.5E+08	**33%**	2.4E+07	**32%**
GSL-pruned-train	ResNet-101	7.68%	3.5E+08	33%	2.4E+07	32%

4.1 Pruned VGG-16

VGG-16 [2] is a well-known architecture based on CNNs, celebrated for its simplicity and efficacy. It comprises 16 weighted layers, consisting of 13 convolutional layers and 3 fully connected layers. VGG-16 is widely recognized for its remarkable depth, boasting a total of 138 million parameters, which empowers it to capture intricate image features with great prowess. This architecture finds application in diverse image recognition tasks, encompassing image classification, object detection, and image segmentation. However, the substantial parameter count and computational complexity of VGG-16 present challenges for deployment on constrained devices, thus highlighting the need for optimization and enhancement. Table 2 provides information regarding the original VGG-16 model as well as the proportions of its components that are pruned.

Table 2. Comparison of the VGG-16 model before and after construction, the last two columns are the parameters of the convolutional layer and the first fully connected layer in the pruned model and the percentage reduction of FLOPs.

layers	Params (before)	Params (after)	FLOPs (before)	FLOPs (after)	Params %	FLOPs %
Conv_1	1.7E+03	8.1E+02	1.8E+06	8.3E+05	53%	53%
Conv_2	3.7E+04	1.3E+04	3.8E+07	1.4E+07	64%	64%
Conv_3	7.4E+04	4.0E+04	1.9E+07	1.0E+07	46%	46%
Conv_4	1.5E+05	8.3E+04	3.8E+07	2.1E+07	44%	44%
Conv_5	2.9E+05	1.5E+05	1.9E+07	9.6E+06	49%	49%
Conv_6	5.9E+05	2.6E+05	3.8E+07	1.7E+07	56%	56%
Conv_7	5.9E+05	3.3E+05	3.8E+07	2.1E+07	44%	44%
Conv_8	1.2E+06	6.2E+05	1.9E+07	9.9E+06	48%	48%
Conv_9	2.4E+06	7.1E+05	3.8E+07	1.1E+07	70%	70%
Conv_10	2.4E+06	5.9E+05	3.8E+07	9.4E+06	75%	75%
Conv_11	2.4E+06	7.3E+05	9.4E+06	2.9E+06	69%	69%
Conv_12	2.4E+06	7.3E+05	9.4E+06	2.9E+06	69%	69%
Conv_13	2.4E+06	6.4E+05	9.4E+06	2.6E+06	73%	73%
Linear1	2.6E+05	1.3E+05	2.6E+05	1.3E+05	50%	50%
Linear2	4.1E+04	4.1E+04	4.1E+04	4.1E+04	0%	0%
Total	1.5E+07	5.1E+06	3.1E+08	1.3E+08	66%	58%

4.2   Pruned ResNet-101

ResNet-101 is a variation of the ResNet (Residual Network) architecture initially proposed by He et al. [29]. It stands out for its remarkable depth, comprising 101 weighted layers. ResNet-101 builds upon the concept of residual learning, which introduces skip connections to enable the network to learn residual mappings instead of directly learning the desired output. This approach helps mitigate the issue of vanishing gradients and facilitates the training of exceptionally deep networks. The building blocks of ResNet-101 consist of basic residual units, which entail two 3×3 convolutional layers along with a skip connection. Notably, ResNet-101 does not employ pooling layers but instead relies on residual connections for downsampling the feature maps. Table 3 provides details regarding the original ResNet-101 model as well as the pruned components.

Table 3. The comparison before and after the construction of the ResNet-101 model, the last two columns are the parameters of the convolutional layer and the first layer of the fully connected layer in the pruned model and the reduction percentage of FLOPs

layers	Params (before)	Params (after)	FLOPs (before)	FLOPs (after)	Params %	FLOPs %
Conv1	9.4E+03	6.6E+03	9.6E+06	6.8E+06	30%	30%
Conv2_x	1.7E+05	1.1E+05	4.4E+07	2.7E+07	39%	39%
Conv3_x	9.8E+05	6.0E+05	6.3E+07	3.8E+07	39%	39%
Conv4_x	2.3E+07	1.5E+07	3.6E+08	2.5E+08	31%	32%
Conv5_x	1.2E+07	7.9E+06	4.7E+07	3.2E+07	33%	33%
Linear	2.0E+04	1.8E+04	2.0E+04	1.8E+04	14%	14%
Total	3.6E+07	2.41E+07	5.3E+08	3.5E+08	32%	33%

5   Conclusions and Future Works

This paper proposes an innovative approach for compressing CNNs by leveraging the relationships among different parameters within network layers. The proposed method involves constructing a graph structure to represent the neural network parameters, which is then inputted into a dual-branch GSL module. During the learning process, two constraints are introduced to update the original graph topology while ensuring consistency between node embeddings and the original network parameters. By employing GSL, the correlations between parameters of CNNs are uncovered, enabling network compression based on the strength of these correlations. This method achieves a significant reduction in the number of parameters and floating-point operations, leading to improved efficiency and reduced model complexity while maintaining prediction accuracy. The effectiveness of the proposed method is demonstrated through evaluations on VGG-16 and ResNet-110, achieving a significant reduction in FLOPs and parameters on CIFAR-10 without substantial loss in accuracy. Comparative analysis with other compression algorithms showcases the accuracy and efficiency of our method. In future research, we plan to expand the dataset and incorporate label distributions of experimental samples to further analyze interactions between nodes within and across layers with larger margins. Additionally, we aim to extend the application of this method to other deep learning models, continually enhancing its practicality and versatility to achieve favorable outcomes in optimizing specialized hardware and adapting to dynamic environments.

Acknowledgements. This work is supported by the Natural Science Foundation of Shandong Province China (NO. ZR2020LZH008, ZR2021MF118, ZR2022LZH003), the Key R&D Program of Shandong Province, China (NO. 2021CXGC010506, NO. 2021SFGC0104) and the National Natural Science Foundation of China (NO. 62101311).

References

1. Wang, B., et al.: SparG: a sparse GEMM accelerator for deep learning applications. In: Meng, W., Lu, R., Min, G., Vaidya, J. (eds.) Algorithms and Architectures for Parallel Processing: 22nd International Conference, ICA3PP 2022, Copenhagen, Denmark, 10–12 October 2022, Proceedings, pp. 529–547. Springer, Cham (2023). https://doi.org/10.1007/978-3-031-22677-9_28

2. Aktas, K., Ignjatovic, V., Ilic, D., Marjanovic, M., Anbarjafari, G.: Deep convolutional neural networks for detection of abnormalities in chest X-rays trained on the very large dataset. Signal Image Video Process. **17**(4), 1035–1041 (2023). https://doi.org/10.1007/s11760-022-02309-w

3. Atakishiyev, S., Salameh, M., Yao, H., Goebel, R.: Explainable artificial intelligence for autonomous driving: a comprehensive overview and field guide for future research directions. CoRR abs/2112.11561 (2021). https://arxiv.org/abs/2112.11561

4. LeCun, Y., Denker, J.S., Solla, S.A.: Optimal brain damage. In: Touretzky, D.S. (ed.) Advances in Neural Information Processing Systems 2, [NIPS Conference, Denver, Colorado, USA, 27–30 November 1989], pp. 598–605. Morgan Kaufmann (1989). http://papers.nips.cc/paper/250-optimal-brain-damage

5. Denil, M., Shakibi, B., Dinh, L., Ranzato, M., de Freitas, N.: Predicting parameters in deep learning. In: Burges, C.J.C., Bottou, L., Ghahramani, Z., Weinberger, K.Q. (eds.) Advances in Neural Information Processing Systems 26: 27th Annual Conference on Neural Information Processing Systems 2013. Proceedings of a Meeting Held 5–8 December 2013, Lake Tahoe, Nevada, United States, pp. 2148–2156 (2013). https://proceedings.neurips.cc/paper/2013/hash/7fec306d1e665bc9c748b5d2b99a6e97-Abstract.html

6. Lin, Y., Wang, C., Chang, C., Sun, H.: An efficient framework for counting pedestrians crossing a line using low-cost devices: the benefits of distilling the knowledge in a neural network. Multim. Tools Appl. **80**(3), 4037–4051 (2021). https://doi.org/10.1007/s11042-020-09276-9

7. Iandola, F.N., Moskewicz, M.W., Ashraf, K., Han, S., Dally, W.J., Keutzer, K.: SqueezeNet: AlexNet-level accuracy with 50x fewer parameters and <1 MB model size. CoRR abs/1602.07360 (2016). http://arxiv.org/abs/1602.07360

8. Li, H., Kadav, A., Durdanovic, I., Samet, H., Graf, H.P.: Pruning filters for efficient convnets. In: 5th International Conference on Learning Representations, ICLR 2017, Toulon, France, 24–26 April 2017, Conference Track Proceedings. OpenReview.net (2017). https://openreview.net/forum?id=rJqFGTslg

9. Zhang, L., Wei, W., Shi, Q., Shen, C., van den Hengel, A., Zhang, Y.: Accurate tensor completion via adaptive low-rank representation. IEEE Trans. Neural Networks Learn. Syst. **31**(10), 4170–4184 (2020). https://doi.org/10.1109/TNNLS.2019.2952427

10. Kang, H.: Accelerator-aware pruning for convolutional neural networks. IEEE Trans. Circuits Syst. Video Technol. **30**(7), 2093–2103 (2020). https://doi.org/10.1109/TCSVT.2019.2911674

11. Shen, W., Wang, W., Zhu, J., Zhou, H., Wang, S.: Pruning-and quantization-based compression algorithm for number of mixed signals identification network. Electronics **12**(7), 1694 (2023)

12. Yuan, C., Agaian, S.S.: A comprehensive review of binary neural network. CoRR abs/2110.06804 (2021). https://arxiv.org/abs/2110.06804

13. Zhao, R., et al.: Accelerating binarized convolutional neural networks with software-programmable FPGAs. In: Greene, J.W., Anderson, J.H. (eds.) Proceedings of the 2017 ACM/SIGDA International Symposium on Field-Programmable Gate Arrays, FPGA 2017, Monterey, CA, USA, 22–24 February 2017, pp. 15–24. ACM (2017). http://dl.acm.org/citation.cfm?id=3021741

14. Li, E., Zeng, L., Zhou, Z., Chen, X.: Edge AI: on-demand accelerating deep neural network inference via edge computing. IEEE Trans. Wirel. Commun. 19(1), 447–457 (2020). https://doi.org/10.1109/TWC.2019.2946140

15. Srivastava, N., Hinton, G.E., Krizhevsky, A., Sutskever, I., Salakhutdinov, R.: Dropout: a simple way to prevent neural networks from overfitting. J. Mach. Learn. Res. 15(1), 1929–1958 (2014). https://doi.org/10.5555/2627435.2670313

16. Han, S., Pool, J., Tran, J., Dally, W.J.: Learning both weights and connections for efficient neural network. In: Cortes, C., Lawrence, N.D., Lee, D.D., Sugiyama, M., Garnett, R. (eds.) Advances in Neural Information Processing Systems 28: Annual Conference on Neural Information Processing Systems 2015 December, pp. 7–12, 2015, Montreal, Quebec, Canada, pp. 1135–1143 (2015). https://proceedings.neurips.cc/paper/2015/hash/ae0eb3eed39d2bcef4622b2499a05fe6-Abstract.html

17. Srinivas, S., Babu, R.V.: Data-free parameter pruning for deep neural networks. In: Xie, X., Jones, M.W., Tam, G.K.L. (eds.) Proceedings of the British Machine Vision Conference 2015, BMVC 2015, Swansea, UK, 7–10 September 2015, pp. 31.1–31.12. BMVA Press (2015). https://doi.org/10.5244/C.29.31

18. Chen, S., Zhao, Q.: Shallowing deep networks: layer-wise pruning based on feature representations. IEEE Trans. Pattern Anal. Mach. Intell. 41(12), 3048–3056 (2019). https://doi.org/10.1109/TPAMI.2018.2874634

19. Fiesler, E., Choudry, A., Caulfield, H.J.: Weight discretization paradigm for optical neural networks. In: Optical Interconnections and Networks, vol. 1281, pp. 164–173. SPIE (1990)

20. Balzer, W., Takahashi, M., Ohta, J., Kyuma, K.: Weight quantization in Boltzmann machines. Neural Netw. 4(3), 405–409 (1991). https://doi.org/10.1016/0893-6080(91)90077-I

21. Gilmer, J., Schoenholz, S.S., Riley, P.F., Vinyals, O., Dahl, G.E.: Neural message passing for quantum chemistry. In: Precup, D., Teh, Y.W. (eds.) Proceedings of the 34th International Conference on Machine Learning, ICML 2017, Sydney, NSW, Australia, 6–11 August 2017. Proceedings of Machine Learning Research, vol. 70, pp. 1263–1272. PMLR (2017)

22. Dai, H., et al.: Adversarial attack on graph structured data. In: Dy, J.G., Krause, A. (eds.) Proceedings of the 35th International Conference on Machine Learning, ICML 2018, Stockholmsmässan, Stockholm, Sweden, 10–15 July 2018. Proceedings of Machine Learning Research, vol. 80, pp. 1123–1132. PMLR (2018). http://proceedings.mlr.press/v80/dai18b.html

23. Zhu, D., Zhang, Z., Cui, P., Zhu, W.: Robust graph convolutional networks against adversarial attacks. In: Teredesai, A., Kumar, V., Li, Y., Rosales, R., Terzi, E., Karypis, G. (eds.) Proceedings of the 25th ACM SIGKDD International Conference on Knowledge Discovery & Data Mining, KDD 2019, Anchorage, AK, USA, 4–8 August 2019, pp. 1399–1407. ACM (2019). https://doi.org/10.1145/3292500.3330851

24. Luo, D., et al.: Learning to drop: robust graph neural network via topological denoising. In: Lewin-Eytan, L., Carmel, D., Yom-Tov, E., Agichtein, E., Gabrilovich, E. (eds.) WSDM 2021, The Fourteenth ACM International Conference on Web Search and Data Mining, Virtual Event, Israel, 8–12 March 2021, pp. 779–787. ACM (2021). https://doi.org/10.1145/3437963.3441734

25. Newman, M.: Networks. Oxford University Press, Oxford (2018)
26. Li, R., Wang, S., Zhu, F., Huang, J.: Adaptive graph convolutional neural networks. In: McIlraith, S.A., Weinberger, K.Q. (eds.) Proceedings of the Thirty-Second AAAI Conference on Artificial Intelligence, (AAAI-2018), the 30th innovative Applications of Artificial Intelligence (IAAI-2018), and the 8th AAAI Symposium on Educational Advances in Artificial Intelligence (EAAI-2018), New Orleans, Louisiana, USA, 2–7 February 2018, pp. 3546–3553. AAAI Press (2018). https://www.aaai.org/ocs/index.php/AAAI/AAAI18/paper/view/16642
27. Preparata, F.P., Shamos, M.I.: Computational Geometry - An Introduction. Texts and Monographs in Computer Science, Springer, Cham (1985). https://doi.org/10.1007/978-1-4612-1098-6
28. Song, L., Smola, A.J., Gretton, A., Borgwardt, K.M., Bedo, J.: Supervised feature selection via dependence estimation. In: Ghahramani, Z. (ed.) Machine Learning, Proceedings of the Twenty-Fourth International Conference (ICML 2007), Corvallis, Oregon, USA, 20–24 June 2007. ACM International Conference Proceeding Series, vol. 227, pp. 823–830. ACM (2007). https://doi.org/10.1145/1273496.1273600
29. He, K., Zhang, X., Ren, S., Sun, J.: Deep residual learning for image recognition. In: 2016 IEEE Conference on Computer Vision and Pattern Recognition, CVPR 2016, Las Vegas, NV, USA, 27–30 June 2016, pp. 770–778. IEEE Computer Society (2016). https://doi.org/10.1109/CVPR.2016.90

Reliability-Aware VNF Provisioning in Homogeneous and Heterogeneous Multi-access Edge Computing

Haolin Liu[1,2] , Zehang Tan[1], Zhetao Li[3(✉)], Saiqin Long[3], Shujuan Tian[1,2], and Xiaoshan Li[1]

[1] School of Computer Science, Xiangtan University, Xiangtan 411105, Hunan, China
[2] Hunan International Scientific and Technological Cooperation Base of Intelligent Network, Xiangtan University, Xiangtan 411105, Hunan, China
[3] College of Information Science and Technology, Jinan University, Guangzhou 510632, Guangdong, China
liztchina@hotmail.com

Abstract. In the emerging network architecture of multi-access edge computing (MEC), virtualized network function (VNF) can be deployed to provide network services to users, thus reducing the cost of the service provider. However, the server is not 100% reliable, and its hardware failure can lead to malfunctions of the VNFs deployed on it, which can affect the service reliability for users. In this paper, we focus on reliability-aware VNF service provisioning in MEC to meet the service reliability requirements of users by deploying redundant replicas of VNF instances to different servers. For that, a profit maximization problem for reliability assurance (PMRA) is first formulated and proved to be an NP-hard problem. Then, an efficient local ratio based algorithm (LRBA) is proposed to solve the PMRA problem in a homogeneous MEC scenario. Meanwhile, for the other PMRA problem in the heterogeneous scenario, a fast benefit-cost ratio preference algorithm (BRPA) is proposed. Finally, we evaluate the proposed algorithms by simulation experiments. The experimental results show that the proposed are promising.

Keywords: Multi-access edge computing · VNF provisioning · Reliability aware · Approximation algorithm

1 Introduction

Multi-access edge computing (MEC) is becoming a promising computing platform in the era of Internet of Everything. It extends cloud computing services to the edge of the mobile network by deploying servers to the user side [1]. Meanwhile, MEC is implemented with the support of Network Functions Virtualization (NFV) technology, which uses instances of Virtual Network Functions (VNFs) instead of dedicated hardware devices. That not only reduces costs but also has the advantage of being able to flexibly adapt services to meet rapidly changing user demands [2].

© The Author(s), under exclusive license to Springer Nature Singapore Pte Ltd. 2024
Z. Tari et al. (Eds.): ICA3PP 2023, LNCS 14488, pp. 147–167, 2024.
https://doi.org/10.1007/978-981-97-0801-7_9

In practical MEC scenarios, users usually require not only specific VNFs, such as firewalls, proxies, and intrusion detection systems (IDSs) but also a reliability requirement for the services [3]. The service reliability is defined as the ability of the network to provide stable services to ensure service level agreement (SLA) in the face of the risk of failure of the underlying network components [4]. Satisfying user service reliability requirement is critical for the service provider, and temporary failure of the service can lead to data loss and endanger the secure integration of services [5]. Traditional carrier-class systems are carefully designed to provide nearly 99.999% (five 9 s) reliable services [6] and are highly fault-tolerant [7]. While edge servers are close to users, it is difficult to compare with traditional cloud computing servers in terms of hardware structure, security assurance, and maintenance timeliness, so the reliability of MEC services is an issue that must be considered to provide stable and high-quality services to users. Since the services in MEC are implemented by VNFs, which are instances running on virtual machines of edge servers, the instances used to implement VNFs may contain errors [7]. Not only that, the server hosting the VNF instance may also malfunction and fail, all of which means that it is more difficult to achieve 100% reliability of the service.

Fault tolerance tries to ensure the continuity of services even when failures occur, and it can more practically improve the service reliability of MEC. Many fault-tolerance mechanisms have been proposed, and checkpointing is one of them [8]. However, both periodic saving and fault recovery in checkpointing mechanism cause significant performance overhead and are not suitable for time-sensitive MEC tasks. Another common fault-tolerance mechanism is replication, i.e., redundant deployment of VNF instances [9], which is very suitable for MEC reliability guarantees because of its simplicity and efficiency. Specifically, the failure of a single VNF instance can be mitigated by deploying other VNF backup instances in the same servers as its primary VNF instance [4]. Nevertheless, the VNF reliability under this single-point deployment scheme is limited by the reliability of the server where the VNF instance resides. If one server fails, all VNF instances in that server will fail. Therefore, the multi-point deployment scheme is often adopted in practice, which uses VNF replicas as backups and redundantly deploys the same VNF instances on multiple servers to satisfy the reliability requirement of users.

A redundant deployment scheme can consume a significant amount of computation resources and increase costs. As a result, it is important to consider how to ensure reliability and profitability for the service provider under limited computation resource constraints. To address this issue, the service provider can reduce the initiation and task execution costs of VNF replicas through task admission and assignment decisions to maximize their profit. For that, we formulate a profit maximization problem for reliability assurance (PMRA). Our goal is to maximize the profit of the service provider by admitting and assigning tasks and their replicas under resource constraints in both homogeneous and heterogeneous network contexts. The main contributions of this paper are summarized as follows.

- We use a multi-point deployment scheme to achieve fault tolerance and construct a PMRA problem that maximizes the profit of the service provider while guaranteeing task reliability. The problem is proved to be an NP-Hard problem.
- We consider the PMRA problem in both homogeneous MEC (Homo-MEC) and heterogeneous MEC (Heter-MEC) network scenarios. All servers in the homogeneous-PMRA (Homo-PMRA) have the same reliability. In contrast, heterogeneous-PMRA (Heter-PMRA) is more generalized, where all servers have different reliability.
- For Homo-PMRA, we propose a local ratio based algorithm (LRBA) that converts the original optimization problem into several identical subproblems to be solved. We prove the approximation ratio of the algorithm to illustrate its effectiveness. benefit-cost ratio preference algorithm (BRPA) prioritizes the assignment of tasks based on the benefit ratio on the high-reliability server and is able to find the approximate optimal solution quickly.
- We conduct simulation experiments for two network scenarios, comparing LRBA and BRPA with the LocalSearch algorithm and random algorithm in the homogeneous scenario, and comparing BRPA with the greedy algorithm and random algorithm in the heterogeneous scenario. The results demonstrate that the proposed LRBA and BRPA outperform the compared algorithms.

2   Related Work

VNF provisioning in MEC has been extensively studied [10–15]. For example, Tian et al. [10] proposed the improved service function graph and approximation algorithm for parallelization and deployment of VNFs on edge nodes to reduce latency. Yang et al. [11] proposed a reinforcement learning approach to handle the random arrival of SFC requests in MEC, with the goal of maximizing the number of allowed requests while satisfying the reliability and latency requirements of allowed requests. Mao et al. [12] developed a online joint SFC placement and flow routing model that takes into account the flow variation of each SFC. To address this issue, they propose a two-stage flow-sensitive scheme. Wang et al. [13] propose a model for a multi-user noncooperative computing offload problem for MEC scenarios and design a payoff function for the participants to evaluate their benefits and achieve maximum utility for the users. Li et al. [14] proposed a metric called coflow age to measure Result freshness. Xu et al. [15] use the random forest to classify flows, and then assign rates and paths to flows based on confidence to reliably predict flow information in advance.

In addition to those studies on resource allocation mentioned above, there are several works that consider service reliability issues in VNF provisioning. For example, Huang et al. [3] propose a new reliability-aware VNF instance configuration problem for network function services in MEC. Efficient approximate and exact algorithms are proposed to allocate network resources to accommodate primary and secondary VNF instances in different servers to meet users' reliability requirements. Li et al. [1] studied the VNF service reliability problem to maximize the service provider's revenue. The authors first considered the provisioning of reliable VNF services by joint considering both the VNF instance reliability (software) and the cloudlet reliability (hardware). Then they proposed online algorithms with provable competitive ratios for the problems under two different VNF placement backup schemes: the on-site and off-site schemes.

Different from the above works, this paper considers the reliability problem brought by server hardware and divides the network into Homo-MEC and Heter-MEC scenarios. Then, we propose a multi-point deployment scheme for distributed redundancy deployment in both scenarios. However, this scheme consumes more computation resources and may not accommodate all task requests. To address this issue, we formulate a PMRA problem to maximize profit from edge servers executing tasks by selectively admitting task requests, thus guaranteeing the reliability requirement of tasks and the revenue of the service provider.

3   Preliminaries

In this section, the system model is described. We first introduce the network model, describing the components of the network. The task model and cost model are then detailed respectively.

3.1   Network Model

There exists a set of servers $S = \{s_1, s_2, ..., s_m, ..., s_M\}$ within the network. Each server $s_m \in S$ can provide K types of VNF instances, expressed as $F = \{f_1, f_2, ..., f_k, ..., f_K\}$. Each of them has an available computation resource capacity of cap_m, which is used to instantiate various VNFs and provide network services to users. The reliability of server s_m is expressed in terms of q_m. A task run on a VNF instance on an edge server, and if the server has hardware failures and crashes, the VNF instances and tasks on it will also be affected and stop working, so q_m can also be considered as the success rate of tasks on server s_m. Figure 1 shows an illustration of the MEC network.

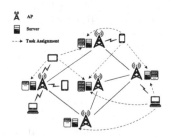

Fig. 1. An illustration of the MEC network.

3.2 Task Model

Suppose there is a set of tasks $T = \{t_1, t_2, ..., t_n, ...t_N\}$ that need to be processed and cannot be partitioned. For task t_n, we assume it contains a set of properties $(d_n, v_{nk}, \varepsilon_n, \gamma_n, pay_n)$, where d_n represents the size of data packets of task t_n (e.g., input parameters and processing codes). Assume that each task replica can only run on a VNF instance of a specific type, denoting the required VNF type by $v_{nk} \in \{0, 1\}(f_k \in F)$, where $v_{nk} = 1$ represents that the VNF type required for task t_n is f_k. To ensure that tasks can be executed independently, ε_n represents the computation resources required for task t_n. The reliability goal of task t_n is represented by γ_n, $0 < \gamma_n < 1$. It is worth noting that the service provider admitting task t_n represents its reliability requirement is satisfied, i.e., the task reliability goal γ_n is achieved through redundant deployment, at this point the service provider gets paid, denoted by pay_n.

3.3 Cost Model

If a replica of a task is assigned to a server, the server needs to launch the VNF instances required for the task. This process is called VNF instance initialization [16], which causes some energy and resource consumption for which costs need to be paid. We assume that VNF instances of the same type have equal initiation costs, denote by $C^i(k)$ the initiation cost of VNF f_k. In addition, task replicas consume computation resources on the server during processing, so there is a cost of computation resource consumption that is related to the type of VNF required for the task. Specifically, denote by δ_{mk} the cost per unit of data size for an instance f_k of the VNF to perform a task on the server s_m. In summary, the total cost of a replica of task t_n executing on the server s_m can be calculated as

$$C^t(n, m) = v_{nk} \cdot d_n \cdot \delta_{mk} + v_{nk} \cdot C^i(k). \tag{1}$$

4 Homogeneous MEC Scenario

In this section, we first consider a Homogeneous MEC (Homo-MEC) scenario. For ease of management and maintenance, the service provider uses servers with the same hardware architecture to provide services to users. This results in all servers having almost identical hardware and parameter settings, leading to a more uniform user experience [17]. In this case, all servers have almost the same hardware reliability and computation capacity. For that Homo-MEC scenario, we first formulate a profit maximization problem and then propose a local ratio based algorithm (LRBA) to solve it.

4.1 Reliability Model

We first consider the reliability model of redundant deployment for the Homo-MEC scenario. For that, we define the service reliability as the probability that the network can serve the admitted tasks, i.e., the probability that at least one replica of the task can be executed properly in the network. Since all servers in the Homo-MEC are of the same type, we can assume that the computation capacity of each server is cap and the reliability is q, i.e., $1 - q$ is the failure rate of the server.

To satisfy the reliability goal γ_n for task t_n, assuming that t_n needs to be redundantly assigned to a_n different servers, then according to the homogeneous property of the network, we can obtain

$$1 - (1 - q)^{a_n} \geqslant \gamma_n, \tag{2}$$

where $(1 - q)^{a_n}$ is the probability that all a_n replicas of task t_n fail so that $1 - (1 - q)^{a_n}$ represents the probability that at least one of the replicas can work successfully. Therefore, the minimum number of redundant deployed replicas a_n required for task t_n can be calculated as

$$a_n = \lceil \log_{1-q} (1 - \gamma_n) \rceil . \tag{3}$$

4.2   Problem Formulation

Because the computation capacity of edge servers is limited compared to cloud servers, and especially when redundant deployment is considered, cost becomes a factor that must be considered. Therefore, the service provider must pursue sufficient profit to ensure that they can provide long-term stable edge computing services. For that, we consider a profit maximization problem of reliability assurance (PMRA) in Homo-MEC, which selects and admits some tasks under the premise of satisfying the task reliability requirements to ensure that the resources of servers do not exceed the limits and can maximize the profit of the server provider. Specifically, we first define a binary decision variable $x_{nm} \in \{0, 1\}$ to indicate whether a replica of task t_n is assigned to server s_m, where $x_{nm} = 1$ if task t_n is assigned to server s_m and $x_{nm} = 0$ otherwise. Let $y_n \in \{0, 1\}$ denote whether task t_n is admitted, where $y_n = 1$ if task t_n is admitted by the service provider and $y_n = 0$ otherwise. Let X and Y denote the set of x_{nm} and y_n, respectively. Finally, the Homo-PMRA problem can be formulated as the following integer programming problem,

$$\text{P1}: \max_{X,Y} \ \sum_{t_n \in T} pay_n \cdot y_n - \sum_{t_n \in T} \sum_{s_m \in S} C^t (n, m) \cdot x_{nm}$$

$$\text{s.t.} \quad \sum_{s_m \in S} x_{nm} \geq y_n \cdot a_n, \quad \forall t_n \in T, \tag{4}$$

$$\sum_{t_n \in T} \varepsilon_n \cdot x_{nm} \leq cap, \quad \forall s_m \in S, \tag{5}$$

$$x_{nm} \in \{0, 1\}, y_n \in \{0, 1\}, \forall t_n \in T, \forall s_m \in S, \tag{6}$$

where constraint (4) denotes that for any task, if it is admitted, then the minimum number of task replicas needs to be satisfied due to the reliability goal.

Constraint (5) indicates that the computation resources required by the task replicas deployed on a server must not exceed the total computation capacity of the server. Constraint (6) denotes that decision variable x_{nm}, y_n is a binary variable.

To solve P1 more conveniently, constraint (4) is reduced to be strictly equal, thus transforming P1 into P2:

$$P2 : \max_{X,Y} \sum_{t_n \in T} y_n \cdot \sum_{s_m \in S} \left(\frac{pay_n}{a_n} - C^t(n, m) \right) \cdot x_{nm}$$

$$\text{s.t.} \quad \cdot \sum_{s_m \in S} x_{nm} = y_n \cdot a_n, \quad \forall t_n \in T, \tag{7}$$

$$\text{Constraints} \quad (5), (6).$$

Theorem 1: P1 and P2 are equivalent.

Proof: We can prove it by induction. When all $y_n = 0$, the objective function of P1 is maximizing $-\sum_{t_n \in T} \sum_{s_m \in S} C^t(n, m) \cdot x_{nm}$, and constraint (4) will also no longer work, so the optimal solution of P1 will be all $x_{nm} = 0$, the same as the solution and result of P2. Therefore the induction hypothesis holds when all $y_n = 0$.

Without loss of generality, we then consider the case where $y_n = 1$ for arbitrary task t_n. For task t_n, the objective function of P1 can be formulated as $pay_n - \sum_{s_m \in S} C^t(n, m) \cdot x_{nm}$. Constraint (4) then becomes $\sum_{s_m \in S} x_{nm} \geqslant a_n$. For P2, the objective function can be formulated as $\sum_{s_m \in S} (pay_n/a_n - C^t(n, m)) \cdot x_{nm}$. Constraint (7) becomes $\sum_{s_m \in S} x_{nm} = a_n$. Therefore, in this case, the objective functions of P1 and P2 are identical. Suppose the results of P1 and P2 are $R(P1)$ and $R(P2)$, respectively. Because the objective functions are identical, but the constraints are different, it is easy to know that $R(P1) \geqslant R(P2)$ when $y_n = 1$ and the other y_n are 0. $R(P1)$ and $R(P2)$ can be strictly equal when the solutions of P1 and P2 are exactly the same, i.e., all x_{nm} are the same. By so on, the above conclusion still holds for other y_n taking the value of 1. Hence, the induction hypothesis follows. P1 and P2 are equivalent.   ∎

Theorem 2: P2 is an NP-hard problem.

Proof: We prove that P2 is an NP-hard problem by a reduction from a well-known NP-complete knapsack problem. The knapsack problem is defined as, given a set of items, each of which has its own weight and price, how to choose the items so that the total price of the items is the highest within a limited total weight. We consider a special case of P2 where there is only one server with capacity cap in the MEC network, and each task requires only one replica that satisfies its reliability requirement, while all tasks arrive at the same time. Then, if the server admits task t_n, it needs to gain profit $pay_n - C^t(n, m)$ at the cost of ε_n computation resources. Therefore, in the special case, the goal of this optimization problem is to maximize the profit by admitting as many high-profit tasks as possible within the limit of the server capacity *cap*. It can be seen that a solution to this profit maximization problem is a solution to the knapsack problem. Thus, according to the reduction theorem [18], the more complex P2 is also an NP-hard problem.   ∎

4.3   Local Ratio Based Algorithm

To ease the solving of the Homo-PMRA problem, we transform P1 into P2, which can be regarded as a generalized assignment problem (GAP). Then, we propose a combinatorial algorithm using local ratio technique [19] to solve the GAP problem. The local ratio technique provides a way to decompose the complex problem into several simple subproblems. Specifically, We denote the feasible set by $\mathbb{R}$, $f(\cdot)$ is the objective function, and $(\mathbb{R}, f(\cdot))$ is the original problem. Through the local ratio technique, $(\mathbb{R}, f(\cdot))$ is decomposed into several $0-1$ knapsack problems $(\mathbb{R}, f_m(\cdot))$, where $f_m(\cdot)$ is the profit function in subproblem m. In general, the proposed LRBA is summarized in the following steps:

1) Clustering: According to (3), the number a_n of redundant deployments required for each task t_n is obtained. Gather tasks with the same a_n into a group. The task groups are sorted in the order of a_n from smallest to largest, and the following steps are sequentially adopted for the assignment of tasks.
2) Decomposition: Generate a profit matrix for the tasks in the current group with the corresponding servers. Decompose the assignment problem of the current task group into multiple subproblems, which leads to M knapsack problems $(\mathbb{R}, f_m(\cdot))$.
3) Solving: Solve the knapsack problem for each server from s_1 to s_M. Denote by $\boldsymbol{O}_m$ the set of tasks assigned to the server s_m found by the knapsack problem algorithm. Then update the profit matrix.
4) Refinement: The tasks in $\boldsymbol{O}_m$ are assigned to each server s_m in the order from server s_M to s_1, and the set of tasks that have been determined to be assigned to server s_m is denoted by $\boldsymbol{R}_m$, where $\boldsymbol{R}_M = \boldsymbol{O}_M$. Tasks that have been assigned are excluded in the next iteration, i.e., $\boldsymbol{R}_m = \boldsymbol{O}_m \setminus \cup_{k=m+1}^{M} \boldsymbol{R}_k$, eliminating the effect of sequentially on the results. Set the profit of the tasks included in $\boldsymbol{R}_m$ in the profit matrix to zero, which also means that these replicas will not be reassigned. Meanwhile, the rows corresponding to tasks that meet the number of replicas deployed and the columns corresponding to servers that can no longer admit any tasks are cleared to zero.
5) Iteration: Determine whether the profit matrix is zero. If not, return to Step 3). Otherwise, continue to the next group and restore the computation resources occupied by tasks that do not meet the reliability goals in the current group. The algorithm will end after all groups have been traversed, and unadmitted tasks will be dropped.

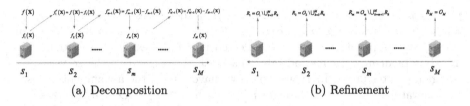

(a) Decomposition   (b) Refinement

Fig. 2. Illustration of Step 2) and Step 4).

Next, we further explain the steps in the algorithm. When generating the profit matrix in Step 2), for ease of understanding, denote a $N \times M$ matrix p the profit that can be obtained by the server. The element $p[n, m]$ indicates the profit available to server s_m for deploying a replica of task t_n, i.e., $p[n, m] = pay_n/a_n - C^t(n, m)$. First, we assume that the tasks are all preadmitted, so the objective function is converted to $f(X) = \sum_{t_n \in T} \sum_{s_m \in S} p[n, m] \cdot x_{nm}$, which shows that $f(X)$ is a linear function on $\{x_{nm}\}$. Let p_m denote the profit matrix before the assignment of tasks for mth server, where $p_1 = p$. Then, We divide p_m into two parts, i.e., $p_m = p_m^1 + p_m^2$, where p_m^1 denotes the modified profit matrix after the assignment for mth server and p_m^2 is the profit matrix used for the $(m + 1)$th server, i.e., $p_m^2 = p_{m+1}$. The calculation of p_m^1 can be given by:

$$p_m^1[n, k] = \begin{cases} p_m[n, m], & \text{if } t_n \in O_m \text{ or } k = m \\ 0, & \text{otherwise,} \end{cases} \tag{8}$$

where O_m is the replicas assignment result for mth server. So we have $p_{m+1} = p_m - p_m^1$.

By constructing the above profit matrix, we can define a new profit function, $f_1(X) = \sum_{t_n \in T} \sum_{s_m \in S} p_1^1[n, m] \cdot x_{nm}$.

Similarly, let: $f_1'(X) = \sum_{t_n \in T} \sum_{s_m \in S} p_1^2[n, m] \cdot x_{nm}$.

Clearly, we have $f(X) = f_1(X) + f_1'(X)$. In the same way, $f_1'(X)$ can also become $f_1'(X) = f_2(X) + f_2'(X)$. Eventually, $f(X)$ can be iteratively expressed as

$$f(X) = f_1(X) + f_2(X) + \cdots + f_M(X), \tag{9}$$

where $f_M'(X) = 0$ due to $p_M^2 = 0$.

From that, we can decompose the optimization problem P2 into M subproblems, each of which can be expressed as $(\mathbb{R}, f_m(X))$. Figure 2a illustrates the process of problem decomposition.

In Step 3), based on the problem decomposition in the previous step, we use a fast approximation knapsack algorithm [20] to solve each subproblem. This step will continue until $m = M$ or $p_m = 0$.

In Step 4), the refinement is performed by backward loop on the above assignment results. In the previous step, each task may be assigned to more than one server when solving knapsack problems on different servers. Since

$p_{m+1} = p_m - p_m^1$, each profit value in the p_{m+1} decreases as the subproblem is solved, then tasks in the p_{m+1} that are less than zero will never be selected in subsequent subproblems. If task t_n selected in subproblem m is selected again in subproblem $m + 1$, this indicates that task t_n assigned to the server s_{m+1} will gain greater profit. In view of that, we have refined the assignment of tasks in backward order. In subproblem m, for each task $t_n \in O_m$, if the task has been assigned to server s_k with $k > m$, then it will be excluded from R_m, i.e., $R_m = O_m \setminus \cup_{k=m+1}^{M} R_k$. Figure 2b shows the process of refinement.

In Step 5), after completing a round of task assignment, a new round of knapsack subproblems will continue to be solved in that group to obtain the results of task redundancy assignment until the tasks meet the redundancy deployment requirement or the servers can no longer admit tasks. The above iterations will then be repeated in the next redundancy number group.

Algorithm 1 shows the pseudocode of the proposed LRBA for the Homo-PMRA problem. Algorithm 2 is called the recursion algorithm and is used to implement the procedures in Steps 3) and 4).

4.4   Algorithm Analysis

To theoretically prove the performance of LRBA, we introduce the approximation ratio.

Algorithm 1: Local Ratio Based Algorithm (LRBA)

  Input: Task set T, server set S
  Output: Assignment results X, Y
1 Initialization: $x_{nm} \leftarrow 0, y_n \leftarrow 1 \, (\forall t_n \in T, s_m \in S)$
2 Calculate the number of redundant deployments a_n for each task
3 Sort all a_n and remove duplicates to generate the set α
4 **for** each a_i in α **do**
5 | Cluster the tasks with redundancy assignment number a_i as T_i
6 | Generate the initial profit matrix p corresponding to server
7 | **while** $p \neq 0$ **do**
8 | | $p_1 \leftarrow p$
9 | | Invoke RA(T_i, S, p_1) to get the task assignment scheme $R = \{R_1, R_2, ..., R_M\}$
10 | | **for** each s_m in S **do**
11 | | | **for** each t_n in R_m **do**
12 | | | | $x_{nm} \leftarrow 1, cap_m \leftarrow cap_m - \varepsilon_n, p[n, m] \leftarrow 0$
13 | | | | **if** $\sum_{s_m \in S} x_{nm} = y_n \cdot a_i$ **then**
14 | | | | | $p[n, k] \leftarrow 0 \, (k = 1, \cdots, M)$
15 | | | **if** s_m *can no longer admit more replicas because of capacity limitations* **then**
16 | | | | $p[k, m] \leftarrow 0 \, (k = 1, \cdots, |T_i|)$
17 | **for** $t_n \in T_i$ **do**
18 | | **if** $\sum_{s_m \in S} x_{nm} \neq y_n \cdot a_i$ **then**
19 | | | $y_n \leftarrow 0$
20 | | | **for** $s_m \in S$ **do**
21 | | | | $x_{nm} \leftarrow 0, cap_m \leftarrow cap_m + \varepsilon_n$

22 **return** X, Y

Definition 1: For a maximization problem, a polynomial time algorithm is λ-approximation for a problem $(\mathbb{R}, f(x))$ if the approximation solution x deliveried by the algorithm satisfies $\lambda \cdot f(x) \geqslant f(x^*)$, where x^* is the optimal solution.

Lemma 1: If $\{X, Y\}$ is a λ-approximation solution of $(\mathbb{R}, f_m(\cdot))$, $m = 1, 2, ..., M$, where $(\mathbb{R}, f_m(\cdot))$ is the mth subproblem decomposed from the original problem $(\mathbb{R}, f(\cdot))$, then $\{X, Y\}$ is also a λ-approximation solution of $(\mathbb{R}, f(\cdot))$.

Proof: From constraint (7), we know that when the decision variable X is fixed, Y is also fixed, so we can consider only X first. Based on (9), we can learn that the decomposition of the objective function $f(X)$ of the original problem can be achieved by using the local ratio technique, i.e., satisfying $f(X) = f_1(X) + f_2(X) + ... + f_M(X)$. Then we first consider the case when $M = 2$, suppose X^*, X_1^* and X_2^* are optimal solutions to problems $(\mathbb{R}, f(X)), (\mathbb{R}, f_1(X))$ and $(\mathbb{R}, f_2(X))$, respectively. Then we can get

$$
\begin{aligned}
f(X^*) = f_1(X^*) + f_2(X^*) &\leq f_1(X_1^*) + f_2(X_2^*) \\
&\leq \lambda \cdot f_1(X) + \lambda \cdot f_2(X) \\
&= \lambda \cdot (f_1(X) + f_2(X)) = \lambda \cdot f(X).
\end{aligned}
\tag{10}
$$

Inequality (10) holds because we assume in advance that $\{X, Y\}$ is a λ-approximation solution of $(\mathbb{R}, f_m(\cdot))$.

The above inequality also holds for any M. Thus we can conclude that the feasible solution $\{X, Y\}$ is a λ-approximation solution of the problem $(\mathbb{R}, f(\cdot))$ [19]. ∎

Lemma 2: LRBA provides a feasible solution (X, Y) for the optimization problem P2.

Proof: First, from Step 3), which solves the subproblem as a knapsack problem, we know that the computation capacity of each server is not violated. Meanwhile, at the end of each round of subproblem decomposition, the remaining capacity of the server is also updated according to the allocation result of each round. Then,

---

Algorithm 2: Recursion Algorithm (RA)

Input: Task set T_i, server set S, profit matrix p_m
Output: Assignment Scheme R

1 Invoke the fast knapsack algorithm to get the task assignment scheme O_m of server s_m
2 Dividing p_m into two parts: p_m^1 and p_m^2, where
$$p_m^1[n, k] = \begin{cases} p_m[n, m] & , \text{if}(t_n \in O_m) \text{ or } (k = m) \\ 0 & , \text{otherwise} \end{cases}, \quad p_m^2 \leftarrow p_m - p_m^1$$
3 **if** $m < M$ **then**
4 $\quad$ $p_{m+1} \leftarrow p_m^2$
5 $\quad$ Recursive invoke RA(T_i, S, p_{m+1})
6 $\quad$ $R_m \leftarrow O_m \setminus \cup_{k=m+1}^M R_k$
7 **else**
8 $\quad$ $R_m \leftarrow O_m$
9 **return** R

in Step 5), a decision is made on whether the number of redundant deployments of the task is satisfied. If the condition is satisfied the task will be admitted, and vice versa, it will be dropped. Thus, the reliability goals of the admitted tasks are satisfied. Therefore, the solution (X, Y) is a feasible solution to the optimization problem P2. ∎

Lemma 3: If the fast knapsack algorithm is a α-approximation algorithm, then $R = \{R_1, R_2, ..., R_M\}$ is a $(1 + \alpha)$-approximation solution to problem $(\mathbb{R}, f_m(X))$.

Proof: We first assume that the profit obtained from the assignment scheme R is denoted by $P(R)$ and then We consider two cases. When $m = M$, $R_M = O_M$ is the solution of $f_M(X)$, where O_M is α-approximation. Since $R_M \in R$, $P(R) \leq P(R_M)$, we can obtain R is a α-approximation of $f_M(X)$. According to the definition of the approximation ratio, we can also know R is a $(1 + \alpha)$-approximation solution.

When $m < M$, we can get

$$f_m(X) = \sum_{t_n \in T} \sum_{s_{m'} \in S} p_m^1[n, m'] x_{nm'}$$
$$= \sum_{t_n \in T} p_m^1[n, m] x_{nm} + \sum_{m' > m} \sum_{t_n \in O_m} p_m^1[n, m'] x_{nm'}, \qquad (11)$$

(11) can be hold due to (8).

It is easy to know that the first term of (11) is the objective function of a knapsack problem, so we can obtain $\sum_{t_n \in T} p_m^1[n, m] x_{nm}^* \leq \alpha P(O_m)$.

The second term of (11) includes only the tasks in O_m, so the profit that can be obtained is at most $P(O_m)$. Thus, by combining the above two terms, we have $f_m(X^*) \leq \alpha P(O_m) + P(O_m) = (1 + \alpha) P(O_m)$.

Due to $R_m = O_m \setminus \cup_{k=m+1}^M R_k$, the tasks in O_m have two cases: 1)$s_n \in O_m$ and $s_n \in R_m$ or 2)$s_n \in O_m$ and $s_n \in R_{m'}(m' > m)$. But in either case, in R, s_n will be assigned, and not only that but also other tasks that are not in O_m will be assigned. So, we finally get $f_m(X^*) \leq (1 + \alpha) P(O_m) \leq (1 + \alpha) P(R)$.

Therefore, R is a $(1 + \alpha)$-approximation solution of $(\mathbb{R}, f_m(X))$. ∎

Theorem 2: LRBA is a $(1 + \alpha)$-approximation algorithm for the optimization problem P1. Its time complexity is $O(a_{max} M N^3)$.

Proof: Accoding to Lemma 3, we know R is a $(1 + \alpha)$-approximation solution of $(\mathbb{R}, f_m(X))$. Again, because of (9) and Lemma 1, we can conclude LRBA is a $(1 + \alpha)$-approximation algorithm.

The time complexity of LRBA is analyzed as follows: In LRBA, tasks are clustered according to the number of redundant deployments, and each task is repeated until the number of redundancies is met, so the process is performed at most $O(a_{max} N)$, where $a_{max} = \max\{a_n\}$. The time complexity of the fast knapsack algorithm is $O(N^2)$ [20]. In RA, the knapsack algorithm is performed recursively up to M times. Therefore, the time complexity of LRBA is $O(a_{max} M N^3)$. ∎

5 Heterogeneous MEC Scenario

In the previous section, we discussed the profit maximization problem in Homo-MEC. However, in practical scenarios, MEC is usually operated by different telecommunication companies, which will result in a clear distinction in the hardware architecture of the servers [17]. These heterogeneous hardware architectures not only have different computation capacities and costs but also have an impact on the reliability of the tasks. This means that even different replicas of the same task, running on servers with different hardware architectures, can have completely different failure rates. This heterogeneous MEC (Heter-MEC) scenario is obviously very different from the Homo-MEC, so the profit maximization problem under Homo-MEC is no longer applicable to Heter-MEC, and its algorithm is also inappropriate. To this end, we propose a new profit maximization problem for Heter-MEC and a heuristic algorithm for solving it.

5.1 Reliability Model

We assume that the reliability rate of server s_m is q_m and that $1 - q_m$ is the failure rate of s_m. Since the servers are heterogeneous, the reliability rates of the servers are independent of each other. To meet the reliability requirement γ_n for task t_n, task t_n needs to be redundantly assigned to multiple different servers with the corresponding number of VNF instances deployed on the servers, so we can get

$$1 - \prod_{s_m \in S} (1 - q_m) \cdot x_{nm} \geq \gamma_n, \tag{12}$$

where $x_{nm} = 1$ represents that task t_n is assigned to server s_m, otherwise $x_{nm} = 0$. The second term $\prod_{s_m \in S} (1 - q_m) \cdot x_{nm}$ in (12) is the probability that all VNF instances of task t_n are faulted. The left side of the inequality, therefore, represents the probability that at least one VNF instance of the task, also known as a replica, will run successfully. Whether the reliability of the task can be satisfied, i.e. whether (12) holds, will be determined by the decision to assign replicas of the task.

5.2 Problem Formulation

As in Homo-PMRA, we consider a profit maximization problem for reliability assurance in Heter-MEC. Similarly, we define two sets of decision variables, X and Y, and formulate the following integer programming problem,

$$P3 : \max_{X,Y} \sum_{t_n \in T} pay_n \cdot y_n - \sum_{t_n \in T} \sum_{s_m \in S} C^t (n, m) \cdot x_{nm}$$

$$\text{s.t.} \left(1 - \prod_{s_m \in S} (1 - q_m) \cdot x_{nm} \right) \geq y_n \cdot \gamma_n, \forall t_n \in T, \tag{13}$$

$$\sum_{s_m \in S} x_{nm} \geq y_n, \quad \forall t_n \in T, \tag{14}$$

$$\sum_{t_n \in T} \varepsilon_n \cdot x_{nm} \leq cap_m, \quad \forall s_m \in S, \tag{15}$$

$$x_{nm} \in \{0,1\}, y_n \in \{0,1\}, \forall t_n \in T, \forall s_m \in S, \tag{16}$$

where Constraint (13) indicates that the admitted task is to meet its reliability requirements by the way of redundant deployment. Constraint (14) represents that if a task is admitted, then it has at least one replica deployed in the network. It is worth noting that the benefit maximization objective of P3 would avoid the possibility of only deploying replicas without admitting tasks. Constraint (15) denotes that replicas deployment to server s_m cannot exceed its computation resource capacity limit. Constraint (16) denotes the decision variables x_{nm}, y_n are binary variables.

Theorem 4: P3 is an NP-hard problem.

Proof: The proof procedure is the same as the proof of Theorem 2, so it is omitted. ∎

5.3   Benefit-Cost Ratio Preference Algorithm

To solve the problem of P3, we propose a heuristic algorithm termed benefit-cost ratio preference algorithm (BRPA). The core idea is to use the greedy method to select the server with higher reliability and prefer the task with a higher benefit-cost ratio when deploying replicas, where the benefit-cost ratio of task t_n on the server s_m is $pay_n/C^t (n, m)$. The specific steps are:

1) We start by ranking the servers in descending order of reliability, prioritizing the more reliable servers for the deployments of task replicas.
2) Split the task set into two sets, T', which already has replicas deployed, and T'', which does not yet have replicas deployed. Each server ranks tasks in decreasing order of benefit-cost ratios. The sorted tasks are pushed into a queue Q. The tasks in T' are given priority so that tasks with replicas are deployed to meet their reliability requirements earlier.
3) Determine sequentially whether each replica of the task in the queue satisfies the computation capacity constraint of the server. If the capacity constraint is met, then deploy a replica of the task to that server and determine whether all replicas of the current task have satisfied the reliability requirements. If satisfied then the task is admitted and it is removed from T' to avoid being involved in subsequent deployments of the servers. If it is the first replica of the task to be deployed, move the task from T'' to T'.

4) Traverse through each task in T. Tasks that are not admitted will have all their deployed replicas removed.

Algorithm 3 shows the pseudocode of our heuristic algorithm for the PMRA problem.

Algorithm 3: Benefit-cost Ratio Preference Algorithm (BRPA)

Input: Task set T, server set S, Benefit-cost ratio matrix B
Output: Assignment results X, Y

1 Initialization: $x_{nm} \leftarrow 0$, $y_n \leftarrow 0$ ($\forall t_n \in T$, $s_m \in S$), $T' \leftarrow \emptyset$, $T'' \leftarrow T$
2 Sort set S in the order of q_m from largest to smallest, and name the sorted set as S'
3 **for** each s_m in S' **do**
4    $Q \leftarrow \emptyset$
5   Sort set T' in the order of $pay_n/C^t (n, m)$ from largest to smallest, and push the sorted set into queue Q
6   Sort set T'' in the order of $pay_n/C^t (n, m)$ from largest to smallest, and push the sorted set into queue Q
7   **while** $Q \neq \emptyset$ **do**
8      $t_n \leftarrow Q.\text{Pop}()$
9   **if** $\varepsilon_n + \sum_{t'_n \in T} \varepsilon_n \cdot x_{n'm} \leq cap_m$ **then**
10        $x_{nm} \leftarrow 1$
11   **if** $(1 - \prod_{s_m \in S} (1 - q_m) \cdot x_{nm}) \geq \gamma_n$ **then**
12          $y_n \leftarrow 1$, $T' \leftarrow T' \setminus \{t_n\}$
13   **if** $t_n \in T''$ **then**
14          $T' \leftarrow T' \cup \{t_n\}$, $T'' \leftarrow T'' \setminus \{t_n\}$

15 **for** $t_n \in T$ **do**
16   **if** $y_n \neq 1$ **then**
17   **for** each s_m in S **do**
18        $x_{nm} \leftarrow 0$

19 **return** X, Y

5.4 Algorithm Analysis

Theorem 5: Given a set of servers $S = \{s_1, s_2, ..., s_m, ..., s_M\}$, and a set of tasks $T = \{t_1, t_2, ..., t_n, ...t_N\}$, there exists an algorithm BRPA that provides a feasible solution to the profit maximization problem for VNF deployments, and the time complexity of BRPA is $O(M \cdot N)$.

Proof: BRPA provides a feasible solution because each admitted task t_n satisfies its reliability requirement and the VNF instances deployment on the server have sufficient computation resources to execute the admitted tasks. Therefore, the BRPA solution is feasible.

The time complexity of BRPA is analyzed as follows: It takes $O(M)$ time to traverse all servers and $O(M \cdot N)$ time to compare the profit of each task assigned to each server. Therefore, the time complexity of BRPA is $O(M \cdot N)$. ∎

6 Simulation Evaluation

In this section, we focus on the impact of important parameters on the performance of the proposed algorithm, including the number of tasks and the number of servers. The simulations are run on a server with Intel i7-9700, 3 GHz CPU, and 32 GB RAM using Matlab.

6.1 Experimental Parameters

We consider a MEC network. There are 10 different types of VNF instances in the network, i.e., $|F| = 10$ and the initiation cost $C^i(k)$ for each VNF instance is set randomly within $[1,3]\$$ [21]. The number of servers M is increased from 10 to 100, and the total computation resources cap_m of each server is set randomly from 2 GHz to 6 GHz [1]. For server s_m, the cost consumed to execute a packet of VNF f_k is drawn within $[0.03, 0.3]\$$. The number of tasks N in the network is set from 100 to 1000, and the number of packets d_n for each task is uniformly distributed within $[5, 15]$ [22]. Each task that is admitted, i.e., satisfies its reliability requirements, will pay the service provider a payment that takes values in the range $[10, 40]\$$. Each task also takes up the computation resource capacity of the server, which is set from 40 MHz to 400 MHz. We assume that the reliability of each server is between 0.99 and 0.99999 [23], while the reliability requirements of the tasks are randomly taken between 0.99999 and 0.999999 [6]. Note that in the Homo-MEC, servers all have equal capacity and reliability. In Heter-MEC, it's the opposite. We execute each point in the simulation 500 times independently and calculate the average as the final result.

Since Homo-MEC can be considered as a special case of Heter-MEC, the BRPA algorithm can be applied to both Homo-MEC and Heter-MEC scenarios, while the LRBA algorithm is only applicable to Homo-MEC. In our experiments, we compare the performance of BRPA, LRBA, and other comparison algorithms in Homo-MEC. For Heter-MEC, we only compare BRPA and the comparison algorithms. The comparison algorithms mainly include the following three, namely:

1) *LocalSearch* [24]: First, the admitted task set is tuned by adding, deleting, and replacing steps. Then a greedy strategy is used to find the minimum cost deployment solution based on the admitted task set. Finally, the set of tasks will be continuously tuned until the benefits no longer increase. The algorithm is applicable to Homo-MEC scenarios.
2) *Greedy* [25]: The tasks with the highest payment are selected first. Tasks are then sequentially assigned to servers in order of cost from smallest to largest until the reliability of the task is satisfied. The algorithm ends if all the computation resources of the server are exhausted or all the tasks are traversed. The algorithm is applicable to Heter-MEC scenarios.
3) *Random* [26]: One task is randomly selected for assignment at a time. Server selection is also random. If the task eventually satisfies the reliability requirements it is admitted, otherwise, it is dropped. The algorithm is applicable to both scenarios.

6.2 Impact of Number of Tasks

The performance of the proposed algorithm against LocalSearch, the random, and the optimal solution *Optimal* is evaluated by varying the number of tasks in a small-scale Homo-MEC network. The solution *Optimal* is obtained by ILOG CPLEX v12.10 [27].

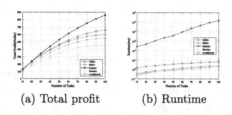

(a) Total profit   (b) Runtime

Fig. 3. Impact of the number of tasks on algorithm performance in the small-scale Homo-MEC scenario

Figure 3 shows the impact of a small number of tasks on the performance of the algorithm, with the number of tasks ranging from 10 to 100, fixing the number of servers to 10. From Fig. 3a, we can see that the total profit of *Optimal* obtained by CPLEX is the largest and the LRBA is nearest to the optimal solution. Figure 3b shows the running time of the algorithms as the number of tasks varies from 10 to 100, and we can see that obtaining the optimal solution to the problem P1 with the CPLEX solver takes a significant amount of solution time. Moreover, with more decision variables, the solution time of the optimal solution becomes further longer, so that the optimal solution is no longer pursued in large-scale networks.

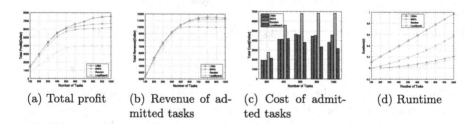

(a) Total profit   (b) Revenue of ad-   (c) Cost of admit-   (d) Runtime
                        mitted tasks   ted tasks

Fig. 4. Impact of the number of tasks on algorithm performance in the Homo-MEC scenario

Next, we vary the number of tasks from 100 to 1000 and fix the number of servers to 100 to study the performance of the algorithm in the case of a Homo-MEC scenario with a large number of tasks. In Fig. 4a, we compare the profits of the four algorithms. From Fig. 4a, we can see that the total profit of the service provider increases with the number of tasks. As can be seen, our proposed algorithms outperform the other two compared algorithms. When the number of tasks is 1000, LRBA obtains 12.3%, 22.5%, and 90.5% higher profit than BRPA, LocalSearch, and the random, respectively. However, it can be seen from Fig. 4d that the running time of LRBA is also significantly higher than that of BRPA and other algorithms, mainly because LRBA iterations and recursions both take longer computation time. To further show the performance of the proposed algorithms, we also compare the total revenue achieved by each algorithm for a varying number of tasks and the cost incurred in deploying these tasks and their replicas. Figure 4b shows the revenue that each algorithm can obtain. It can be seen that as the number of tasks increases, the revenue obtained by each algorithm will not continue to increase. This is because the total capacity of the server is fixed and there is a limit to the number of tasks and their replicas that can be admitted. However, LRBA and BPRA can obviously admit more high-revenue tasks. Figure 4c illustrates the cost of admitting these tasks. As can be seen, LRBA and BRPA costs are kept at a low level. Combining Fig. 4b and 4c, it is clear that the reason why LRBA and BRPA are more profitable is the ability to admit more high-revenue tasks while minimizing the cost of deploying their replicas. It is worth noting that the LocalSearch first obtains the set of admitted tasks, then goes on to find the lowest cost deployment result on that set, and finally adjusts the set according to the profit. This comparison algorithm, therefore, allows for very low costs but limits the revenue of admitted tasks.

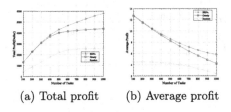

(a) Total profit   (b) Average profit

Fig. 5. Impact of the number of tasks on algorithm performance in the Heter-MEC scenario.

We also investigate the performance of the BRPA algorithm in the Heter-MEC scenario by varying the number of tasks from 100 to 1000 and fixing the number of servers to 100. From Fig. 5a, we can see that the total profit of our proposed BRPA outperforms that of the random and greedy algorithms. In particular, due to the limited total computation resources of the servers, the growth trend of each algorithm gradually slows down after the number of tasks reaches 500, but BRPA can still admit more high-revenue tasks. In Fig. 5b, we show the average profit of the tasks admitted by the three algorithms for varying numbers of tasks, which allows us to obtain that BRPA prefers to admit tasks with high profits.

6.3   Impact of Number of Servers

The performance of the proposed algorithms, as well as the LocalSearch algorithm, the greedy algorithm, and the random algorithm is evaluated by varying the number of servers in two network scenarios, where the number of servers is varied from 10 to 100 and the number of tasks is fixed to 1000.

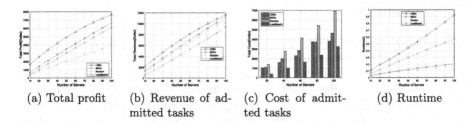

(a) Total profit   (b) Revenue of admitted tasks   (c) Cost of admitted tasks   (d) Runtime

Fig. 6. Impact of the number of servers on algorithm performance in the Homo-MEC scenario.

In the Homo-MEC scenario, it can be seen from Fig. 6a that the proposed algorithm LRBA always outperforms BRPA and the other two compared algorithms as the number of servers increases. When $M = 100$, LRBA's profits are 14.5% higher than BRPA, 25.6% higher than LocalSearch, and 95.7% higher than the random respectively. From Fig. 6c, we can see that the total cost increases with the number of servers, and compared to LocalSearch, the total cost of LRBA and BRPA is higher. However, as can be seen from Fig. 6b, the LRBA and BRPA yield higher revenues and thus higher profits. This is also because as the number of servers increases, so do the resources available to admit more tasks, which in turn bring more revenue. LRBA and BRPA are able to admit more high-revenue tasks even though they come at a higher cost. As can be seen from Fig. 6d, LRBA has the longest running time, while BRPA is just above the random, which can also show the advantage of BRPA.

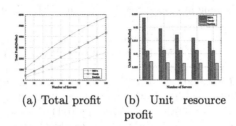

(a) Total profit   (b) Unit resource profit

Fig. 7. Impact of the number of servers on algorithm performance in the Heter-MEC scenario.

In the Heter-MEC scenario, we can see from Fig. 7a that the performance of each algorithm grows linearly as the number of servers increases, with BRPA outperforming the other two compared algorithms. We have also compared the profits that can be brought by a unit of resources. Figure 7b shows that the profit per unit of the resource decreases as the number of servers increases. However, the BRPA is higher than the other two algorithms, which also indicates that BRPA makes more efficient use of resources.

7 Conclusion

In this paper, we studied the optimization problems of VNF provisioning in the Homo- and Heter-MEC scenarios. The formulated optimization problem is an integer programming problem that ensures service reliability through redundant task deployment with the objective of maximizing the benefit of the admitted tasks. Then, an efficient approximation algorithm and a fast heuristic algorithm are proposed and the approximation is proved. Finally, the proposed algorithms are evaluated by simulation experiments, and the simulation results show that the proposed algorithms have some advantages compared with the greedy, the random, and the LocalSearch algorithm.

Acknowledgements. This work was supported in part by the National Natural Science Foundation of China under Grants 62372396, 62172350, and 61902336, in part by the Natural Science Foundation of Hunan under Grant 2023JJ30595.

References

1. Li, J., Liang, W., Huang, M., et al.: Reliability-aware network service provisioning in mobile edge-cloud networks. IEEE Trans. Parallel Distrib. Syst. **31**(7), 1545–1558 (2020)
2. Alonso, R.S., Sittón-Candanedo, I., Casado-Vara, R., et al.: Deep reinforcement learning for the management of software-defined networks in smart farming. In: 2020 International Conference on Omni-layer Intelligent Systems (COINS), pp. 1–6 (2020)
3. Huang, M., Liang, W., Shen, X., et al.: Reliability-aware virtualized network function services provisioning in mobile edge computing. IEEE Trans. Mobile Comput. **19**(11), 2699–2713 (2020)
4. Sarrigiannis, I., Ramantas, K., Kartsakli, E., et al.: Online VNF lifecycle management in an MEC-enabled 5G IoT architecture. IEEE Internet Things J. **7**(5), 4183–4194 (2020)
5. Gill, P., Jain, N., Nagappan, N.: Understanding network failures in data centers: measurement, analysis, and implications. In: Proceedings of the ACM SIGCOMM 2011 Conference, pp. 350–361. ACM, Toronto (2011)
6. Han, B., Gopalakrishnan, V., Kathirvel, G., et al.: On the resiliency of virtual network functions. IEEE Commun. Mag. **55**(7), 152–157 (2017)
7. Fan, J., Jiang, M., Qiao, C.: Carrier-grade availability-aware mapping of Service Function Chains with on-site backups. In: 2017 IEEE/ACM 25th International Symposium on Quality of Service (IWQoS), pp. 1–10 (2017)
8. Wang, L., Liu, J., He, Q.: Concept drift-based checkpoint-restart for edge services rejuvenation. IEEE Trans. Serv. Comput. **16**, 1713–1725 (2022)

9. Kang, R., Zhu, M., He, F., et al.: Implementation of virtual network function allocation with diversity and redundancy in kubernetes. In: 2021 IFIP Networking Conference (IFIP Networking), pp. 1–2 (2021)
10. Tian, F., Liang, J., Liu, J.: Joint vnf parallelization and deployment in mobile edge networks. IEEE Trans. Wireless Commun. **22**, 8185–8199 (2023)
11. Yang, L., Jia, J., Lin, H., et al.: Reliable dynamic service chain scheduling in 5g networks. IEEE Trans. Mobile Comput. **22**(8), 4898–4911 (2023)
12. Mao, Y., Shang, X., Yang, Y.: Provably efficient algorithms for traffic-sensitive SFC placement and flow routing. In: IEEE INFOCOM 2022 - IEEE Conference on Computer Communications, pp. 950–959 (2022)
13. Wang, Y., Lang, P., Tian, D., et al.: A game-based computation offloading method in vehicular multiaccess edge computing networks. IEEE Internet Things J. **7**(6), 4987–4996 (2020)
14. Li, W., Yuan, X., Qu, W., et al.: Efficient coflow transmission for distributed stream processing. In: IEEE INFOCOM 2020 - IEEE Conference on Computer Communications, pp. 1319–1328 (2020)
15. Xu, R., Li, W., Li, K., et al.: Darkte: towards dark traffic engineering in data center networks with ensemble learning. In: 2021 IEEE/ACM 29th International Symposium on Quality of Service (IWQOS), pp. 1–10 (2021)
16. Gao, T., Li, X., Wu, Y., et al.: Cost-efficient VNF placement and scheduling in public cloud networks. IEEE Trans. Commun. **68**(8), 4946–4959 (2020)
17. Lu, W., Wu, W., Xu, J., et al.: Auction design for cross-edge task offloading in heterogeneous mobile edge clouds. Comput. Commun. **181**, 90–101 (2022)
18. Zou, P., Zhou, Z.: A multilevel reduction algorithm to TSP. J. Softw. **14**(1), 35–42 (2003)
19. Bar-Yehuda, R., Even, S.: A local-ratio theorem for approximating the weighted vertex cover problem. In: North-Holland Mathematics Studies, vol. 109, pp. 27–45. Elsevier (1985)
20. Zhao, C., Li, X.: Approximation algorithms on 0–1 linear knapsack problem with a single continuous variable. J. Comb. Optim. **28**(4), 910–916 (2014)
21. Ma, Y., Liang, W., Wu, J.: Online NFV-enabled multicasting in mobile edge cloud networks. In: 2019 IEEE 39th International Conference on Distributed Computing Systems (ICDCS), pp. 821–830 (2019)
22. Li, Y., Xuan Phan, L.T., Loo, B.T.: Network functions virtualization with soft real-time guarantees. In: IEEE INFOCOM 2016 - The 35th Annual IEEE International Conference on Computer Communications, pp. 1–9 (2016)
23. Li, J., Liang, W., Huang, M., et al.: Providing reliability-aware virtualized network function services for mobile edge computing. In: 2019 IEEE 39th International Conference on Distributed Computing Systems (ICDCS), pp. 732–741 (2019)
24. Yang, S., Li, F., Trajanovski, S., et al.: Recent advances of resource allocation in network function virtualization. IEEE Trans. Parallel Distrib. Syst. **32**(2), 295–314 (2021)
25. Li, J., Liang, W., Ma, Y.: Robust service provisioning with service function chain requirements in mobile edge computing. IEEE Trans. Netw. Service Manag. **18**(2), 2138–2153 (2021)
26. Chen, M., Hao, Y.: Task offloading for mobile edge computing in software defined ultra-dense network. IEEE J. Sel. Areas Commun. **36**(3), 587–597 (2018)
27. IBM: IBM ILOG CPLEX optimizer (2011). http://www-01.ibm.com/software/integration/optimization/cplex-optimizer/

Approximate Query Processing Based on Approximate Materialized View

Yuhan Wu[1], Haifeng Guo[1], Donghua Yang[1(✉)], Mengmeng Li[1], Bo Zheng[2], and Hongzhi Wang[1]

[1] Harbin Institute of Technology, Harbin, China
yang.dh@hit.edu.cn
[2] ConDB, Beijing, China

Abstract. In the context of big data, the interactive analysis database system needs to answer aggregate queries within a reasonable response time. The proposed AQP++ framework can integrate data preprocessing and AQP. It connects existing AQP engine with data preprocessing method to complete the connection between them in the process of interaction analysis.

After the research on the application of materialized views in AQP++ framework, it is found that the materialized views used in the two parts of the framework both come from the accurate results of precomputation, so there's still a time bottleneck under large scale data. Based on such limitations, we proposed to use approximate materialized views for subsequent results reuse. We take the method of identifying approximate interval as an example, compared the improvement of AQP++ by using approximate materialized view, and trying different sampling methods to find better time and accurate performance results.

By constructed larger samples, we compared the differences of time, space and accuracy between approximate and general materialized views in AQP++, and analyzed the reasons for the poor performance in some cases of our methods.

Based on the experimental results, it proved that the use of approximate materialized view can improve the AQP++ framework, it effectively save time and storage space in the preprocessing stage, and obtain the accuracy similar to or better than the general AQP results as well.

Keywords: Approximate materialized view · Materialized views reuse · AQP++ optimization · Approximate query processing

1 Introduction

With the increasing amount of data, practical applications have higher requirement in query. The query results within the precise or error threshold should

This paper was supported by The National Key Research and Development Program of China (2020YFB1006104) and NSFC grant (62232005).

be returned in the acceptable time. Sampling based approximate query processing and aggregate precomputation are the two methods proposed in the past to try to solve this problem. During interactive queries, the database will produce a large number of materialized views, and the queries on certain data sets are usually concentrated in practice, which makes the query results reusable.

AQP++ framework integrates AQP with aggregate precomputation. AQP usually transforms query conditions in a reasonable way to get approximate results, data preprocessing focuses on the reasonable organization of data set for efficient query. However, the reuse results of the two parts of AQP++ all come from the precise answer of precomputation, so there is still a time bottleneck in the case of huge data scale.

Considering the reusability and reduction of the materialized views, we proposed a new perspective of approximate materialized view. That is to say, data preprocessing is based on sampling set, using general methods to generate approximate materialized views for reuse in the subsequent query process.

Using an approximate materialized view will approximate the results twice, so we are faced with the challenge of how to choose a way of view generation and AQP method. We use the precomputing method based on data aggregation on the results of simple random sampling to generate the interval results as the approximate materialized view, and identify the approximate intervals on them, so as to realize the approximate query.

We make the following contributions in this paper:

- Sampling and aggregation precomputing of data sets, and generate approximate materialized view on the intervals.
- Identifying the approximate interval and get the approximate result.
- Comparison and improvement of sampling methods: select different sampling methods, or combine multiple sampling methods to find the relatively optimal method according to their respective time and accuracy performance.

The remaining sections of this paper are organized as follows. In Sect. 2, we make overview for this paper. In Sect. 3, we introduce the framework and steps of using approximate materialized view. In Sect. 4, the experimental results indicate the performance of our method and we carry out comparative experiments and error analysis. In Sect. 5, we study the influence of parameters. In Sect. 6, we survey related work for this paper. In Sect. 7, we provide the conclusions and give a brief overview of our future work.

2   Overview

For the range query of attributes with continuous values on the dataset, the data cube based method for data preaggregation is a way of AQP. We will apply this method to sample sets to generate approximate materialized views. In this section, we will give the definition of the problem and review the interval partition method.

2.1   Problem Definition

We consider aggregate queries on continuous attributes. Let the value range of the query attribute be $[x, y]$, aggregate query is defined as the number of samples whose values belong to the query interval $[a, b]$. Suppose through data preaggregation, we have obtained the approximate aggregate value of the attribute on k different intervals: $[a, x_1], [x_1, x_2, ..., [x_{k-1}, b]$, these results will be reused as our materialized views. Take the age query on census data as an example, the new range query is:

SELECT age, COUNT(*) FROM census WHERE $a \le age \le b$

We aim to use one or more views as approximate solutions of actual query intervals. For example, use the sum of aggregate values of interval $[x_{t-1}, x_t]$ and $[x_t, x_{t+1}]$ to replace $[a, b]$ if $\|a - x_{t-1}\| \le \epsilon$ and $\|b - x_{t+1}\| \le \epsilon$.

2.2   Data Aggregation

This paper uses the preaggregation and approximate interval recognition methods based on interval partition provided by previous strudies. It finds some partition points and approximate point of query according to interval evaluation.

Partition Evaluation. The current partition points on the ordered dataset is $x_1, x_2, ..., x_n$. For any query point x, I_x and H_x are represented the first partition point less than x and the first partition point greater than x respectively. The interval $[I_x, H_x]$ is also divided into two parts $\overline{L}_x$ and L_x as shown in Fig. 1.

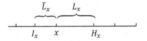

Fig. 1. Interval Partition

Approximate method will select I_x or H_x to replace the query point x, so the error comes from the smaller value between $\overline{L}_x$ and L_x. Define error as:

$$error_x = \min\{\frac{\lambda N}{\sqrt{n}}\sqrt{var(A_{L_x})}, \frac{\lambda N}{\sqrt{n}}\sqrt{var(A_{\overline{L}_x})}\} \qquad (1)$$

If i and j are the two points with the largest error in the whole dataset, then the upper bound of the interval error will be $error_i + error_j$.

Data Aggregation and Precomputation. Interval Partition use the adaptive climbing method, starting from the initial state, trying to move the partition point to a better position according to the evaluation (1). The specific steps of the algorithm will be given in Sect. 3.

Aggregation Recognition. After interval pre partition, for query in interval $[a, b]$, there are four most relevant intervals: $[I_a, I_b], [I_a, H_b], [H_a, I_b], [H_a, H_b]$. Calculate confidence on these candidate intervals:

$$\lambda N \sqrt{\frac{var(cond(A = 0))}{n}} \tag{2}$$

where $cond(A = 0)$ means $\{x|x \in data, x \notin A\}$. Calculate the confidence of candidate interval on subsampling, and choose one of them with the minimum confidence as the approximate result.

3 Reuse of Approximate Materialized View

In this section, we divide the reuse of approximate materialized view into two parts: view generation and query approximate interval recognition. In the part of view generation, we find the partition points of the sampling data according to the method described in Sect. 2 and store the aggregation results of the partition intervals. Note that these aggregation results are based on the sampling set, so they are approximate. In the part of interval recognition, based on the method described in Sect. 2, we find four candidate values corresponding to the query interval, and choose the best one according to their confidence. That is to achieve the approximate treatment of approximate materialized view.

3.1 Aggregation and Precomputing

In this part, we need to sample the dataset, and find a resonable partition of the sampled data, use the aggregation results of each partition as a materialized view for subsequent reuse.

Firstly we use simple random sampling, assume that the sampled data has been sorted, and a parameter k is given to represent the number of partition points. Starting with an initial partition, each iteration calculates the upper bound of the current partition error, only when the new iteration reduces the upper bound can the partition points move. When it need to move the partition point, we move the point with the minimum moving error to the position with the maximum error in the dataset. The details will be introduced with the description of the pseudo code in Algorithm 1.

The pseudo code of partition points generation is shown in Algorithm 1. We first sample and sort the dataset (Line 1). Then we initialize the partition points, here we find k points evenly on the sample set (Line 2). We define three conditions to stop the algorithm: (a)iterations exceeds the threshold, (b)new iteration cannot reduce the upper bound of error, (c)no better partition points can be found. We initialize the above three conditions (Line 3–5).

Algorithm 1: Interval Partition

Input: dataset D, number of partition points k, maximum iteration I_{max}
Result: partition points set $P = \{p_{1,2}, ..., p_k\}$

1   $S \leftarrow Sample\_Sort(D)$;
2   $P \leftarrow EqualPartition(S, k)$;
3   $iterator \leftarrow 0$;
4   $upper \leftarrow +\infty$;
5   $stop \leftarrow False$;
6   **while** *not stop and* $ErrorBound(P) \leq upper$ *and* $iterator \leq I_{max}$ **do**
7      $i_1 \leftarrow \arg\max_i error(S)$;
8      $i_2 \leftarrow \arg\max_{i \neq i_1} error(S)$;
9   **if** $i_1, i_2 \in P$ **then**
10        $stop \leftarrow True$;
11   **else**
12        $t \leftarrow \arg\min_{t \in [1,k]} \{\max_{r \in S \wedge r \in [p_{t-1}, p_{t+1}]} error(S_r)\}$;
13   **if** $i_1 \notin P$ **then**
14          $p_t \leftarrow i_1$;
15   **else**
16          $p_t \leftarrow i_2$;
17   **end**
18   **end**
19 **end**

In an iteration, we find two points i_1, i_2 with the largest error in the sampling set (Line 7–8). We should move the partition points to i_1 or i_2 to reduce the upper bound of error. If i_1, i_2 are already partition points, means algorithm has found the partition result (Line 9–10). Otherwise, we need to select a partition point to move to the location of i_1 or i_2. Suppose three consecutive partition points are p_{t-1}, p_t, p_{t+1}, after moving p_t, only the query on $[p_{t-1}, p_{t+1}]$ will be affected by p_t and may choose other approximate points. So we define the movement error of p_t as the maximum error of interval $[p_{t-1}, p_{t+1}]$, find the partition point with the minimum moving error to move (Line 12). Note that error of i_1 is greater than that of i_2, so we move to the location of i_1 preferentially (Line 13–17).

At the end of the algorithm, we get the set of partition points on the sampling set, the aggregation values on the intervals according to these points will be used as approximate materialized view.

3.2 Approximate Interval Recognition

In Sect. 3.1, we store approximate aggregation values on several intervals. We aim to identify the available partition intervals and find the approximate results when a new query comes. We have reviewed the principle of approximate identification in Sect. 2.2, details and the pseudo code will be explained in Algorithm 2.

The pseudo code of finding approxiamte result is shown in Algorithm 2. We first find the four closest partition points to the upper and lower bounds a and b

---

Algorithm 2: Interval Recognition

Input: sample set S, partition points set P, approximate aggregate values
   $V = \{V_i | sum(p_{i-1} \leq x \leq p_i) \wedge x \in S\}$, query interval $[a, b]$
Result: approximate aggregation value $\hat{Q}$

1 $I_a, I_b, H_a, H_b \leftarrow LowAndHighPoint(P, a, b)$;
2 $C \leftarrow \{[I_a, I_b], [I_a, H_b], [H_a, I_b], [H_a, H_b]\}$;
3 $c \leftarrow +\infty$;
4 $interval \leftarrow \emptyset$;
5 $S' \leftarrow Subsample(S)$;
6 **for** t *in* C **do**
7 | $tmp \leftarrow confidence(t, S')$;
8 | **if** $tmp < c$ **then**
9 | | $c \leftarrow tmp$;
10 | | $interval \leftarrow t$;
11 | **end**
12 **end**
13 $\hat{Q} \leftarrow AggregationValue(interval, V)$;

---

of the query (Line 1), according to these four points constructed four candidate intervals (Line 2). Initialize to find the candidate interval (Line 3–4). In order to improve the computational efficiency of confidence, the set is subsampled once (Line 5). Calculate the confidence of candidate interval in subsample set S' and find the minimum (Line 6–12). Finally, the approximate result can be obtained by simple calculation according to the stored aggregated values V and the found interval (Line 13).

Example 3.2. We want to find out the age distribution in the census data, and there comes a query:

SELECT age, COUNT(*) FROM census WHERE $23 \leq age \leq 67$.

Assume that the age distribution on the census is between $[10, 100]$, and after Algorithm 1 we got a set of 6 partition points as $P = \{22, 35, 47, 61, 85, 100\}$. So we know $I_a = 22, I_b = 61, H_a = 35, H_b = 85$ and get the four candidate intervals as $[22, 61], [22, 85], [35, 61], [35, 85]$. After confidence calculation, we get that the interval $[22, 61]$ is the best approximate result, then we can get the aggregation result by adding the precomputed values of $[10, 22], [22, 35], [35, 47], [47, 61]$.

3.3   Analysis for Twice Approximations

We use the following theorem to prove that the error of twice approximation methods using approximate materialized views is controllable.

Result Estimation. Given dataset D and the sample set S on D, for some aggregate queries q, the answer can be estimated from the sample:

$$q(D) \approx \hat{q}(S) \tag{3}$$

AQP++ uses samples to estimate the difference between query results and pre calculated aggregate values. Let q denote user's query, pre denote precomputed aggregate quary on complete dataset, i.e.

q: SELECT f(A) FROM D WHERE Condition
q: SELECT f(A) FROM D WHERE Condition

AQP++ estimates the differences in query results as follows:

$$q(D) - pre(D) \approx \hat{q}(S) - \hat{pre}(S)q(D) \approx pre(D) + \hat{q}(S) - \hat{pre}(S) \qquad (4)$$

In the general case, after data pre aggregate, the $pre(D)$ obtained by AQP++ is a known constant. Moreover, AQP++ can estimate the results of user's query $\hat{q}(S)$ and the precomputed results $\hat{pre}(S)$ by formula (3). Bring these two estimates into formula (4) to get the final approximate results. When data pre aggregation is based on samples, we use Δ to represent the difference between sample set S and dataset D, i.e. $pre(D) = pre(S) + \Delta$. Substitute it into formula (??), there will be:

$$q(D) \approx pre(S) + \Delta + \hat{q}(S) - \hat{pre}(S) \qquad (5)$$

where $pre(S)$ has been precomputed. Similarly, we can get $\hat{q}(S)$ and $\hat{pre}(S)$ by formula (3).

Error Estimation of Twice Approximation. We use the approximate materialized views for estimation, formula (5) is further approximated by rounding off Δ:

$$q(D) \approx pre(S) + \hat{q}(S) - \hat{pre}(S) \qquad (6)$$

Compare formula (4) and (6), the difference is only the constant value $pre(D)$ and $pre(S)$ in the precomputation part on the right side of the formula, so Lemma 1 still holds.

Lemma 1. *For any aggregation function f, if AQP can anwer queries like: SELECT $f(A)$ FROM D WHERE Condition, then AQP++ can also answer the query.*

Because the constant value from $pre(D)$ to $pre(S)$, when AQP gets unbiased estimation, the answer returned by twice approximation is likely to be biased. If AQP get the unbiased answer, i.e. $q(D) = E[\hat{q}(S)]$ and $pre(D) = E[\hat{pre}(S)]$. Based on formula (6) we get the result $pre(S) + \hat{q}(S) - \hat{pre}(S)$, calculate its expectation:

$$\begin{aligned}
& E[pre(S) + \hat{q}(S) - \hat{pre}(S)] \\
&= E[pre(S)] + (E[\hat{q}(S)] - E[\hat{pre}(S)]) \\
&= pre(S) + (q(D) - pre(D)) \qquad (7) \\
&= q(D) + (pre(S) - pre(D)) \\
&= q(D) - \Delta
\end{aligned}$$

When precomputation is based on the complete datasetm, i.e. $pre(S)$ equals $pre(D)$, AQP++ get unbiased estimation. But when the precomputation is based on the sample set, the expectation difference after twice approximation only comes from the difference Δ. Then we consider the variance of twice approximation:

$$
\begin{aligned}
D[pre(S) &+ \hat{q}(S) - p\hat{r}e(S)] \\
&= E\{[pre(S) + \hat{q}(S) - p\hat{r}e(S) - (pre(S) + \hat{q}(S) - p\hat{r}e(S))]^2\} \\
&= E\{[pre(S) + \hat{q}(S) - p\hat{r}e(S) - (q(D) + (pre(S) - pre(D)))]^2\} \\
&= E\{[pre(D) - p\hat{r}e(S)]^2\} \\
&= E\{pre^2(D) - 2pre(D)p\hat{r}e(S) + p\hat{r}e^2(S)\} \\
&= pre^2(D) - 2pre(D)p\hat{r}e(S) + D(p\hat{r}e(S)) + E^2[p\hat{r}e(S)] \\
&= D[p\hat{r}e(S)]
\end{aligned}
\tag{8}
$$

Formula (8) explain that the variance of the result estimation is completely from the error of sample estimation in the pre aggregation stage. Therefore, it can be inferred from the results of formula (7) and (8) that a reasonable sampling method can make the preprocessing result on the sample dataset approximate to that on the complete dataset. It means the error of twice approximation is controllable.

4   Experimental Results

In this seciton, we exprimentally study the proposed algorithms.

4.1   Experimental Setup

We will introduce the hardware, datasets and some important parameters before we describe and analyze the experimental results.

Hardware All the experiments were conducted on a laptop with an Intel Core i7 CPU with 2.20 GHz clock frequency and 16 GB of GAM.

Datasets We use UCI's public dataset: *Adult Data Set*. This dataset is extracted from census data, with 15 attributes and 32561 instances. Attributes include numerical (age, income, etc.) and discrete (gender, nationality, etc.) The attribute $fnlwgt$ is the weight control value, because it has no important meaning in the queries, we ignored it in the experiment. Each row of the dataset has an instance, the attributes are arranged in order and separated by commas.

Parameters We execute experiments on attribute *age*, which is a integer attribute and has the value range $[17, 75]$. The default number of interval partition points $k = 5$.

4.2   Accuracy

The sampling methods are simple random sampling and stratified sampling based on attribute *age*, about 10% of the data are extracted from both sampling methods. We analyze the accuracy of approximate materialized views reuse from two perspectives:

- The difference between the approximate query results on the approximate materialized view and the exact query results on the complete dataset: compare absolute error, X and Y axis are the upper and lower bounds of recognition interval, the error is taken as Z axis. The mark lines for X and Y axis represent the results of interval partition in preprocessing.
- The difference between the approximate query results on the approximate materialized view and the approximate query results on the complete dataset: the interval partitions are obtained on the sampling sample and the complete dataset respectively. Then get the approximate results according to the partition points. We compare the difference between them and show it in the form of triangular surface graph. The difference is represented by $Q_C - Q_{AMV}$, where Q_C is the results on complete dataset, Q_{AMV} is the result based on approximate materialized views. The blue and red mark line of X and Y aixs represent the repective results of interval partitions.

Simple Random Sampling. Using simple random sampling, the results are shown in Fig. 2. The processing time to generate the partition is about 627 s.

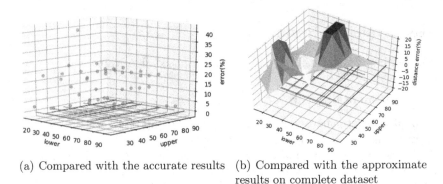

(a) Compared with the accurate results   (b) Compared with the approximate results on complete dataset

Fig. 2. Accuracy of simple random sampling

Figure 2(a) indicates the errors are concentrated in the region of 0% and 15%, few points (only one in Figure) is more than 35%. It shows that the error can be stablized within 15% by using our method.

Analyze the results in Fig. 2(b), it can be roughly divided into three different areas: high-rise area, smooth area with error difference about 0% and concave area. The smooth area represents the approximate query result based on the approximate materialized view is similar to that using the exact materialized view. The high-rise area indicates that our method performs better, and concave area means it is not good as using materialized view from the complete dataset. Since most of the regions in the graph are smooth and high-rise, it can be shown that the approximate query based on approximate materialized views has better performance in accuracy on simple random sampling.

Attribute Stratified Sampling. We set every 10 years as a layer for stratified sampling and the results shown in Fig. 3. The processing time to generate the partition is about 624 s.

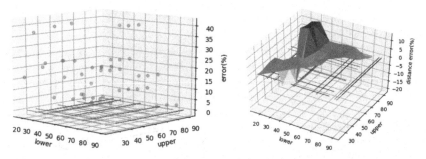

(a) Compared with the accurate results   (b) Compared with the approximate results on complete dataset

Fig. 3. Accuracy of stratified sampling

Errors in Fig. 3(a) are also concentrated in the region of 0% and 15%. But compared with Fig. 2(a), there are more points with large error (more than 35%), indicating that stratified sampling is not as accurate as simple random sampling in reusing approximate materialized views. Similarly divide Fig. 3(b) into three areas. Compared with Fig. 2(b), there is a significant difference in the scale of Y axis, which indicates that stratified sampling for approximate materialized view reuse can get a closer result than directly reusing materialized view. And most of the areas in Fig. 3(b) are above 0%, shows that stratified sampling can get better accuracy results.

Error Analysis. Based on the precomputation of Fig. 2(a) and 3(b), we take multiple perspectives of the result graph to analyze the error. We use red arrows and red dotted line box to identify the parts with lower error in the result of the diagram, and use black dashed line to mark the line *lower = upper* on the *lower − upper* plane. The results of two sampling methods are shown in Fig. 4 and 5.

Notice the red marked parts in Fig. 4 and 5, the lower errors are concentrated near the blue border on the *lower − upper* plane. It shows that the error of approximate query is lower when query interval approaches the partition interval. This is corresponding to with the theory, i.e. when query interval is replaced by the partition intervals, there has lower error. When query interval moves to the inside of the blue line, the error trends to rise, and reach the peak in the most

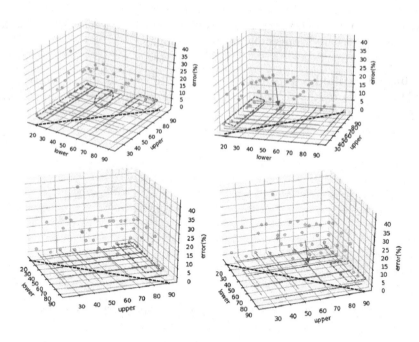

Fig. 4. Accuracy on simple random sampling. (Color figure online)

central areas, which also indicates that the query accuracy is lower when query interval distance from partition intervals.

By observing the scattered points projected near the black dotted line, the query error varies from high to low, showing an obvious fluctuation trend. The possible reason is: when the upper and lower bounds of query are close, the approximate query interval is easily changed to $[p_i, p_i]$ (p_i is a partition point), so the returned result will be 0. Once the actual data is distributed on $[p_i-\varepsilon, p_i+\varepsilon]$, the approximate result is far from the actual result.

Analyze the Fig. 2(b) and 3(b) more specifically. Use red arrows and red dotted line box the mark the high-rise part, and use the black dotted line to mark the line $lower = upper$ on the $lower - upper$ plane. The results are shown in Fig. 6.

The red mark in Fig. 6 indicates that the error of using approximate materialized view is lower than that of using general views. Moreover, these intervals show a trend of a lower bound and higher upper bound, that is, when the interval span is larger, using approximate materialized views can improve the effect obviously. The projection area near the black dotted line is basically at 0%, which indicates that when intervals have similar lower and upper bounds, the approximate materialized view has almost the same accuracy as the general method. It can be seen that this type of query interval is prone to high and low query error, which is the original limitation of using data aggregation method, and it can not be improved by using approximate materialized view.

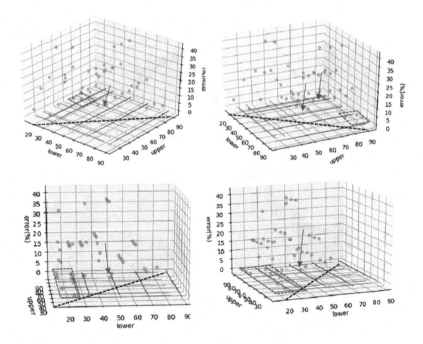

Fig. 5. Accuracy on stratified sampling. (Color figure online)

4.3   The Summary of Experimental Results

We summarize the experimental results as follows:

- In most cases, the performance of reusing approximate materialized view is better than that of reusing general materialized view. It can not only save preprocessing time and storage space, but also have better accuracy.
- The error of reusing approximate materialized view is mainly limited by the method of data pre aggregation.

5   Influence of Parameters

In this section, we modified the experimental parameters to study their influence on the results. These parameters include sample size and the number of partition points k.

5.1   Sample Size

We stay the other experimental conditions the same, using simple random sampling, selected 10%, 20%, 30% and 50% from the complete dataset to repeat the above experiments. The results are shown in Fig. 7. Figure 7 indicates that under different sample size, the error of using approximate materialized view is about

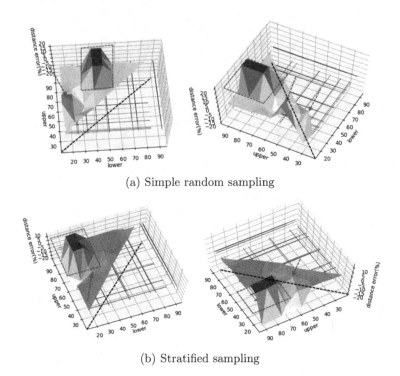

(a) Simple random sampling

(b) Stratified sampling

Fig. 6. Accuracy of stratified sampling

15%, and the increase of sample size has little effect on the accuracy. We think the possible reason is: the query processing use the data aggregation recognition method, query error of this method mainly comes from the difference between the pre partition points and the real query interval. The change of sample size will not lead to a huge difference of interval partition, so it has no obvious impact on the accuracy.

We compare the interval partitions under different sample size as shown in Fig. 8. Figure 8(a) indicates there is no significant difference in the time of interval partition within the increase of sample size, and Fig. 8(b) shows the results of interval partition are similar. So we can infer that the sample size has little effect on improving the pretreatment time and accuracy of reusing approximate materialized view. But considering the space occupation, lower sample size has better space performance.

5.2   Number of Partition Points

We stay the other experimental conditions the same, using simple random sampling to take about 10% samples, set the number of partition points as 3, 5, 8, 10 to repeat these experiments. The accuracy are shown in Fig. 9.

Figure 9 shows the error concentration area decreases significantly with the increase of k, so the parameter k can improve the approximate materialized

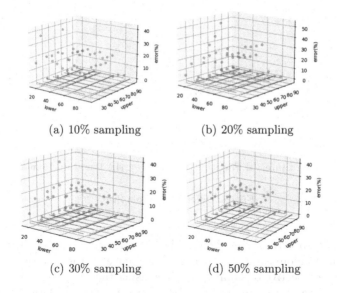

(a) 10% sampling   (b) 20% sampling

(c) 30% sampling   (d) 50% sampling

Fig. 7. Accuracy under different sampling scale

view. Then we compare the time-consuming of interval partition and the reduced preprocessing time by using sampling under different k, results are shown in Fig. 10.

From the results of Fig. 10, the preprocessing time is positively correlated with number of partition points k, and the use of sampling can effectively reduce the time of interval partition. Moreover, this performance improvement increases with the increase of k. It shows that our method can improve the performance greatly when k is large.

Finally we compared the difference between using approximate materialized view and general view under different parameter k, as shown in Fig. 11.

According to previous description, the area above 0% in Fig. 11 indicates that our approximate materialized view have better performance. Figure 11(a) shows when k is too small, there is almost no high-rise area. As the number of partition points from $k = 5$ to $k = 8$, high-rise area increases obviously, and the depression area of $k = 8$ is smaller than that of $k = 5$. It shows that the increase of k makes the use of approximate materialized view more efficient than general views. When k becomes large ($k = 10$), the increase of high-rise area is no longer obvious, but the peak value decrease, which indicates that the difference between our method and general method is reduced. However, most of the areas are still above the horizontal plane, so using approximate materialized view still has the advantage of accuracy.

5.3   The Summary of Parameter Setting

We summarize the influence of parameters as follows:

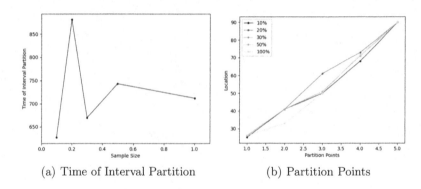

(a) Time of Interval Partition   (b) Partition Points

Fig. 8. Different Sample Size

- From the perspective of sampling method, single attribute stratified sampling and simple random sampling have little influence on the results.
- From the perspective of sample size, increasing the sampling scale can effectively improve the accuracy of AQP. However, the too large sample size is no longer obviously improve the performance, and the space occupation increases. The 30% sampling ratio in the experiment can get the best results.
- From the selection of parameter k, it directly affect the accuracy of AQP. When the number of partition is increased, the accuracy will be improved but the preprocessing time will be increased too. In our experiment, choosing the number of partition points with $k = 8$ can get the best comprehensive result.

6 Related Work

Tranditional database management systems execute query paradigm based on blocking. In order to deal with the challenge of interactive analysis database system answering aggregate query in a reasonable time under large scale of data, the technology of pre aggregation (materialized view, data cube) can significantly reduce query latency. But they need a lot of preprocessing, there are dimension bottlenecks and the cost of storing a complete data cube is usually very expensive. The methods that try to overcome these problems (such as imMens and NanoCubes) usually limit the number of attributes that can be filtered at the same time, and limit the possible exploration paths.

In order to achieve low latency query, the system for interactive data exploration must rely on AQP, which provide query result estimation with bounded error. Most AQP systems conduct research on sampling, including using some form of biased sampling (such as AQUA, BlinkDB [2], DICE), and study how to generate better hierarchical samples [2–6], or trying to supplement samples with auxiliary index [3,7]. However, these methods usually need a lot of preprocessing time to obtain prior knowledge, which is still insufficient in the face of unknown queries. In addition to sampling based AQP, some non sampling techniques are

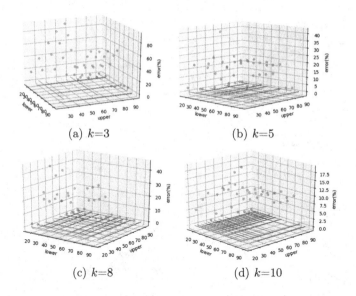

Fig. 9. Accuracy under different k

also proposed in the study [8,9]. They provide certainty by using indexes rather than samples. But for interactive query types, their effectiveness is not as good as sampling based AQP.

Based on the combination of preprocessing and approximate query, the proposed AQP++ framework [10,11] is used to connect any existing AQP engine with aggpre for the connection of AQP and aggregate precomputation of interaction analysis.

In many AQP systems, the ueusability of materialized views is brought into full play [12]. And there are many sample collection methods to help us study the generation of approximate materialized view [13,14].

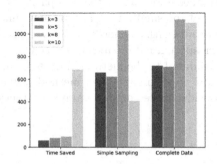

Fig. 10. Time Performance under different k

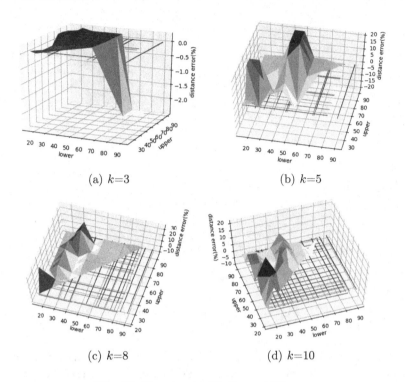

(a) $k=3$ (b) $k=5$

(c) $k=8$ (d) $k=10$

Fig. 11. Compared with General Views

7 Conclusion

In this paper, we proposed to generate approximate materialized views on sample datasets and combine it with AQP++ framework to improve its performance in the data preprocessing stage. The experimental results show that sampling can reduce the preprocessing time based on data aggregation, thus greatly improving the overall time performance and reduces the occupation of storage space. And when the sample size is not too small, using approximate materialized views can get approximate query results with smaller error than using general view in more cases. Generate a better approximate materialized view is still a problem to be studied. Our research is limited to the method of data pre aggregation. We plan to horizontally select more approximate query methods using materialized views for comparison, and consider more complex sample extraction methods to test our views.

References

1. Gray, J., et al.: Data cube: a relational aggregation operator generalizing group-by, cross-tab, and sub totals. In: Data Mining and Knowledge Discovery, pp. 29–53 (1997)

2. Agarwal, S., Mozafari, B., Panda, A., Milner, H., Madden, S., Stoica, I.: BlinkDB: queries with bounded errors and bounded response times on very large data. In: EuroSys (2013)
3. Chaudhuri, S., Das, G., Datar, M., Motwani, R., Narasayya, V.R.: Overcoming limitations of sampling for aggregation queries. In: ICDE (2001)
4. Acharya, S., Gibbons, P.B., Poosala, V.: Congressional samples for approximate answering of group-by queries. ACM SIGMOD Rec. **29**(2), 487–498 (2000)
5. Chaudhuri, S., Das, G., Narasayya, V.R.: A robust, optimization-based approach for approximate answering of aggregate queries. In: SIGMOD (2001)
6. Ganti, V., Lee, M., Ramakrishnan, R.: ICICLES: self-tuning samples for approximate query answering. In: VLDB (2000)
7. Moritz, D., Fisher, D., Ding, B., Wang, C.: Trust, but verify: optimistic visualizations of approximate queries for exploring big data. In: CHI (2017)
8. Cao, Y., Fan, W.: Data driven approximation with bounded resources. PVLDB **10**(9), 973–984 (2017)
9. Potti, N., Patel, J.M.: DAQ: a new paradigm for approximate query processing. PVLDB **8**(9), 898–909 (2015)
10. Peng, J., Zhang, D., Wang, J., et al.: AQP++: connecting approximate query processing with aggregate precomputation for interactive analytics. In: The 2018 International Conference. ACM (2018)
11. Wang, Y., Xia, Y., Fang, Q., et al.: AQP++: a hybrid approximate query processing framework for generalized aggregation queries. J. Comput. Sci. **26**, 419–431 (2017)
12. Galakatos, A., Crotty, A., Zgraggen, E., et al.: Revisiting reuse for approximate query processing. Proc. VLDB Endow. **10**(10), 1142–1153 (2017)
13. Gibbons, P.B., Matias, Y.: New sampling-based summary statistics for improving approximate query answers. ACM SIGMOD Rec. **27**(2), 331–342 (1998)
14. Babcock, B., Chaudhuri, S., Das, G.: Dynamic sample selection for approximate query processing. In: The 2003 ACM SIGMOD International Conference on Management of Data. ACM (2003)

Schema Integration on Massive Data Sources

Tianbao Li[1], Haifeng Guo[1], Donghua Yang[1(✉)], Mengmeng Li[1], Bo Zheng[2], and Hongzhi Wang[1]

[1] Harbin Institute of Technology, Harbin, China
`yang.dh@hit.edu.cn`
[2] ConDB, Beijing, China

Abstract. As the fundamental phrase of collecting and analyzing data, data integration is used in many applications, such as data cleaning, bioinformatics and pattern recognition. In big data era, one of the major problems of data integration is to obtain the global schema of data sources since the global schema could be hardly derived from massive data sources directly. In this paper, we attempt to solve such schema integration problem. For different scenarios, we develop batch and incremental schema integration algorithms. We consider the representation difference of attribute names in various data sources and propose ED Join and Semantic Join algorithms to integrate attributes with different representations. Extensive experimental results demonstrate that the proposed algorithms could integrate schemas efficiently and effectively.

Keywords: Information integration · Schema mapping · Schema integration

1 Introduction

In the era of big data, the integration of massive data sources from the web is crucial, but existing techniques face challenges due to the absence of a global schema.

Traditional information integration systems rely on predefined global schemas and schema mapping techniques to unify heterogeneous data. However, in the case of massive data sources, defining a global schema is impractical because users can't grasp the entirety of all data sources.

Schema integration, creating a global schema and mapping it to local schemas, becomes essential in this context, posing two main challenges. Firstly, integrating schemas can be complicated by synonyms, homonyms, and misspellings in attribute names across schemas, impacting the quality of the integrated schema. Secondly, handling large sets of schemas with potentially billions of attributes efficiently is demanding.

Supported by The National Key Research and Development Program of China (2020YFB1006104).

To address these challenges, this paper explores schema integration for millions or billions of attributes, considering both efficiency and effectiveness. For effectiveness, an approximate matching algorithm is introduced, utilizing a knowledge base to evaluate semantic similarity and identify relationships between attributes. For efficiency, the paper develops efficient algorithms, adapting traditional join operations and optimizing disk I/O by clustering related data in continuous blocks.

The contributions of this paper include:

- Pioneering research on schema integration for massive data sources in information integration.
- Proposing an efficient and effective schema integration framework leveraging a knowledge base.
- Designing batch and incremental integration algorithms for scalability and effectiveness.
- Conducting extensive experiments validating the proposed methods' performance.

The paper is organized as follows: Preliminaries and backgrounds are introduced in Sect. 2. An overview of the framework is presented in Sect. 3. Join algorithms for schema integration are explained in Sect. 4. Detailed solutions for batch integration are provided in Sect. 5. Experimental results and analyses are discussed in Sect. 6. A comparison with related work is presented in Sect. 7, and the paper concludes in Sect. 8.

2   Preliminary

In this section, we introduce the backgrounds and definitions of the problem studied in this paper. At first, we give a brief introduction to knowledge base and edit distance. Then we define the problem and related symbols.

2.1   Knowledge Base

Incorporating a knowledge base helps assess semantic similarity among attributes, addressing synonyms and homonyms in attribute names. Knowledge bases like Freebase, WordNet, Probase, and YAGO follow a graph structure. Concepts serve as nodes, and relationships as edges, forming a graph (G). Nodes have a 3-tuple (id, name, type), e.g., "Pies" and "Sweet pies" are $(id1,$ "Pies", "wikicategory") and $(id2,$ "Sweet pies", "wikicategory"). Some knowledge bases use an "is a" hierarchy, representing subclass relationships as edges (set S).

Definition 1 (edge). $\exists\ a,b \in G,$ *if a "is a" b, then edge (a,b) $\in$ S.*

For an attribute like a, if a concept c_a in the knowledge base has the closest literal distance to a within a threshold ε_t, it's considered as the representation of a. This enables the computation of the semantic distance between attributes

a and b based on their corresponding concepts in the knowledge base. In other words, $dis(a, b) \leq \varepsilon$ if $dis_r(c_a, c_b) \leq \gamma$, $dis_t(a, c_a) \leq \varepsilon_t$, and $dis_t(b, c_b) \leq \varepsilon_t$.

For example, let's consider two attributes: a = "Sweet pies" and b = "meat pie," along with their corresponding knowledge base concepts, c_a = "Sweet pies" and c_b = "meat pie." Setting $\varepsilon_t = 2$ and verifying that the literal distances between the attributes and concepts meet the threshold, we only need to check if the semantic distance between the concepts satisfies γ. In this case, both "Sweet pies" and "meat pie" share the common concept "pie," resulting in $dis_r(c_a, c_b)$ = 2. Since $dis(a, b)$ falls within the threshold, "Sweet pies" and "meat pie" are considered related attributes.

In situations where no literal matches are found for concepts in the knowledge graph for either a or b, we rely solely on the literal distance between them to measure their dissimilarity. For instance, if "Abraham Lincoln" is misspelled as "Abrehan Lincon" and the calculated dis_t between them is 3, and ε_t is set to be less than 3, then the comparison of two attribute names like "Abrehan Lincon" and "Abraham Robinson" is based solely on literal difference, resulting in $dis(a, b) = dis_t(a, b)$. In the provided example, $dis(a, b)$ would be 6.

It's important to note that even with the knowledge base and edit distance, accurately determining attributes for integration can be challenging due to semantic complexities. Take, for instance, the attributes "import" and "export." They share a strong literal similarity, and a path "import" - "commodity" - "export" in the knowledge base between them has a length of 2, which seems relatively small. However, these attributes are actually antonyms and not semantically similar. Resolving such semantic ambiguities often necessitates human intervention, as semantic understanding remains a challenging issue in schema matching. In this paper, while we strive for automated processing, we acknowledge that addressing false positives may require further human verification, as discussed in Sect. 3.4.

2.2   Distance Function

Based on the knowledge base, we define the semantic distance between two attributes in the schemas as follows.

Definition 2 (semantic distance). $\exists\ a, b \in G$, *s.t.* $(a,\ b) \in S$, *a semantic distance means the length of the path between a and b, denoted as $dis_r(a, b)$.*

According to this definition, the smaller dis_r is, the more similar a and b are, as described in Sect. 2.1. Then we use a threshold γ to constrain whether two concepts are similar enough. Two concepts are regarded similar with the distance under the given threshold γ. For example, if we define $\gamma = 2$ when dis_r ("Sweet pies", "pie")= 1, then we regard "Sweet pies" and "pie" as related concepts.

With misspellings, an attribute name may not be found in the knowledge base. Thus, we should consider literal difference between attributes and concepts in the knowledge base. In this paper, we use edit distance [4], a commonly-used distance function for strings to represent the literal distance between attribute names and concepts, denoted by dis_t. Utilization of edit distance will be discussed in detail in Sect. 2.3.

With these considerations, we define following constraints of the determination whether attributes could be matched in schema integration.

Distance Constraint

$$dis(a,b) \leq \varepsilon =$$
$$\begin{cases} dis_r(c_a, c_b) \leq \gamma \wedge dis_t(a, c_a) \leq \varepsilon_t \wedge dis_t(b, c_b) \leq \varepsilon_t, \\ \quad \exists c_a, c_b, dis_t(a, c_a) \leq \varepsilon_t \wedge dis_t(b, c_b) \leq \varepsilon_t \\ dis_t(a, b) \leq \varepsilon_t, \\ \quad \forall c_a, c_b, dis_t(a, c_a) \geq \varepsilon_t \vee dis_t(b, c_b) \geq \varepsilon_t \end{cases}$$

When dealing with an attribute name like a, if a concept c_a within the knowledge base has the closest literal distance to a below the ε_t threshold, we designate c_a as the representation of a. This enables us to compute the semantic distance between attributes a and b based on their respective knowledge base concepts. To clarify, $dis(a,b) \leq \varepsilon$ holds when $dis_r(c_a, c_b) \leq \gamma$, $dis_t(a, c_a) \leq \varepsilon_t$, and $dis_t(b, c_b) \leq \varepsilon_t$.

In cases where there are no literal matches for concepts in the knowledge graph for either a or b, we rely solely on the literal distance between them to measure their dissimilarity. For instance, if "Abraham Lincoln" is mistakenly spelled as "Abrehan Lincon," and the calculated dis_t between them is 3, and ε_t is set to a value less than 3, we treat two attribute names like "Abrehan Lincon" and "Abraham Robinson" as a and b. In such instances, the distance is determined solely by literal differences, leading to $dis(a,b) = dis_t(a,b)$, resulting in $dis(a,b)$ being 6.

Even with the knowledge base and edit distance, accurately identifying attributes for integration can be challenging due to the inherent complexity of semantics. For example, consider the attributes "import" and "export." They exhibit strong literal similarity, and addressing such semantic ambiguities often necessitates human intervention. In this paper, while we strive for automated processing, addressing false positives may require additional human verification, as discussed in Sect. 3.4.

2.3   Edit Distance and Q-Gram

Edit distance, a metric for measuring the literal dissimilarity between two strings[1], has been a focus of extensive research by scholars like [5,6], and [9]. In our context, we employ $dis_t(a,b)$ to denote the edit distance between the strings a and b, with a predefined threshold of ε_t.

Edit distance computation primarily relies on manipulating the q-gram structure of strings. A q-gram refers to a substring of length q within a string. Notably, strings with a small edit distance should share numerous common q-grams. A commonly used filtering condition, as proposed in [2], is presented below.

---

[1] http://en.wikipedia.org/wiki/Edit_distance.

Count filtering means that a and b must share at least LB_{ab} common q-grams.

$$LB_{ab} = (\max(|a|, |b|) - q + 1) - q * \varepsilon_t$$

To enforce constraints, we use LB_{ab} and apply count filtering to alternative schemas within the ε_t threshold.

In schema integration for big data, exhaustive concept scanning in the knowledge base is unfeasible. To address this, we employ an inverted list structure to organize our index.

For a word w, q-gram splitting yields segments w_1, w_2, ..., w_k ($1 \leq k \leq |w| - q + 1$). Each w_i ($1 \leq i \leq k$) is represented as (h_i, v_i), where h_i is the hash value of w_i, and v_i is a set of words containing w_i. Disk storage uses h_i as the index for w_i.

Literal difference assessment relies on count filtering. When matching a string s to word set W, s is divided into q-grams: s_1, s_2, ..., s_k ($1 \leq k \leq |s| - q + 1$). Hashing s_i ($1 \leq i \leq k$) and matching them with attribute set A yield mapped entries: s_1, s_2, ..., s_j ($j \leq k$). We then scan v_1, v_2, ..., v_j, counting word occurrences. If an attribute a_i appears no less than $|s| - q + 1 - \varepsilon_t * q$ times, we consider s and a_i as literally similar, merging them into the integrated set.

2.4   Problem Definition

Given a schema set Σ, each schema s_i is represented as (id_i, n_i, A_i). Schema integration aims to create a global schema S_g from Σ, mapping attributes in each s_i to S_g while ensuring that for each attribute $t \in A_i$, its corresponding attribute a_t in A_g satisfies $dis(t, a_t) \leq \varepsilon$. In practical terms, this means identifying and including similar attributes like "Savory pies," "Tiropita," "meat pie," and "tourtiere" during integration.

3   Overview

In this section, we outline schema integration, a process that combines semantic and literal distance measures to assess attribute similarity. We discuss initializing the knowledge base in Sect. 3.1.

To cater to different scenarios, we've devised two schema integration algorithms: batch integration and incremental integration. Batch integration is suited for cases where multiple schemas require simultaneous integration, while incremental integration is tailored for updating existing schemas with small-sized input. These algorithms will be discussed further in Sect. 3.2 and Sect. 3.3, respectively.

To streamline processing, we've introduced a specialized data structure known as a cluster set. This data format is used for both operands and output in subsequent functions. Henceforth, in this paper, we will refer to this data structure simply as the "cluster set."

Definition 3 (cluster set). *With S as the concepts set of the knowledge base, a cluster set is a set of pairs $\{U, S_U\}$, where U is a set of attributes and $S_U = \{(r, d) | d = min_{\forall t \in U}\{dis(t, r)\} \wedge r \in S\}$. The function is the combination of both literal and semantic distance, as defined in Sect. 2.*

3.1 Initialization

As highlighted in Sect. 2.1, our system relies on the "is a" relationship between concepts, represented as triples (id, name, type). Here, id serves as the concept's index, name as the identifier, and type denotes the knowledge base section of origin. Consequently, relationships between concepts are described using six-tuples (subId, subName, subType, superId, superName, superType).

3.2 Batch Integration

Batch integration processes multiple schemas in a batch by clustering their attributes and merging them into an integrated schema. This involves two types of similarity join operations: ED Join and Semantic Join. ED Join identifies pairs of attributes with edit distances smaller than a threshold, while Semantic Join finds pairs with semantic similarities greater than a threshold based on the knowledge base. Merged pairs from these joins are integrated into the schema. However, transitivity issues can arise during this process due to the nature of the similarity function.

To address transitivity problems, a further step called "resolve" is employed. The algorithm for batch integration is outlined in Algorithm 1. Initially, all attributes from the input schemas are added to a set U (Lines 1–3). Subsequently, ED Join is performed on U to merge literally similar attributes (Line 4), followed by Semantic Join to merge semantically similar attributes (Line 5). This results in attribute pairs forming clusters. However, transitivity issues may persist in the clusters, which are resolved through the Resolve(U) function (Line 6). Further details on the resolve process are discussed in Sect. 5.2.

Algorithm 1: Batch Integration

Input: schema batch W
Output: integration set U
1 **foreach** $w \in W$ **do**
2 | $U \leftarrow U \cup A_w$;
3 **end**
4 $U \leftarrow U -$ EDJoin(U,U);
5 $U \leftarrow U -$ SemanticJoin(U,S);
6 $U \leftarrow$ Resolve(U);
7 **return** U;

3.3 Incremental Integration

Incremental integration adds schemas to the existing global schema one at a time, ideal for integrating new data sources. It employs a cluster set U containing all global schema attributes to streamline the process. When integrating a new schema K, each attribute a in K is compared to U for both literal and semantic similarity. If a lacks similar attributes in U, it's inserted, and U is updated accordingly, with an additional verification step to reduce false positives.

Algorithm 2: Incremental Integration

Input: inserting schema K, integration set U
Output: integration set U' after insert
1 $T \leftarrow$ EDJoin(K,U);
2 $R \leftarrow$ Verify(T,U);
3 $V \leftarrow K - R$;
4 $V \leftarrow$ EDJoin(V,S);
5 $V \leftarrow$ SemanticJoin(V,S);
6 $U \leftarrow U \cup V$;
7 $U \leftarrow$ Resolve(U);
8 **return** U;

Both incremental and batch schema integration share common operations: (1) ED Join, a similarity join based on edit distance, (2) Semantic Join, which leverages semantic similarity from the knowledge base, and (3) Resolve, which involves verification and cluster partitioning. It's worth noting that the Verify() function in Algorithm 2 corresponds to a part of the Resolve() function in Algorithm 1.

3.4 Verification

To address potential false positives in attribute integration, we employ a verification approach comprising value verification and manual verification.

Value Verification: This method relies on attribute values to validate integration results. It assumes that if two attributes are similar, their values should exhibit similarity or identity. Value verification employs structural analysis and predefined rules for judgment:

- **Type:** Attributes have specific data types (e.g., integer, string, list). Similar attributes often share the same data type, especially for complex structures. For example, if attributes contain string sets of names, they may be considered similar, like a list of football team names. Conversely, attributes with dissimilar data types (e.g., mixing strings and integers) are less likely to be similar. Type serves as the primary judgment criterion.
- **Affix:** Examining common prefixes and suffixes in attribute values can reveal specific word structures. For instance, shared prefixes or suffixes like "..." might indicate attributes related to cost or financial records. Thus, affixes are employed as a secondary judgment criterion.

These rules help identify false positives. If attributes initially considered similar fail to meet these value-based criteria, they are reevaluated as false positives, and the proposed relationship is rejected.

Manual Verification: While value verification is effective in some cases, it may be insufficient for attributes lacking distinct structures or those without values. For enhanced accuracy, manual verification involving human assessment is utilized. Crowdsourcing is employed in specific high-accuracy domains to further validate and refine integration results.

4 Join Schema Integration

According to the definition of cluster set, the operators of ED Join and Semantic Join are defined as follows.

Definition 4 (ed join). *Given two families of cluster sets, R and T, and a threshold d, two elements (U_1, S_1) and (U_2, S_2) from R and T, respectively, are ED joined if they satisfy one of the following constraints.*

1. $\min\limits_{r_1 \in U_1, r_2 \in U_2} dist_t(r_1, r_2) \leq \varepsilon_t$
2. $\exists (r, d) \in S_2, \min\limits_{r_1 \in U_1} dist_t(r_1, r) \leq \varepsilon_t - d$
3. $\exists (r, d) \in S_1, \min\limits_{r_2 \in U_2} dist_t(r_2, r) \leq \varepsilon_t - d$

The ED Join result of (U_1, S_1) and (U_2, S_2) is a pair (U, S_U), where $U = U_1 \cup U_2$ and $S_U = \{(r, d) | r \in S \land d = \min_{t \in U}\{dis(r, t)\}\}$.

Definition 5 (semantic join). *Given two families of cluster sets R, T, and a threshold d, two elements (U_1, S_1) and (U_2, S_2) are from R and T, respectively are semantically joined if they satisfy one of the following constraints.*

1. $\min\limits_{r_1 \in U_1, r_2 \in U_2} dis_r(r_1, r_2) \leq \gamma$
2. $\exists (r, d) \in S_2, \min\limits_{r_1 \in U_1} dis_r(r_1, r) \leq \gamma - d$
3. $\exists (r, d) \in S_1, \min\limits_{r_2 \in U_2} dis_r(r_2, r) \leq \gamma - d$

The result of Semantic join on (U_1, S_1) and (U_2, S_2) is a pair (U, S_U), where $U = U_1 \cup U_2$ and $S_U = \{(r, d) | r \in S \land d = \min_{t \in U}\{dis(r, t)\}\}$.

ED Join and Semantic Join are employed to integrate sets of attributes based on edit distance and semantic similarity, respectively. These operations involve specific conditions for attributes to be considered similar and are described in Definition 4 and Definition 5.

For ED Join, it primarily relies on direct attribute similarity, defined as the first constraint in the mentioned definitions. If the distance between $r_1 \in U_1$ and $r_2 \in U_2$ is within one of the specified thresholds, these attributes are considered similar, enabling the integration of the respective cluster sets. For instance, attributes "Sander" and "Sunder" have an edit distance of 1, making them similar, and thus, U_1 and U_2 can be joined.

Moreover, both ED Join and Semantic Join take into account conditions 2 and 3 as per the definition of the cluster set. If the distance between r_1 in U_1 and r in S_2 is within $\gamma - d$, they are regarded as similar, allowing the integration of the cluster sets containing r_1 and r. This judgment process is applicable to both ED Join and Semantic Join.

Algorithm 3: Pair Join

    Input: two cluster pairs (U_1, S_1) and (U_2, S_2)
    Output: joined pair (U, S)
1 $U \leftarrow U_1 \cup U_2$;
2 **foreach** $(r, d) \in S_1 \cup S_2$ **do**
3   **if** $dist_t(r, v) \leq d$ **then**
4 | $S \leftarrow S \cup (r, dist_t(r, v))$;
5   **end**
6   **else**
7 | $S \leftarrow S \cup (r, d)$;
8   **end**
9 **end**
10 **return** (U, S);

Based on the Pair Join solution, ED Join and Semantic Join are two crucial steps to finish batch integration and incremental integration. We will discuss ED Join and Semantic Join respectively in Sect. 4.1 and Sect. 4.2.

4.1   ED Join

ED Join focuses on merging cluster sets with attributes that are literally similar. This operation resembles the similarity join on sets of strings based on edit distance, as discussed in [4,5]. To efficiently achieve this, we adapt q-gram-based methods for ED Join.

Our fundamental data structure is the inverted list, with each q-gram serving as an entry. Attributes within the cluster sets are indexed using q-grams. Given inputs R and T, the q-gram-based inverted lists for their attribute sets are denoted as X_R, X_T for U, and Z_R, Z_T for S_U, respectively. In ED Join, two cluster sets can be joined based on the three constraints outlined in Definition 4. Consequently, q-gram-based similarity joins are performed according to index pairs X_R and X_T, X_R and Z_T, as well as X_T and Z_R.

Within the ED Merge function, the input H represents the index list, while the output K consists of pairs that have been joined. Each gram g in the list H is associated with a list $v_1, v_2, \ldots$ containing attributes that contain g. The function begins by counting the frequency of appearance for each attribute v in different parts of the list H, such as $X_R \cap X_T$ (Line 10), and initializes the answer set K as an empty set (Line 11). For each v that appears more than $|v| - q + 1 - \varepsilon_t * q$ times, indicating the presence of similar attributes in H (as discussed in Sect. 2.3), the cluster set containing v is merged into K using Pair

Join (Lines 12–16). The Locate(v) function is used to determine the cluster set to which v belongs. Ultimately, the resulting cluster set M contains integrated attributes for ED Join.

Algorithm 4: ED Join

Input: two cluster sets R and T, threshold ε_t and d
Output: joined cluster sets M including pairs (U, S)

1 $X_R \leftarrow$ q-gram($R.U_i$);
2 $X_T \leftarrow$ q-gram($T.U_i$);
3 $Z_R \leftarrow$ q-gram($R.S_i$);
4 $Z_T \leftarrow$ q-gram($T.S_i$);
5 $M \leftarrow M\cup$ EDMerge ($X_R \cap X_T$);
6 $M \leftarrow M\cup$ EDMerge ($X_R \cap Z_T$);
7 $M \leftarrow M\cup$ EDMerge ($X_T \cap Z_R$);
8 **return** M;
9 **function EDMerge**
        Input: q-gram H
        Output: set of joined pairs K
10   Count($v \in g \in H$);
11      $K \leftarrow \emptyset$;
12   **foreach** $v \in g \in H$ **do**
13   **if** $count[v] \geq |v| - q + 1 - \varepsilon_t * q$ **then**
14 | $K \leftarrow$ PairJoin(K, Locate(v));
15   **end**
16   **end**
17   **return** K;
18 **end**

4.2 Semantic Join

To accelerate join processing, we maintain a hash table for k-hop neighbors of all concepts, denoted as H_k. Such table is used to find required pre-processed neighbors relationship within O(1) time complexity. Thus, running time is saved by turning the paths in the knowledge base into accessing a series of segments in hash tables. We define neighbor table as first.

Definition 6 (neighbor table). t *is an attribute and* P *is the set of all paths in the knowledge base.* $H_k(t)$ *is a table on the disk indexed by hash value of string* t, *s.t.*

$$H_k(t) = \{a_i | (t, a_i, d) \in P \wedge d = k\}$$

The hash function's primary objective is to cluster attributes closely together on the disk. To achieve this, we divide the table into multiple buckets, ensuring that attributes accessed together are placed in the same bucket. To illustrate, let's consider a hash seed of 13, a bucket length of 10,000, and a base offset for this bucket at 1,000,000. The offsets for each attribute are detailed in Table 1.

Table 1. Example of Bucket Hash Offset

Attribute	Offset in Bucket	Total Offset
Name	9277	1009277
Speed	5109	1005109
Amount	2380	1002380
Streetname	2708	1002708

The algorithm is shown in Algorithm 5.

Algorithm 5: Bucket Hash

Input: one set of attributes of A, offset base value R_0
Output: set K of (hashkey, attribute)
1 $K \leftarrow \emptyset$;
2 **foreach** $a \in A$ **do**
3      $k \leftarrow 0$;
4   **foreach** $s \in a$ **do**
5          $k \leftarrow (k * hash\_seed + s)$;
6          $s \leftarrow next(s)$;
7   **end**
8      $k \leftarrow k\%bucket\_length$;
9      $k \leftarrow k + R_0$;
10     $K \leftarrow (k, a)$;
11 **end**
12 **return** K;

In this algorithm, we take a set of attributes A as input, along with the base offset R_0 indicating the starting point of the bucket on the disk. For each attribute a, we calculate its hash value k (Lines 2–7). To ensure that these attributes are grouped together in a single bucket for reduced disk access time, we compute the offset for each attribute by adding the bucket offset k to the base offset R_0 (Lines 8–9). All pairs of hash keys and attributes are then added to set K as the output.

Join Algorithm. The logical process of Semantic Join under the semantic threshold γ can be described as follows:

$$(R) \cup (R \bowtie E) \cup (R \bowtie^2 E) \cup \cdots \cup (R \bowtie^{\gamma-1} E) \cup (R \bowtie^\gamma E)$$

During this join operation on the knowledge base, we establish connections between paths of varying lengths. Each connection requires matching between the end nodes of one path and the starting nodes of another. To optimize computation speed, we introduce the concept of a "path set," which is a sophisticated data structure designed to group paths with the same end node into the same hash bucket. This data structure is defined as follows:

Definition 7 (path set). P_a *is a path set, all paths in which share the same end node* a, *s.t.*

$$P_a = \{(start, k)|\exists start \in H_k(a)\}$$

The Semantic Join algorithm, as outlined in Algorithm 6, operates based on the provided data structure. It takes as input the target cluster set R, the threshold γ, and the knowledge base with hash-based storage, producing the joined cluster sets M.

Algorithm 6: Semantic Join

Input: one cluster set R, semantic threshold γ and 1-hop neighbor table H_1
Output: joined cluster sets M

1 $P \leftarrow \emptyset$;
2 $M \leftarrow \emptyset$;
3 **foreach** $w \in U_i|(U_i, S_i) \in R$ **do**
4 $\quad$ $P_h \leftarrow P_h \cup \{(w, 1)|h \in H_1(w)\}$;
5 $\quad$ $P \leftarrow P \cup P_h$;
6 **end**
7 **for** $i{=}1$ **to** $\gamma - 1$ **do**
8 $\quad$ **foreach** $P_i \in P$ **do**
9 $\quad\quad$ $P_j \leftarrow P_j \cup \{(start, len + 1)|j \in H_1(i)\}$;
10 $\quad\quad$ $P \leftarrow P \cup P_j$;
11 $\quad\quad$ **if** $j \in U_i|(U_i, S_i) \in R$ **then**
12 $\quad\quad\quad$ **foreach** $start \in P_j$ **do**
13 $\quad\quad\quad\quad$ $|$ $M \leftarrow M \cup PairJoin(atCluster(start), (U_i, S_i))$;
14 $\quad\quad\quad$ **end**
15 $\quad\quad$ **end**
16 $\quad\quad$ **if** $(j, d) \in S_i|(U_i, S_i) \in R \wedge len + 1 + d < \gamma$ **then**
17 $\quad\quad\quad$ **foreach** $start \in P_j$ **do**
18 $\quad\quad\quad\quad$ $|$ $M \leftarrow M \cup PairJoin(atCluster(start), (U_i, S_i))$;
19 $\quad\quad\quad$ **end**
20 $\quad\quad$ **end**
21 $\quad$ **end**
22 **end**
23 **return** M;

Our goal is to identify concepts within a subgraph where paths have a length no greater than the specified threshold.

After adding 1-hop neighbor information into P during step 2, we proceed to expand from the end concepts of each path set by performing joins with H_1.

For the Semantic Join algorithm, the expansion step relies on P obtained from the initialization step. Based on the 1-path set to obtain a γ-path subgraph, the number of expansion loop iterations is $\gamma - 1$ (Line 7). For each path set P_i in the set P, the first task is to identify paths that the end concept i can connect to. Utilizing the 1-hop neighbor table H_1, we filter out all the paths $(i, j, 1)$ and link them to i, thereby expanding the paths to reach j. Subsequently, we insert $(start, len + 1)$ into P_j (Lines 8–10).

When executing the operation $P_j \leftarrow P_j \cup (start, len + 1)|j \in H_1(i)$ (Line 9) to expand paths in the knowledge base, P_j gets updated with the insertion of new paths. However, P_j often ends up containing duplicated concepts with the same start and end concepts.

According to the definition of semantic distance in Definition 2, the difference between two concepts is represented by the path length in the algorithm. Therefore, the smaller the length, the more similar the two concepts are. Under a given threshold γ, a shorter path (length $= d$) has a higher chance of joining with other concepts ($\gamma - d$ times, which is greater). For paths that share the same start and end concept, the shorter path is preferred for integration.

Finally, the end concept j in Line 9 is used to determine whether cluster sets (U_i, S_i) should be merged. If the end concept j belongs to any U_i, then the cluster set where the path started from, denoted as $(atCluster(start))$, is merged with (U_i, S_i) and added to M (Lines 11–15). Additionally, if the end concept j is within any S_i and the distance to the corresponding concept in U_i is d such that $len + 1 + d < \gamma$ (indicating that the distance between $start$ and any concept in U_i is no more than γ), then the cluster set $(atCluster(start))$ is also merged with (U_i, S_i) and added to M (Lines 16–20).

Implementations. When the threshold becomes larger, scanning H_1 multiple times (specifically $|\gamma| - 1$ times) can become costly. According to the definition of the neighbor table, the concepts found by H_k and $H_{k_1} \bowtie H_{k_2} \bowtie \cdots \bowtie H_{k_m}$ (where $k_1 + k_2 + \cdots + k_m = k$) are the same. To address this, the main idea is to use neighbor tables with higher k values to reduce the number of times we access H_{k_i}. Additionally, generating neighbor tables becomes more expensive as k increases, so constructing a neighbor table for every possible k is impractical.

Here, we utilize integer powers of 2 (e.g., 1, 2, 4, 8, 16, 32, 64, etc.) to balance the initialization cost and cover the threshold of the join algorithm by adding up some of these numbers. This approach enhances the efficiency of Semantic Join by reducing the number of times we need to access neighbor tables, while still covering a wide range of threshold values.

According to the analysis above, the time cost of the proposed algorithm is unrelated to the size of knowledge base. Without accessing the knowledge base too many times, the algorithm saves much time and can be easily adopted in problems on the knowledge bases in various sizes. The time cost is related to the sets of input and output. We can save time by controlling the threshold to diminish the size of these sets and finally save time.

5   Batch Integration

In this section, we discuss batch integration implementation in detail. We first introduce the steps to construct the cluster set for batch integration in Sect. 6.1. How to resolve subset consisting unrelated schemas efficiently is provided in Sect. 6.2.

5.1   Flow of Batch Integration

The batch integration algorithm, as outlined in Sect. 3.2, consists of four main steps: initialization, ED Join, Semantic Join, and resolve. In the initialization step, all attributes are added to set U as the center attributes for the join operations. Self-ED-Join is then performed on U to group literally similar attributes together.

Following that, Semantic Join is applied to U to aggregate semantically related attributes. This step merges cluster sets that share similar semantic attributes, resulting in cluster sets containing both literally and semantically similar attributes. To achieve accurate schema integration, a resolve process is introduced to break down the merged cluster sets into smaller ones, eliminating extraneous information.

5.2   Resolving

After the merge operation by the join algorithm, similar attributes are grouped together in cluster sets. However, as discussed in Sect. 5.1, the cluster set doesn't always meet the predetermined closure constraint. Here, we propose a simple and efficient solution that is functionally sufficient to resolve this issue. Let's illustrate this solution with an example.

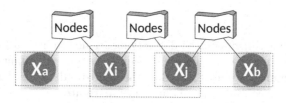

Fig. 1. Example of Resolving

Figure 1 illustrates some relational structures within cluster sets. Suppose we perform a join operation under a given threshold. The dashed line box contains attributes within the threshold. The distances $dis(X_a, X_i)$, $dis(X_i, X_j)$, and $dis(X_j, X_b)$ are all within the threshold, indicating that these are similar attributes. Consequently, the attributes X_a, X_i, X_j, X_b in the figure can be integrated into one cluster set.

However, it's essential to note that X_a and X_b have no meaningful relationship, and their distance $dis(X_a, X_b)$ exceeds the threshold.

There are multiple valid approaches to resolve this situation, such as partitioning into X_a, X_i, X_j, X_b or X_b, X_i, X_j, X_a, depending on the specific context. Additionally, it's worth considering whether it's possible to tolerate X_a and X_b being in the same set to reduce the number of resolution steps, especially when $dis(X_a, X_b)$ is not significantly large.

6   Experiments

6.1   Experimental Settings

Environment. The experiments were performed on a Windows 10 64-bit computer with an Intel Core i7 2.4 GHz processor and 8 GB of memory.

Data Sets. In order to test the accuracy and efficiency of our algorithm, we use real data from the Internet. For knowledge base, we choose *Freebase* to provide knowledge linkage between concepts with special initialization mentioned following. Other knowledge base can be used as well.

For attributes of database tables to be integrated, we choose open data set from NYC OpenData[2] and SF OpenData[3]. These open data sets are sourced from real data. Such data cover various fields and test our algorithm comprehensively.

Parameters. Threshold ε is a key parameter of join schema integration, which decides how many concepts are accessed during join. For two algorithms ED Join and Semantic Join, we set threshold ε_t to describe how much misspelling can be tolerated while γ determine the similarity degree. The values of these two thresholds are determined by the schema of attributes and the chosen knowledge base. To achieve a better performance and low cost, default value of these thresholds are $\varepsilon_t = 1$ and $\gamma = 3$.

6.2   Accuracy

We conducted experiments on the datasets mentioned in Sect. 6.1 to assess how closely the algorithm's results align with human judgments. Manual integration results on the input attributes served as the gold standard for this experiment. While the datasets contain a large number of attributes and concepts, we opted to focus on a smaller dataset to facilitate analysis and manual judgment.

For batch integration, the experimental results for different target attributes are presented in Table 2. The metrics of recall and precision indicate that join schema integration performs relatively well for different target words. The average recall and precision are 0.9266862 and 0.7431666, respectively.

Table 2. Batch Integration Accuracy

Word	$|S_A|$	$|S_T|$	$|S_T \cap S_A|$	Recall	Precision
name	76	61	57	0.934426	0.750000
year	93	64	58	0.906250	0.617021
type	73	58	53	0.913793	0.726027
number	79	68	65	0.955882	0.822785
category	12	13	15	0.923077	0.800000

---

[2] https://data.cityofnewyork.us/.
[3] https://data.sfgov.org/.

6.3 Efficiency

For large-scale data, efficiency is extremely important. Therefore, we test the efficiency. From the algorithm, the running time is influenced by data size, target, threshold and the existence of cluster set. Hence, we show the experimental results and analyze their impact. In this section, we run the experiments on a real piece of knowledge from *Freebase* containing 9,471,476 items.

The Impact of Data Size. To evaluate the impact of the size of data, we assign the size of input attributes at different values. The size of result attributes also change with it. The experimental result is shown in Table 3 for batch integration. It is easy to find running time is increasing with data growth. As for the result, we focus on the increasing rate of running time.

Table 3. Time Cost VS Data Size (batch)

input set No.	input size	result size	running time
1	1	47	0.015
2	1	267	0.021
3	1	67476	0.672
4	5	12073	0.453
5	5	201529	5.625
6	10	106	0.084
7	10	19207	0.582
8	20	252163	51.13
9	20	84962	2.573
10	30	99243	12.39
11	30	177027	20.979
12	40	188034	30.185
13	40	376257	66.327
14	50	189247	18.009
15	50	204929	30.384

For batch integration, we perform experiment on variation of input attribute size and result size. As observed from Table 3, the result size has larger influence on running time than the input data size. For example, although the input set size is the same, the running time for input set 3 is larger than the input set 2 due to larger result size of the third. Even input set 5 has a larger running time than set 6 and 7, even though set 5 has a smaller input set. The reason is that integrating the attributes with many related concepts in the knowledge base means more time while accessing to the disk. Some input sets with less related knowledge cost less time. In conclusion, running time has no specific relationship with input size, but the result size makes much sense. The larger the result set is, the more information we can get form integration. By choosing a suitable threshold, we can limit the size of integrated attributes to save time while satisfying needed accuracy.

The Impact of Target. Since how much knowledge around a concept is unknown, we conduct experiments on concepts of different parts in the knowledge base to testify the impact of the target. Here, we select different attributes to conduct the experiment, both attributes with many neighbors and neighborless attributes. The running time is shown in Table 4 for batch integration and Fig. 2 for incremental integration. The running time differs quite a lot from these two types of attributes.

Table 4. Time Cost VS Targets (batch)

input set no.	input size	result size	running time
1	1	75	0.031
2	1	3924	0.069
3	1	67476	0.672
4	5	377	0.108
5	5	12073	0.453
6	5	201529	5.625
7	10	106	0.084
8	10	19207	0.582
9	10	128659	4.463

From the table, we know that although input data and threshold are same, the running time can differ a lot. From the analysis in the experiment of data size, we know that result size has much influence. The experiment in this section

Table 5. Time Cost VS Threshold (batch)

input size	threshold	result size	running time
1	2	66	0.024
1	3	213	0.129
1	4	69862	36.641
2	2	89	0.039
2	3	4145	0.304
2	4	108493	101.54
3	2	730	0.103
3	3	7086	3.481
3	4	111161	131.434

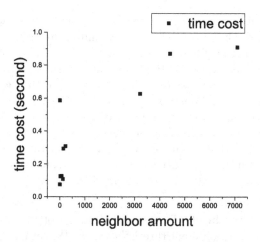

Fig. 2. Time Cost VS Targets (incremental)

verifies it as well. By analyzing the knowledge base, we can make sure why the variance of running time happens. In the knowledge base, the neighbor amount differs a lot among concepts. Some concepts belong to a small subgraph of neighbors, so there is a little knowledge to be dealt with during integration. On the other hand, if an attribute shares a lot of relationship with others, the problem can be very complex to enlarge the cluster set. For example, the input set 3 "living people" has much more neighbors (67476) that the input set 1 "cancer" (75) under threshold of 2 in Table 4. In conclusion, treating attributes with too much knowledge means much time to cost. To avoid large running time, when we foreknow the input attributes with many related attributes, we set a lower threshold, as discussed in Sect. 6.3.

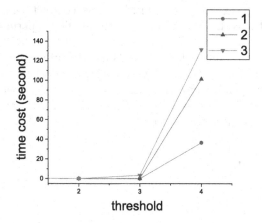

Fig. 3. Time Cost VS Threshold (batch)

For incremental integration, we test the running time when the target attribute is inserted. Based on the structure and cluster set mentioned in Definition 3, if the added attributes are in the generated set S, it is unnecessary to insert and time can be saved. Here, we only consider the target with none relationship in S and to be inserted to U. As observed in Fig. 2, running time varies a lot when the neighbor amount is changed. Here, we select a part of the knowledge base and count the amount of 1-hop neighbor for each concept. The neighbor amount indicates that the concept comes from a dense subgraph or not. Figure 2 shows that when the neighbor amount raises, the running time usually increases with it. Some outliers are caused by some 2-hop or more distant neighbor-rich concepts which are not presented by the neighbor size in Table 2.

The Impact of Threshold. To assess the impact of the threshold, we conducted experiments with different threshold values. We set the minimum threshold to 2 to ensure a sufficient number of integrated attributes for meaningful analysis. We investigated how the running time changes with varying threshold values in batch integration. The experimental results are presented in Table 5 and Fig. 3.

From the results of batch integration, we can deduce that the choice of threshold value is a crucial factor. A larger threshold leads to the integration of more attributes, which, in turn, increases the processing time or even becomes unnecessary. As the path range expands rapidly and the path set P grows larger, the time required for each iteration of path expansion also increases. Notably, we observe that the running time increases more than linearly when the threshold surpasses a certain value. In other words, there is an acceptable threshold limit beyond which the increase in running time becomes less practical.

The Necessity of Cluster Set. For incremental integration, for cluster set defined in Definition 3, the generation of S can save much time when the inserted attribute is related to existing ones in it. Here, we conduct some experiments to verify that the set S really works for acceleration. The experimental results are shown in Fig. 4.

According to the results, we know that if one attribute exists in S, the running time for integrating such attribute keeps low. However, new attributes for the cluster set usually spends more time than those in S. Therefore, in this way, we can observe that the proposed structure, cluster set, can save time when the attributes appear for more than once.

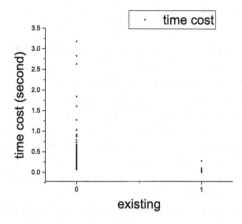

Fig. 4. Time Cost VS Cluster Set Necessity (incremental)

Conclusion. As is stated above, running time of join schema integration is complicated affected by data size, target, threshold and the existence of cluster set. To decrease the running time, it is necessary to balance these factors according to the requirement. For a certain problem, one good solution is to decrease the threshold as low as possible to limit the answer and save time.

7   Related Work

As a basic but crucial technique in database, schema integration has been discussed for many years. In old days, schema integration using similarity metric such as Jaccard similarity could not deal with semantic relation. Later, one marvelous work [8] concludes many approaches to finish the work of schema mapping and integration. In this paper, the authors made classification for existing methods of schema integration and schema mapping, using techniques such as linguistic ways. For methods applied to schema integration, DIKE [7] and ARTEMIS [1] lead the ways. These two methods both computes the relationship between objects or attributes while our proposed algorithm use existing knowledge base. At most cases, relationship in knowledge base extracted from Web is in closer proximity to human's mind.

Recently, Microsoft has done some research [3] on schema integration. In this paper, precision and recall of integration has a high value. Compared to our schema-level algorithm, much instance information is used in their SEMA-JOIN. As the database tables have too many rows storing details, it is not possible to bring them all during the integration. For the efficiency, here, we choose to discard the instance information. What's more, there are quite a lot databases with less maintenance that have even no value for some attributes, integration in schema-level can be more widely used.

8 Conclusions and Future Work

In this paper, we study a novel problem of schema integration on big data. To process this problem, we propose batch and incremental integration algorithms for different scenarios. The former is suitable for a set of attributes needed to be integrated, and the latter is used to insert information of newly adding attributes to the existing integrated cluster set. For effectiveness issues, we involve both semantics and syntactic similarity during integration. The semantics similarity is computed according to the knowledge based, and the syntactic similarity is based on the edit distance. For efficiency issues, we propose ED Join and Semantic Join algorithms. Experimental results show that our approaches could integrate schema efficiently and effectively.

Considering that current knowledge base actually cannot provide all needed information, our future work is to develop novel transformation rules discovery algorithms and weight determination algorithms for the knowledge base to achieve high accuracy for integration.

References

1. Castano, S., De Antonellis, V.: Global viewing of heterogeneous data sources. IEEE Trans. Knowl. Data Eng. **13**(2), 277–297 (2001)
2. Gravano, L., et al.: Using q-grams in a DBMS for approximate string processing. IEEE Data Eng. Bull. **24**(4), 28–34 (2001)
3. He, Y., Ganjam, K., Chu, X.: SEMA-JOIN: joining semantically-related tables using big table corpora. VLDB Endow. **8**, 1358–1369 (2015)
4. Levenshtein, V.I.: Binary codes capable of correcting deletions, insertions, and reversals. In: Soviet Physics Doklady, vol. 10, pp. 707–710 (1966)
5. Li, L., Wang, H., Li, J., Gao, H.: ED-SJOIN; an optimal algorithm for similarity joins with edit distance constraints. J. Comput. Res. Dev. **46**, 319–325 (2009)
6. Lin, X.M., Wang, W.: Set and string similarity queries: a survey. Jisuanji Xuebao (Chin. J. Comput.) **34**(10), 1853–1862 (2011)
7. Palopoli, L., Saccá, D., Ursino, D.: An automatic technique for detecting type conflicts in database schemes. In: Proceedings of the Seventh International Conference on Information and Knowledge Management, pp. 306–313. ACM (1998)
8. Rahm, E., Bernstein, P.A.: A survey of approaches to automatic schema matching. VLDB J. **10**(4), 334–350 (2001)
9. Xiao, C., Wang, W., Lin, X.: Ed-join: an efficient algorithm for similarity joins with edit distance constraints. Proc. VLDB Endow. **1**(1), 933–944 (2008)

A Hybrid Few-Shot Learning Based Intrusion Detection Method for Internet of Vehicles

Yixuan Zhao[1], Jianming Cui[1], and Ming Liu[2](✉)

[1] School of Information Engineering, Chang'an University, ShaanXi 710064, China
cjianming@chd.edu.cn
[2] National Computer Network Emergency Response Technical Team/Coordination Center of China, Beijing 100029, China
liuming@cert.org.cn

Abstract. With the rapid development of vehicle networks technologies, cyber security threats in the Internet have gradually penetrated into the Internet of vehicles. In view of the risks and challenges, this paper proposed a hybrid Meta-Learning based intrusion detection method, which core task is to distinguish normal and network flow samples as its learning task. By constructing a feature extraction network based on 3D-CNN, the characteristic values of network flow classification are learned, and then the constructed feature comparison network is used for learning and discrimination. It should be emphasized that the model can obtain enough prior knowledge to realize lightweight intrusion detection by constructing few-shot sample task training. In the experimental section, we first selected Car-Hacking dataset to evaluate the performance of the proposed method and analyze the accuracy, detection rate, precision, false positive rate and F-Score, etc., and extended the testing to the ICSX2012 dataset. The experimental results show that, the method proposed can effectively implement network intrusion detection in few-shot sample scenarios, and has the expansibility in cyber security applications.

Keywords: Intrusion detection · Internet of Vehicles · Convolutional neural network · Few-shot learning

1 Introduction

With the development of in-vehicles communication technology, Connected Vehicles (CV) and Autonomous Vehicle (AV) are becoming more and more popular [12,17]. As the main communication facility of emerging technologies, IoV connects other IoV entities (such as smart devices, infrastructure, and pedestrians) through wireless communication technology [4]. IoV is mainly composed of in-vehicle network and out-of-vehicle network [9]. The in-vehicle network implements various functions through a variety of Electronic Control Unit (ECU),

---

This work is financially supported by the National Natural Science Foundation of China under Grant 62106060.

Z. Tari et al. (Eds.): ICA3PP 2023, LNCS 14488, pp. 207–220, 2024.
https://doi.org/10.1007/978-981-97-0801-7_12

and all ECUs are connected through the Controller Area Network (CAN) bus to transmit information [14]. The external network connects modern vehicles with the external environment through vehicle to everything (V2X) technology, and realizes the function of communicating with other vehicles, roadside infrastructure and road users. The improved functionality and connectivity of modern smart vehicles has also brought about the security of the Internet of Vehicles.

Deep learning has been widely used in IoV intrusion detection, which can improve some security issues encountered by IoV [2,8,13]. Due to the good feature learning and generalization ability of convolutional neural network (CNN), CNN has become a powerful tool for feature learning and data analysis [11]. Whether 1D-CNN [6] or 2-D-CNN [6] networks are used, there are problems such as inability to capture information beyond two-dimensional space and weak feature representation. Aiming at the limitations of 1D-CNN and 2D-CNN, Ji et al.proposed 3D-CNN [7]. Using 3D-CNN can not only extract spatial features, but also capture the time relationship between IoV traffic packets, making full use of CNN's feature learning ability.

Traditional IoV intrusion detection methods need to train a large number of labeled samples to identify known attacks. However, there will always be new attacks in the IoV environment, and vehicle safety agencies cannot quickly produce new data sets. Aiming at the above problems, an IoV intrusion detection method based on meta-learning is proposed. This method is applied to IoV intrusion detection in few-shot scenarios through feature extraction and detection, and the effectiveness of IoV intrusion detection is verified. After experimentation, the following work was done:

- 3D-CNN is used to extract features and classify the attack types of IoV traffic, and the prior knowledge in the meta-learning algorithm is used to detect samples that are not heavily trained and only based on a limited number of labels.
- A data conversion method is proposed to convert vehicle network traffic data into image format data suitable for input to CNN, making it easier to distinguish various attack modes.
- It is evaluated on two public IoV safety datasets suitable for in-vehicle and out-of-vehicle network data, and compared with the results of other state-of-the-art methods.

2   Design of Intrusion Detection Model for IoV

This section introduces the overall framework and design process of a meta-learning based intrusion detection method for IoV. The overall architecture of the method is shown in Fig. 1, including data collection, dataset construction, feature extraction, feature comparison, and classification. The design process of each module in the framework is detailed below.

2.1   Data Pre-processing

ISCX2012 is an unsplit PCAP file. First, we need to use the pkt2flow tool [3] to split the original traffic data into multiple corresponding streams. The PCAP

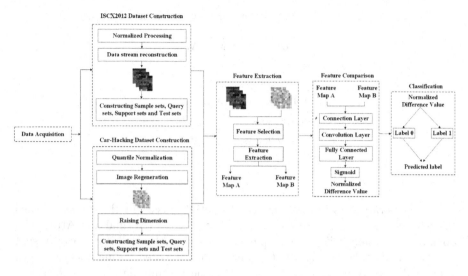

Fig. 1. The framework diagram of the intrusion detection method for connected vehicles based on meta-learning.

packet is reconstructed according to the five tuples: source IP, source port, destination IP, destination port and protocol. However, the PCAP file is still large after extraction. In order to speed up the data reading process, the pickle tool in Python is also used to package the traffic. The difference between each type of attack comes from the header and payload of the packet. The packet is processed according to the uniform length of the *packet_len* byte, that is, if the packet length is greater than *packet_len*, the byte is intercepted, otherwise filled with 0. In this experiment, *packet_len* takes 256 [23]. The length of the data stream is also not fixed. Take *flow_len* packets to represent this data stream. In this experiment, *flow_len* takes 16. Finally, each packet is folded into a single video stream, arranged in the corresponding order, and then transmitted to 3D-CNN.

The original Car-Hacking data set is table data, the original network data should be converted into image format and then input into the 3D-CNN. Since the Car-Hacking dataset contains nine important features: CAN ID, DATA[0]-DATA [7], these features are converted into blocks after data standardization. Therefore, each block of 27 consecutive samples with 9 features is converted into an image with a shape of 9×9×3. Each transformed image is a square color image composed of three channels of red, green and blue. Because the input size of the model is the same, the dimensionality reduction process will lead to a reduction in the number of features. Therefore, the Car-Hacking dataset is increased from 9×9×3 to 16×16×16, which is consistent with the ISCX2012 dataset. The hyperparameter values for the network are shown in the Tab. 1, with reference to literature [10] and empirical rules:

After converting both data sets into a sample format suitable for input into 3D-CNN, it is necessary to divide the training and test set. Meta-learning method

Table 1. The value of hyperparameter in intrusion detection method

Hyper-parameter	Value
Episode	100
Batch size per class	15
Activation functions	ReLU
Optimizers	Adam
Dropout rate	0.4
K	5,10

needs to generate a meta-training set of multiple tasks. In this research, each task is to classify two different categories of instances (Normal and Attack) by providing a classifier with K examples for each class. We call it a 2-way K-shot task, K usually takes 5 and 10. The task set constructed by meta-learning can be divided into Train Task and Test Task. The training data in Train Task is called Sample Set, and the test data is called Query Set. The training data in Test Task is called Support set, and the test data is called Test Set.

2.2 Feature Extraction

The feature extraction part is based on the data packet as a two-dimensional tensor, and the data stream is used as a three-bit tensor, which is input into the 3D-CNN in chronological order, so that the three-dimensional convolution operation can extract the features between different data packets. Therefore, the feature extraction part is a two-way processing of three-bit tensor convolutional neural network [22]. The feature extraction part extracts different features for different data sets. For the Car-Hacking dataset, we mainly extract nine important features of the original vehicle network traffic through the feature extraction part: CAN ID and DATA[0]-DATA [7]. For the ISCX2012 dataset, we mainly extract the header and load of the original network traffic data. The input of the feature comparison part is the feature map output by the feature extraction part. Due to the difficulty of manual design, the feature comparison part is created to learn this comparison function. It obtains a learnable sample comparator through network training, which uses the fully connected layer to combine the features found in the feature extraction part into an overall prediction of which category the input is most likely to belong to.

Through the data preprocessing part, the ISCX2012 data set and the Car-Hacking data set are converted into $16 \times 16 \times 16$ normalized three-dimensional images. Then, using the feature extraction network (Left of Fig. 2), a 128-dimensional network traffic feature vector is extracted from the image data through four convolutional layers (The size of the convolution kernel is $2 \times 2 \times 2$), batch normalization and activation function ReLU. Next, two data sets are used for meta-training set and meta-testing set respectively. In the meta-training set, the sample feature vectors belonging to each category are averaged to obtain the prototype vector of each category. In the training phase, the prototype vector of the meta-training set is spliced with the sample feature vector of the meta-testing set, and input into the feature comparison network (Right of Fig. 2) for classification. The feature comparison network contains two convolution layers, batch standardization, pooling layer and fully connected layer, and outputs the category prediction of network traffic data.

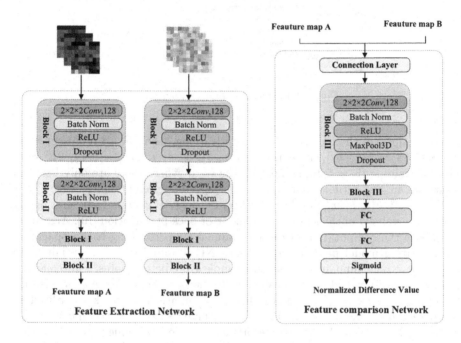

Fig. 2. Network structure of feature extraction and feature comparison.

3   Meta-learning Training Network Process

Through the training of multi-component training set tasks, the model learns meta-knowledge unrelated to specific tasks but related to general discrimination ability. The meta-test set is used to simulate the tasks that the model ultimately needs to deal with, that is, to identify unknown malicious traffic through K samples. In the meta-training phase, the samples in the query set and the sample set are compared one by one, and the average NDV of the Normal and Attack samples in the query set and the sample set is calculated. The prediction label of the sample in the final query set is the sample label with the smallest average NDV difference compared with the Sample Set. The meta-testing phase is the same, and the classification process is shown in Fig 3.

Algorithm 1 uses F and C to represent feature extraction and feature comparison network respectively. Algorithm 1 gives an example of how to train

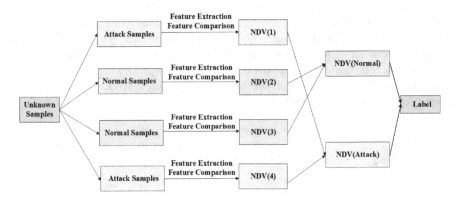

Fig. 3. Classification process diagram.

3D-CNN through the task based on meta-learning, using Ss to represent Sample Set, $Ss = \{(x_1, y_1), (x_2, y_2), ..., (x_K, y_K)\}$, $x_i \in R^d, y_i \in \{0, 1\}$, $2K$ samples for training, $K = 5, 10$, simulating the scene where there are only ' several ' samples in the actual environment. Use Qs to represent Query Set, $Qs = \{(x_1^q, y_1^q), (x_2^q, y_2^q), ..., (x_B^q, y_B^q)\}$, $x_i \in R^d, y_i \in \{0, 1\}$, B sample tests, and finally output the value J of the loss function for error back propagation [10]. Algorithm 1 employs the computation of the loss value J for performing backpropagation and updating model parameters. Initially, NDV0 and NDV1 are initialized to 0, serving to accumulate the current sample's predicted scores. Subsequently, K samples are iterated over to calculate the feature maps A0, A1, and B* for Ss and Qs, respectively. Then, the feature comparison network is utilized to compute the predicted scores. These K predicted scores are accumulated into NDV0 and NDV1, and the average predicted score is obtained. By comparing the magnitudes of NDV0 and NDV1, the final predicted class for the samples in Qs is selected as the one with a smaller average NDV when compared to Ss. The mean squared error between the predicted results and the true labels is computed, and the result is added to the overall loss J. Ultimately, the algorithm outputs the total loss J during the training process, which is used to update the model parameters. Through repeated iterations of this training loop, the model progressively learns and optimizes its predictive capabilities, thereby achieving more accurate classification of unknown samples. As the final output of the 3D-CNN is NDV, the Mean Squared Error (MSE) function is used as the loss function in the training process. This function can calculate the degree of closeness between the predicted value $\hat{y}_l$ and the true label y_i, as shown in Formula (1):

$$MSE = \frac{\sum_{i=1}^{n}(y_i - \hat{y}_l)^2}{n} \tag{1}$$

Algorithm 1. Training 3D-CNN through meta-learning.

```
 1: for i  in{1, 2, ..., B} do
 2:     NDV0 ← 0
 3:     NDV1 ← 0
 4:     for j  in{1, 2, ..., K} do
 5:         Calculate feature map A0 ← F(Ss(0))
 6:         Calculate feature map A1 ← F(Ss(1))
 7:         Calculate feature map B∗ ← (x_i^q)
 8:         NDV0 ← NDV0 + C(A0, B∗)
 9:         NDV1 ← NDV1 + C(A1, B∗)
10:     end for
11:     NDV0 ← NDV0/K
12:     NDV1 ← NDV1/K
13:     if NDV0 < NDV1 then
14:         Predict = 0
15:     else
16:         Predict = 1
17:     end if
18:     J ← J + MSE(Predict, y_i^q)
19: end for
```

The time complexity of algorithm 1 depends on the time complexity of feature extraction network and feature comparison network. For the feature extraction network, the time complexity of each convolution layer is $O(KS^3 * D * H * W * C)$, where KS is the size of the convolution kernel, D, H and W are the depth, height and width of the input graph, and C is the number of channels. For the feature comparison network, the time complexity of the fully connected layer is $O(M^2)$, where M is the input size of the fully connected layer, so the time complexity of the feature comparison network is $O(KS^3 * D * H * W * C + M^2)$. For Algorithm 1, KS, D, H, W, etc. are constants, so the time complexity of the feature extraction network and the feature comparison network can be regarded as $O(1)$. Since the main loop of algorithm 1 executes B * K times, the time complexity of Algorithm 1 is $O(B*K)$, where B is the batch size and K is the number of samples. The space complexity of algorithm 1 depends on the space complexity of feature extraction network and feature comparison network. The Batch Normalization layer of the feature extraction network needs to save the mean and variance on each channel, so the space complexity of the Batch Normalization layer is $O(C)$. Therefore, the overall space complexity of feature extraction network is $O(C)$. The parameters of the fully connected layer in the feature comparison network need to be saved, so the space complexity of the fully connected layer is $O(M^2)$. Therefore, the overall space complexity of feature comparison network is $O(M^2)$. The overall algorithm needs to save the parameters of feature extraction network and feature comparison network, so the overall space complexity of the algorithm is $O(C + M^2)$, which can be simplified to $O(1)$.

4   Simulation and Results Evaluation

In order to protect privacy, many vehicle companies do not publish in-vehicle intrusion detection data. Therefore, we use the well-known IoV open source dataset-Car-Hacking dataset [16] to represent in-vehicle traffic data, which is generated by transmitting CAN data packets to the CAN bus of a real vehicle. The Car-Hacking dataset involves four main types of attacks: Denial of Service (DoS) Attack, Fuzzy Attack, Gear Spoofing Attack and RPM Spoofing Attack. There are many off-board network traffic datasets for intrusion detection, such as NSL-KDD, UNSW-NB15, KDD99, etc. Some researchers have noted that the types of attacks considered in traditional intrusion detection datasets are now outdated [20]. In contrast, the attack type of ISCX2012 is more modern and closer to reality. In addition, the percentage of attack traffic is about 2.8%, which makes ISCX2012 similar to real-world datasets [5]. Therefore, the ISCX2012 data set is used to represent the network traffic outside the car [18]. This dataset contains seven days of raw network traffic data, including normal traffic and four types of attack traffic: BruteForce SSH (BF) Attack, DDoS using an IRC Botnet (DDoS) Attack, HTTP Denial of Service (HttpDoS) Attack and Infiltrating the network form inside (Infiltrating) Attack.

4.1   Metrics

The classification accuracy (ACC), detection rate (DR), precision (PRE), false alarm rate (FPR) and F-Score are expressed as:

$$Accuracy = \frac{TP + TN}{TP + TN + FN + FP} \tag{2}$$

$$DetectionRate = \frac{TP}{TP + FN} \tag{3}$$

$$Precision = \frac{TP}{TP + FP} \tag{4}$$

$$DetectionRate = \frac{FP}{FP + TN} \tag{5}$$

$$F - Score = \frac{2 \times PRE \times DR}{PRE + DR} \tag{6}$$

In the field of IoV intrusion detection, attack samples are defined as positive samples, and normal samples are defined as negative samples. The results of the classification are shown in Table 2:

Table 2. Classified results

Predicted label	Real label	
	Positive samples	Negative samples
Positive samples	Ture Positive, TP	False Positive, FP
Negative samples	False Negative, FN	Ture Negative, TN

4.2 Experimental Results and Comparative Analysis

The experiments were conducted on the following software and hardware platforms: the hardware environment included an Intel(R) Xeon(R) Silver 4110 CPU @2.10 GHz and an NVIDIA Quadro P4000 graphics card; the software environment included Ubuntu 16.04LTS, CUDA10.2, cuDNN7.5, and PyTorch3.7.1.

Type I Experiments. Type I experiments were performed on Car-Hacking and ISCX2012 datasets. Taking Car-Hacking as an example, three of the four types of attacks on the dataset are taken as the data source of the meta-training set, and the remaining one is the data source of the meta-testing set. Since the training and test categories are rotated in turn, there are four groups of parallel experiments. Each group of experiments is repeated for 2000 times to take the average, so as to construct enough few-shot training tasks and test tasks to ensure the fairness of the experiment.

Table 3. Experimental results on Car-Hacking

Type	k = 5					k = 10				
	Metrics (%)									
	ACC	DR	FPR	PRE	F-Score	ACC	DR	FPR	PRE	F-Score
RPM	99.755	99.904	0.283	99.613	99.732	99.755	99.904	0.283	99.613	99.732
Gear	98.321	98.781	1.295	98.404	98.177	98.321	98.781	1.295	98.404	98.177
DoS	96.175	98.014	3.435	94.882	95.829	96.175	98.014	3.435	94.882	95.829
Fuzzy	98.963	99.270	0.942	98.768	98.884	98.963	99.270	0.942	98.768	98.884
Overall	98.310	98.992	1.489	97.917	98.155	98.310	98.992	1.489	97.917	98.155

Table 4. Experimental results on ISCX2012

Type	k = 5					k = 10				
	Metrics (%)									
	ACC	DR	FPR	PRE	F-Score	ACC	DR	FPR	PRE	F-Score
Infiltrating	96.335	96.230	3.407	96.581	96.369	96.397	96.124	3.101	96.901	96.749
HTTPDoS	97.834	99.191	3.354	96.562	97.831	97.702	98.952	3.252	96.655	97.742
DDoS	97.081	99.527	3.045	94.631	96.981	96.887	99.146	4.960	94.728	96.816
Brute Force	97.559	99.759	4.317	96.299	97.492	97.206	99.476	4.691	94.969	97.100
Overall	97.199	98.676	4.307	96.018	97.168	97.048	98.425	4.001	95.813	97.102

The experimental results of Type I experiments on the Car-Hacking dataset are shown in Table 3, and the experimental results on the ISCX2012 dataset are shown in Table 4. The following conclusions can be drawn:

(1) Comparing Table 3 and Table 4, the model is not sensitive to the number of samples K. K is 5 or 10 is a few-shot scenario, and the number of samples is doubled, but the experimental results of the proposed method are not much different under different sample sizes.

(2) Comparing Table 3, it can be seen that regardless of the number of samples, the model used achieves a higher evaluation index on RPM Spoofing Attack. When $K = 5$, the model achieves a detection rate of 99.904% on RPM Spoofing Attack. By comparing Table 4, it can be seen that the evaluation indicators of the four types of attacks, Infiltration, HTTPDoS, DDoS and Brute Force, mostly fluctuate within 1%, with no significant changes.

(3) Comparing Table 3 and Table 4, the evaluation index of the method on Car-Hacking is better than ISCX2012. This shows that Car-Hacking is a more suitable data set for meta-learning than ISCX2012.

Type II Experiments. Type II experiments are based on Type I experiments, using two data sets for cross experiments. For example, three types of attacks in Car-Hacking are selected as the data source of the meta-training set to detect four types of attacks in ISCX2012. Since these two datasets come from networks composed of different hardware and software environments, the types of attacks are different and the detection is difficult, which can fully test the proposed IoV intrusion detection ability based on meta-learning.

Table 5. Experimental results of the training task for ISCX2012, test task for Car-Hacking

	k = 5					k = 10				
	Metrics(%)									
Type	ACC	DR	FPR	PRE	F-Score	ACC	DR	FPR	PRE	F-Score
RPM	93.083	94.916	8.761	93.809	93.384	93.204	95.785	6.683	91.624	92.708
Gear	89.646	96.177	12.855	83.644	88.301	90.457	97.394	11.784	84.055	88.735
DoS	97.971	98.149	1.041	98.795	98.078	97.924	97.748	0.570	99.520	98.276
Fuzzy	90.977	92.176	7.319	93.698	91.156	95.609	96.106	3.456	95.833	95.529
Overall	92.919	95.355	7.494	92.479	92.730	94.298	96.758	5.598	92.758	93.812

Table 6. Experimental results of the training task for Car-Hacking, test task for ISCX2012.

	k = 5					k = 10				
	Metrics(%)									
Type	ACC	DR	FPR	PRE	F-Score	ACC	DR	FPR	PRE	F-Score
Infiltrating	99.746	99.903	0.286	99.602	99.737	99.029	99.604	0.903	98.579	98.918
HTTPDoS	98.470	98.917	1.264	98.467	98.406	97.231	97.684	1.872	98.234	97.536
DDoS	96.071	97.870	3.292	95.134	95.676	94.152	95.837	4.491	94.085	94.210
Brute Force	99.110	99.320	0.812	99.025	99.025	97.848	98.443	1.700	97.590	97.778
Overall	98.349	99.003	1.413	98.057	98.211	97.065	97.892	2.2425	97.122	97.111

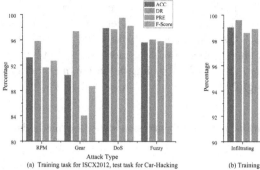 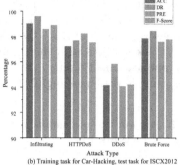

(a) Training task for ISCX2012, test task for Car-Hacking   (b) Training task for Car-Hacking, test task for ISCX2012

Fig. 4. Experimental results of type II experiments($K = 10$).

The type II experiments are experiments that cross-use the Car-Hacking and ISCX2012 datasets. Trained on ISCX2012, the experimental results tested on Car-Hacking are shown in Table 5 and Fig. 4(a). Trained on Car-Hacking, the experimental results tested on ISCX2012 are shown in Table 6 and Fig. 4(b). All the test classes of type II experiments are not involved in the training, so the ability of IoV intrusion detection based on meta-learning can be well tested. From type II experiments, the following conclusions can be drawn:

(1) Although ISCX2012 and Car-Hacking datasets are collected through experimental networks composed of different software and hardware environments, they essentially belong to the same category of IoV, and the network traffic also contains certain commonalities, which verifies that the proposed meta-learning-based IoV intrusion detection method is universal and not limited to a single dataset or attack type.

(2) Comparing Fig. 4, it can be seen that whether K is 5 or 10, training on Car-Hacking, testing on ISCX2012 is better than training on ISCX2012, and testing on Car-Hacking. The evaluation indicators are better, further indicating that Car-Hacking is a data set more suitable for meta-learning than ISCX2012.

Model Comparison and Analysis. IoV intrusion detection is still a relatively new research field in the few-shot scenario, and only a few related research results can be referenced. Therefore, we summarize several recent IoV intrusion detection research results using Car-Hacking and ISCX2012 datasets. The results show that these methods use hundreds of thousands or even millions of samples for training, as shown in Table 7.

Most of the methods and models used in the table can only detect existing attacks through training data. Only the MTH-IDS [24] model is similar to the work of this paper. It effectively detects known and unknown attacks by combining signature-based IDS and anomaly-based IDS. From Table 7, it can be concluded that the method used in this paper achieves a smaller and higher average DR than the MTH-IDS model sample. A large number of research results on IoV intrusion detection are based on a large number of samples. However, in the real network environment, it is difficult for vehicle safety agencies to obtain enough attack samples in a short time, and it is more difficult to make data sets for publication. Therefore, the IoV intrusion detection method based on meta-learning is very meaningful.

Table 7. Performance evaluation of VANET intrusion detection methods on Car-Hacking and ISCX2012)

Model	Datasets	Detect the type of attack	number of samples	ACC (%)	DR (%)	FPR (%)	PRE (%)	F-Score (%)
DCNN [19]	Car-Hacking	known	1,162,188	99.93	99.84	0.16	–	99.91
MTH-IDS [24]	Car-Hacking	known	5,138,977	–	99.999	0.0006	–	99.999
MTH-IDS [24]	Car-Hacking	unknown	5,138,977	–	93.740	0.128	–	96.307
SVM [1]	Car-Hacking	known	3,665,771	96.5	95.7	4.8	–	93.3
KNN [1]	Car-Hacking	known	3,838,860	97.4	96.3	5.3	–	93.4
CNN [21]	ISCX2012	known	915,695	99.69	96.91	0.22	–	–
CNN+LSTM [15]	ISCX2012	known	3,320,315	99.09	99.08	2.27	99.10	99.09
3D-CNN	Car-Hacking	known	5	98.310	98.992	1.489	97.917	98.155
3D-CNN	Car-Hacking	known	10	99.038	99.565	0.906	98.710	98.957
3D-CNN	Car-Hacking	unknown	5	92.919	95.355	7.494	92.479	92.730
3D-CNN	Car-Hacking	unknown	10	94.298	96.758	5.598	92.758	93.812
3D-CNN	ISCX2012	known	5	97.199	98.676	4.307	96.018	97.168
3D-CNN	ISCX2012	known	10	97.048	98.425	4.001	95.813	97.102
3D-CNN	ISCX2012	unknown	5	98.349	99.003	1.413	98.057	98.211
3D-CNN	ISCX2012	unknown	10	97.065	97.892	2.2425	97.122	97.111

5   Conclusion

In order to enhance the network security of IoV, an IoV intrusion detection method based on meta-learning and 3D-CNN is proposed. This method combines normal and attack IoV traffic into a pair of data streams, and then learns a pair of eigenvalues of traffic classification through the feature extraction network composed of 3D-CNN, and compares the two pairs of traffic samples as the basic task. Due to the training of several few-shot tasks, sufficient prior knowledge can be obtained to detect unknown traffic types with fewer samples.

In order to evaluate the proposed detection method, two open source IoV intrusion detection datasets are used to convert them into sample formats suitable for input into 3D-CNN, and two types of experiments are carried out. The experimental results of training and testing on the same data set show that the performance of this method is close to the working type in the large sample scenario, reaching 98.992% average DR; the experimental results of cross-use data sets for training and testing show that this method can identify unknown IoV intrusion detection traffic by training known network traffic and achieve an average DR of 99.003%.

References

1. Alshammari, A., Zohdy, M.A., Debnath, D., Corser, G.: Classification approach for intrusion detection in vehicle systems. Wirel. Eng. Technol. **9**(4), 79–94 (2018)
2. Chatzidakis, M., Hadjiefthymiades, S.: A trust change detection mechanism in mobile ad-hoc networks. Comput. Commun. **187**, 155–163 (2022)
3. Chen, X.: A simple utility to classify packets into flows (2017)
4. Cui, J., Ma, L., Wang, R., Liu, M.: Research and optimization of GPSR routing protocol for vehicular ad-hoc network. China Commun. **19**(10), 194–206 (2022)
5. Ghanem, W.A.H., et al.: Cyber intrusion detection system based on a multiobjective binary bat algorithm for feature selection and enhanced bat algorithm for parameter optimization in neural networks. IEEE Access **10**, 76318–76339 (2022)
6. Hossain, M.D., Inoue, H., Ochiai, H., Fall, D., Kadobayashi, Y.: An effective in-vehicle can bus intrusion detection system using cnn deep learning approach. In: GLOBECOM 2020–2020 IEEE Global Communications Conference, pp. 1–6. IEEE (2020)
7. Ji, S., Xu, W., Yang, M., Yu, K.: 3d convolutional neural networks for human action recognition. IEEE Trans. Pattern Anal. Mach. Intell. **35**(1), 221–231 (2012)
8. Kaur, G., Kakkar, D.: Hybrid optimization enabled trust-based secure routing with deep learning-based attack detection in vanet. Ad Hoc Netw. **136**, 102961 (2022)
9. Khan, I.A., Moustafa, N., Pi, D., Haider, W., Li, B., Jolfaei, A.: An enhanced multistage deep learning framework for detecting malicious activities from autonomous vehicles. IEEE Trans. Intell. Transp. Syst. **23**(12), 25469–25478 (2021)
10. Kingma, D.P., Ba, J.: Adam: a method for stochastic optimization. arXiv preprint arXiv:1412.6980 (2014)
11. Ma, W., Zhang, Y., Guo, J., Yu, Q.: Few-shot abnormal network traffic detection based on multi-scale deep-capsnet and adversarial reconstruction. Int. J. Comput. Intell. Syst. **14**(1), 195 (2021)
12. Mabrouk, A., Naja, A.: Intrusion detection game for ubiquitous security in vehicular networks: a signaling game based approach. Comput. Netw. 109649 (2023)
13. Mchergui, A., Moulahi, T., Zeadally, S.: Survey on artificial intelligence (AI) techniques for vehicular ad-hoc networks (vanets). Veh. Commun. **34**, 100403 (2022)
14. Naqvi, I., Chaudhary, A., Rana, A.: Intrusion detection in vanets. In: 2021 9th International Conference on Reliability, Infocom Technologies and Optimization (Trends and Future Directions)(ICRITO), pp. 1–5. IEEE (2021)
15. Pektaş, A., Acarman, T.: A deep learning method to detect network intrusion through flow-based features. Int. J. Netw. Manag. **29**(3), e2050 (2019)
16. Seo, E., Song, H.M., Kim, H.K.: Gids: gan based intrusion detection system for in-vehicle network. In: 2018 16th Annual Conference on Privacy, Security and Trust (PST), pp. 1–6. IEEE (2018)
17. Shams, E.A., Rizaner, A., Ulusoy, A.H.: Flow-based intrusion detection system in vehicular ad hoc network using context-aware feature extraction. Veh. Commun. 100585 (2023)
18. Shiravi, A., Shiravi, H., Tavallaee, M., Ghorbani, A.A.: Toward developing a systematic approach to generate benchmark datasets for intrusion detection. Comput. Secur. **31**(3), 357–374 (2012)
19. Song, H.M., Woo, J., Kim, H.K.: In-vehicle network intrusion detection using deep convolutional neural network. Veh. Commun. **21**, 100198 (2020)

20. Suthishni, D.N.P., Kumar, K.S.: A review on machine learning based security approaches in intrusion detection system. In: 2022 9th International Conference on Computing for Sustainable Global Development (INDIACom), pp. 341–348. IEEE (2022)
21. Wang, W., et al.: Hast-ids: learning hierarchical spatial-temporal features using deep neural networks to improve intrusion detection. IEEE Access **6**, 1792–1806 (2017)
22. Xu, C., Shen, J., Du, X.: A method of few-shot network intrusion detection based on meta-learning framework. IEEE Trans. Inf. Forensics Secur. **15**, 3540–3552 (2020)
23. Yang, J., Li, H., Shao, S., Zou, F., Wu, Y.: FS-IDS: a framework for intrusion detection based on few-shot learning. Comput. Secur. **122**, 102899 (2022)
24. Yang, L., Moubayed, A., Shami, A.: MTH-IDS: a multitiered hybrid intrusion detection system for internet of vehicles. IEEE Internet Things J. **9**(1), 616–632 (2021)

Noise-Robust Gaussian Distribution Based Imbalanced Oversampling

Xuetao Shao and Yuanting Yan$^{(\boxtimes)}$ (iD)

Artificial Intelligence Institute, School of Computer Science and Technology,
Anhui University, Hefei 230601, Anhui, People's Republic of China
ytyan@ahu.edu.cn

Abstract. Imbalanced data classification has become one of the hot topics in the field of data mining and machine learning. Oversampling is one of the mainstream methods to solve the imbalance problem by synthesizing new samples to balance the data distribution. However, due to the limited sample local information, the data synthetic process is risky in deteriorating the class overlap phenomenon, showing a vulnerable robustness with respect to data noise. In this paper, we propose a noise robust gaussian distribution based imbalanced oversampling (NGOS). NGOS first determines the neighborhood radius based on the global information, and then assigns sampling weights to minority class samples based on the density and the distance information within each of the neighborhoods. Finally, NGOS generates new samples with a Gaussian distribution model. We validate the effectiveness of our proposed method on the 38 KEEL datasets, DT classifier and eleven comparison methods. Experimental results show that our method outperforms the other compared methods in terms of Fmeasure, AUC, Gmean. The codes of NGOS are released in https://github.com/ytyancp/NGOS.

Keywords: Imbalanced data classification · Oversampling · Noise · Gaussian distribution

1 Introduction

Imbalanced data classification has become a hot topic in the fields of data mining and machine learning. Data imbalance poses a great challenge to the robustness of traditional classification algorithms. And they are widespread in fields such as fraud detection [12], network intrusion monitoring [17], software detect prediction [6]. Researchers have proposed a variety of methods for learning imbalanced data, which can be roughly divided into two categories: data-level methods, algorithm-level methods [8]. Algorithm-level methods mainly adapt classifiers specifically designed for imbalanced data or improve traditional classifiers to

This work was supported in part by the National Natural Science Foundation of China under Grant 62376002.

make them suitable for imbalanced data. Data-level methods mainly resample the imbalanced dataset by adding minority class samples (oversampling) or removing majority class samples (undersampling) to balance the dataset.

Data-level methods have become the mainstream method for solving imbalanced data classification problems due to their simplicity, efficiency, and independence of subsequent classifiers [16]. Data-level methods are mainly divided into undersampling, oversampling, and hybrid sampling [8]. Undersampling achieves balance by removing some of the majority class samples. Oversampling synthesizes minority class samples to balance the dataset. Hybrid sampling combines the above two strategies to achieve better learning results. Recent studies have shown that oversampling methods are significantly superior to undersampling methods on traditional classifiers because they provide a higher proportion of safe samples while reducing the proportion of non-safe samples [7].

SMOTE [4] is the most classic oversampling method, but its mechanism of randomly selecting the nearest neighbors of minority class samples for linear interpolation to generate new samples ignores the distribution information of the samples. To address the shortcomings of SMOTE, researchers have proposed many oversampling methods in recent years. These include Borderline-SMOTE [9], which emphasizes synthesizing samples in the boundary region, Safe-Level-SMOTE [3], which emphasize synthesizing samples in the safe region. Unlike the above-mentioned method that only utilizes minority class information, GDO [20] utilizes the density and distance information of both the majority and minority classes to weight the minority class samples, simultaneously. However, these methods either overemphasize synthesizing samples in specific areas, leading to overfitting, or overemphasize preserving the original data distribution and ignore the adverse effects of noisy samples on the classification model.

To address these problems, this paper proposes a noise-robust gaussian distribution based imbalanced oversampling (NGOS). NGOS uses an adaptive neighborhood determination method to mine sample neighborhood information and introduces information entropy to measure the uncertainty of different sample distributions within the neighborhood to reduce the sampling rate of highly overlapping samples (even noise samples) and reduce the risk of introducing additional class overlap and noise samples. To avoid oversampling of the minority class being too concentrated in the boundary region, the method combines the distance information between the minority and majority classes in the neighborhood to expand the potential space for synthesizing minority class samples.

The main contributions of this paper are summarized as follows:

- A noise-robust oversampling method (NGOS)based on Gaussian distribution for imbalanced data is proposed.
- NGOS enhances the robustness of the minority oversampling model to noise by introducing a fixed neighborhood information mining method and information entropy, and reduces the risk of introducing additional class overlap and noise samples by reducing the sampling rate of highly overlapping samples (even noise samples).

- NGOS expands the potential space for synthesizing minority class samples by combining the distance information between the minority and majority classes in the neighborhood, and properly synthesizes new samples in the safe region to avoid overfitting problems in the boundary region.
- We evaluate the performance of NGOS on 38 KEEL datasets by comparing it with 11 data-level methods. The experimental results show that we achieved the best performance in terms of Fmeasure, AUC, and Gmean.

The rest of this paper is organized as follows: Sect. 2 introduces related work and the GDO algorithm. Section 3 proposes the NGOS algorithm. Section 4 presents experimental comparisons and analyses. Section 5 concludes the paper.

2   Related Work

2.1   Resampling Methods

Undersampling methods balance the dataset by removing some majority class samples. SDUS [22] uses a supervised constructive process to learn majority-class local patterns in terms of sphere neighborhoods (SPN) to maintain the distribution pattern of original data in selecting majority-class sample subsets from different perspectives. RUS [2] randomly removes majority class samples to balance the dataset. It may discard important information. Tomek [11] links identify Tomek pairs where a minority class sample and a majority class sample are mutual nearest neighbors, and remove the majority class samples. ENN [19] removes majority class samples that have mostly minority class samples among their k nearest neighbors. However, they do not explicitly specify the number of samples for removing which may lead to undesired level of data imbalance. CC [15] first clusters the minority class samples and then selects either the centroid or the majority class sample closest to the centroid of each cluster. In addition, RBU [13] performs undersampling by calculating inter-class potentials, which reflect the amount of information contained in the majority class. However, it requires iterative steps, making it slower.

Oversampling methods balance the dataset by synthesizing minority class samples. SMOTE [4] synthesizes minority class samples by randomly selecting seed samples and applying linear interpolation. It may generate a large number of new noisy samples. To address this, researchers started to restrict the selection of seed samples in SMOTE. Borderline-SMOTE [9] confines the seed sample selection to the boundary region, considering samples at the classification boundary more difficult to classify. In contrast, Safe-Level-SMOTE [3] argues that minority class samples located in safe regions are better suited as seed samples, because synthesizing samples in the boundary region is more likely to introduce noisy and overlapping samples. ADASYN [10] adaptively assigns weights for seed sample selection based on the density of sample distributions. MWMOTE [1] combines location and density factors and integrates data clustering to assign weights for minority class samples. These methods are all derived from SMOTE [4], and their synthesis methods use linear interpolation. Therefore, overgeneralization issues may arise during sample synthesis. GDO [20] samples and proposes a new sample synthesis method based on Gaussian models. As mentioned above, it overly emphasizes majority class samples in the weighted selection of seed samples, which may result in the synthesis of noisy samples.

Hybrid sampling methods balance the dataset by combining oversampling methods and undersampling methods, which combines the advantages of both. Most of these methods use SMOTE [4] as the main oversampling process and then combine it with different undersampling methods to balance the dataset. SMOTE+TL [2] and SMOTE+ENN [2] combine SMOTE with Tomek links and ENN, respectively. They first use SMOTE to oversample, and then use Tomek links and ENN for undersampling. However, using SMOTE for synthesis can lead to overgeneralization issues. LDAS [21], which is different from the traditional oversampling-then-undersampling process mentioned above, first cleans the overlap area using undersampling methods, and then synthesizes minority class samples using oversampling methods.

2.2   Gaussian Distribution Based Oversampling (GDO)

GDO [20] believes that different minority classes carry different information, so it considers both density and distance information to assign different weights for selecting seed samples to different minority classes.

The sample selection weight factor of GDO is shown in Eq. (1). Where $C(x_i)$ represents the proportion of majority class samples in the K nearest neighbors, and $D(x_i)$ represents the proportion of the distance between majority class samples and the total distance in the K nearest neighbors.

$$I(x_i) = C(x_i) + D(x_i) \tag{1}$$

Then, the weights are normalized as shown in Eq. (2):

$$\widehat{I}(x_i) = \frac{I(x_i)}{\sum_{i=1}^{|N^{min}|} I(x_i)} \tag{2}$$

Where $|N^{min}|$ represents the number of minority class samples.

Let o be the origin of the coordinates, and for any seed sample x_i, a random vector $\overrightarrow{ov}$ is generated. Then, $\overrightarrow{x_iv}$ is the direction vector, and the newly synthesized sample point is on this direction.

$$\overrightarrow{x_iv} = \overrightarrow{ov} - \overrightarrow{ox_i} \tag{3}$$

Next, the length of vector $\overrightarrow{x_ix'}$ is determined, which follows the Gaussian distribution:

$$|\overrightarrow{x_ix'}| = d_i \sim N(\mu_i, \alpha\sigma_i) \tag{4}$$

Where $\mu_i = 0$ and σ_i is the Euclidean distance between the seed sample x_i and its nearest same-class sample.

Therefore, the vector form of the newly synthesized sample is:

$$\overrightarrow{ox'} = \overrightarrow{ox_i} + \frac{|\overrightarrow{x_i x'}|}{|\overrightarrow{x_i v}|} \cdot \overrightarrow{x_i v} \tag{5}$$

3 Proposed Method

3.1 Analysis of the GDO Algorithm

The GDO algorithm relies on K-nearest neighbor (KNN) calculation to obtain local distribution information. However, in imbalanced data, the majority class is the dominant one in the sample space. Therefore, the decision process based on K is prone to bias towards the majority class, and the method has poor robustness to noise. In addition, to achieve better performance for different data distributions, it is usually necessary to find suitable parameter K, which makes the algorithm less adaptable.

As shown in Fig. 1, when the parameter K in the K-nearest neighbor calculation is set to 5, the weight of sample A calculated by Eq. (1) is 2 (each sample can obtain the maximum weight value). Therefore, sample A has the highest probability of being selected as the seed sample, but synthesizing minority samples based on sample A as the seed sample will further increase the difficulty of classification. In addition, GDO believes that samples in the safe region are easier to identify and ignores these samples. This causes the sampling process to concentrate too much on the boundary area, which may cause overfitting.

As shown in Fig. 1, both B and C have a weight of 0 (they will not be selected as seed samples). However, compared with C, it can be seen that sample B is clearly closer to the decision boundary. Synthesizing samples based on B can strengthen the classification boundary to a certain extent and expand the potential generation space of synthesized samples, which can avoid the potential overfitting problem caused by synthesizing too many samples in the boundary area and improve the subsequent learning performance.

To address the above issues, this paper proposes an improved Gaussian sampling method, NGOS, which uses a fixed-radius neighborhood partition method and an information entropy-based neighborhood information measurement method to enhance the performance of imbalance learning.

3.2 Local Information of Samples

The KNN method measures the local distribution information of samples by finding their K-nearest neighbors. As shown in Fig. 1(a), this method cannot effectively characterize the differences in sample distributions, as it only considers the relationship between the target sample and its nearest neighbors, ignoring the local distribution information of its neighbors. To address this issue, this

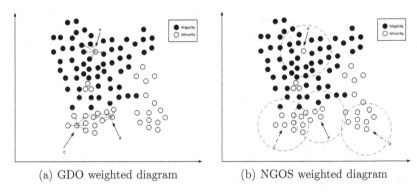

(a) GDO weighted diagram   (b) NGOS weighted diagram

Fig. 1. Comparison of the weighting schemes between GDO and NGOS

paper utilizes the global information of sample distributions to achieve adaptive determination of neighborhoods.

$$R = 2 \sum_{x_i, x_j \in X_{train}} dist(x_i, x_j)/(|N_{train}|(|N_{train}| - 1)) \tag{6}$$

Where $|N_{train}|$ is the number of samples in the training set, and $dist(x_i, x_j)$ is the Euclidean distance between sample x_i and x_j. It can be seen that this radius R considers the global distribution information of the samples. With R as the radius and the target sample x_{min} as the center, we obtain a subset X_{candi} of all samples whose distance to x_{min} is less than R.

$$X_{candi} = \{x_p \in X_{train}, dist(x_{min}, x_p) < R\} \tag{7}$$

Based on X_{candi}, it is easy to evaluate the data distribution within the sample neighborhood.

3.3   Estimation of Weight for Minority Class Sample Selection

In classification tasks, samples located in different regions have different impacts on the classification model [18]. To characterize the impact of different samples on the classification model, GDO uses the KNN method to characterize the importance of minority class samples. Although this method assigns high weights to boundary samples, it also overly emphasizes minority class samples located in dense majority class areas, which leads to the synthesis of a large number of potential noise samples. In other words, the GDO is not robust to noise samples. To address this issue, this paper uses information entropy [5] and distribution density to measure the samples and enhance the robustness of the model to noise.

Specifically, we first measure the distribution differences within the local neighborhood of a sample using the Eq. (8):

$$p_i = \left| N_{i_{candi}}^{maj} \right| / (\left| N_{i_{candi}}^{maj} \right| + \left| N_{i_{candi}}^{min} \right|) \tag{8}$$

Where $\left| N_{i_{candi}}^{min} \right|$ represents the number of minority class samples in the candidate set of sample x_i, and $\left| N_{i_{candi}}^{maj} \right|$ represents the number of majority class samples in the candidate set of sample x_i.

$$E(x_i) = -p_i \log{_2}p_i - (1 - p_i) \log{_2}(1 - p_i) \tag{9}$$

From Eq. (9), it can be seen that when $p_i = 1/2$, Eq. (9) obtains the maximum value of 1, and gradually decreases as p_i decreases or increases. When $p_i = 0$ or 1, we set the value to 0. In other words, the closer a sample is to the decision boundary, the greater its weight, and the further it goes into the majority class area, the smaller its weight. As shown in Fig. 1(b), all the neighboring samples of sample A are of different classes (sample A is a noise sample), and synthesizing samples at this position will increase the difficulty of training the classifier. Therefore, we set its weight to 0 according to Eq. (9). Similarly, sample D located in the safe area has neighboring samples that are all of the same class and can be easily identified by the classifier. Therefore, we also set its weight to 0.

However, using Eq. (9) alone will overly focus on samples with high uncertainty in the boundary area, in other words, it will assign higher weights to samples with higher uncertainty, which may lead to overemphasizing such samples and causing overfitting problems. Therefore, NGOS introduces distance information of the samples within the neighborhood, appropriately expands the selection range of seed samples, increases the synthesis space of potential synthesized samples, and enhances the robustness of the model. To achieve this, NGOS proposes the following distance measurement method:

$$D^{'}(x_i) = \frac{\frac{\sum_{x_j \in X_{i_{candi}}^{maj}} dist(x_i, x_j)}{\left| N_{i_{candi}}^{maj} \right|}}{\frac{\sum_{x_j \in X_{i_{candi}}^{maj}} dist(x_i, x_j)}{\left| N_{i_{candi}}^{maj} \right|} + \frac{\sum_{x_j \in X_{i_{candi}}^{min}} dist(x_i, x_j)}{\left| N_{i_{candi}}^{min} \right|}} \tag{10}$$

When $\left| N_{i_{candi}}^{maj} \right| = 0$ or $\left| N_{i_{candi}}^{min} \right| = 0$, it means that the candidate set of the sample only contains majority class samples (such as sample A in Fig. 2) or minority class samples (such as sample D in Fig. 2). In these cases, the weight $D^{'}(x_i)$ of the sample is set to 0. It can be seen that Eq. (10) uses distance information between samples to select seed samples for minority class synthesis that are farther away from the decision boundary for unstable samples (i.e., samples with different classes in their neighborhoods), thus avoiding overfitting.

By considering both density and distance factors, the following method is proposed to calculate the weight of each sample:

$$I(x_i) = D^{'}(x_i) + E(x_i) \tag{11}$$

From Eq. (11), it can be seen that the weight of minority samples that are deep in the majority class area is 0, while the weight of samples located at or close to the decision boundary is relatively high.

3.4   Probabilistic Seed Sample Selection and Time Complexity

After calculating the weight of each minority sample using Eq. (11), we normalize the weights using Eq. (2) to convert them into probabilities. The selection of seed samples and generation of new samples follow the iterative process below: at each iteration, a seed sample is chosen based on its probability $\widehat{I}$, and new minority samples are synthesized based on the chosen seed sample. This process continues until the number of minority samples is equal to the number of majority samples. The number of samples to be synthesized is determined by the Eq. (12).

$$G = \left|N^{maj}\right| - \left|N^{min}\right| \tag{12}$$

The process of NGOS is described in Algorithm 1. To calculate Eq. (9) and Eq. (10), we first need to compute the candidate set of samples x_i, which has a time complexity $O(|N_{train}|)$. Then, each minority class sample needs to be calculated, resulting in a time complexity $O(|N_{train}||N^{min}|)$ for the minority class weighting process (lines 2–8). The data generating process (lines 14–18), the minority class instances are resampled G times and the time complexity is $O(|G|)$. Because G is smaller than N_{train}, the time complexity of Algorithm 1 is $O(|N_{train}||N^{min}|)$.

Algorithm 1 . NGOS(α)

Input: the original dataset D , scaling factor α;
Output: balanced dataset S;
 1: Divide into minority class D_{min} and majority class D_{maj};
 2: for x_i in D_{min}:
 3:    Calculate the radius R; Eq. (6)
 4:    Calculate the candidate set X_{candi} of x_i; //Eq. (7)
 5:    Obtain the density factor weight $E(x_i)$; Eqs. (8) and (9)
 6:    Obtain the distance factor weight $D^{'}(x_i)$; //Eq. (10)
 7:    Calculate the information weight $I(x_i)$; //Eq. (11)
 8: end for
 9: for x_i in D_{min}:
10:    Calculate the normalized weight $\widehat{I}(x_i)$; //Eq. (2)
11: end for
12: Calculate the number of samples needed for balance G; //Eq. (12)
13: Initialize the number of minority class samples to be synthesized $n = 0$;
14: *while $n < G$:*
15:    Synthesizing samples with using Eqs. (3)–(5);
16:    Add the synthesized sample to $D^{'}_{min}$;
17:    $n = n + 1$;
18: end while
19: $S = D \cup D^{'}_{min}$;

---

4   Experiments and Analysis

To validate the effectiveness of our proposed NGOS algorithm, we designed a three-stage experimental study. First, we will briefly introduce the evaluation metrics and settings used in our experiments. Then, we analyzed the influence of algorithm parameters on its performance. Finally, we compare our proposed method with other state-of-the-art resampling methods on the KEEL dataset.

4.1   Experimental Settings

Evaluation Metrics. We use *Fmeasure, Gmean, AUC* (the area under the ROC curve) [16] which are the most frequently used metircs in imbalance learning were applied in this study.

Datasets. Table 1 provides detailed information about the datasets, including the dataset name, the abbreviation of the dataset (Abbr), the number of attributes (Atts), the size of the dataset (Size), the number of samples in the minority class (Min), and the imbalance ratio (IR).

Classifiers. In our experiments, we use Decision Tree (DT) classifiers provided by the scikit-learn library in Python with default parameters. To ensure the correctness of the experimental results, we used 5-fold cross-validation with 10 repetitions for the training and test set split.

Comparison Methods. In our experiments, we compared our proposed NGOS algorithm with 11 other resampling methods, including SMOTE(SMO), Borderline-SMOTE (BSM), ADASYN(ADA), MWMOTE (MWO), SMOTE Tomek links (STL), SMOTE ENN (SENN), GDO, CC, ROS, RUS, RBU.

4.2   Experimental Results and Analysis

Parameter Analysis. In NGOS, when performing oversampling, the length of the synthetic minority class mode d is derived from $N(\mu_i, \alpha\sigma_i)$, where α is a scaling factor to control the sampling density of the seed sample. To investigate the influence of the parameter α on NGOS under different data distributions, we selected 10 datasets. The best value is highlighted in bold.

Table 2 shows the AUC values and their average values for 10 datasets under different parameter values for the DT classifier. The average values indicate that NGOS performs best when α is set to 1.5, with D02, D03, D05, and D12 datasets achieving the best performance at $\alpha = 1.5$. The D14, D24 and D33 datasets achieve the best performance at $\alpha = 1.4$, 6 out of 10 datasets perform best around these values. Therefore, we recommend setting the α to 1.5.

Table 1. Description of KEEL Datasets

Dataset	Abbr	Size	Atts	Min	IR	Dataset	Abbr	Size	Atts	Min	IR
abalone19	D01	4173	9	32	129.41	newthyroid1	D20	214	6	35	5.11
abalone918	D02	730	9	41	16.8	newthyroid2	D21	214	6	35	5.11
car-good	D03	1727	7	69	24.03	pb134	D22	471	11	28	15.82
car-vgood	D04	1727	7	65	25.57	page-blocks0	D23	5471	11	559	8.79
cleveland04	D05	176	14	13	12.54	pima	D24	767	9	267	1.87
dermatology6	D06	357	35	20	16.85	p86	D25	1476	11	17	85.82
e013726	D07	280	8	7	39	p97	D26	243	11	8	29.38
e01	D08	219	8	77	1.84	segment0	D27	2307	20	329	6.01
flare-F	D09	1065	12	43	23.77	s25	D28	3315	10	49	66.65
glass1	D10	213	10	76	1.8	scvc	D29	1828	10	123	13.86
glass5	D11	213	10	9	22.67	vehicle0	D30	845	19	198	3.27
haberman	D12	305	4	81	2.77	vehicle2	D31	845	19	218	2.88
iris0	D13	149	5	49	2.04	vowel0	D32	987	14	89	10.09
kgpvs	D14	1641	42	52	30.56	wr35	D33	690	12	10	68
krivb	D15	2224	42	22	100.09	wr4	D34	1598	12	53	29.15
krvkzovd	D16	2900	7	104	26.88	wisconsin	D35	682	10	239	1.85
kvkzvf	D17	2192	7	27	80.19	yeast1	D36	1483	9	429	2.46
l024567891	D18	442	8	37	10.95	yeast6	D37	1483	9	35	41.37
lnf	D19	147	19	6	23.5	zoo-3	D38	100	17	5	19

Comparison with Other Resampling Methods. This section compares NGOS with 11 resampling methods in Sect. 4.1, which include 6 oversampling methods, 3 undersampling methods, and 2 hybrid methods.

Due to space limited, we only provide the AUC for each dataset. From Table 3, it can be seen that NGOS performs the best overall compared to other comparison methods, achieving the best average values for AUC. For easily classified datasets such as D13, its evaluation metrics also reach 1, like other comparison methods. Additionally, NGOS achieves the best performance on 12 datasets for AUC. SENN achieves the best performance on 13 datasets for AUC. It can be seen that SENN is the biggest competitor of NGOS, although it achieves the best performance on one more dataset than NGOS for AUC, its overall average performance is not as good as NGOS.

Therefore, we use Bayesian analysis [14] to further compare the performance of NGOS and other comparison methods (especially SENN). Unlike other testing methods, Bayesian analysis does not fall into the pitfalls of black and white thinking and could estimate the probability that the performance of two classifiers is different(or equal). Figure 2 shows the corresponding results of Bayesian testing.

As shown in Fig. 2(a) and (c), on the DT classifier, the probability that NGOS outperforms all other comparison methods except SENN is close to 100%, and

Table 2. Influence of parameter α on DT in terms of the AUC metric

Dataset	1.0	1.1	1.2	1.3	1.4	1.5	1.6	1.7	1.8	1.9	2.0
D02	0.6501	0.6533	0.6788	0.6576	0.6567	**0.6856**	0.6460	0.6465	0.6591	0.6794	0.6623
D03	0.9286	0.9227	0.9216	0.9134	0.9137	**0.9390**	0.9240	0.9298	0.9347	0.9273	0.9313
D05	0.7453	0.7472	0.7686	0.7226	0.7498	**0.7853**	0.7313	0.7808	0.7659	0.7738	0.7746
D10	0.7305	**0.7496**	0.7326	0.7418	0.7448	0.7483	0.7071	0.7361	0.7379	0.7251	0.7172
D12	0.5747	0.5763	0.5971	0.5754	0.5858	**0.5974**	0.5842	0.5675	0.5725	0.5886	0.5876
D14	0.9882	0.9861	0.9899	0.9919	**0.9957**	0.9862	0.9919	0.9922	0.9942	0.9884	0.9859
D24	0.6761	0.6712	0.6746	0.6789	**0.6832**	0.6747	0.6729	0.6765	0.6736	0.6727	0.6709
D33	0.6475	0.6848	0.6689	0.6640	**0.7177**	0.6909	0.6586	0.6283	0.7072	0.6587	0.7094
D35	0.9427	0.9434	0.9376	0.9430	0.9370	0.9384	0.9370	0.9431	**0.9443**	0.9383	0.9422
D38	0.6593	0.6432	**0.7201**	0.6379	0.6622	0.6982	0.6591	0.6835	0.7043	0.6863	0.6633
Avg	0.7285	0.7341	0.7432	0.7276	0.7396	**0.7506**	0.7227	0.7328	0.7438	0.7375	0.7376

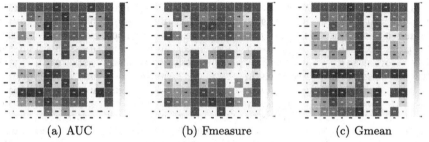

|(a) AUC|(b) Fmeasure|(c) Gmean|

Fig. 2. The value in the i-th row and j-th column represents the probability that the i-th method performs better than the j-th method.

the probability of outperforming SENN is as high as 95%. Although NGOS has one less best dataset than SENN on AUC, its performance on all 38 datasets far exceeds SENN. From Fig. 2(b), it can be seen that the performance of NGOS on Fmeasure is very outstanding, significantly better than other comparison methods, with a probability of almost 100% of outperforming other comparison methods, including SENN, even though, SENN is a hybrid resampling method.

5   Conclusion

This paper proposed the NGOS to addresses several issues of the GDO: 1) GDO emphasizes the majority class in local regions, resulting in the generation of too many synthetic samples around minority class samples deep in the majority class region, introducing more difficult-to-learn samples that hinder the training of the learning model. 2) GDO regards that samples in safe regions are easier to recognize, thus ignoring these samples, but this can lead to oversampling being too concentrated on the boundary region, which increases the risk of overfitting. 3) Both density and distance information in the GDO method rely on the KNN

Table 3. AUC results on KEEL datasets obtained by DT

	NGOS	GDO	ADA	BSM	CC	MWO	ROS	RUS	SMO	SENN	STL	RBU
D01	0.5226	0.5220	0.5460	0.5426	**0.7073**	0.4944	0.5188	0.6448	0.5402	0.5816	0.5739	0.5237
D02	0.6856	0.6643	0.6879	**0.6968**	0.6674	0.6940	0.6111	0.6870	0.6495	0.6920	0.6616	0.6624
D03	0.9390	0.9108	0.7764	0.7994	0.8492	0.8419	0.9549	0.9567	0.8192	0.8214	0.7943	**0.9660**
D04	0.9858	0.9870	0.9920	0.9905	0.8629	0.9890	0.9792	0.9780	**0.9935**	0.9788	0.9766	0.9774
D05	0.7853	0.7533	0.7779	0.7899	0.7088	0.6471	0.6971	0.7800	**0.8051**	0.7746	0.7638	0.7460
D06	**0.9985**	0.9832	0.9835	0.9635	0.9546	0.9885	0.9835	0.9619	0.9885	0.9885	0.9885	0.9544
D07	0.8501	0.8590	0.8294	0.7387	0.8380	0.6087	0.5905	0.7059	**0.8879**	0.8372	0.8198	0.7205
D08	0.9758	0.9663	0.9708	0.9789	0.9732	0.9724	0.9647	0.9695	0.9668	**0.9846**	0.9815	0.9702
D09	0.6227	0.5887	0.5916	0.6100	0.6239	0.6556	0.6214	0.7526	0.5818	**0.7564**	0.6467	0.6167
D10	0.7483	0.7456	0.7186	0.7361	0.7258	**0.7510**	0.7385	0.7345	0.7364	0.7052	0.7365	0.7330
D11	0.9020	0.8551	0.9376	0.8476	0.9293	0.8451	0.9076	0.8507	0.8476	**0.9756**	0.8576	0.8388
D12	0.5971	0.5781	0.5420	0.5633	0.5281	0.5791	0.5726	0.5828	0.5297	**0.6177**	0.5941	0.5437
D13	**1.0000**	**1.0000**	**1.0000**	**1.0000**	**1.0000**	**1.0000**	**1.0000**	**1.0000**	**1.0000**	**1.0000**	**1.0000**	**1.0000**
D14	0.9976	0.9832	0.9979	0.9980	0.9978	0.9978	0.9979	0.9927	0.9978	0.9977	**0.9999**	0.9979
D15	**1.0000**	0.9736	**1.0000**	**1.0000**	0.9787	**1.0000**	**1.0000**	0.9886	**1.0000**	**1.0000**	**1.0000**	0.9992
D16	0.9635	0.9538	0.9446	0.9525	0.9238	0.9579	0.9463	0.9646	0.9526	**0.9883**	0.9586	0.9470
D17	**1.0000**	0.9997	0.9927	**1.0000**	0.9651	**1.0000**	**1.0000**	0.9877	**1.0000**	0.9960	0.9853	0.9982
D18	**0.9043**	0.9038	0.8956	0.8758	0.8619	0.8908	0.8436	0.8485	0.8929	0.8788	0.9021	0.8620
D19	**0.8860**	0.8173	0.7909	0.7900	0.7354	0.5379	0.6539	0.7103	0.8322	0.8250	0.8003	0.5779
D20	0.9530	0.9425	0.9710	0.9607	**0.9767**	0.9574	0.9355	0.9410	0.9356	0.9464	0.9516	0.9210
D21	0.9582	0.9442	**0.9688**	0.9682	0.9617	0.9648	0.9139	0.9506	0.9482	0.9556	0.9453	0.9413
D22	**0.9977**	0.9901	0.9955	0.9837	0.9114	0.9977	0.9784	0.9562	0.9977	0.9644	0.9898	0.9636
D23	0.9252	0.9138	0.9296	0.9194	0.8959	0.9236	0.9065	0.9371	0.9238	**0.9378**	0.9325	0.8888
D24	0.6832	0.6702	0.6756	0.6724	0.6388	0.6782	0.6739	0.6769	0.6621	**0.7057**	0.6690	0.6624
D25	0.9380	**0.9761**	0.7578	0.6379	0.5253	0.5436	0.5071	0.5999	0.7409	0.6634	0.6898	0.6021
D26	**0.8188**	0.6484	0.5426	0.6932	0.6020	0.4907	0.5370	0.6099	0.6041	0.5956	0.6719	0.7427
D27	0.9908	0.9883	0.9863	0.9896	0.9692	0.9887	0.9905	0.9796	**0.9917**	0.9885	0.9882	0.9864
D28	**1.0000**	**1.0000**	**1.0000**	**1.0000**	**1.0000**	**1.0000**	**1.0000**	0.9976	**1.0000**	**1.0000**	**1.0000**	**1.0000**
D29	**1.0000**	0.9996	**1.0000**	**1.0000**	**1.0000**	**1.0000**	**1.0000**	0.9998	**1.0000**	**1.0000**	**1.0000**	**1.0000**
D30	0.9316	0.9133	0.9145	0.9079	**0.9329**	0.9127	0.9076	0.9264	0.9064	0.9164	0.8998	0.9024
D31	0.9527	0.9434	0.9533	**0.9552**	0.9358	0.9472	0.9499	0.9403	0.9394	0.9428	0.9490	0.9526
D32	0.9725	**0.9813**	0.9564	0.9522	0.9547	0.9674	0.9223	0.9359	0.9650	0.9668	0.9618	0.9557
D33	**0.7177**	0.6514	0.5362	0.5216	0.6014	0.5129	0.5453	0.6339	0.4905	0.6614	0.5097	0.5103
D34	**0.6500**	0.5919	0.5599	0.5752	0.5887	0.5698	0.5376	0.6303	0.5808	0.6255	0.5531	0.5653
D35	0.9443	0.9392	0.9366	0.9330	0.9293	0.9363	0.9362	0.9458	0.9325	**0.9480**	0.9335	0.9404
D36	0.6616	0.6312	0.6654	0.6433	0.6479	0.6663	0.6493	0.6473	0.6524	**0.6835**	0.6616	0.6295
D37	0.8012	0.7248	0.7577	0.7502	0.7403	0.7620	0.7110	**0.8075**	0.7300	0.7816	0.7539	0.6260
D38	0.7201	0.6960	0.6676	**0.7800**	0.5032	0.7150	0.7147	0.6358	0.6045	0.6905	0.6963	0.6846
Avg	**0.8679**	0.8471	0.8355	0.8347	0.8196	0.8154	0.8131	0.8381	0.8323	0.8519	0.8368	0.8187

algorithm, which requires setting an appropriate K value for different datasets, reducing the adaptability of the algorithm. Experimental results on 38 KEEL datasets demonstrate that our method outperforms GDO in terms of the average rank of all evaluation metrics. Moreover, compared to other state-of-the-art resampling methods, our method also achieves the best performance.

References

1. Barua, S., Islam, M.M., Yao, X., Murase, K.: MWMOTE-majority weighted minority oversampling technique for imbalanced data set learning. IEEE Trans. Knowl. Data Eng. **26**(2), 405–425 (2012)
2. Batista, G.E., Prati, R.C., Monard, M.C.: A study of the behavior of several methods for balancing machine learning training data. ACM SIGKDD Explor. Newsl. **6**(1), 20–29 (2004)
3. Bunkhumpornpat, C., Sinapiromsaran, K., Lursinsap, C.: Safe-Level-SMOTE: safe-level-synthetic minority over-sampling technique for handling the class imbalanced problem. In: Theeramunkong, T., Kijsirikul, B., Cercone, N., Ho, T.-B. (eds.) PAKDD 2009. LNCS (LNAI), vol. 5476, pp. 475–482. Springer, Heidelberg (2009). https://doi.org/10.1007/978-3-642-01307-2_43
4. Chawla, N.V., Bowyer, K.W., Hall, L.O., Kegelmeyer, W.P.: SMOTE: synthetic minority over-sampling technique. J. Artif. Intell. Res. **16**, 321–357 (2002)
5. Chen, Y., Wu, K., Chen, X., Tang, C., Zhu, Q.: An entropy-based uncertainty measurement approach in neighborhood systems. Inf. Sci. **279**, 239–250 (2014)
6. Folino, G., Pisani, F.S., Sabatino, P.: An incremental ensemble evolved by using genetic programming to efficiently detect drifts in cyber security datasets. In: Proceedings of the 2016 on Genetic and Evolutionary Computation Conference Companion, pp. 1103–1110 (2016)
7. García, V., Sánchez, J.S., Marqués, A., Florencia, R., Rivera, G.: Understanding the apparent superiority of over-sampling through an analysis of local information for class-imbalanced data. Exp. Syst. Appl. **158**, 113026 (2020)
8. Haixiang, G., Yijing, L., Shang, J., Mingyun, G., Yuanyue, H., Bing, G.: Learning from class-imbalanced data: review of methods and applications. Exp. Syst. Appl. **73**, 220–239 (2017)
9. Han, H., Wang, W.-Y., Mao, B.-H.: Borderline-SMOTE: a new over-sampling method in imbalanced data sets learning. In: Huang, D.-S., Zhang, X.-P., Huang, G.-B. (eds.) ICIC 2005, Part I 1. LNCS, vol. 3644, pp. 878–887. Springer, Heidelberg (2005). https://doi.org/10.1007/11538059_91
10. He, H., Bai, Y., Garcia, E.A., Li, S.: ADASYN: adaptive synthetic sampling approach for imbalanced learning. In: 2008 IEEE International Joint Conference on Neural Networks (IEEE World Congress on Computational Intelligence), pp. 1322–1328. IEEE (2008)
11. Ivan, T.: Two modifications of CNN. IEEE Trans. Syst. Man Commun. (SMC) **6**, 769–772 (1976)
12. Jurgovsky, J., et al.: Sequence classification for credit-card fraud detection. Exp. Syst. Appl. **100**, 234–245 (2018)
13. Koziarski, M.: Radial-based undersampling for imbalanced data classification. Pattern Recogn. **102**, 107262 (2020)
14. Krawczyk, B., Koziarski, M., Woźniak, M.: Radial-based oversampling for multiclass imbalanced data classification. IEEE Trans. Neural Netw. Learn. Syst. **31**(8), 2818–2831 (2019)
15. Lin, W.C., Tsai, C.F., Hu, Y.H., Jhang, J.S.: Clustering-based undersampling in class-imbalanced data. Inf. Sci. **409**, 17–26 (2017)
16. López, V., Fernández, A., García, S., Palade, V., Herrera, F.: An insight into classification with imbalanced data: empirical results and current trends on using data intrinsic characteristics. Inf. Sci. **250**, 113–141 (2013)

17. Rodriguez, D., Herraiz, I., Harrison, R., Dolado, J., Riquelme, J.C.: Preliminary comparison of techniques for dealing with imbalance in software defect prediction. In: Proceedings of the 18th International Conference on Evaluation and Assessment in Software Engineering, pp. 1–10 (2014)
18. Vuttipittayamongkol, P., Elyan, E., Petrovski, A.: On the class overlap problem in imbalanced data classification. Knowl. Based Syst. **212**, 106631 (2021)
19. Wilson, D.L.: Asymptotic properties of nearest neighbor rules using edited data. IEEE Trans. Syst. Man Cybern. **3**, 408–421 (1972)
20. Xie, Y., Qiu, M., Zhang, H., Peng, L., Chen, Z.: Gaussian distribution based oversampling for imbalanced data classification. IEEE Trans. Knowl. Data Eng. **34**(2), 667–679 (2022)
21. Yan, Y., Jiang, Y., Zheng, Z., Yu, C., Zhang, Y., Zhang, Y.: LDAS: local density-based adaptive sampling for imbalanced data classification. Exp. Syst. Appl. **191**, 116213 (2022)
22. Yan, Y., Zhu, Y., Liu, R., Zhang, Y., Zhang, Y., Zhang, L.: Spatial distribution-based imbalanced undersampling. IEEE Trans. Knowl. Data Eng. **35**, 6376–6391 (2023)

LAST: An Efficient In-place Static Binary Translator for RISC Architectures

Yanzhi Lan[1,2], Qi Hu[1,2], Gen Niu[1,2], Xinyu Li[1,2], Liangpu Wang[1,2], and Fuxin Zhang[1,2(✉)]

[1] State Key Lab of Processors, Institute of Computing Technology,
Chinese Academy of Sciences, Beijing, China
{lanyanzhi22b,huqi20s,niugen18z,lixinyu20s,wangliangpu21s,
fxzhang}@ict.ac.cn
[2] University of Chinese Academy of Sciences, Beijing, China

Abstract. The lack of software has been a persistent issue for emerging instruction set architecture (ISA). To overcome this challenge, binary translation has emerged as a widely adopted solution, enabling programs written for older ISA to run on new ones. In the past, dynamic binary translation (DBT) was commonly utilized for software migration, but this technique required dynamic translation and often suffered from suboptimal efficiency. In contrast, Static binary translation (SBT) is an offline technique for translating binary code without runtime translation overhead. Existing SBT systems always employ address mapping tables to handle the address relocation problem, but this approach introduces performance overhead and leads to issues with indirect jump correctness. To address these limitations, we propose a novel static in-place instruction translation method for reduced-instruction set computing (RISC) architectures. This method ensures that the address of the guest program remains unchanged after translation, leveraging the regular length of various RISC instructions. We have implemented this method in a portable SBT tool called LAST, specifically designed to run MIPS or RISCV programs on the LoongArch platform. Based on the SPEC CPU2000 benchmark results, LAST achieves over 80% performance compared to the native LoongArch program, demonstrating its effectiveness and efficiency.

Keywords: Static Binary Translation · In-place instruction translation · instrumentation

1 Introduction

Binary translation enable software of one architecture to execute on a hardware platform of another architecture. This technology has a wide range of application scenarios, such as fast software simulation [4,9,13,22], program runtime analysis [6,10,20], debugging [11,14,21], and dynamic optimization [5].

Supported by The National Key Research and Development Program of China under grant number 2022YFB3105103.

Dynamic binary translation has played a prominent role in software migration endeavors in the past decades. As diverse instruction sets continue to evolve, a multitude of exceptional dynamic binary translation systems have emerged, showcasing the advancements in this field. Notable examples include IA-32 EL, which enables the execution of IA-32 applications on IA-64 processor family systems [2]. And Rosetta2, which facilitates the migration of x86 executables to the ARM platform [1]. Additionally, QEMU is a fast and portable dynamic translator, which support multiple guest and host ISAs [4]. However, it is important to note that dynamic binary translation often incurs additional performance and memory overhead.

Static binary translation not only does not need to translate at execution time but also can use larger-scale optimization methods, thus it can often achieve higher execution efficiency. SBT can often achieve higher execution efficiency because no real-time translation is required, which can be used to complete software migrations efficiently.

The Address relocation problem is caused by instruction expansion during translation, breaking the original indirect jump relationship, critically affecting the efficiency of static binary translators. The correctness of the jump relationship in the guest program is guaranteed by the compiler of the guest platform, but some address information will be lost during the compilation process, which makes it particularly difficult for the binary translator to reconstruct the jump relationship of the translated program. For direct jumps, the translator can easily calculate the new jump target through the offset value in the instruction. But for indirect jumps, their targets are unknown during translation. So address mapping tables is used to look up the targets by guest address at the runtime. However, this lookup table can introduce extra overhead, we will discuss the overhead in Sect. 2.2.

Instruction instrumentation technology is widely employed in various binary analysis tools, enabling the modification of the execution flow of the original program. These tools incorporate instrumentations into the program to facilitate statistical analysis and program debugging [23]. However, it should be noted that these tools are typically limited to programs within the same ISA. One factor contributing to this limitation is the variability in instruction lengths across different ISAs. For instance, the X86 instruction set utilizes variable-length instructions, whereas the MIPS instruction set adheres to a fixed-length format.

Over the past few decades, RISC architecture has witnessed remarkable advancements, giving rise to prominent instruction sets such as ARM, MIPS, RISCV, and LoongArch. These instruction sets have regular instruction encoding, which facilitates efficient addressing and decoding of instructions. Introduced by Loongson Technology in 2020, LoongArch is a new RISC instruction set that incorporates state-of-the-art advancements in instruction system design [26]. It offers a favorable environment for the development of low-power, high-performance CPUs [17–19,27]. With a fixed instruction length of 32 bits and a regular encoding format, LoongArch ensures simplicity and ease of instruction handling. The design of LoongArch also prioritized software compatibility, its basic software, such as the Linux kernel, GCC compiler, and QEMU simulator, has already been successfully integrated into the community.

In this paper, we propose a novel approach called in-place instruction translation, which enables the preservation of address relationships in the translated guest program, eliminating the need for address relocation. This innovative technique is implemented in a new SBT tool named LAST, which currently supports the translation from MIPS or RISCV to LoongArch, called LASTM and LASTR. Through extensive evaluation using the SPEC CPU2000 benchmark, LASTM demonstrates remarkable performance. It achieves over 80% of the program efficiency of the native LoongArch platform. These results showcase the effectiveness and efficiency of our proposed in-place instruction translation approach in the context of static binary translation.

The main contributions of this paper are as follows:

- We propose an in-place instruction translation method in binary translation systems that can efficiently solve the address relocation problem encountered in the mutual translation of RISC architectures. This method can effectively reduce runtime translation overhead and improve system performance.
- We implement and evaluate LAST, demonstrating the effectiveness of the in-place instruction translation method. Our experiments show that in-place instruction translation can solve the address relocation problem in binary translation with minimal overhead.
- This paper discuss the potential of this instruction translation method to be efficiently applied to the mutual translation of other RISC architectures, further expanding its applications beyond LoongArch.

The organization of the rest of this paper is as follows. Section 2 introduces static binary translation, address relocation problems, and overhead in SBT. Section 3 describes the design of LAST, and Sect. 4 shows the implementation details of LAST. Section 5 analyses some experimental results. Section 6 concludes this paper.

2   Background

This section provides a concise overview of static binary translation and highlights its key challenges, focusing on the main performance overheads associated with this technique.

2.1   Static Binary Translation

Binary translation systems fall into two general categories: static binary translator and dynamic binary translator. Figure 1 shows the difference between dynamic binary translation and static binary translation. The dynamic binary translator dynamically translates instructions during the execution of the native program. While the static binary translator converts the original program to the target program offline, and the translated program executes without the assistance of the binary translator. Compared to DBT, SBT is more convenient and efficient.

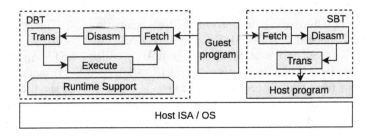

Fig. 1. The Difference between DBT and SBT

UQBT [8] is a versatile SBT tool that utilizes an intermediate language called HRTL to translate source binaries. HRTL can be further transformed into target binary assemblers, enabling compatibility with multiple platforms. However, UQBT still relies on a runtime interpreter to handle indirect register calls that cannot be determined during static translation.

LLBT [24,25] is a portable SBT tool specifically designed for translating ARM binaries to various target platforms. It employs LLVM IR (Intermediate Representation) as an intermediate representation and leverages the LLVM compiler infrastructure to retarget the LLVM IR to different ISAs. This approach significantly enhances the quality of the generated code. Nonetheless, LLBT still requires an address mapping table to effectively handle indirect jumps.

SBT can cause the address of the guest program to change due to instruction expansion during translation, breaking the original indirect jump relationship. This leads to the address relocation problem, which is described in detail in Sect. 2.2.

2.2   Address Relocation Problem in SBT

The best situation is that SBT can complete the instruction translation without any code expansion, thus avoiding updating the branch target address [28]. Nevertheless, it is obvious that one-to-many translations always exist, which requires extra space to store the extra instructions. So the branch instructions, especially indirect branch instructions, need to change their target address to avoid the incorrect execution flow.

Figure 2 show the Address Relocation Problem. Because the range of immediate numbers that can be used in the LoongArch is smaller than that in the MIPS, two additional instructions are required in Fragment L1. But the address of LABEL L3 is not changed, which may misleads some branch instructions into jumping to the wrong address, like instruction jr ra.

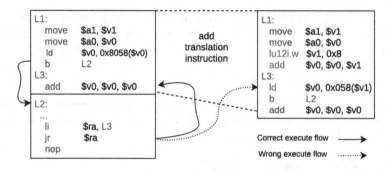

Fig. 2. The Address Relocation Problem

These problems can be easily solved for direct branches, but updating the target address for indirect branches sometimes is difficult [16]. To tackle this problem, Killian's Pixie uses a translation table in which all indirect branches need to find their target addresses [7].

Currently, the predominant approach for addressing this issue in static binary translation systems is to employ an address hash table [25]. However, challenges related to code discovery and performance persist. Please see Sect. 2.3 for further details on these challenges.

2.3   Performance Overhead in SBT

The performance of binary translation is significantly influenced by the disparities between the Guest ISA and the Host ISA. To achieve optimal performance in SBT tools, two critical issues must be addressed, as they have a substantial impact on the translation process.

- Instruction expansion. The semantics of instructions on different architectures directly leads to the inevitable instruction expansion during binary translation, which increases the number of dynamic instructions and affects efficiency.
- Indirect branch handling overhead. Caused by binary code expansion, one-to-many translations can affect the address of the original instruction. Therefore, the translator needs to correctly handle the modified jump relationships, especially indirect branches.

Instruction expansion is a common outcome of disparities between instruction sets, but the overhead associated with indirect jumps can be minimized. Numerous remarkable studies have been conducted to mitigate the impact of indirect jumps. For instance, Amanieu extensively examined various types of indirect jumps and implemented optimizations tailored to specific contextual scenarios in dynamic binary translation [12].

However, static binary translators do not actively participate in program execution and, as a result, lack the capability to dynamically handle indirect jumps during runtime. To address this limitation, a common approach is to employ a address mapping table within static binary translation systems [3]. However, this method necessitates hash table queries, thereby introducing additional overhead.

Algorithm 1: Lookup Indirect JMP Target

---

input : GPC

hash = HASH(*GPC*);
HPC = Address_Mapping_Table[hash];
if *HPC* ! = *NULL* **then**
  | **jmp** to HPC   // hit the target;
else
  | **return** error   // Lookup error;
end

The address mapping table serves as a repository where the guest program counter address (GPC) is stored as the key, and the corresponding host program counter address (HPC) is stored as the value. The lookup process is outlined in Algorithm 2. During runtime, the GPC is hashed to generate an index that corresponds to the HPC stored in the address mapping table. This HPC was previously stored during the static binary translation (SBT) process. Consequently, the HPC associated with the given GPC can be swiftly retrieved based on the hash. Additionally, it is crucial to compare the guest addresses within the lookup table to ensure that the hash algorithm does not generate any conflicts.

There is no doubt that Algorithm 2 is reliable and efficient. However, the production of instructions required to calculate the address hash, fetch data from the lookup table and determine if a hit has taken place still imposes a considerable performance cost. In SPEC2000, as shown in Fig. 3, indirect jumps account for approximately 1.27% of the total number of instructions, resulting in approximately 15% performance loss due to its 1 : 12 instruction inflation.

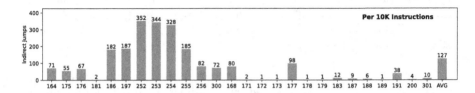

Fig. 3. Number of indirect jumps per 10K instructions

Furthermore, indirect jumps pose another huge challenge for static binary translation systems, as those jump entries can be difficult to fully identify. If the address entries, such as the switch-case jump table, cannot be recognized fully, some parts of the program are not being executed correctly because the address mapping table will lose some entry mappings.

3 Design

3.1 Overview

The design of the LAST architecture is shown in Fig. 4, which mainly includes the following modules:

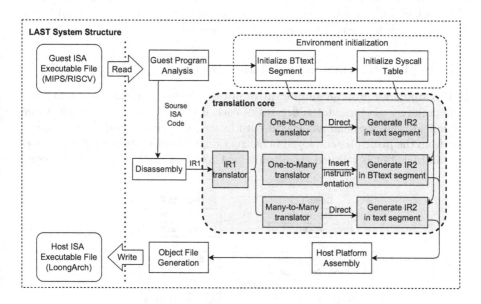

Fig. 4. The overall design of the LAST

Guest Program Analysis Module. This module accepts the origin binary executable file, parses the ELF format and maps each program segment. Then the segments will be recorded and reordered, making it easier for translator to analyze, manage, and provide essential information to other modules.

Environment Initialization Module. This module configures the basic structure related to translation, including initialization translated code segment and system call table. The translated code segment is the location for the one-to-many translated code, which will be written to the translated executable file. The system call table is used to convert different system call numbers between original and translated programs.

Disassembly Module. This module disassembles the original code segments and converts each instruction into internal IR-GUEST data.

Translation Module. This module will use the IR-GUEST data to classify different types of instructions, and then the corresponding translation function will translate them to the IR-HOST data.

In the translator core, the instructions are divided into three categories: *one-to-one*, *one-to-many* and *many-to-many* translation. Each translator puts the translated instruction in the original position, and in addition, one-to-many translator will place the extra instructions in the translated code segment.

Host Platform Assembly Module. This module integrates and assembles the IR-HOST data to generate the binary code and store it in the memory.

Object File Generation. This module will reorganize the segments (segment of the original file and the new segments generated by the translator), fill in the necessary ELF file header and write to the target file.

3.2 In-place Instruction Translation Design

In this study, we propose an innovative in-place instruction translation method that preserves the original addresses of each instruction in the program while minimizing the overhead associated with branches. This method draws inspiration from the principles employed in instruction instrumentation binary transformation tools but is tailored specifically for Cross-ISA static binary translation.

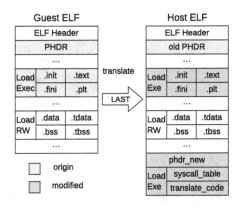

Fig. 5. Modifications of LAST to ELF Sections

LAST will generate an additional code segment to store translated binary codes and system call table. Figure 5 shows the difference between original and translated ELF file.

For the translation of each instruction, we need to consider two cases, the *one-to-one* and *one-to-many* translation. We also consider some optimization by using *many-to-many* translation.

- **One-to-one translation.** If the guest platform instruction can be converted into host platform instructions one-to-one, it is simple to place the translated instruction at the original address.
- **One-to-many translation.** Some instructions are complex and need to use multiple instructions to translate them. We put these translated binary codes at the additional code segment and then replace the original instruction with a direct jump instruction which can jump to the translated fragment.
- **Many-to-many translation.** Some instructions are used as a pattern on the guest platform to achieve a certain function. Usually, the host platform also has an instruction pattern with equal length to achieve the same function. In this case, the host instruction pattern can equivalently replace the guest instruction pattern.

The proposed tool, LAST, effectively addresses the challenges posed by indirect jumps through its in-place instruction translation mechanism. Section 4.2 of the paper will provide comprehensive implementation details, offering a deeper understanding of how LAST successfully mitigates the issues related to indirect jumps.

3.3   System Call Design

To solve the issue of system call incompatibility, LAST employs a system call table for system call conversion.

Unlike binary translators that "wrap" system calls by modifying call numbers and handling different structures [15], LAST uses the system call table to handle these issues, which is particularly useful in static in-place binary translation where dynamic interception and processing of system calls is challenging.

Algorithm 2: Handle Syscall By Syscall Table

---

input : Guest Syscall Number

hash = **HASH**(*Guest Syscall Number*);
Host Syscall entry = Address_Mapping_Table[hash];
if *Host Syscall entry* != *NULL* **then**
> **jmp** to Host Syscall entry   // **hit the target**;
> **covert** Guest Arguments to Host Arguments;
> **do** Host Syscall;
> **covert** Host return Arguments to Guest return Arguments;
> **return**
else
> **return** error   // **Unsupport Syscall**;
end

During the translation process, the initial step of LAST involves the insertion of the system call table into the translation code segment, as depicted in Fig. 5. The role and functionality of the system call table are further elucidated in

Algorithm 2. It facilitates several key functions: locating the appropriate table item based on the Host Syscall number, performing the necessary conversion from Guest ABI to Host ABI to enable system call invocation and kernel entry, and addressing any variations in return values that may arise between different architectures, ensuring a seamless transition back to the normal execution flow. For a more comprehensive understanding of the implementation specifics, please refer to Sect. 4.3.

3.4   ISA-Level Support

LAST utilizes some of the binary translation optimizations provided by the LoongArch instruction set, which are optional. These optimizations are provided as optional enhancements, and their detailed descriptions are presented below.

To address two efficiency issues in translation, LoongArch has designed binary translation support into its ISA. The first issue is register shortage, which may occur due to register mapping during binary translation. To solve this, LoongArch adds four new scratch registers (SCRs) alongside the 32 general-purpose registers (GPRs). These SCRs can interact with GPRs through data move instructions and are used as temporary registers. The second issue is the jump range limit. In-place instruction translation, as used in LAST, requires jumping to the translated code block. To address the problem of the limited jump range of the direct jump instruction, a jump-and-link instruction using SCRs is also added. By setting the value in one of the SCRs as the address of the translated code block, it is possible to jump to the translated code block effectively and return to the original instruction conveniently.

In LAST, the SCR is primarily used in the following situations: first, when the number of registers is insufficient during instruction translation, the SCR register is used as a temporary register instead of using a virtual register in memory. Second, when LAST needs to interrupt the current execution flow, but cannot jump for a long distance, the SCR is used as the address register for a long-distance indirect jump, thus reducing the storage and recovery of the source register.

If the host system does not support SCRs, LAST employs a strategy to identify the least frequently used registers in the guest program. These registers are then transformed into memory accesses, allowing the freed registers to be utilized for the same functionalities as the SCR registers mentioned earlier. As a result, even in the absence of SCR support on the host, the aforementioned challenges can still be addressed using additional instructions. However, it is important to note that this approach may lead to a potential loss of efficiency.

4   Implementation

4.1   Register Mapping

Register mapping, which binds the guest platform's registers to the host platform's registers, is an important part of binary translation, directly affecting the execution efficiency of binary translators.

Most binary translators, such as QEMU, use translation blocks (TBs) as base units and perform dynamic register allocation in each TB. This dynamic register allocation is convenient for design but tends to cause data transfer overhead. Regarding the implementation of LAST, it adopts a global static register mapping approach. Specifically, we illustrate this with LASTM as an example, and the corresponding mapping rules are presented in Table 1.

Table 1. Register mapping in LASTM

MIPS		LoongArch		LASTM
num	name	num	name	
0	zero	0	zero	zero
1	at	19	t7	tr_at
2, 3	v0, v1	17, 18	t5, t6	tr_v0, tr_v1
4–11	a0–a7	4–11	a0–a7	tr_a0–tr_a7
12–15	t0–t3	12–15	t0–t3	tr_t0–tr_t3
16–23	s0–s7	23–30	s0–s8	tr_s0–tr_s7
24, 25	t8, t9	16, 20	t4, t8	tr_t8, tr_t9
26, 27	k0, k1	21, 22	tp, fp	tmp
28	gp	2	gp	tr_gp
29	sp	3	sp	tr_sp
30	s8/fp	31	s8	tr_s8
31	ra	1	ra	tr_ra
hi, lo	hi, lo	-	scr2, scr3	tr_hi, tr_lo
-	-	-	scr0, scr1	tmp

The main reason for using these mapping rules is the difference in the definition of the ABI and the differences in hardware. For example, register 31 is used as the Return Address (RA) in MIPS, while the RA in LoongArch is register 1. In addition, when the branch predictor supports RAS, it is important to map the RA register of target planform to the host RA register, reducing the overhead of function returns. Also, four scratch registers (SCRs) are designed in the LoongArch which provide additional temporary registers for binary translation. LAST maps SCR2 and SCR3 to the HI and LO registers in MIPS, and SCR0 and SCR1 are used as temporary registers.

In addition, LAST takes into consideration the scenario where the number of guest registers exceeds that of the host registers. In this case, the translator can still implement the translation process by using memory as virtual registers. The translator keeps track of the frequency of register usage, and saves and restores the less frequently used registers, treating them as temporary registers. These marked temporary registers can be used when necessary by loading them from memory, where they were saved earlier. It is worth noting that this load/restore overhead is unavoidable. This case $NumRegs(Host) >= NumRegs(Guest)$ requires additional loading and restoring in any translator.

4.2   In-place Instruction Translation

Instruction translation is the critical part of LAST, which is related to the efficient execution of the translated program.

LAST's translator disassembles each input instruction and stores the detailed information in the IR. Then the translator will identify the classification of the instruction from the IR and translate them using different translation functions. LAST classifies instruction translations into three types.

One-to-One Translation. Both the host ISA and the guest ISA are RISC architectures, the behavior of the instructions is relatively similar. So some guest instructions can be translated into host instructions one-to-one. For such instructions, LAST can replace them at the original addresses.

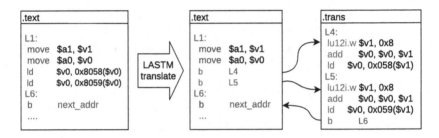

Fig. 6. One-to-many translation in LAST

One-to-Many Translation. Because of the differences between the host ISA and the guest ISA, some instructions require one-to-many translation, such as system call instruction and some instructions with 16-bit immediate. In this case, the translator will translate this instruction and put these translated codes in an additional code segment. At the same time, the original instruction will be replaced by a direct jump whose target address is this additional code segment. For example, Fig. 2 will be translated into the case of Fig. 6. The instruction *ld $v0,0x8058($v0)* is replaced by *b L4*, and the segment *.trans* is used to store these "one-to-many translated code". At the end of the translated code, a branch instruction *b L6* will be added to return to the original control flow.

In addition, if multiple consecutive instructions require one-to-many translation, they need to jump to the translated code segment only once. Thus, *b L4* need not to return next instruction.

Many-to-Many Translation. Compilers often combine several instructions as patterns. If we translate these instructions one by one, it may result in multiple *one-to-many translations*. However, if we use *many-to-many translation* to translate this instruction pattern, we often get good results. LAST can recognize these instruction patterns and translate them into instruction sequences with the same function.

For instance, in the case of MIPS, the instruction sequence *Lui +Ori* is utilized to load 32-bits immediate values. Conversely, in LoongArch, the corresponding instruction has a different immediate width. Nonetheless, both architectures provide instructions capable of loading 32-bits immediate values. In such scenarios, LAST treats the combination of *Lui+Ori* as a unit and replaces them with suitable instructions from the host instruction set.

4.3   System Call Handling

The translated programs cannot execute on the platform directly due to the differences between the host and guest operating systems. So, System calls must be handled in binary translation.

LAST stores a system call table in the translation code segment, which is used to handle system call differences between the guest Kernel and the host Kernel. Figure 7 shows the execution flow of the system call. Whenever the guest program needs to run a system call, the program first jumps to the header of the system call table, where a piece of code is stored for preprocessing. Then, LAST will use the guest system call number as an index to find the handler's entry in the system call table and jump over to execute the handler function. In these handlers, LAST will convert the system call parameters, including system call numbers and some structures in memory.

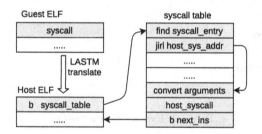

Fig. 7. Syscall Execution Flow in LAST

Another thing we need to consider is that there may be some special system calls that exist only in the guest platform and cannot be implemented by the host system calls. It is difficult to implement a non-existent system call in user mode. To handle those system calls, LAST uses system call simulation by employing other host system calls to mimic the function of the guest system call. For instance, the CLONE system call in MIPS can be emulated by the FORK system call in LoongArch.

4.4   Delay Slot

Due to historical reasons, the MIPS has designed the delay slot. However, delay slots are not available in the LoongArch, which leads to additional processing in LASTM.

In general, the instruction in the delay slot and the branch instruction have no data dependencies and can be easily translated by swapping the positions of two instructions.

When data dependencies exist, the relationship between the instructions needs to be handled with care, as depicted in Fig. 8. The daddiu instruction is an example of a data-dependent operation instruction, and simply changing the order of its execution can cause the beq instruction to execute incorrectly, resulting in program errors. To solve this error, LASTM need to translate according to the following steps. First, the value in the dependent register needs to be copied to a temporary register. Then, the delay slot instruction is executed. Finally, the branch instruction where the dependent register is replaced by the temporary register will be executed.

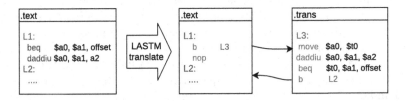

Fig. 8. Data Dependencies Delay Shot Handling

5 Evaluation

5.1 Evaluation Setup

Table 2. Evaluation Platform

	Loongson 3A4000	Loongson 3A5000
Architecture	GS464V	GS464V
ISA	**MIPS64 Release 5**	**LoongArch**
Compiler	gcc 8.3	gcc 8.3 (LoongArch)
Options	-Ofast -static	-Ofast -static
Frequency	**1.8 GHz**	**2.3 GHz**
L1 cache I/D	64 KiB	64 KiB
L2 cache	256 KiB	256 KiB
LLC	**8 MiB**	**16 MiB**

LAST offers the capability to convert MIPS or RISCV programs into LoongArch programs, which are referred to as LASTM and LASTR, respectively, for clarity. To evaluate its performance, we conducted tests in two distinct environments.

The term *MIPS* denotes the execution of native MIPS programs on the Loongson 3A4000 platform, while *LA* represents the execution of LoongArch programs on the Loongson 3A5000 platform. Although LAST supports RISCV-to-LoongArch translation, our evaluation was limited to the simulation environment for RISCV. Therefore, we were unable to perform actual chip tests for RISCV. The detail information about the experimental environment is shown in Table 2.

For evaluation purposes, we utilized the Coremark and SPEC CPU2000 benchmark testing programs. Coremark is specifically designed to assess the fixed-point performance of CPUs, and although it has a relatively small test scale, it serves as a suitable tool for evaluating the performance of binary translation. In addition to Coremark, we employed SPEC CPU2000 to obtain more detailed insights into the performance of binary translation. SPEC CPU2000 encompasses both fixed-point and floating-point testing, enabling a more comprehensive evaluation of performance details. All benchmarks were compiled by GCC version 8.3 with -Ofast as the optimization level.

The 3A5000 is an evolution of the 3A4000 and they share the same microarchitecture. There are three main differences, ISA, frequency, and LLC capacity. Other microarchitectural features are unchanged, so the programs behave very similarly on the 3A4000 and 3A5000. Therefore, for LAST, it is significant to compare the original MIPS programs on 3A4000 with the LAST-translated programs on 3A5000.

5.2 Performance

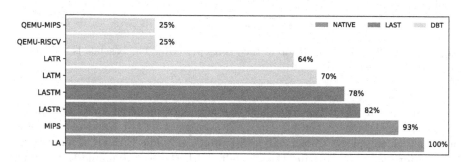

Fig. 9. Coremark Relative Ratio of the Scores-per-GHz The baseline is the scores-per-GHz of the *native LA*.

Figure 9 presents a performance comparison of various translators and native programs, using the score of the native LA as the baseline. It is obvious that LAST has significant performance advantages over other dynamic binary translator.

QEMU [4] is a commonly used binary translator in the industry that supports mutual translation of multiple architectures, but its efficiency is low due to the use of TCG as the intermediate code for translation. LATR and LATM are dynamic translators developed in the research group that use one-to-one instruction translation and special optimization for indirect jump, resulting in higher efficiency than QEMU. LASTR and LASTM represent LAST's translation of RISC and MIPS programs, respectively, and show much higher efficiency than dynamic translators, because In-place static translators do not generate indirect jump overhead and do not require translation time.

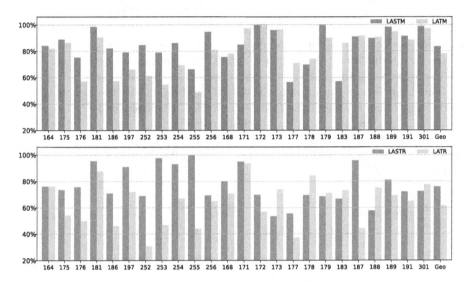

Fig. 10. SPEC2000 Relative Ratio of the Scores-per-GHz. The baseline is the scores-per-GHz of the *native LA*.

Figure 10 illustrates the SPEC2000 performance of LAST, with the native LA program serving as the baseline. On average, LASTM achieves over 80% performance compared to the *native LA* program, while LASTR achieves over 75% performance. Comparing these results with those of dynamic translators, it is evident that LAST demonstrates superior performance across most sub-items. The key factor contributing to LAST's higher performance is its ability to address the overhead of indirect jumps through the translation of interpolated instructions, without incurring the performance loss associated with dynamic translation techniques.

We further analyze the relevance between the instruction expansion ratio and the actual performance in LASTM. Figure 11 shows the relationship between the instruction expansion ratio and the execution time. Due to our translation rules, the number of *LASTM*'s dynamic executed instructions is always bigger than the number of *native MIPS*'s. So the value in the x-axis is always greater than 1.

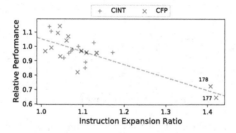

Fig. 11. Instruction Expansion Ratio vs Relative Performance. The x-axis is the ratio of the dynamic instruction count, which stands for the instruction expansion ratio. The y-axis is the ratio of the number of execution cycles, which stands for the relative performance.

The y-axis is the ratio of the number of execution cycles. We divide the number of *native MIPS*'s execution cycles by the number of *LASTM*'s.

Note that there are two points at the bottom-right corner. They are 177.mesa and 178.galgel, whose performance is less than 80%. The reason why their performance is such low is that their instruction expansion ratio is too high. They contain many instructions that can not be translated by the one-to-one translator.

5.3 Translation vs. Compilation

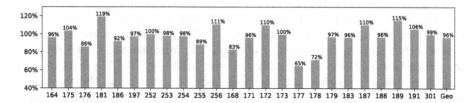

Fig. 12. SPEC2000 Relative Ratio of the Scores-per-GHz. The baseline is the scores-per-GHz of the *native MIPS*.

Figure 12 shows the SPEC2000 performance of LASTM. The result of *native MIPS* is the baseline. On average, the performance for SPEC benchmarks is 96%. Note that there are only two benchmarks whose performance is lower than 80%. And they both belong to the floating-point benchmark. In all, we can conclude from the result that LASTM almost does not lose performance compared to the native MIPS program. As shown in Fig. 10, the SPEC2000 performance based on the LA program is only 84%.

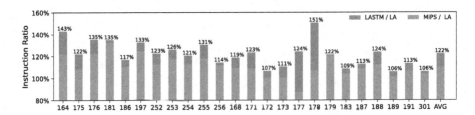

Fig. 13. Relative Ratio of Instruction Count. The baseline is the instruction count of *native LA*

However, this result is heavily influenced by compilers on different platforms. Figure 13 shows the relative ratio of the instruction count compared to the instruction count of the *native LA*. On average, for integer benchmarks, the instruction count of *native MIPS* is 17% more than that of *native LA* and *LASTM* is 27%. Since LASTM is translating MIPS instructions into LoongArch instructions, a sizeable proportion of the instruction expansion comes from the difference in the compiler.

In fact, the microarchitecture of 3A4000 and 3A5000 is nearly the same except for the size of the LLC. Comparing the performance of LASTM in 3A5000 with the performance of MIPS in 3A4000 while considering the difference in the hardware platform is a better way to describe the actual efficiency of our binary translator.

6   Conclusion

In conclusion, this paper introduces a novel approach, in-place static binary translation, which effectively addresses the challenge of address relocation and significantly improves the efficiency of SBT. The implementation of LAST on the LoongArch platform successfully converts MIPS or RISCV programs into the LA program. Experimental results demonstrate that while the translated program may experience a slight increase in direct jumps, it has minimal impact on efficiency. In fact, the translated program achieves over 80% performance compared to the native LA program, confirming the high efficiency of this approach. These findings highlight the effectiveness and potential of in-place static binary translation for efficient program translation and execution.

References

1. Apple: Rosetta. https://support.apple.com/en-us/HT211861
2. Baraz, L., et al.: IA-32 execution layer: a two-phase dynamic translator designed to support IA-32 applications on Itanium/spl reg/-based systems. In: 2003 Proceedings of the 36th Annual IEEE/ACM International Symposium on Microarchitecture, MICRO-36, pp. 191–201 (2003). https://doi.org/10.1109/MICRO.2003. 1253195

3. Bauman, E., Lin, Z., Hamlen, K.W.: Superset disassembly: statically rewriting x86 binaries without heuristics. In: 25th Annual Network and Distributed System Security Symposium, NDSS 2018. Internet Society, Reston (2018). wOS:000722005800038. https://doi.org/10.14722/ndss.2018.23300

4. Bellard, F.: QEMU, a fast and portable dynamic translator. In: Proceedings of the FREENIX Track: 2005 USENIX Annual Technical Conference, 10–15 April 2005, pp. 41–46. USENIX, Anaheim (2005). https://www.usenix.org/events/usenix05/tech/freenix/bellard.html

5. Bruening, D., Garnett, T., Amarasinghe, S.: An infrastructure for adaptive dynamic optimization. In: 2003 International Symposium on Code Generation and Optimization, CGO 2003, pp. 265–275 (2003). https://doi.org/10.1109/CGO.2003.1191551

6. Bruening, D., Garnett, T., Amarasinghe, S.: An infrastructure for adaptive dynamic optimization. In: Proceedings of the International Symposium on Code Generation and Optimization: Feedback-Directed and Runtime Optimization, CGO 2003, pp. 265–275. IEEE Computer Society, USA (2003)

7. Chow, F.C., Himelstein, M.I., Killian, E., Weber, L.: Engineering a RISC compiler system. In: COMPCON, pp. 132–137 (1986)

8. Cifuentes, C., Van Emmerik, M.: UQBT: adaptable binary translation at low cost. Computer 33(3), 60–66 (2000). https://doi.org/10.1109/2.825697

9. Cota, E.G., Bonzini, P., Bennée, A., Carloni, L.P.: Cross-ISA machine emulation for multicores. In: Proceedings of the 2017 International Symposium on Code Generation and Optimization, CGO 2017, pp. 210–220. IEEE Press (2017)

10. Cota, E.G., Carloni, L.P.: Cross-ISA machine instrumentation using fast and scalable dynamic binary translation. In: Proceedings of the 15th ACM SIGPLAN/SIGOPS International Conference on Virtual Execution Environments, VEE 2019, pp. 74–87. Association for Computing Machinery, New York (2019). https://doi.org/10.1145/3313808.3313811

11. Cunha, M., Fournel, N., Pétrot, F.: Collecting traces in dynamic binary translation based virtual prototyping platforms. In: Proceedings of the 2015 Workshop on Rapid Simulation and Performance Evaluation: Methods and Tools, RAPIDO 2015, Association for Computing Machinery, New York (2015). https://doi.org/10.1145/2693433.2693437

12. d'Antras, A., Gorgovan, C., Garside, J., Luján, M.: Optimizing indirect branches in dynamic binary translators. ACM Trans. Archit. Code Optim. 13(1) (2016). https://doi.org/10.1145/2866573

13. Dehnert, J.C., et al.: The Transmeta Code Morphing™ software: using speculation, recovery, and adaptive retranslation to address real-life challenges. In: Proceedings of the International Symposium on Code Generation and Optimization: Feedback-Directed and Runtime Optimization, CGO 2003, pp. 15–24. IEEE Computer Society, USA (2003)

14. Eyolfson, J., Lam, P.: Detecting unread memory using dynamic binary translation. In: Qadeer, S., Tasiran, S. (eds.) RV 2012. LNCS, vol. 7687, pp. 49–63. Springer, Heidelberg (2013). https://doi.org/10.1007/978-3-642-35632-2_8

15. Federico, A.D., Agosta, G.: A jump-target identification method for multi-architecture static binary translation. In: 2016 International Conference on Compliers, Architectures, and Synthesis of Embedded Systems (CASES), pp. 1–10 (2016)

16. Horspool, R., Marovac, N.: An approach to the problem of detranslation of computer-programs. Comput. J. 23(3), 223–229 (1980). WOS:A1980KD91500005. https://doi.org/10.1093/comjnl/23.3.223

17. Hu, W., et al.: Godson-3B: a 1GHz 40W 8-core 128GFLOPS processor in 65nm CMOS. In: 2011 IEEE International Solid-State Circuits Conference, pp. 76–78 (2011). https://doi.org/10.1109/ISSCC.2011.5746226

18. Hu, W., Yang, L., Fan, B., Wang, H., Chen, Y.: An 8-core MIPS-compatible processor in 32/28 nm bulk CMOS. IEEE J. Solid-State Circ. **49**(1), 41–49 (2014). https://doi.org/10.1109/JSSC.2013.2284649

19. Hu, W., et al.: Godson-3B1500: a 32nm 1.35GHz 40W 172.8GFLOPS 8-core processor. In: 2013 IEEE International Solid-State Circuits Conference Digest of Technical Papers, pp. 54–55 (2013). https://doi.org/10.1109/ISSCC.2013.6487634

20. Luk, C.K., et al.: Pin: building customized program analysis tools with dynamic instrumentation. In: Proceedings of the 2005 ACM SIGPLAN Conference on Programming Language Design and Implementation, PLDI 2005, pp. 190–200. Association for Computing Machinery, New York (2005). https://doi.org/10.1145/1065010.1065034

21. Molnar, I.: Performance counters for Linux (2009). https://lwn.net/Articles/337493/. Accessed 23 Feb 2022

22. Niu, G., Zhang, F., Li, X.: Eliminate the overhead of interrupt checking in full-system dynamic binary translator. In: Proceedings of the 15th ACM International Conference on Systems and Storage (2022)

23. Prasad, M.: A binary rewriting defense against stack-based buffer overflow attacks. In: 2003 USENIX Annual Technical Conference, USENIX ATC 03, San Antonio, TX, June 2003. USENIX Association (2003). https://www.usenix.org/conference/2003-usenix-annual-technical-conference/binary-rewriting-defense-against-stack-based

24. Shen, B.Y., Chen, J.Y., Hsu, W.C., Yang, W.: LLBT: an LLVM-based static binary translator. In: International Conference on Compilers, Architecture, and Synthesis for Embedded Systems (2012)

25. Shen, B.Y., Hsu, W.C., Yang, W.: A retargetable static binary translator for the arm architecture. ACM Trans. Archit. Code Optim. **11**(2) (2014). https://doi.org/10.1145/2629335

26. Loongson Technology: Loongarch documentation (2022). https://loongson.github.io/LoongArch-Documentation/

27. Weiwu, H., et al.: Loongson instruction set architecture technology. J. Comput. Res. Dev., 1–22 (2022)

28. Wenzl, M., Merzdovnik, G., Ullrich, J., Weippl, E.: From hack to elaborate technique - a survey on binary rewriting. ACM Comput. Surv. **52**(3), 1–37 (2020)

Personalized Privacy Risk Assessment Based on Deep Neural Network for Image Sharing on Social Networks

Hongyun Cai[1,2], Ao Zhao[1,2]([✉]), Shiyun Wang[1,2], Meiling Zhang[1,2], and Yu Zhang[1,2]

[1] School of Cyber Security and Computer, Hebei University, Baoding 071000, Hebei, China
chosen_ao@163.com
[2] Key Laboratory on High Trusted Information System in Hebei Province, Hebei University, Baoding 071000, Hebei, China

Abstract. With the extensive usage of social networks, many users get used to share images with their friends frequently without thinking carefully about the private information in the images, which may cause the leakage of user private information. To help users to improve their privacy awareness, in this paper, we propose a two-stage personalized privacy risk assessment framework based on images sharing history between users. In the first stage, the privacy information of the shared images are identified by using the faster R-CNN, based on which, the user-image privacy vector is generated. In the second stage, we predict the user behavior for image sharing using deep neural network and calculate the risk leakage probability of user privacy for the sharing image. The experimental results show the effectiveness of the proposed method, which can reach a prediction accuracy of 98% and the average time of 0.47 s.

Keywords: social networks · image sharing · image privacy · presonalized requirement vector · deep neural network

1 Introduction

At present, there are 4.2 billion social media users in the world, and image sharing has become the common interactive behavior between users on social networks. For example, on Facebook, Wechat, et al., users frequently exchange various kinds of image information about learning, life, and work. Sharing images is a great convenience to social network users, but also brought the privacy risk. Through the analysis of sharing images, it is possible to obtain very detailed personal information including location, social relationship, and property, etc., which will undoubtedly cause many risks and bring huge losses to users [9]. For avoiding privacy disclosure, many social network platforms allow users to set their fine-grained privacy preference. However, research in [10] has shown that

Z. Tari et al. (Eds.): ICA3PP 2023, LNCS 14488, pp. 255–274, 2024.
https://doi.org/10.1007/978-981-97-0801-7_15

only 37% of users have personalized privacy settings on Facebook, which means that many users are unaware of privacy and their potential risks. Moreover, in the process of sharing images, users may misjudge the privacy information in images even they use the personalized privacy settings provided by social networking sites [12]. Therefore, it is very important to analyze the privacy information in the sharing image accurately and predict their privacy leakage risk for different users effectively, which will help enhance awareness of privacy for social network users.

To predict privacy risk for users on social networks, some methods have been proposed to analyze the image privacy or calculate the privacy score. The existing image privacy prediction methods can classify the sharing image into private class or public class according to the image itself, while those methods of calculating privacy score usually focused on the user attributes. However, the concept and definition of privacy are very subjective and different people may have different views on privacy [17]. For example, introverted users and extroverted users usually have different privacy requirements for the same personal privacy information. Therefore, the privacy leakage risk of the sharing image should be related to the behavior preference of image owner [25]. In addition, privacy policy may change with the subject, context and time, and leakage path of private images is complex.

To address the above limitations, we propose a personalized privacy risk assessment model for image sharing on social networks, which is called PPRAS. The proposed approach consists of two stages, i.e. generating user-image privacy vector and analyzing image shared behavior between users. In the first stage, we use faster R-CNN to identify the privacy information in the shared image and generate the privacy attribute vectors, the privacy vector is generated by combining the generated privacy attribute vectors with the personalized user privacy requirement vector. In the second stage, we adopt deep neural network to predict the image shared behavior between users based on their historical sharing behavior, and calculate the risk leakage probability of user privacy for the sharing image. The main contributions of this paper are as follows:

- The proposed PPRAS takes into account both the personalized privacy requirement of each user and privacy information in the shared image, which will be more aligned with the characteristic of privacy.
- PPRAS adopts a deep neural network to model the shared behavior between users, and the privacy leakage risk for shared images is calculated based on the prediction probability, which can minimize the subjective impact of the evaluation results.
- PPRAS does not need provide privacy setting by themselves, which can greatly reduce the difficulty of extracting information and enhance the universal application of privacy risk assessment on different social networks. The experiments results demonstrate the effectiveness of the proposed model.

The rest of this paper is organized as follows: Sect. 2 reviews some related work. Section 3 details the proposed model including problem description, definition and the framework of PPRAS. Section 4 introduces the experimental results and evaluation. Section 5 concludes the paper.

2 Related Work

For image sharing on social networks, the identification of private information in images is an important research direction. Hong et al. [6] proposed a privacy attributes-aware message passing neural network framework, which can effectively deal with multiple privacy attributes, and use the message passing algorithm to model the relationship between the privacy attributes in the image. Zhang et al. [3] calculated the attention value of each pixel in the image through the visual attention mechanism, which improved the accuracy of image privacy attribute recognition. Yang et al. [22] proposed dynamic region-aware graph convolutional network based on feature extraction and dynamic region selection mechanism, which can model the correlation between different elements to detect privacy-leaking images. Tonge et al. [16] identified and predicted the sensitive content of images by dynamic deep multi-modal fusion. These studies focused on measuring whether an image contains private information or the type of private information.

For evaluating the leakage risk of user privacy on social networks, Li et al. [7–9] used structural similarity, attribute correlation and behavioral intimacy to calculate the risk of privacy leakage. In order to reduce the influence of subjective factors, they improved the calculation method using SGC neural network. However, their methods omit some factors that cause the leakage of private information. Dan et al. [10] developed a probabilistic model and coordination strategy based on the social user's image sharing history. This method does not take into account the hidden relationships such as intimacy between users. Oukemeni et al. [13] analyzed the internal and external environment of the system from the perspective of the social network platform and calculated the privacy level by combining the influence of privacy and security requirements, accessibility, and information extraction difficulty. This method statically evaluates the system as a whole, but it does not consider the individual requirement of different users. The work in [2,5,11] explored the risk factors affecting user privacy based on the privacy behavior and potential privacy threats of social network users.

In addition, there also have some related researches on evaluating the leakage risk of privacy information in other background. In the mobile phone system and application, Wang et al. [20] calculated the risk value of privacy information leakage by quantifying the influence of authority through association rule learning. Wang et al. [18] introduced the TF-IDF model to calculate the privacy leakage risk of the app and used domain name similarity, ip address similarity, and tcp channel similarity optimization models. However, this method ignores the disclosure of private information caused by app service providers. Chen et al. [1] proposed a semantic-aware privacy risk assessment framework. They took whether the information leaves the local device as an important basis for privacy information leakage, and combined data transmission paths, permissions, and user intentions to qualitatively (risk level) and quantitatively (risk score) privacy leakage risks of apps. However, it is difficult to distribute these data extraction and analysis tasks on terminal devices. Zhang et al. [24] constructed a hierarchy model to evaluate the user's privacy leakage risk in mobile commerce applications, which is based on information entropy and the Markov chain. However, the classification of risk categories and levels is subjective. The above methods focused on the risk of privacy leakage from the analysis and modeling of the information propagation path in the App. In text data, Wu et al. [19] utilized information entropy to represent the amount of information containing original information in differential privacy publishing data, thus quantifying the risk of privacy leakage. Xiong et al. [21] used the word2vec algorithm and bag-of-words model to capture the embedding distance between words as the impact and contribution, i.e., correlation, thus calculating the sensitivity of words and obtaining the privacy risk value of text. These methods modeled and analyzed the risk of privacy leakage from the perspective of text content.

3   The Framework of PPRAS

3.1   Problem Description and Definition

Users often share image on social networks, which may involve a large amount of private information about users, thereby causing leakage of users' private information. For example, u_0 and u_1 are friends, and u_1 also has other friends, e.g., u_2, u_3, et al. For the images shared by u_0 to u_1, u_1 usually shares those interesting images to u_2, some images about life to u_3, and images about work to u_4. This different sharing feature can also imply the relationship or intimacy between u_1 and other friends. Facing the same privacy category, users have different privacy requirements for different friends. In addition, due to different personalities and views on private information, the sharing behavior of u_1 may inadvertently leak the private information of u_0. Therefore, our goal is to evaluate the privacy leakage risk brought by the sharing images between social network users. Our work is developed on the basis that social network service providers are fully aware of their network structure and all the images sharing records between users.

Definition 1. *Image Privacy Attribute Vector (IPAV). For each image, the image privacy attribute vector is defined as $IPAV = (a_1, a_2, ..., a_n)$, where a_i represents the ith privacy category extracted from the image, n represents the number of privacy categories in the system. If there exists the ith privacy category in the image, the attribute $a_i = 1$; otherwise, $a_i = 0$.*

Definition 2. *Personalized Privacy Requirement Vector (PPRV). For each user u, let $p_{v,i}^u$ be an indicator function that represents the privacy attitude of user u towards friends v in the ith privacy category. If user u believes that the ith category contains private information, $p_{v,i}^u = 1$; otherwise, $p_{v,i}^u = 0$. Therefore, the personalized privacy requirement vector of user u towards friend v is defined as $PPRV_v^u = (p_{v,1}^u, p_{v,2}^u, ..., p_{v,n}^u)$.*

Definition 3. *User-Image Privacy Information Vector (UIPIV). For each user and shared image, let v_i be an indicator function that represents the relationship between the user and the shared image. If the private image shared by the user contains the ith information and the user believes the ith category contains private information, $v_i = 1$; otherwise, $v_i = 0$. And the user-image privacy information vector is defined as $UIPIV = (v_1, v_2, ..., v_n)$.*

Definition 4. *Image Privacy Policy. The image privacy policy is defined as a quadruple $Pol = (sen, rec, img, PR)$, where sen is the sender of the private image, rec is the receiver of the private image, and PR represents the set of other potential receivers for the private image img shared by user rec.*

The work in this paper aims to predict the risk probability (as shown in Definition 5) that a shared private image may be inadvertently propagated to other recipients by the recipient user of the private image. The reasons for this risk include differences in the degree of intimacy of the relationship between users and their different perceptions and concepts of privacy. Therefore, personalized privacy risk assessment needs to collect users' privacy views on different types of privacy information (as shown in Definition 2).

Definition 5. *Privacy Disclosure Risk of Shared Image. For each non-zero user-image privacy information vector $UIPIV_{u,i}$, a specific image privacy policy $Pol_0 = \{sen_0, rec_0, img_0^l, PR_0\}$, img_0^l represents the private image containing l kinds of sensitive private information, let $user_t \in PR_0$, then the privacy leakage risk of the private image img_0^l is recorded as the leakage probability $P_{rec_0 \to user_t}^{sen_0}$ after sharing and dissemination by the sender sen_0 to other potential recipients $user_t$.*

3.2   The Framework of PPRAS

The basic framework of PPRAS is shown in Fig. 1. For all users, they can customize specific privacy requirements on different privacy features and generate personalized privacy requirement vectors. For each user, if his/her personalized privacy requirement vector is not a zero vector, we use faster R-CNN to identify privacy information on each shared images. Combining the personalized privacy requirement vector and the image privacy feature vector, the user-image privacy information vector can be constructed. According to the user-image privacy information vector, it is understood whether the image to be shared contains the privacy information that the user cares about. If the image does not contain the sensitive privacy information related to that user, indicating that this user can freely share the corresponding image with his/her friends; otherwise, the deep neural network is used to predict the risk probability based on the history of privacy image sharing behaviors between the receiver and its friends.

Construction of User-Image Privacy Information Vector. Everyone has different views on various privacy categories, and their privacy attitudes towards

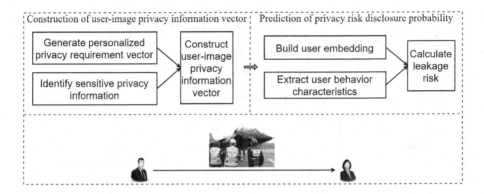

Fig. 1. Framework of PPRAS.

different individuals within the same category also different. So the privacy requirements of users towards others are also various. PPRAS uses the privacy requirement vector $PPRV_o^u$ to represent the privacy requirement of user u towards others. For his/her each shared image, PPRAS needs to identify all privacy regions in the image by using Faster R-CNN [14], which has more excellent results on various object detection than R-CNN and Fast R-CNN. The usage of cumbersome privacy settings brings many inconveniences to users. Furthermore, the dynamics of users' relationships with their friends, such as changes in intimacy levels, mean that privacy configurations with the same friend are not set in stone. These dynamic changes undoubtedly present great challenges for personalized privacy settings. Consequently, we explore a personalized privacy configuration algorithm that dynamically modifies PPRV, which is illustrated in Algorithm 1.

As shown in Algorithm 1, lines 1–5 calculate the frequency of images shared by a user containing different types of private information, which is used to represent the user's sharing tendencies towards different types of private information. Lines 6–8 calculate the frequency at which a user shares images containing the same type of private information with different friends, representing the user's sharing attitude towards different friends under a particular type of private information. Based on these factors, Algorithm 1 obtains PPRV, which represents a user's attitudes towards sharing different types of private information with different friends.

Inspired by the previous work in prediction image privacy category, we define the 20 feature categories identified by Faster R-CNN on the VOC2007 dataset [14] as sensitive private information research. The main process of generating the user privacy information vector is depicted in Fig. 2, which consists of four steps:

Step1: we adopt the ResNet50 convolutional neural network to extract the features of the image and generate the feature map;

Step2: The feature map is sent to the Regional Proposal Network (RPN), and a 3×3 sliding window is used to slide on the feature map to generate

---

Algorithm 1. Personalized Privacy Configuration Algorithm

Input: the friends list V_u and image sharing records D_u of user u.
Output: $PPRV_v^u$
1: $N \leftarrow$ the number of shared privacy images in D_u
2: **for** the ith privacy category **do**
3:    $S_i \leftarrow$ the set of shared images containing the ith private category in D_u
4:    $N_i \leftarrow |Si|$
5:    $\theta_i \leftarrow \frac{N_i}{N}$
6:    **for** each v in V_u **do**
7:       $N_{i,v} \leftarrow$ the number of private images that user u shared with v in S_i
8:       $P_{i,v} \leftarrow \frac{N_{i,v}}{N_i}$
9:       **if** $P_{i,v} > \theta_i$ **then**
10:          $PPRV_v^u(i) \leftarrow 0$
11:       **else**
12:          $PPRV_v^u(i) \leftarrow 1$
13:       **end if**
14:    **end for**
15: **end for**
16: **return** $PPRV_v^u$

---

the corresponding candidate frame, the target candidate frame containing the privacy feature area is screened out, and the corresponding feature matrix is obtained according to the projection relationship between the generated target candidate frame containing the privacy feature and the feature map;

Step3: Each feature matrix is converted into a feature map with a fixed spatial range (e.g., 7×7) through the ROI pooling layer, and then the classifier and regressor through two fully connected layers (FC1, FC2) are used to calculate the privacy feature vector and determine the location of the privacy feature.

Step4: The user-image privacy information vector is constructed by combining the generated image privacy attribute vector and the user personalized privacy requirement vector, i.e., $UIPIV_o^u = PPRV_o^u \otimes IPAV = (v_1, v_2, ..., v_n)$, where

$$v_j = \begin{cases} 1, if \text{ the } i\text{th elements in } PPRV_o^u \text{ and } IPAV \text{ are the same non-zero} \\ 0, otherwise \end{cases},$$

o represents the recipient of the image shared by user u.

Prediction of Privacy Risk Disclosure Probability. A shared image may contain a variety of sensitive private information. In our daily life, we should consider carefully whether it is safe when we share an image containing private information with others. It is obvious that the privacy risk is more if the receiver tends to share this kind of privacy images with others more frequently. Therefore, for user u, image i and the corresponding receiver, PPRAS predicts the privacy leakage risk based on the shared history between the receiver and its other friends. In social networks, users usually take images of their life, work, and other privacy-related content and share them with some friends. To learn

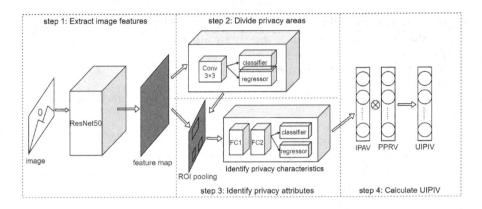

Fig. 2. Construction of user-image privacy information vector.

the user's historical sharing behavior, we introduce deep neural network. It is worth noting that although this model can predict the potential recipient users, only the sharing probability results will be known to the sender users because the recipient user's friendship and sharing behaviors are part of the recipient user's privacy.

The detailed process of predicting privacy risk in this section is shown in Fig. 3, which consists of four parts.

Historical behavior records. For each user, let $Seq = \{\mathrm{Pol}_1, \mathrm{Pol}_2, \ldots, \mathrm{Pol}_N\}$ be the set of image privacy policies, where N is the number of images shared by the user, and the user's ith sharing behavior $Pol_i = \{sen_i, rec_i, img_i^l, PR_i\}$ indicates the dissemination policy of the privacy image img_i^l shared by user rec_i from user sen_i.

Embedding Layer. Embedding is a commonly used technology to convert high-dimensional sparse feature vectors into low-dimensional dense feature vectors. After Embedding, the ith user can be represented as $u_i \in R^{N_M \times dim}$, where N_M represents the number of sparse features of user ids, and dim represents the dimension of embedding. The privacy image img_i^l can be represented by $img^l \in R^{N_l \times dim}$, where N_l represents the number of sparse features of private information. Then the user behavior sequence is represented as $q^l = p_1, p_2, \ldots, p_n \in R^{N \times sz \times dim}$, sz is the size of the input, and p_i represents the embedding of the ith behavior.

User behavior feature extraction. The behavioral characteristics of users sharing images will change with time and the intimacy of the user's friend relationship. For example, in certain periods, users will frequently share information with other users who usually interact less in the past, and this phenomenon is particularly obvious in holidays. As a variant of the LSTM [23], GRU [4] can not only capture and learn long-term time series changes well, but also GRU has fewer parameters than LSTM. Therefore, in this paper, on the basis of the embedded user historical behavior sequence Seq, we use GRU to extract user's

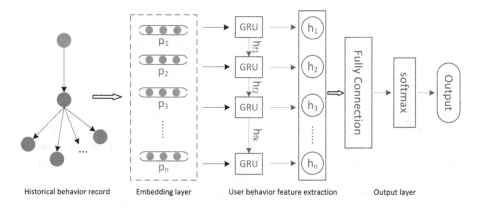

Fig. 3. The detailed process of predicting privacy risk.

behavior characteristics. The reset gate and update gate can be calculated as follows:

$$R_t = \sigma \left(p_t W_{pr} + H_{t-1} W_{hr} + b_r \right) \tag{1}$$

$$Z_t = \sigma \left(p_t W_{pz} + H_{t-1} W_{hz} + b_z \right) \tag{2}$$

where $\sigma(\cdot)$ is the sigmoid function, R_t is the reset gate, Z_t is the update gate, p_t is the user's historical behavior record, H_{t-1} is the hidden state of the previous time step, and the shapes of the weight matrices W_{pr}, W_{hr}, W_{pz}, and W_{hz} are represented by subscripts, b_r and b_z are offset vectors. The candidate hidden states and hidden states at time step t can be calculated by:

$$\widetilde{H}_t = tanh \left(p_t W_{ph} + \left(R_t \odot H_{t-1} \right) W_{hh} + b_h \right) \tag{3}$$

$$H_t = Z_t \odot H_{t-1} + \left(1 - Z_t \right) \odot \widetilde{H}_t \tag{4}$$

Where $\widetilde{H}_t$ is the status of the candidate hidden layer, H_t is the final hidden state, $\odot$ represents element-wise multiplication, W_{ph} and W_{hh} refer to the weight matrices, and b_h represents the offset vector.

Output layer. The perceptron receives the flattened user behavioral features as input for further processing. Finally, the softmax function is used to calculate the probability that the recipient may share the private image with his friends. An image may contain multiple category that are considered to be sensitive information, and the disclosure of any one category is likely to result in the compromise of other sensitive information within the same image. As a result, the risk of an image being leaked is defined as the maximum probability of leakage among all sensitive information contained within the image. Furthermore, when an image is shared with multiple recipients, the sender's risk is defined as the maximum probability of leakage among these recipients.

$$P_{img^l} = max \left\{ \frac{\exp(W_i H + b_i)}{\sum_i^M \exp(W_i H + b_i)} \right\} \tag{5}$$

$$P_{rec \to user}^{sen} = max \left\{ P_{img_j^l} \right\} \tag{6}$$

where P_{img^l} represents the risk value of the private image img^l containing sensitive private information l being leaked, $P_{rec \to user}^{sen}$ represents the risk value of private image leakage, M represents the number of friends of the recipient user, H is the hidden layer state, W_i represents the weight matrix with the corresponding shape, and b_i represents the offset vector.

We calculate the cross-entropy loss of each privacy category in the privacy image and shared friends to optimize the risk prediction model:

$$L = -\sum_i^M y_i \cdot log \hat{y}_i \tag{7}$$

where y_i is the true label of the ith shared friend, and $\hat{y}_i$ represents the corresponding predicted probability.

4   Experimental Results and Evaluation

4.1   Dataset

In this paper, privacy information refers to sensitive information set by users, such as faces, mobile phones, cars, and other objects in images. In the dissemination of social images, users usually pay more attention to this sensitive information. For the extraction of sensitive private information, we use the Pascal VOC2012 dataset to train the model of privacy information recognition. The Pascal VOC2012 dataset is one of the commonly used datasets in object detection and image segmentation tasks. To facilitate research, we define its twenty categories as sensitive private information. In the model of privacy risk prediction in social networks, the related research is lagging behind and there is no suitable public datasets because the dataset of privacy assessment involves user privacy. On one hand, the existing datasets in privacy assessment research mainly come from data mining, recommender systems, and other fields. However, these datasets are collected for their specific fields and do not contain the content related to the user image-sharing history required in this paper. On the other hand, the existing publicly published social network datasets usually only contain information such as user node characteristics and social network graph structures, or other published texts. For example, Facebook and Twitter datasets contain the nodes and edges of the network graphs; Reddit dataset consists of Reddit forum posts; Epinions dataset is collected from online product review sites that contains reviewers' ratings of products and reviewers' trustworthiness of other reviewers. Therefore, we simulate user image-sharing scenarios and generate a simulated dataset to conduct our experiments, while the relationship between users is based on the existing real social network dataset (e.g., Facebook dataset). The details of Facebook dataset [15] is listed in Table 1.

Table 1. Details of Facebook Dataset.

Dataset	Social circles: Facebook
Node	4039
Edge	88234
Average degree	43
Max degree	1045
Min degree	1

According to the degree of user nodes, we generate different number of close friends and private image-sharing records for them. The user with the largest degree has 1045 friends, while the user with the smallest degree only has one friend. The average number of friends is 43 for all users. Since different user nodes have different degree, users with very few friends have single behavioral characteristics and are mostly zombie users, so we do not consider users with only one friend. For other users, we randomly select a subset of his/her friends as intimate users, and randomly select a group of privacy features for each user in the subset to generate a random number of images containing the same privacy features, simulating the private image records received by the user. Then, for the private images received by the user, we randomly select a subset of his/her friends to share these images, and the maximum size for the subset comprising intimate users is set to 20, whereas the maximum size for the subset encompassing friends involved in information propagation is set to 60. In this way, the user's behavior of forwarding private images is simulated. For each user's private image-sharing records, we choose the top 80% of the records as the training set and the rest as the validation set. Since the sharing behavior characteristics of each user usually remain stable, and it is difficult for randomly generating records to simulate the characteristics of users in the time series, our validation set also includes the test set. It is worth noting that although these private image-sharing history records are simulated and synthesized, the social network topology is real, and the simulation process covers a wide range of possible sharing behaviors, so it can be used to conduct the experiments and verify the effectiveness of our proposed model.

4.2 Evaluation Metrics and Baseline Methods

Evaluation Metrics. Users may forward a variety of private images from different users to other users, and the number of forwarding will be quite different. To evaluate the performance of the proposed model, we use the metrics of macro-F1, weighted-F1, and the accuracy, which are calculated as follows.

$$macroF1 = \frac{1}{n}\sum_{i=1}^{n}\frac{2 \times TP_i}{2 \times TP_i + FP_i + FN_i} \tag{8}$$

$$weightedF1 = \frac{1}{n}\sum_{i=1}^{n}\frac{w_i \times 2 \times TP_i}{2 \times TP_i + FP_i + FN_i} \tag{9}$$

$$accuracy = \frac{\sum_i TP_i}{N} \tag{10}$$

Where n represents the number of friends of the user, w_i represents the ratio of the number of private image records shared by the user with the ith friend to the number of shared records owned by the user, TP_i represents the number of historical behaviors correctly predicted that the user shared the private image with user i, FP_i represents the number of historical behaviors misclassified as the sharing behavior from the user to the ith friend, FN_i represents the number of incorrectly predicted historical behaviors in the set of all sharing behavior from the user to the ith friend, N represents the total number of behavior records of users sharing private images.

For the metrics, the higher the value, the better the model performance. Among them, macro-F1 is based on the weighted average of the user's friends. Each friend has the same weight, which can take into account the influence of friends who share a small number of private image records, while weighted-F1 gives each friend a different weight according to the proportion of users sharing private image records with each friend. The more records are shared, the greater the weight, which can take into account the influence of the close relationship between users and friends. The metric of accuracy is used to represent the ability of the model to correctly predict the recipients of private image-sharing. It is based on the weighted average of the number of private image sharing history records. Each record has the same weight, reflecting the overall performance of the model. Furthermore, although the model does not return the predicted recipient user directly, this does not affect the evaluation of model performance.

Variant Models. Due to the lack of related researches on the risk of privacy image leakage, we designed two additional variants PPRAS-LSTM and PPRAS-BILSTM as baselines, which adopt LSTM and BiLSTM instead of GRU in PPRAS, respectively.

4.3   Train the Neural Network

For the identification of private features in images, PPRAS uses the faster R-CNN model, which can achieve more accurate extraction of high-precision, multi-scale and small object image information. The training epoch is set to 15, the learning rate is dynamically adjusted and its initial value is set to 0.01. After every 3 epochs, the learning rate is adjusted to $1/3$ of the original. In the common metrics of the PASCAL VOC dataset, the Intersection over Union (IoU) of the predicted bounding box and the real bounding box is set to 0.5. When the mean average precision (mAP) reaches 0.8, our training goal is reached, which can meet the needs of PPRAS. The variation curve of training loss and learning rate with epoch is shown in Fig. 4(a), and the variation curve of mAP is shown in Fig. 4(b). As seen from Fig. 4(a), the loss value gradually begins to flatten and the model gradually stabilizes with the adjustment of the learning rate after the 6th epoch. Also, we can see from Fig. 4(b) that the mAP value reaches 0.8 and it tends to be stable after the 6th epoch. In addition, in order to better evaluate the performance of the prediction model, we set all components in the user privacy requirement vector to 1 in the experiment, indicating that the user has the strongest privacy demand.

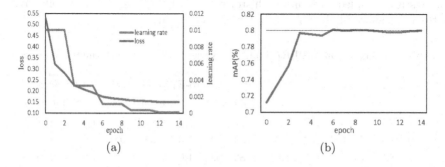

Fig. 4. Curve of loss, learning rate, and mAP versus epoch.

In the module of privacy risk prediction, the dimension of embedding vector is set to 8, the size of hidden layer in the recurrent neural networks is 128, the training epoch is set to 30, the batch size is set to 128, and the learning rate is set to 0.001. All parameters setting are all optimal on basis of many experiments. The above models are all trained on a Linux system equipped with NVIDIA GeForce RTX 3080.

4.4 Comparison of PPRAS and Two Variants in Three Performance Metrics

We have trained a personalized privacy risk prediction model for each user. Since the privacy image-sharing records of each user is different and the performance is also different, we analyze and verify the effectiveness of the model proposed in this paper from the minimum, maximum and average values. 75 users in the dataset only have one friend, it can be considered that they do not unintentionally forward their friend's images to cause the disclosure of their friend's private information. In addition, there is only one user's private image history, which makes it difficult for the model to learn its private image sharing behavior characteristics. Therefore, we trained the remaining 3963 users in the dataset. Table 2 lists the comparison of PPRAS and two variants in metrics of accuracy, macro-F1, and weighted-F1.

As shown in Table 2, the PPRAS and PPRAS-BiLSTM have a higher average score than PPRAS-LSTM. The worst model has an accuracy of 0, and the best model has an accuracy of 1. This is due to the fact that some users in the Facebook dataset have fewer friends. In today's social network environment, we can think that these users do not use the Facebook social platform frequently, so the number of historical records generated by the simulation is small, resulting in a large range of changes in accuracy. Accordingly, the range of changes in macro-F1 score and weighted-F1 score is also large. For example, in a model with an accuracy of 0, the user with id 129 has 7 friends, but the user has only 5 shared records of sensitive private information, and the shared recipients are two different friends. Due to the difference in recipients used during model training and testing, coupled with limited historical records, so it is difficult to effectively extract the shared behavior characteristics of this user.

Table 2. Comparison of PPRAS and two variants in three performance metrics.

	accuracy			macro-F1 Score			weighted-F1 Score		
	min	max	mean	min	max	mean	min	max	mean
PPRAS-LSTM	0.0	1.0	0.981	0.0	1.0	0.973	0.0	1.0	0.979
PPRAS-BiLSTM	0.0	1.0	0.982	0.0	1.0	0.976	0.0	1.0	0.980
PPRAS	0.0	1.0	**0.982**	0.0	1.0	**0.977**	0.0	1.0	**0.980**

However, this does not mean that the number of friends is small and the prediction accuracy is low. For example, there are 2563 instances with an accuracy rate of 1, and 761 of them correspond to users with less than 10 friends, and the maximum number of friends is 547. There are at least 2 shared records of sensitive private information and 3483 at most. The number of shared records has an impact on the performance of risk prediction models, because the number of records can reflect the complexity of users' sharing behavior to some extent. In instances with accuracy of 0 or 1, the average number of shared records of the corresponding users is about 887. But in other instances, the average number of shared records of corresponding users is about 2060.

On the metrics of macro-F1 score and weighted-F1 score, PPRAS still have the same results or exceed the results of two variants. PPRAS-BiLSTM can learn more fine-grained features through the combination of forward LSTM and backward LSTM, while PPRAS improves performance by optimizing the gate structure and hidden state, both of which are better than PPRAS-LSTM. The macro-F1 score gives each friend of the user the same weight, so when the difference is large in the number of shared history records between different friends, the value can fully consider the influence of users with fewer private image sharing records. On the contrary, weighted-F1 score takes into account the number of sensitive private information shared by the user with each friend. The more records of shared sensitive private information, the greater the weight of the friend. This is also similar to the real social network. Different friends have different degrees of intimacy, and the frequency and amount of sensitive privacy information they interact with are also different. This is why the weighted-F1 score of the three models is significantly higher than the macro-F1 score. Based on these experimental results, we can conclude that PPRAS outperforms the other two variant models in three metrics on the Facebook dataset.

4.5 Comparison of PPRAS and Two Variants in Time Utility

The training time of the model is affected by various factors such as the processor, batch size, and other model parameters. For the privacy risk prediction models of all users, the results of training time and prediction time are listed in Table 3, where every result is the average of ten experiments based on the time required for 3963 users.

Table 3. Comparison of PPRAS and two variants in time utility.

	Training time (s)			Prediction time (s)		
	min	max	mean	min	max	mean
PPRAS-LSTM	1.99	4.39	2.88	0.55	0.61	0.57
PPRAS-BiLSTM	3.04	6.36	4.49	1.05	1.14	1.09
PPRAS	1.87	9.65	2.85	0.44	0.51	0.47

As shown in Table 3, the time efficiency of the PPRAS is significantly less than that of the PPRAS-LSTM and PPRAS-BiLSTM models in training and in the prediction process. This is because PPRAS-BiLSTM improves the learning ability of the model by combining forward and backward LSTM, but also greatly increases the time cost of training and prediction. By optimizing the internal structure of PPRAS-LSTM, PPRAS not only improves accuracy, but also reduces the amount of computation.

In addition, due to the lack of a real private image-sharing record dataset, the number of users' private image-sharing records is difficult to accurately estimate. The time efficiency of training and prediction using the simulation dataset is the same effect as the real dataset. The experimental results show that among 1138 users with more than 2000 sensitive privacy information sharing records, the training time of the PPRAS model has the shortest 2.54 s and the longest 9.65 s, and the average training time of each user is 3.36 s. Among the 1059 users with no more than 200 private image-sharing records, the shortest training time is 1.92 s, the longest is 2.99 s, and the average training time is 2.34 s. Therefore, we can conclude that PPRAS is the best among three methods in time utility.

4.6 Utility Analysis

In order to minimize the impact of system operation on the sharing experience of social network users, the running time should be reduced as much as possible. The time for PPRAS mainly includes two parts, namely privacy information detection and risk assessment calculation. In this paper, Faster R-CNN is used to detect image privacy information. In our experimental setup (refer to Sect. 4.3), it takes an average of 0.41 s to detect 10 images continuously. Therefore, when the image shared by the user does not contain private information, the delay for the user to share the image will take about 0.41 s longer than when this function is not enabled. When the images shared by users contain sensitive privacy information, the risk prediction model takes about 0.47 s to compute (refer to Table 3). PPRAS will cause the delay of sharing images to increase by about 0.88 s.

4.7 Parametric Analysis

To verify the sensitivity of hyperparameters, we conducted some experiments to analyze the impact of different parameter selections on model performance. For intuitive display, we only compare the average values of various indicators of the model.

The Number of Hidden Neurons in the GRU. The number of hidden neurons is selected from {32, 64, 128, 256, 512, 1024}, and its influence is shown in Table 4. As the number of hidden neurons increases, the effect of PPRAS gradually improves, and the experimental effect reaches the optimum when number of units is 512. When the number of units continues to increase, the performance of the model does not improve, but the training time and prediction time increase rapidly, indicating that the size of the hidden layer is sufficient to learn the user's sharing characteristics. Therefore, the number of units is set to 512 in experiment.

Table 4. The effect of the number of hidden neurons.

Units	Accuracy (mean)	Macro-F1 (mean)	Weighted-F1 (mean)	Training time (mean)	Predict time (mean)
32	0.97	0.95	0.97	2.78	0.39
64	0.98	0.96	0.97	2.74	0.39
128	0.98	0.97	0.98	2.76	0.39
256	0.98	0.97	0.98	2.75	0.40
512	0.98	0.98	0.98	2.85	0.47
1024	0.98	0.98	0.98	3.26	0.91

Number of GRU Network Layers. The number of network layers of GRU is selected from {1, 2, 3}, and the performance of different layers is shown in Table 5. It can be seen from Table 5 that the increase in the number of layers does not bring about a performance improvement. This is because the expressive ability of a 1-layer GRU network can express the user's sharing behavior, and too much superposition will lead to a decrease in time performance. Therefore, the number of network layers of GRU in this paper is 1.

Table 5. The impact of the number of GRU network layers.

Number of layers	Accuracy (mean)	Macro-F1 (mean)	Weighted-F1 (mean)	Training time (mean)	Predict time (mean)
1	0.98	0.98	0.98	2.85	0.47
2	0.98	0.98	0.98	4.34	0.87
3	0.98	0.98	0.98	5.87	1.27

Dimensions of User Feature Representation. Dimensions are selected from {8, 16, 32, 64, 128, 256}, and the performance of different values is shown in Table 6. It can be seen from the Table 6 that when the dimension reaches 128, the experimental effect reaches the best, which can express the sharing behavior characteristics of users. When the dimension increases to 256, the model performance does not change, and the time performance decreases. Therefore, we set the dimension of user feature representation to 128.

Table 6. The influence of the dimensionality of the user's feature representation.

Dim	Accuracy (mean)	Macro-F1 (mean)	Weighted-F1 (mean)	Training time (mean)	Predict time (mean)
8	0.97	0.94	0.96	2.81	0.47
16	0.97	0.96	0.97	2.83	0.48
32	0.98	0.97	0.98	2.84	0.49
64	0.98	0.97	0.98	2.81	0.50
128	0.98	0.98	0.98	2.85	0.47
256	0.98	0.98	0.98	2.84	0.48

5   Conclusions

For improvement of user privacy awareness in social networks, this paper proposes a personalized privacy risk leakage prediction model of sharing image on social networks, which consists of two main parts including generating user-image privacy information vector and predicting image privacy leakage risk score. In the stage of generating user-image privacy information vector, starting from the user's personalized privacy requirements, PPRAS identifies the privacy regions of the shared image by using faster R-CNN, and then constructs user-image privacy information vectors. In the stage of predicting image privacy leakage risk score, PPRAS adopts GRU to predict image shared behaviors between users and calculates the leakage probability. The proposed model does not require the quantitative statistical information such as user attribute information, personal configuration, and other factors, and can provide more accurate personalized privacy services.

Due to insufficient data in related research, it is difficult to measure the characteristics of proposed model in the time dimension, and the training of the model is a huge and arduous task. Therefore, in future research, we will consider simulating real social networks and collecting corresponding data. At the same time, we will focus on solving the problems of diversity and temporal validity of private information in private images, improving the model to reduce training and prediction time to achieve better performance. It is worth noting that after the further development of related research on the identification and extraction of private information, the model proposed in this paper can be used for dealing with other types of information such as text, audio, and video for privacy risk assessment.

Acknowledgement. This research is funded by Science and Technology Project of Hebei Education Department (ZD2022105), Hebei Natural Science Foundation (F2020201023), and the high-level personnel starting project of Hebei University (521100221089).

References

1. Chen, J., et al.: Semantics-aware privacy risk assessment using self-learning weight assignment for mobile apps. IEEE Trans. Dependable Secur. Comput. **18**(1), 15–29 (2021). https://doi.org/10.1109/TDSC.2018.2871682
2. Chen, R., Kim, D.J., Rao, H.R.: A study of social networking site use from a three-pronged security and privacy threat assessment perspective. Inf. Manag. **58**(5), 103486 (2021). https://doi.org/10.1016/j.im.2021.103486
3. Chen, Z., Kandappu, T., Subbaraju, V.: PrivAttNet: predicting privacy risks in images using visual attention. In: 25th International Conference on Pattern Recognition, ICPR 2020, Virtual Event, Milan, Italy, pp. 10327–10334. IEEE (2020). https://doi.org/10.1109/ICPR48806.2021.9412925
4. Dey, R., Salem, F.M.: Gate-variants of gated recurrent unit (GRU) neural networks. In: IEEE 60th International Midwest Symposium on Circuits and Systems, MWSCAS 2017, Boston, MA, USA, pp. 1597–1600. IEEE (2017). https://doi.org/10.1109/MWSCAS.2017.8053243
5. Heravi, A., Mubarak, S., Choo, K.R.: Information privacy in online social networks: uses and gratification perspective. Comput. Hum. Behav. **84**, 441–459 (2018). https://doi.org/10.1016/j.chb.2018.03.016
6. Hong, H., Bao, W., Hong, Y., Kong, Y.: Privacy attributes-aware message passing neural network for visual privacy attributes classification. In: 25th International Conference on Pattern Recognition, ICPR 2020, Virtual Event, Milan, Italy, pp. 4245–4251. IEEE (2020). https://doi.org/10.1109/ICPR48806.2021.9412853

7. Li, X., Xin, Y., Zhao, C., Yang, Y., Chen, Y.: Graph convolutional networks for privacy metrics in online social networks. Appl. Sci. **10**(4), 1327 (2020)
8. Li, X., Xin, Y., Zhao, C., Yang, Y., Luo, S., Chen, Y.: Using user behavior to measure privacy on online social networks. IEEE Access **8**, 108387–108401 (2020). https://doi.org/10.1109/ACCESS.2020.3000780
9. Li, X., Zhao, C., Tian, K.: Privacy measurement method using a graph structure on online social networks. ETRI J. **43**(5), 812–824 (2021)
10. Lin, D., Steiert, D., Morris, J., Squicciarini, A.C., Fan, J.: REMIND: risk estimation mechanism for images in network distribution. IEEE Trans. Inf. Forensics Secur. **15**, 539–552 (2020). https://doi.org/10.1109/TIFS.2019.2924853
11. Liu, C., Zhu, T., Zhang, J., Zhou, W.: Privacy intelligence: A survey on image privacy in online social networks. ACM Comput. Surv. **55**(8), 161:1–161:35 (2023). https://doi.org/10.1145/3547299
12. Orekondy, T., Schiele, B., Fritz, M.: Towards a visual privacy advisor: understanding and predicting privacy risks in images. In: IEEE International Conference on Computer Vision, ICCV 2017, Venice, Italy, pp. 3706–3715. IEEE Computer Society (2017). https://doi.org/10.1109/ICCV.2017.398
13. Oukemeni, S., Rifà-Pous, H., Puig, J.M.M.: IPAM: information privacy assessment metric in microblogging online social networks. IEEE Access **7**, 114817–114836 (2019). https://doi.org/10.1109/ACCESS.2019.2932899
14. Sharma, V., Mir, R.N.: Saliency guided faster-RCNN (SGFr-RCNN) model for object detection and recognition. J. King Saud Univ. Comput. Inf. Sci. **34**(5), 1687–1699 (2022). https://doi.org/10.1016/j.jksuci.2019.09.012
15. Su, X., et al.: A comprehensive survey on community detection with deep learning. CoRR abs/2105.12584 (2021)
16. Tonge, A., Caragea, C.: Dynamic deep multi-modal fusion for image privacy prediction. In: Liu, L., et al. (eds.) The World Wide Web Conference, WWW 2019, San Francisco, CA, USA, pp. 1829–1840. ACM (2019). https://doi.org/10.1145/3308558.3313691
17. Wang, Y., Nepali, R.K.: Privacy impact assessment for online social networks. In: 2015 International Conference on Collaboration Technologies and Systems, CTS 2015, Atlanta, GA, USA, pp. 370–375. IEEE (2015). https://doi.org/10.1109/CTS.2015.7210451
18. Wang, Z., He, K., Wang, X., Niu, B., Li, F.: Traffic characteristic based privacy leakage assessment scheme for android device. J. Commun. **41**, 155–164 (2020)
19. Wu, N., Peng, C., Mou, Q.: Information entropy metric methods of association attributes for differential privacy. Acta Electonica Sinica **47**(11), 2337 (2019)
20. Xinyu, W., Ben, N., Fenghua, L.I., Kun, H.E.: Risk assessing and privacy-preserving scheme for privacy leakage in app. J. Commun. **40**, 13–23 (2019)
21. Xiong, P., Liang, L., Zhu, Y., Zhu, T.: PriTxt: a privacy risk assessment method for text data based on semantic correlation learning. Concurr. Comput. Pract. Exp. **34**(5), e6680 (2022). https://doi.org/10.1002/cpe.6680
22. Yang, G., Cao, J., Sheng, Q., Qi, P., Li, X., Li, J.: DRAG: dynamic region-aware GCN for privacy-leaking image detection. In: Thirty-Sixth AAAI Conference on Artificial Intelligence, EAAI 2022 Virtual Event, pp. 12217–12225. AAAI Press (2022). https://doi.org/10.1609/aaai.v36i11.21482
23. Yu, Y., Si, X., Hu, C., Zhang, J.: A review of recurrent neural networks: LSTM cells and network architectures. Neural Comput. **31**(7), 1235–1270 (2019). https://doi.org/10.1162/neco_a_01199

24. Zhang, T., Zhao, K., Yang, M., Gao, T., Xie, W.: Research on privacy security risk assessment method of mobile commerce based on information entropy and Markov. Wirel. Commun. Mob. Comput. 2020, 8888296:1–8888296:11 (2020). https://doi.org/10.1155/2020/8888296
25. Zhong, H., Squicciarini, A.C., Miller, D.J., Caragea, C.: A group-based personalized model for image privacy classification and labeling. In: Sierra, C. (ed.) Proceedings of the Twenty-Sixth International Joint Conference on Artificial Intelligence, IJCAI 2017, Melbourne, Australia, pp. 3952–3958. ijcai.org (2017). https://doi.org/10.24963/ijcai.2017/552

A Pipelined AES and SM4 Hardware Implementation for Multi-tasking Virtualized Environments

Yukang Xie⬤, Hang Tu, Qin Liu$^{(\boxtimes)}$⬤, and Changrong Chen

Key Laboratory of Aerospace Information Security and Trusted Computing,
Ministry of Education, School of Cyber Science and Engineering, Wuhan University,
Wuhan 430072, China
{bathtub,tuhang,qinliu,chenchangrong}@whu.edu.cn

Abstract. Virtualization techniques are becoming increasingly prevalent and are driving trends in hardware development to offer parallelization support for multi-tasking. Existing works on hardware designs of the Advanced Encryption Standard (AES) and SM4 encryption algorithms have primarily focused on optimizing metrics such as throughput and area, but have not fully addressed the demands in virtualized environments. In this article, we propose innovative optimization schemes that partition the resources in AES and SM4 cipher modules into smaller, independent units that can execute tasks from different guests in parallel. Such designs can improve hardware utilization efficiency and enhance the user experience in virtualized environments. Our FPGA-validated designs achieve comparable circuit performance in terms of throughput/area efficiency to existing work. Experiments show that in virtualized environments lacking block-wise parallelism (e.g., cipher block chaining (CBC) mode), our approach reduces context switches over 50% and decreases average task pending time around 75% with similar hardware needs.

Keywords: Hardware acceleration · Parallelism · Virtualization · AES · SM4

1 Introduction

Modern computer architecture practices commonly employ dedicated hardware to offload compute-intensive tasks, such as graphics processing and tensor processing, from CPUs [8,17]. Cryptographic operations lend themselves well to hardware implementation due to their routine patterns of execution, along with their demands for considerable resources. To address various use cases ranging from server-level acceleration to co-processing on embedded devices, industry professionals and researchers have made significant efforts to optimize the hardware implementation of widely-used encryption algorithms. These optimizations have targeted several metrics, including throughput, circuit area, and countermeasures against side-channel attacks. The rise of virtualization techniques

Z. Tari et al. (Eds.): ICA3PP 2023, LNCS 14488, pp. 275–291, 2024.
https://doi.org/10.1007/978-981-97-0801-7_16

[1,14], in which multiple tasks from different guests are executed on a shared physical infrastructure, has introduced new optimization requirements for hardware designers. In this paper, we propose an innovative optimization scheme for implementing two widely-used symmetric-key encryption algorithms, Advanced Encryption Standard (AES) and SM4, to enhance the designs' suitability for virtualized environments.

Our motivation stems from an observation of existing hardware implementations of AES and SM4, particularly those with pipelined structures. While these implementations leverage additional register resources to maximize circuit frequency and achieve higher throughput, this can result in resource wastage when the pipeline is not fully utilized during operation. Such issue becomes especially pronounced when the guest employs a block operation mode that necessitates sequential processing of data blocks, meaning that the hardware can only handle one data block at a time throughout the entire process. In our designs, resources within an AES or SM4 cipher module can be subdivided into smaller units based on the number of pipeline stages. From the perspective of guests, each resource unit operates independently and can be combined with a specific user key, allowing the module to execute tasks from different guests in parallel. As a result, hardware utilization rate is enhanced in these serial-styled block operation modes. Moreover, since there are more independent resource units available, hypervisors or operating systems can switch contexts less frequently between different guest tasks in the cipher hardware, thereby improving overall system performance.

In this paper, we adopt this optimization strategy for both AES and SM4 implementations, developing four-staged pipeline structures capable of processing tasks from four different guests in parallel. Our designs are open-sourced and have been validated and evaluated on a Zynq UltraScale+ and a Kintex-7 FPGA device. Synthesis and implementation results demonstrate that our designs achieve comparable results in circuit efficiency relative to existing studies. We also notice that previous studies have devoted less attention to the practical methods for incorporating cryptographic hardware into larger systems. To fill this gap, we have developed a prototype cryptosystem based on our cryptographic modules and discussed its key design principles. Experiments conducted on the prototype system reveal that, in virtualized environments, our designs reduce context switches by over 50% and decrease average task pending time by approximately 75% in operation modes lacking block-wise parallelism. Our designs achieve these improvements while maintaining similar hardware consumption compared to classic architectures.

The remainder of this paper is organized as follows: Sect. 2 discusses the background of both AES and SM4 algorithms, as well as related work in their hardware implementations. Section 3 and 4 presents the details of the proposed designs, while Sect. 5 includes evaluation and experimental data. Section 6 provides a discussion on our designs, and the final section concludes the paper.

2   Background

This section presents a concise overview of the AES and SM4 algorithms, as well as a survey of prior research on their hardware implementations.

2.1   Overview of AES and SM4

AES is a widely adopted symmetric-key encryption algorithm that operates on blocks of 128 bits [6]. As shown in Fig. 1, this algorithm relies on iterative rounds of transformations-including substitution, permutation, and mixing operations-to encrypt and decrypt input data. The number of transformation rounds varies between 10, 12, or 14, depending on the key length, which is either 128, 192, or 256 bits. In each round, the intermediate state is XORed with a 128-bit round key, which is generated by the AES key schedule algorithm using the main cipher key. The encryption and decryption processes of the AES algorithm are inverses of each other, and their datapaths can be partially combined in specific implementations.

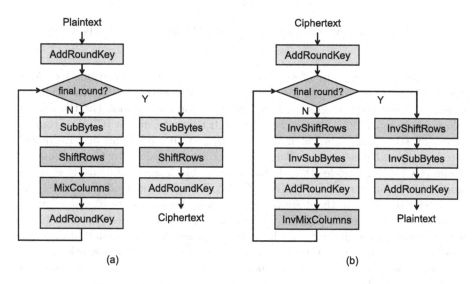

Fig. 1. AES (a) encryption and (b) decryption process.

SM4, released by the Office of State Commercial Cryptography Administrator of China, is a standard algorithm in ISO/IEC 18033-3:2010/Amd 1:2021 [11]. It operates on 128-bit blocks and uses a fixed key length of 128 bits. Like AES, this algorithm comprises an encryption/decryption process and a key scheduling process. Figure 2 illustrates that SM4 employs 32 rounds of permutations, with each round making use of a 32-bit round key. The round permutation in SM4 is less complex than the round transformation in AES, and only updates a 32-bit word in each iteration. The datapaths for encryption and decryption in the SM4 algorithm are identical, with the only difference being the order in which the round keys are used.

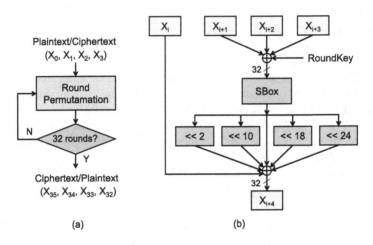

Fig. 2. (a) Overall process and (b) round permutation of SM4.

To process input data of arbitrary size, block cipher algorithms like AES and SM4 must be employed in specific modes of operation. Various modes of operation have been developed, including Electronic Codebook (ECB), Cipher Block Chaining (CBC), Output Feedback (OFB), and Counter (CTR). From an efficiency standpoint, current mainstream operation modes can be categorized into two types: block-wise parallel support mode, exemplified by Galois/Counter Mode (GCM), and modes that do not support block-wise parallelism, as represented by Counter with CBC-MAC (CCM). In operation modes that support block-wise parallelism, there are no dependencies between data blocks, allowing them to be processed in parallel. In contrast, operation modes that lack block-wise parallelism require data blocks to be processed serially as each block's processing result depends on the previous one. Since both types of operation modes are widely adopted, hardware designers must consider their characteristics and the impact they bring.

2.2   Related Works and Discussion

Optimized hardware implementations for both AES and SM4 have been extensively studied. Various designs have been proposed in prior works, targeting different optimization metrics based on application requirements. Numerous research efforts have investigated the trade-off between circuit area and hardware throughput. For example, the authors in [4] presented an ultra-compact design for AES on FPGA, utilizing only 184 logic slices and no Block RAM

(BRAM). Another work in [20] introduced a low-area SM4 design costing only 164 slices on a Virtex-6 FPGA device.

Conversely, to achieve higher throughput, unrolled and pipelined designs have been proposed, such as those in [3,12,13,19]. Although these designs can achieve throughputs of several tens of gigabits per second, they also consume considerable resources. Moreover, the theoretical throughput of these unrolled and pipelined designs can only be achieved when operating in a mode that supports block-wise parallelism, such as ECB and GCM. This limitation is discussed in [21], which abandons pipelined design and presents a round-based architecture with a highly compressed datapath structure.

However, our analysis indicates that hardware resource wastage in pipelined structures is not necessarily inevitable in operation modes that do not support block-wise parallelism, such as CBC and CCM. In these block-chaining operation modes, idle resources in the pipeline can still serve tasks from other guests, which is particularly beneficial for scenarios involving virtualization. Building upon this observation, we propose rolled, four-staged pipeline structures for both AES and SM4 algorithms in this paper. This design supports task-wise parallelism for up to four guests, effectively raising the hardware utilization rate and addressing the limitations of existing approaches.

3   Proposed AES and SM4 Architecture

In this section, we introduce the core of our work, including the implementation of the four-staged encryption and decryption modules for both AES and SM4 algorithms.

Our design is implemented in Chisel [2], a modern hardware description language with rich parameterization and modularity features [18]. The architectures of both the AES and SM4 modules adopt a round-based, four-staged pipeline structure. The key idea for enabling guest-level parallelism is to maintain a set of control information for each running task in the pipeline. This allows the circuit to index the corresponding round key when conducting key-related operations (e.g., AddRoundKey in AES). With this measure, each pipeline stage can be regarded as an independent resource from the user's perspective.

We made several crucial design decisions in our implementation:

(a) Although using lookup tables to implement the S-box is common in high-throughput designs, pipelined combinational logic circuits usually show better throughput/area efficiency. We employed the S-box presented by Maximov and Ekdahl in [15] because it is a state-of-the-art design that can be evenly partitioned without introducing excessively large registers.

(b) We chose a round-based architecture because it is less area-consuming and can adapt to more usage scenarios. When higher throughput is required, users can instantiate multiple modules to achieve that. We have determined the number of pipeline stages to be four, based on the objective of balancing register usage and maximizing circuit frequency.

(c) Since the key scheduling module is used much less frequently than the encryption/decryption modules in most cases, we decided to separate their implementation. The prototype cryptosystem proposed in Sect. 4 combines a key scheduling module with multiple encryption/decryption modules.

3.1   AES Encryption and Decryption Architecture

In our envisioned usage scenarios, multiple cryptographic modules will be instantiated to handle the workload from a large number of guests. Consequently, we focus on simplifying the functionality of each individual module, opting not to merge the encryption and decryption modules. Users can configure and generate the necessary number of encryption and decryption modules in cryptosystems based on their application needs, leveraging the flexibility provided by Chisel.

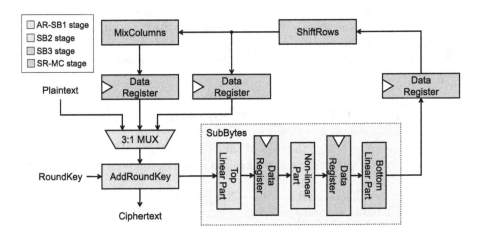

Fig. 3. Proposed AES encryption architecture.

The proposed four-staged architectures for AES encryption and decryption are illustrated in Fig. 3 and Fig. 4. For both processes, two pipeline registers are inserted before and after the non-linear transformation part of the S-box to reduce the critical path in the SubBytes/InvSubBytes module. Additionally, two other pipeline registers are inserted before and after the MixColumns/InvMixColumns module, separating its datapath from the linear transformation part of the S-box.

Here we use the AES encryption architecture as an example to explain the involved pipeline stages. The input data is first processed in the AR-SB1 stage, which contains the AddRoundKey module and the top linear part of the S-box. The second and third stages, SB2 and SB3, contain the non-linear part and bottom linear part of the S-box, respectively. The last stage of the round transformation is SR-MC, which includes the ShiftRows and MixColumns modules. There is an extra register bypassing the MixColumns module since the final round of

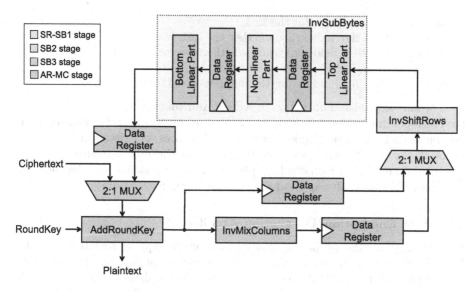

Fig. 4. Proposed AES decryption architecture.

AES encryption does not contain it. Figure 5 depicts a timing diagram for AES-128 encryption, illustrating how distinct input blocks are processed through the pipeline. Note that a round transformation of AES in our design takes four cycles, therefore an AES-128 encryption operation requires 40 cycles. Given that the four-staged pipeline can process four tasks in parallel, the proposed architecture can accomplish four AES-128 encryptions in 43 cycles.

	Time													
	1	2	3	4	5	6	7	8	//	39	40	41	42	43
Guest 1	AR-SB1	SB2	SB3	SR-MC	AR-SB1	SB2	SB3	SR-MC		SB3	SR-MC			
Guest 2		AR-SB1	SB2	SB3	SR-MC	AR-SB1	SB2	SB3		SB2	SB3	SR-MC		
Guest 3			AR-SB1	SB2	SB3	SR-MC	AR-SB1	SB2	//	AR-SB1	SB2	SB3	SR-MC	
Guest 4				AR-SB1	SB2	SB3	SR-MC	AR-SB1		SR-MC	AR-SB1	SB2	SB3	SR-MC

Fig. 5. Example pipeline timing diagram for AES-128 encryption.

In our design, the data register incorporated in the pipeline contains not only a 128-bit AES state but also a set of control information. Each control information set is associated with a specific task, indicating the status of the AES state for that task. The subfields of the control information include:

- `taskID`: a 2-bit specifier for the task.
- `isIdle`: a 1-bit flag indicating whether the task is valid.
- `keyLength`: a 2-bit field indicating whether the task is running in AES-128, AES-192, or AES-256.
- `rounds`: a 4-bit field indicating the number of rounds for the task.

The only key-related operation in AES is AddRoundKey. This module reads the corresponding round key from an external source based on the control information of its input task. Unlike unrolled architectures, our design natively supports all three AES variants of key length. From the user's perspective, each instance of our design includes four physical resources that can independently perform AES tasks, offering good flexibility in virtualized environments.

3.2   SM4 Encryption/Decryption Architecture

Figure 6 displays the proposed SM4 encryption/decryption architecture. The SM4 algorithm employs an unbalanced Feistel network structure, updating a 32-bit word in each round. SM4 encryption and decryption share the same datapath and only differ in the order of applying round keys. Similar to the AES architectures, two pipeline registers are allocated in the S-box, before and after the non-linear transformation part. Two additional two pipeline registers are placed before and after the linear transformation logic outside the S-box. Furthermore, the unchanged part in each round is preserved by shift registers.

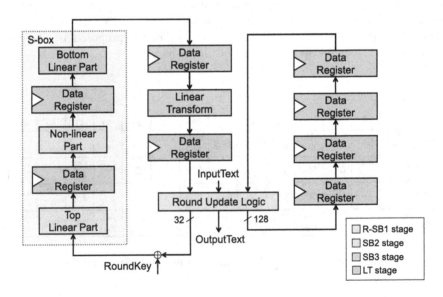

Fig. 6. Proposed SM4 encryption/decryption architecture.

Our proposed architecture for SM4 utilizes the same control information set as that used by the AES architecture, except for the absence of the `keyLength` subfield since SM4 only supports a fixed key length of 128 bits. Although it is possible to add an `isEnc` subfield to the control information set to enable both encryption and decryption tasks to run within the same module, we chose not to implement this feature to avoid increasing the complexity of the external controller.

4 A Prototype Cryptosystem for Virtualization

As a reference for applying our design in virtualized environments, we further construct a prototype cryptosystem comprising four instances of the cryptographic modules discussed in Sect. 3. In this section we present the design of our prototype system, as well as its programming sequence for users.

4.1 Prototype Cryptosystem Architecture

Figure 7 depicts the structure of the prototype cryptosystem. As each cryptographic module is capable of processing four guest tasks in parallel, this prototype system can support a maximum of sixteen guest tasks concurrently. Each guest task can operates either in ECB mode or CBC mode. The prototype system employs the AXI interface for easy integration into FPGA designs.

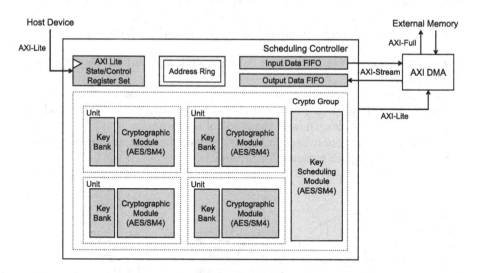

Fig. 7. The multi-instance prototype cryptosystem.

In our design, each cryptographic module is combined with a key bank to form a "unit". The input and output dataflows of a single cryptographic module are managed by the logic within the unit. Our implementation adopts the idea of another Chisel design proposed in [9], enabling users to easily generate the desired combination of AES/SM4 encryption/decryption units through simple configuration. Since the key scheduling module is less utilized than the cryptographic module in typical cases, we have a single key scheduling module serving four units. The round keys generated by this scheduling process are stored in key banks for future use. This combination is referred to as a "crypto group" in our prototype. The control logic within the crypto group manages the plaintext, ciphertext, and key dataflows. The crypto group is then wrapped into a scheduling controller, which receives task configuration sent by a host device through the AXI-Lite bus.

Maximizing the utilization of cryptographic acceleration hardware is contingent on ensuring efficient data transfer between the accelerator and external devices. Therefore, we employed a DMA IP in our prototype system to move plaintext/ciphertext between external memory and our design. To ensure even parallelization of tasks from different guests within the module, the scheduling controller must issue DMA commands from various tasks alternately. This is achieved through an address ring structure within the scheduling controller.

4.2   Programming Sequence of the Prototype Cryptosystem

As previously mentioned, from the user's perspective, our prototype system offers sixteen independent cryptographic resources capable of performing encryption or decryption operations. Typically, the hypervisor or OS allocates a single cryptographic resource to each guest. Figure 8 illustrates the programming sequence for utilizing cryptographic resources, which consists of the following steps:

- keySet: The hypervisor/OS sets a guest's main key, initiating the key scheduling process.
- taskAssign: An encryption or decryption task from the guest begins. The task configuration specifies the source and target addresses of input and output data, the data length, and the mode of operation.
- keyDestroy: The round keys in the key storage bank are destroyed, freeing up the cryptographic resource for future use.

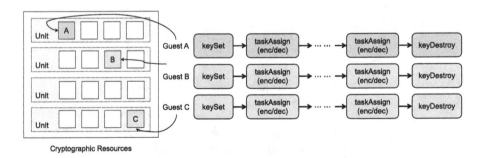

Fig. 8. Programming sequence of the prototype cryptosystem.

As the manager of hardware resources, the hypervisor or OS is responsible for maintaining the mapping between guests and cryptographic resources. Once a cryptographic resource is assigned to a guest, it cannot serve other guests until the hypervisor or OS sends a keyDestroy signal to release the resource. In situations where all cryptographic resources are occupied by various guests, and new guest requests arrive, context switches are triggered. The hypervisor or OS must free up some cryptographic resources and reallocate them to the incoming guests. The overhead associated with context switches arises from both software and hardware aspects. On the software side, the hypervisor or OS must update the mapping between guests and resources. Meanwhile, on the hardware side, before conducting encryption or decryption operations, the module must perform the key scheduling process again.

5   Experimental Results

This section presents a comparison of circuit performance across proposed designs and existing works, along with experiments evaluating the efficiency gains in multi-guest virtualized environments afforded by our designs.

5.1   Circuit Performance Comparison

The designs proposed in Sect. 3 are thoroughly verified on a Zynq UltraScale+ (xczu7eg-ffvc1156) and a Kintex-7 (xc7k325t-ffg900) FPGA device using Xilinx Vivado 2022.1 for synthesis and implementation.

Table 1. Comparison of AES Implementation

Designs	Platforms	Slices	Max. freq. (MHz)	Throughput (Gbps)	Efficiency (Mbps/slice)
[4]	XC3S50-5	184	45.64	0.037	0.2
[13]	XC7VX690T	3436	516.8	66.1	19.2
[16]	XC6VLX240T	4830	617.63	79	16.36
[19]	XC5VLX	5974	622.4	79.7	13.3
[10]	XC7V585TFF	1355+80BRAMs	374	47.8	-
This work	XC7K325T	371	414.3	5.3	14.29
	XCZU7EG	430	535.3	6.85	15.93

The synthesis results of both the AES and SM4 encryption/decryption architectures are compared with previous works in Tables 1 and 2. These tables report hardware utilization, maximum frequency, throughput, and efficiency. Since our round-based designs strike a balance between throughput and hardware utilization, we have selected works focusing on throughput maximization and area minimization for comparison. The proposed AES implementation achieves a maximum frequency of 414.3 MHz and a throughput of 5.3 Gbps, occupying 371 slices. The corresponding figures for the proposed SM4 implementation are 415.3 MHz,

Table 2. Comparison of SM4 Implementation

Designs	Platforms	LUTs	FFs	Slices	Max. freq. (MHz)	Throughput (Gbps)	Efficiency (Mbps/slice)	Efficiency (Mbps/LUT+FF)
[20]	XC6VLX240T	-	-	164	253	0.25	1.54	-
[7]	EP4SE230F29	687	448	-	210.26	0.82	-	0.72
		7667	5438	-	212.13	27.1	-	2.07
[3]	XCZU7EV	8655	10071	-	923.36	118.19	-	6.31
This work	XC7K325T	575	816	236	415.3	1.66	7.03	1.19
	XCZU7EG	574	816	266	547.3	2.19	8.23	1.58

1.66 Gbps, and 236 slices, respectively. The results show that our designs surpass existing compact designs of AES and SM4 in both throughput and efficiency and achieve fair results in efficiency compared to existing high-speed designs. On the one hand, our designs are lightweight enough to be deployed on resource-constrained platforms where virtualization techniques are applied. On the other hand, due to the good throughput/area efficiency of our design, users can always choose to instantiate multiple modules to achieve higher throughput, which is essentially equivalent to the unrolled structure.

It is important to note that, despite the lack of full authority in the comparative data due to differences in FPGA platforms and the randomness of synthesis, placing, and routing optimization algorithms in Vivado, our design successfully maintains competitive hardware performance while introducing support for parallelism of multiple guest tasks. This is reasonable because our designs introduce relatively little additional control information.

5.2   Multi-guest Efficiency Improvement

To validate the performance improvement of the proposed design in multi-tasking virtualization scenarios, we conducted experiments based on the prototype system described in Sect. 4, integrated with the Zynq UltraScale+ MPSoC. The results demonstrate that our design is well-suited for virtualized environments.

Two critical performance metrics in virtualized environments are the average pending time for each task and context switching rate. We compared these two metrics between the classic design and the proposed architecture in the experiment. In the classic design, guest-level parallelism is not exploited, and each guest utilizes the full resources of a cryptographic hardware instance. However, in our design, each guest only occupies a single pipeline stage of the instance and is processed in parallel. To avoid potential unfairness introduced by differences in hardware utilization and given that most existing works are not open-sourced, we decided to simulate the classic architecture using our own design with slight modifications. This is possible because the working method of the classic architecture can be considered a subset of that of our proposed design. There are four AES encryption modules initialized in the prototype system.

In the application for the experiment, a randomly generated task sequence, containing tasks belonging to random guests, was read in by the scheduling function, which allocates cryptographic resources to guest tasks and records their pending time. A context switch occurs when idle resources exist but no resources have been occupied by the incoming guest. We fix the total length of tasks in the task sequence to 4 GB, and the length of each task is randomly determined between 4 MB and 64 MB. The number of guests is set to 8, 16, 32, and 64. Additionally, two operation modes, CBC and ECB, are tested in the experiment, representing block-wise parallelism unsupported and block-wise parallelism supported modes, respectively. We conducted twenty runs for each setup, and each task sequence is reused for setups with the same guest number configuration.

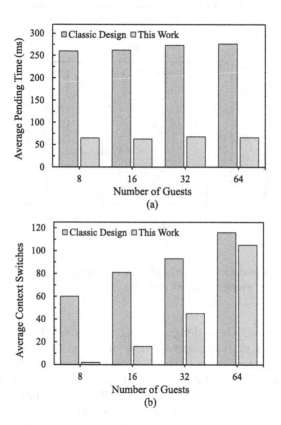

Fig. 9. Multi-guest performance evaluation in CBC mode in terms of (a) task average pending time and (b) average context switches.

Figure 9 and 10 summarize the average task pending time and the count of context switches for both the classic design and our architecture. The experimental results show that the proposed design can significantly reduce context switches when the guest number is within a certain range. When the number of guests is 32 or less, our architecture can reduce context switches by over 50%. This is because our design contains sixteen independent resources, which is four times that of the classic one. However, when the number of guests greatly exceeds

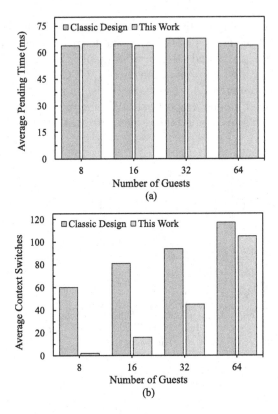

Fig. 10. Multi-guest performance evaluation in ECB mode in terms of (a) task average pending time and (b) average context switches.

the number of resources, our design no longer has a significant advantage, indicating that the user must instantiate more instances to further reduce context switches. In the operation mode that does not support block-wise parallelism, our design shows a clear performance advantage over the classic architecture in terms of pending time, reducing it by 75%. Classic pipelined designs typically perform poorly in this situation as they cannot fully utilize all the resources in the pipeline. In contrast, when operating in a mode that supports block-wise parallelism, both architectures exhibit similar performance in terms of average pending time. It is worth noting that in practical applications, the advantages of our design may be more pronounced because, in actual hypervisors or operating systems, context switches have greater software overhead.

6   Discussion

In multi-tasking virtualized environments, a critical job of the hypervisor/OS is to schedule tasks coming from different guests with limited hardware resources, where context switching is inevitable. Frequent context switching could cause a loss in performance and, for cryptographic hardware, could potentially introduce security implications since sensitive information, including user keys, is conveyed more regularly via the bus. Designers of large-scale systems thus generally instantiate a substantial number of dedicated hardware to meet demand. Our work provides an alternative optimization perspective, demonstrating that this issue can be addressed by dividing resources into finer granularity at the hardware level.

The key to our design is increasing the available hardware resources from the user's perspective by making the pipeline stages independent. The independent pipeline stage brings advantages in two ways. First, it offers great flexibility to the hypervisor/OS level. The hypervisor/OS can apply different scheduling strategies, such as allowing each guest to occupy at most one resource in situations with a large number of guests to reduce context switching, or temporarily allowing a superior user to occupy all the resources of an instance in priority scheduling, which makes our design equivalent to the classic architecture. Second, it enhances the hardware utilization rate when the hardware operates in a mode that does not support block-wise parallelism.

Although the proposed structure is round-based, our design can be easily applied to an unrolled or partially unrolled structure to obtain a higher throughput, since the additional control information register in each pipeline stage is not a significant overhead compared to the 128-bit AES/SM4 state itself. Furthermore, our design strategy is not limited to specific cryptographic algorithms and can potentially be applied to more cryptography implementations, as long as it does not take excessive overhead to maintain the control information for key-related operations. An interesting example would be Ascon [5], the new lightweight cryptography standard selected by NIST, which adopts a sponge structure and only involves key-related operations at its initialization and finalization stage. This feature makes it natural for Ascon implementation to have independent pipeline structure.

7   Conclusion

This paper presented optimized hardware implementations for AES and SM4 algorithms, targeting multi-tasking virtualization scenarios. Evaluation on FPGAs shows that our designs can significantly reduce the average pending time and context switch demands in virtualized environments compared to classic architectures, while achieving competitive throughput/area efficiency. A prototype cryptosystem was further developed to showcase the practicality of our designs, indicating their scalability and potential applications in various scenarios. By enhancing hardware utilization rate and decreasing pending time in virtualized environments, our work demonstrates optimization strategies useful for a wider range of metrics and applications.

The source codes of the AES and SM4 modules are available on GitHub at https://github.com/bathtub-01/cluster-AES.

Acknowledgements. The work was supported by the National Natural Science Foundation of China (No. 62272348).

References

1. Alam, I., et al.: A survey of network virtualization techniques for internet of things using SDN and NFV. ACM Comput. Surv. (CSUR) **53**(2), 1–40 (2020). https://doi.org/10.1145/3379444
2. Bachrach, J., et al.: Chisel: constructing hardware in a Scala embedded language. In: Proceedings of the 49th Annual Design Automation Conference, DAC 2012, pp. 1216–1225. Association for Computing Machinery, New York (2012). https://doi.org/10.1145/2228360.2228584
3. Chen, Y., et al.: Exploring the high-throughput and low-delay hardware design of SM4 on FPGA. In: 2022 19th International SoC Design Conference (ISOCC), pp. 211–212 (2022). https://doi.org/10.1109/ISOCC56007.2022.10031393
4. Chu, J., Benaissa, M.: Low area memory-free FPGA implementation of the AES algorithm. In: 22nd International Conference on Field Programmable Logic and Applications (FPL), pp. 623–626 (2012). https://doi.org/10.1109/FPL.2012.6339250
5. Dobraunig, C., Eichlseder, M., Mendel, F., Schläffer, M.: Ascon v1.2: lightweight authenticated encryption and hashing. J. Cryptol. **34**(3), 1–42 (2021). https://doi.org/10.1007/s00145-021-09398-9
6. Dworkin, M.J., et al.: Advanced encryption standard (AES) (2001). https://doi.org/10.6028/NIST.FIPS.197
7. Guan, Z., Li, Y., Shang, T., Liu, J., Sun, M., Li, Y.: Implementation of SM4 on FPGA: trade-off analysis between area and speed. In: 2018 IEEE International Conference on Intelligence and Safety for Robotics (ISR), pp. 192–197 (2018). https://doi.org/10.1109/IISR.2018.8535613
8. Gui, C.Y., et al.: A survey on graph processing accelerators: challenges and opportunities. J. Comput. Sci. Technol. **34**, 339–371 (2019)
9. Guo, X., El-Hadedy, M., Mosanu, S., Wei, X., Skadron, K., Stan, M.R.: Agile-AES: Implementation of configurable AES primitive with agile design approach. Integration **85**, 87–96 (2022)
10. Harb, S., Ahmad, M.O., Swamy, M.N.S.: A high-speed FPGA implementation of AES for large scale embedded systems and its applications. In: 2022 13th International Conference on Information and Communication Systems (ICICS), pp. 59–64 (2022). https://doi.org/10.1109/ICICS55353.2022.9811140
11. Information technology - Security techniques - Encryption algorithms - Part 3: Block ciphers - Amendment 1: SM4. Standard, ISO/IEC 18033–3:2010/Amd 1:2021 (2021)
12. Kumar, T.M., Reddy, K.S., Rinaldi, S., Parameshachari, B.D., Arunachalam, K.: A low area high speed FPGA implementation of AES architecture for cryptography application. Electronics **10**(16) (2021). https://doi.org/10.3390/electronics10162023. https://www.mdpi.com/2079-9292/10/16/2023
13. Liu, Q., Xu, Z., Yuan, Y.: A 66.1 GBPS single-pipeline AES on FPGA. In: 2013 International Conference on Field-Programmable Technology (FPT), pp. 378–381 (2013). https://doi.org/10.1109/FPT.2013.6718392

14. Mansouri, Y., Babar, M.A.: A review of edge computing: features and resource virtualization. J. Parallel Distrib. Comput. **150**, 155–183 (2021). https://doi.org/10.1016/j.jpdc.2020.12.015.   https://www.sciencedirect.com/science/article/pii/S0743731520304317
15. Maximov, A., Ekdahl, P.: New circuit minimization techniques for smaller and faster AES SBoxes. IACR Trans. Crypt. Hardw. Embed. Syst. **2019**(4), 91–125 (2019). https://doi.org/10.13154/tches.v2019.i4.91-125. https://tches.iacr.org/index.php/TCHES/article/view/8346
16. Oukili, S., Bri, S.: High speed efficient advanced encryption standard implementation. In: 2017 International Symposium on Networks, Computers and Communications (ISNCC), pp. 1–4 (2017). https://doi.org/10.1109/ISNCC.2017.8071975
17. Peccerillo, B., Mannino, M., Mondelli, A., Bartolini, S.: A survey on hardware accelerators: taxonomy, trends, challenges, and perspectives. J. Syst. Architect. **129**, 102561 (2022). https://doi.org/10.1016/j.sysarc.2022.102561
18. Rautakoura, A., Hämäläinen, T.: Does SOC hardware development become agile by saying so: a literature review and mapping study. ACM Trans. Embed. Comput. Syst. **22**(3) (2023). https://doi.org/10.1145/3578554
19. Shahbazi, K., Ko, S.B.: High throughput and area-efficient FPGA implementation of AES for high-traffic applications. IET Comput. Digit. Tech. **14**(6), 344–352 (2020)
20. Shang, M., Zhang, Q., Liu, Z., Xiang, J., Jing, J.: An ultra-compact hardware implementation of SMS4. In: 2014 IIAI 3rd International Conference on Advanced Applied Informatics. 86–90 (2014). https://doi.org/10.1109/IIAI-AAI.2014.28
21. Ueno, R., et al.: High throughput/gate AES hardware architectures based on datapath compression. IEEE Trans. Comput. **69**(4), 534–548 (2020). https://doi.org/10.1109/TC.2019.2957355

Blockchain-Assisted Privacy-Preserving Public Auditing Scheme for Cloud Storage Systems

Wenyu Xiang[1], Jie Zhao[1], Hejiao Huang[1(✉)], Xiaojun Zhang[2], Zoe Lin Jiang[1], and Daojing He[1]

[1] School of Computer Science and Technology, Harbin Institute of Technology, Shenzhen, Guangdong 518055, China
xwyhit@163.com, zhaojswpu2017@163.com,
{huanghejiao,zoeljiang,hedaojing}@hit.edu.cn, zhangxjdzkd2012@163.com
[2] School of Computer Science, Research Center for Cyber Security, Southwest Petroleum University, Chengdu 610500, China

Abstract. Public auditing mechanism can delegate a third-party auditor (TPA) to check the remote data integrity on behalf of data owners. However, the TPA, as an idealized and benefit-oriented entity, may not provide correct auditing results on time. To date, a large number of public auditing schemes utilize the booming blockchain technique to resist dishonest TPA, but most of them are vulnerable to malicious miners who attempt to manipulate the randomness of auditing challenge generation. In this paper, we propose a novel Blockchain-assisted Privacy-preserving Public Auditing scheme, named BPPA. The BPPA scheme utilizes a smart contract deployed on the Ethereum blockchain to replace the TPA. To eliminate the impact of malicious miners, the smart contract employs unpredictable hash values of the nearest Ethereum blocks to generate the index locators. These locators segmentally produce index subsets of challenged data blocks, ensuring the unpredictability of auditing challenge messages. Meanwhile, BPPA achieves conditional identity anonymity for data owners through the employment of identity-based public key cryptography and key exchange technique. We prove the security of our scheme based on the computational Diffie-Hellman assumption and the discrete logarithm assumption. Furthermore, we analyze the performance from theoretical and experimental aspects, and the evaluation results demonstrate that our auditing scheme is effective and efficient.

Keywords: Cloud storage system · Remote data integrity · Ethereum blockchain · Malicious miner · Conditional identity anonymity

1 Introduction

Cloud storage service [18,24,25], as one of the most promising technologies, has been a practical tool for cloud users to alleviate the heavy burden caused by the explosive growth of data. While the cloud user takes advantage of these services, the ensuing security threats have aroused great concern [6]. Particularly, the

Z. Tari et al. (Eds.): ICA3PP 2023, LNCS 14488, pp. 292–310, 2024.
https://doi.org/10.1007/978-981-97-0801-7_17

integrity of outsourced storage data is one of the most important security issues. As ownership of the data is transferred from the cloud user to cloud service providers (CSPs), users lose physical control of these data. Then they always worry about whether outsourced data is lost/damaged due to hardware failures, software bugs, or human errors. In fact, the cloud server is a semi-trusted entity. The cloud server may delete some data blocks that are rarely accessed to save storage space, and hide these events to maintain a great reputation. Even worse, to obtain additional benefits, the CSPs may tamper with the user's sensitive data information, e.g., patient's medical data or network reporting data, which puts lives at risk. Therefore, it is very necessary to periodically check the integrity of outsourced storage data.

To realize the integrity verification of outsourced storage data, numerous remote data auditing mechanisms have been proposed [8,15,19]. They mainly entrust third-party auditors (TPA) to periodically verify the integrity of outsourced storage data on behalf of users. However, TPA is a benefit-oriented entity, which may betray the established auditing protocol. (1) TPA may be extremely curious about the sensitive storage data and endeavor to derive the original content from these data. (2) TPA may output a valid auditing result without performing the validation process indeed thus avoiding the auditing verification costs. (3) TPA may collude with the cloud server for economic profit and convince users the remote data is maintained intact. Accordingly, a full-trusted TPA may suffer from a single point of failure and is difficult to be found in a real-world scenario. In addition, traditional auditing schemes fail to consider the identity privacy protection of cloud users.

Blockchain technology [9] with data non-tamperability and transaction transparency can provide a good solution for the above TPA issue. Hence, many blockchain-based auditing schemes [7,20–22] were proposed. In these schemes, TPA extracts the time-independent random value from the latest block to sample challenges and records the verification information locally allowing cloud users to supervise the TPA's behavior. However, they are vulnerable to attacks from dishonest miners who can be bribed by adversaries. These miners are inclined to mine a satisfying block whose random value would not generate the challenge message locating to the corrupted storage data in the cloud server. Consequently, the adversary could bias the probability distribution of the target block, thereby manipulating the final auditing results. Besides, storing the verification information of numerous auditing tasks leads to excessive storage costs. The majority of existing schemes lack an analysis of these problems. Furthermore, the identity privacy of the cloud user is as important as the confidentiality of outsourced data. Most of them utilize the user's real identity for identification in the system, potentially exposing the user's real identity to other curious entities [21]. While unconditional anonymity will incur obscure problems during dispute resolution, it should be able to achieve the traceability of real identity from the pseudonym. Once a malicious user contaminates the cloud server for extra profit, a trustworthy authority can efficiently reveal his real identity and impose legal penalties. Therefore, it is crucial to realize conditional identity anonymity [23].

To address the aforementioned issues, we propose an efficient Blockchain-assisted Privacy-preserving Public Auditing scheme for cloud storage, named BPPA. The major contributions of this paper are summarized as follows.

- We propose a novel privacy-preserving cloud storage auditing scheme, which deploys a smart contract in Ethereum [17] as a substitute for the TPA. We employ unpredictable hash values of the nearest Ethereum blocks to generate the index locators which segmentally determine index subsets of challenged data blocks, so as to keep the randomness of the auditing challenge message.
- To protect users' identity privacy and reveal the malicious user at the same time, we design an identity anonymity mechanism by integrating key exchange technique into symmetric encryption algorithm. Even if an adversary breaks the ciphertext indistinguishability [3], it cannot deduce who is the original owner of these data. Instead, if a misbehaved user abuses the cloud storage system, its real identity will be revealed by a full-trusted KGC.
- We provide a detailed security analysis to prove that the proposed BPPA is provably secure, including the storage correctness guarantee, and malicious-miner resistance. We also present a comprehensive performance evaluation to demonstrate that BPPA has a higher performance than existing related auditing schemes.

The rest of this paper is organized as follows. We introduce the related work in Sect. 2. In Sect. 3, we present the background knowledge and define our system model, adversary model, as well as design goals. Next, we elaborate on the construction of the proposed BPPA in Sect. 4. Then we present a correctness and security analysis for BPPA in Sect. 5. In Sect. 6, a comprehensive performance evaluation is provided. Finally, we draw a conclusion for this work in Sect. 7.

2   Related Work

Cloud storage technology enables individuals and enterprises to enjoy convenient data outsourcing at a low cost. To check the integrity of outsourced data, Ateniese et al. [1] proposed a provable data possession (PDP) mechanism based on the asymmetric cryptosystem, which allows the legitimate user that has storage data at an untrusted cloud server to verify that the server possesses the original data without retrieving the whole data set. In the same year, Juels et al. [5] presented a proofs of retrievability (POR) mechanism, which adds an erasure code technique to ensure that the damaged data blocks are recovered. It should be noted that the basic idea of the above two data integrity auditing mechanisms is based on the probabilistic inspection method of random sampling. On the basis of literature [1,5], Shacham et al. [13] proposed the first provable secure PoR scheme, which is called compact proof of retrievability (CPoR). This not only indicates that a practical data auditing scheme needs to implement semantic security, but also that any credible verifier could conduct a secure auditing challenge verification process on remote data. Wang et al. [15] innovatively introduced a third-party auditor (TPA) to replace the user for the integrity verification of remote data, thus proposing an outsourced storage data privacy-preserving public auditing scheme with a random masking technique to resist the curious TPA. Later, with distinct features, massive data privacy-preserving public auditing schemes are proposed [10,19]. Particularly, Xu et al. [19] put forward novel proxy-oriented outsourcing with public auditing for cloud-based medical cyber-physical systems by employing the identity-based signature algorithm, which supports secure proxy-oriented data outsourcing between the original data owner and proxy. Scheme [10] combines lattice-assisted linear homomorphic signature with an identity-based data outsourcing public auditing scheme in clouds to achieve post-quantum security.

However, most of the existing data privacy-preserving auditing schemes [8,15,19] regard TPA as a full-trusted entity. In fact, a lazy TPA may not check the remote data integrity on time but conduct the centralized execution of the data auditing tasks, or even directly call the previous auditing result to prevaricate users. Furthermore, a malicious TPA may collude with cloud servers and directionally verify the integrity of those uncorrupted storage data. Zhang et al. [22] proposed a novel public integrity verification scheme for cloud storage, which integrates the booming blockchain technique into the PDP mechanism, so as to resist the procrastinating public auditor. After that, Zhang and Zhao [21] proposed a conditional anonymity privacy-preserving public auditing scheme for wireless body area networks, which could record each auditing result into the Ethereum blockchain as a transaction, thereby resisting TPA's malicious auditing behaviors. Wang et al. [16] proposed an innovative concept of practical private PDP based on RSA and blockchain, thus addressing the inefficient PDP implementation issue. To overcome the drawback of a non-manager group, Huang et al. [4] proposed an incentive public auditing scheme, with the ring signature and a novel blinding technology. Shu et al. [14] analyzed the adverse impact of malicious miners and explored the concept of the decentralized autonomous organization (DAO) while delegating a smart contract to complete auditing tasks. Nevertheless, this scheme is cumbersome and costly in the data auditing challenge-verification process, which is not conducive to deployment in the actual environment. Therefore, how to design a secure and effective privacy-preserving integrity verification is imperatively demanded.

3 Technical Preliminaries

3.1 Bilinear Pairing and Hardness Problem

Given two multiplicative cyclic groups $\mathbb{G}$ and $\mathbb{G}_T$ with the same prime order q, where $\mathbb{G}$ is generated by some generator g. A bilinear pairing $\tilde{e} : \mathbb{G} \times \mathbb{G} \to \mathbb{G}_T$ is a map which satisfies the following properties:

- Bilinearity: For all $A, B \in \mathbb{G}$, and $a, b \in \mathbb{Z}_q^*$, it has $\tilde{e}(A^a, B^b) = \tilde{e}(A, B)^{ab}$.
- Computability: For $\tilde{e}(A, B)$, it can be efficiently calculated, where $A, B \in \mathbb{G}$.
- Non-Degeneracy: For the generator g in group $\mathbb{G}$, it satisfies $\tilde{e}(g, g) \neq 1_{\mathbb{G}_T}$.

Discrete Logarithm (DL) Problem: Given a multiplicative cyclic group $\mathbb{G} =< g >$ with the order q, where g is a generator of $\mathbb{G}$. For any element $A \in \mathbb{G}$, the goal of the DL problem is to find the random value $a \in \mathbb{Z}_q^*$ to satisfy $A = g^a$ within a probabilistic polynomial-time (PPT) algorithm.

Computational Diffie-Hellman (CDH) Problem: For a multiplicative cyclic group $\mathbb{G} =< g >$ with the order q, where g is a generator of $\mathbb{G}$. Given the tuple $(g^a, g^b) \in \mathbb{G}^2$, the goal of the CDH problem is to compute g^{ab} within a PPT algorithm, where $a, b \in \mathbb{Z}_q^*$ are selected randomly.

3.2 Homomorphic Hash Function

A homomorphic hash function [2] not only inherits all the advantages of general hash functions, but also owns the following two outstanding features:

- Homomorphism: Given any two messages m_1, m_2 and scalars ϱ_1, ϱ_2, it has

$$H(\varrho_1 m_1 + \varrho_2 m_2) = H(m_1)^{\varrho_1} \times H(m_2)^{\varrho_2}.$$

  Note that if $\varrho_1 = \varrho_2 = 1$, then $H(m_1 + m_2) = H(m_1) \times H(m_2)$.
- Collision Resistance: For an unknown tuple $(m_1, m_2, \varrho_1, \varrho_2)$, there is no probabilistic polynomial-time adversary that can find the message m_3 to satisfy

$$H(m_3) = H(m_1)^{\varrho_1} \times H(m_2)^{\varrho_2},$$

if and only if $m_3 \neq m_1 \varrho_1 + m_2 \varrho_2$.

3.3 System Model

As depicted in Fig. 1, BPPA is made up of four entities: key generation center (**KGC**), data owner (**DO**), cloud server (**CS**), and blockchain.

- **DO:** The DO is an original data owner. Before uploading the outsourced storage data to the CS, it needs to execute a series of cryptographic operations for the original data. Besides, each DO needs to deploy a smart contract in the blockchain to check the integrity of remote data.
- **CS:** The CS is managed by the cloud server provider, which provides massive data storage space and powerful information processing capabilities for cloud users. After receiving a challenge message sent by the smart contract, CS generates the response auditing proof information.
- **KGC:** The KGC is a full-trusted authority, whose responsibility is to determine the system public parameters and the master private key. Also, it generates an anonymous identity and corresponding private key for each legal user in the system.
- **Blockchain:** The blockchain stores the metadata uploaded by DO as a transaction. The smart contract, which executes in the blockchain unbias, undertakes the duty of TPA. In the challenge-verification process, the smart contract extracts hash values from the latest block headers to generate the challenge message segmentally and sends it to CS. Later it verifies the response proof sent by CS utilizing the metadata, further outputs a fair auditing result.

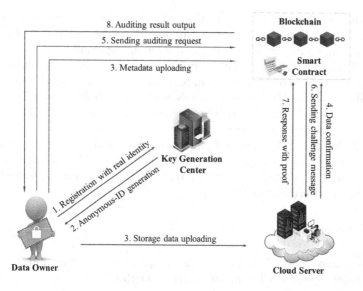

Fig. 1. A brief system model of BPPA

3.4   Adversary Model and Design Goals

The security threats of the system generally come from four aspects, including semi-trusted CS, misbehaved DO, malicious miners, and curious external attackers.

- **Semi-Trusted CS**: The semi-trusted CS may cover the data loss to maintain its reputation. Worse still, it may directly delete the key data blocks of some file and forge the corresponding signature tags, trying to pass the auditing challenge-verification process of the smart contracts.
- **Misbehaved DO**: The misbehaved DO may deliberately accuse CS of the correct behaviors. More specifically, a malicious DO sends correct metadata to the blockchain, and then uploads the wrong/incomplete storage data to the CS intentionally, thereby claiming CS for compensation.
- **Malicious Miners**: The malicious miners attempt to control the decisive block of the blockchain, thereby breaking the randomness and unpredictability of the auditing challenge message.
- **Curious External Attackers**: The external attacker may be curious about the content of the outsourced storage data and the real identity of the DO.

To address the aforementioned security threats, and further present a secure and efficient public cloud auditing scheme, the underlying design goals should be achieved.

- **Storage Correctness Guarantee.** The smart contract in the blockchain could correctly check the integrity of outsourced storage data on behalf of the DO.

- **Malicious-Miner Resistance.** Even if the malicious miner colludes with malicious CS, it cannot break the randomness of the generation of auditing challenge message, thus affecting the final auditing result.
- **Data Privacy Preservation.** Any adversary (including the cloud server) cannot recover the original data content from the ciphertext of DO.
- **Conditional Identity Privacy.** DO's identity should be anonymized. Meanwhile, once a DO in the system has malicious behavior, it should track and reveal its real identity information.
- **High Efficiency.** The proposed public auditing scheme should be as lightweight as possible, including the communication overhead between any two logical entities, the signature computation cost, and the auditing computation costs of the smart contract.

4   Our Protocol

Now, we construct a novel blockchain-assisted privacy-preserving public auditing scheme, which consists of the following six algorithms.

System Initialization: Input a security parameter 1^λ, the KGC generates the system's public parameters and the secret master keys below.

1. Let $\tilde{e}$ be a bilinear pairing: $\tilde{e} : \mathbb{G} \times \mathbb{G} \to \mathbb{G}_T$, and choose g as the generator of $\mathbb{G}$, where $\mathbb{G}$ and $\mathbb{G}_T$ denote two multiplicative cyclic groups with same prime order q.
2. Select a random value $x \leftarrow \mathbb{Z}_q^*$ as the system's master private key, and compute $X = g^x$ as its corresponding master public key.
3. Define five anti-collision hash functions: $H : \mathbb{Z}_q^* \to \mathbb{G}$, $h_1 : \mathbb{G} \times \{0,1\}^* \to \mathbb{Z}_q^*$, $h_2 : \mathbb{G} \to \mathbb{Z}_q^*$, $h_3 : \mathbb{Z}_q^* \times \mathbb{G} \times \mathbb{G} \to \mathbb{Z}_q^*$, $h_4 : \mathbb{Z}_q^* \times \{0,1\}^* \times \{0,1\}^* \times \mathbb{Z}_q^* \to \mathbb{Z}_q^*$, respectively. Also, pick a symmetric encryption/decryption security algorithm pair (ENC, DEC).
4. Set a pseudo-random generator $Prg : SK_{prg} \times I \to \mathbb{Z}_n^*$, and a pseudo-random function $Prf : SK_{prf} \times \mathbb{Z}_q^* \times I \to \mathbb{Z}_q$, where SK_{prg} and SK_{prf} indicate the keysets for Prg and Prf respectively, and I denotes the set of sort location of each data block.
5. Issue all of the public parameters $Para = \{(\tilde{e}, \mathbb{G}, \mathbb{G}_T, g, q), X, (ENC, DEC), Prg, Prf, H, h_1 \sim h_4\}$, and save $\{x, SK_{prg}, SK_{prf}\}$ secretly.
6. CS and each DO are required to initial an external account on the Ethereum blockchain and deposit sufficient assets in it. Subsequently, the DO creates a smart contract that is responsible for the integrity verification of outsourced storage data.

Anonymous-ID Generation and Key Extraction: With unique and real identity information $RID \in \mathbb{Z}_q^*$, each DO needs to interact with KGC to generate its private key. The detailed algorithm steps are as follows.

1. Select a random value $\omega \leftarrow \mathbb{Z}_q^*$, and compute $W = g^\omega$. The DO sends the user registration information (RID, W) to KGC by a secure channel.

2. When receiving (RID, W) from DO, KGC selects a random value $r \leftarrow \mathbb{Z}_q^*$ and computes $R = g^r$. Then it computes the symmetric secret key $\kappa = h_1(W^x, Time)$, and generates the DO's pseudonym $PSID = ENC_\kappa(RID \oplus h_2(W))$, as well as its corresponding private key

$$sk = r + x \cdot h_3(PSID, X, R) \mod (q-1), \tag{1}$$

where $\|$ indicates the concatenation of each portion, and $Time$ denotes the valid time period of $PSID$. Finally, KGC adds R into the public parameters $Para$, and sends the tuple $(sk, Time)$ to the DO by a safe channel.

3. Upon receiving the tuple $(sk, Time)$ from the KGC, the DO recomputes $\kappa = h_1(X^\omega, Time)$, and $PSID = ENC_\kappa(RID \oplus h_2(W))$. After that, it verifies whether the verification equation is correct:

$$g^{sk} \overset{?}{=} R \cdot X^{h_3(PSID, X, R)} \mod q. \tag{2}$$

If the above verification equation does not hold, the DO asks the KGC to regenerate the signature private key by repeating the above algorithm steps. Otherwise, the DO accepts $(sk, Time)$.

Data Processing and Outsourcing: The DO achieves the secure processing and outsourcing of data file as follows.

1. Given a data file $M = \{m_1, m_2, \cdots, m_n\} \in \mathbb{Z}_q^n$ with an identifiable file name $Name \in \{0,1\}^*$, the DO picks a one-time value $\varpi \leftarrow \mathbb{Z}_q^*$, and computes the file tag $Tag = h_4(PSID, Name, n, \varpi)$.
2. Select a secure secret key $sk_{prf} \in SK_{prf}$ for pseudo-random function Prf. The DO encrypts each data block m_i as follows.

$$m_i^* = \eta_i \oplus m_i \in \mathbb{Z}_q, \tag{3}$$

where $\eta_i \leftarrow Prf_{sk_{prf}}(Tag, i) \in \mathbb{Z}_q$, and $i \in [1, n]$, thus $M = \{m_1, m_2, \cdots, m_n\} \in \mathbb{Z}_q^n$ is encrypted as $M^* = \{m_1^*, m_2^*, \cdots, m_n^*\} \in \mathbb{Z}_q^n$. Moreover, the DO computes $HF = H(Tag \| M^*)$.
3. For $i \in [1, n]$, choose a random value $\beta_i \leftarrow \mathbb{Z}_q^*$ and compute $H(\beta_i)$. Then the DO computes the authenticator for each ciphertext block m_i^* below.

$$\delta_i = (H(Tag) \times H(m_i^* + \beta_i))^{sk} \in \mathbb{G}. \tag{4}$$

4. Generate a transaction T_x^Λ for metadata $\Lambda = \{PSID, HF, (H(\beta_1), H(\beta_2), \cdots, H(\beta_n))\}$, and send it to Ethereum blockchain. Once the T_x^Λ is written into the blockchain, the Ethereum immediately returns the hash value $\hbar(T_x^\Lambda)$ of the T_x^Λ to the DO, where $\hbar(\cdot)$ shows the secure hash function of Ethereum such as Keccak-256. Finally, the DO uploads the storage data $\Omega = \{PSID, Name, \varpi, (M^*, \delta), \hbar(T_x^\Lambda)\}$ to CS, where $\delta = \{\delta_1, \delta_2, \cdots, \delta_n\}$.
5. Upon receiving the storage data Ω, the CS firstly retrieves HF by sending the transaction digest $\hbar(T_x^\Lambda)$ to Ethereum. Then it checks whether the following verification equation is established:

$$H(h_4(PSID, Name, n, \varpi) \| M^*) \overset{?}{=} HF. \tag{5}$$

If it holds, the CS accepts Ω by emitting *Sucess*; Otherwise, it refuses Ω by emitting *Error*.

Challenge Message Generation: The DO sends auditing task request $Req = \{PSID, Name, \hbar(T_x^\Lambda)\}$ to the smart contract. Once the Req from DO is successfully received, the smart contract in Ethereum generates the following auditing challenge message on behalf of the DO.

1. Retrieve the BlockNumber of the current newest block by using the chain search interface (e.g., *block.number*), and obtain the corresponding Block-Height Bl. Then it seriatim extracts the hash value of ϑ blocks backward according to the logical order, thus outputting $\{\hbar(\nabla)_{Bl-\vartheta+1}, \hbar(\nabla)_{Bl-\vartheta+2}, \cdots, \hbar(\nabla)_{Bl-1}, Bl\}$. Here, the $\hbar(\nabla)_{Bl-1}$ denotes the header hash value of the previous block of the current newest block, and $\vartheta \geq 20$.
2. Compute $k_1 = \hbar_1(\hbar(\nabla)_{Bl-\vartheta+1}||\hbar(\nabla)_{Bl-\vartheta+2}||\cdots|| \hbar(\nabla)_{Bl-\lceil\vartheta/2\rceil}) \in SK_{prg}$, and $k_2 = \hbar_2(\hbar(\nabla)_{Bl-\lceil\vartheta/2\rceil+1} ||\hbar(\nabla)_{Bl-\lceil\vartheta/2\rceil+2}||\cdots||\hbar(\nabla)_{Bl-1}||Bl) \in SK_{prg}$, where $\hbar_1(\cdot)$ and $\hbar_2(\cdot)$ are two secure hash functions defined on smart contract.
3. Send the auditing challenge message $Chal = \{c, k_1, k_2\}$ to CS through the inline event trigger, where c denotes the number of challenged data block.

Proof Information Generation: Once receiving $Chal$ from the smart contract, the CS generates the response auditing information proof below.

1. Compute $i_{\theta_1} = Prg_{k_1}(\theta_1)$, $i_{\theta_2} = Prg_{k_2}(\theta_2)$, and set $i_\theta = i_{\theta_1} \cup i_{\theta_2}$, where $\theta_1 = 1, 2, \cdots, \lceil c/2 \rceil$, $\theta_2 = \lceil c/2 \rceil + 1, \cdots, c$, and $i_{\theta_1} \cap i_{\theta_2} = \emptyset$. For each $\theta = 1, 2, \cdots, c$, the CS computes $\nu_\theta = Prf_{k_2}(h_4(PSID, Name, n, \varpi), \theta)$.
2. With the data block locator (i_θ, ν_θ), CS computes the combined data blocks $\xi = \sum_{\theta=1}^{\theta=c} \nu_\theta m_{i_\theta}^* \bmod q$, and aggregates the corresponding authenticators $\psi = \prod_{\theta=1}^{\theta=c} H(h_4(PSID, Name, n, \varpi))^{\nu_\theta}$, and $\delta = \prod_{\theta=1}^{\theta=c} \delta_{i_\theta}{}^{\nu_\theta}$.
3. Generate a new transaction on the response auditing proof information $Proof = \{\xi, \delta, \psi\}$, and transmit it to the smart contract.

Proof Verification: Upon receiving the $Proof$ from CS, the smart contract checks the integrity of the outsourced storage data by executing the following algorithm steps.

1. Recompute the data block locator (i_θ, ν_θ) by the same way of CS, where $\theta = 1, 2, \cdots, c$.
2. According to the data auditing request $Req = \{PSID, Name, \hbar(T_x^\Lambda)\}$ and (i_θ, ν_θ), the smart contract acquires these precisely parsed hash values $\{H(\beta_{i_1}), H(\beta_{i_2}), \cdots, H(\beta_{i_c})\}$ by performing transaction data query function (e.g., $Oraclize_{query}$), and further verifies the integrity of the storage data by checking the following auditing verification equation:

$$\tilde{e}(\delta, g) \stackrel{?}{=} \tilde{e}(\psi \cdot H(\xi) \cdot \prod_{\theta=1}^{\theta=c} H(\beta_{i_\theta})^{\nu_\theta}, R \cdot X^{h_3(PSID, X, R)}). \tag{6}$$

3. If the above auditing verification equation holds, the smart contract outputs the auditing result as $True$ and sends $\{True, ST\}$ to the DO; Otherwise, it outputs the auditing result as $False$ and sends $\{False, ST\}$ to the DO, where ST denotes the timestamp of the auditing result output.

5 Evaluation of the Proposed Mechanism

5.1 Correctness

The correctness of verification Eq. (2) is proved below.

$$
\begin{aligned}
g^{sk} &= g^{(r+x \cdot h_3(PSID,X,R)) \mod (q-1)} \mod q \\
&= g^r \cdot (g^x)^{h_3(PSID,X,R)} \mod q \\
&= R \cdot X^{h_3(PSID,X,R)} \mod q.
\end{aligned}
$$

The correctness of auditing verification Eq. (6) on the smart contract side is elaborated below.

$$
\begin{aligned}
\tilde{e}(\delta, g) &= \tilde{e}(\prod_{\theta=1}^{\theta=c} \delta_{i_\theta}{}^{\nu_\theta}, g) \\
&= \tilde{e}(\prod_{\theta=1}^{\theta=c} (H(Tag) \times H(m_{i_\theta}^* + \beta_{i_\theta}))^{\nu_\theta}, g^{sk}) \\
&= \tilde{e}(\prod_{\theta=1}^{\theta=c} (H(h_4(PSID, Name, n, \varpi)))^{\nu_\theta} \times \prod_{\theta=1}^{\theta=c} H(\nu_\theta(m_{i_\theta}^* + \beta_{i_\theta})), g^{sk}) \\
&= \tilde{e}(\psi \cdot H(\sum_{\theta=1}^{\theta=c} \nu_\theta m_{i_\theta}^*) \cdot H(\sum_{\theta=1}^{\theta=c} \nu_\theta \beta_{i_\theta}), g^{sk}) \\
&= \tilde{e}(\psi \cdot H(\xi) \cdot \prod_{\theta=1}^{\theta=c} H(\beta_{i_\theta})^{\nu_\theta}, R \cdot X^{h_3(PSID,X,R)}).
\end{aligned}
$$

5.2 Security Analysis

In this section, we prove that the proposed BPPA scheme is secure in terms of storage correctness guarantee, and malicious-miner resistance.

Theorem 1. *In BPPA, it is computationally infeasible for an adversary $\mathcal{A}$ (including the malicious CS) to falsify a valid response auditing proof information to pass the verification process of smart contract, as long as the hardness assumptions of DL and CDH problems hold.*

Proof. Now, we demonstrate that if there exists an adversary $\mathcal{A}$ (including the malicious CS) breaking the storage correctness guarantee of our auditing scheme with non-negligible probability ε in terms of **Game 1** and **Game 2**, we can construct a challenger $\mathcal{C}$ with corresponding probability ε' to win **Game 1** and **Game 2** by running $\mathcal{A}$ as a subroutine.

Game 1: In this game, a well-trained $\mathcal{A}$ tries to replace or falsify some damaged/lost data block, and further generate the response auditing proof to pass the integrity verification process. The details are described below.

Upon receiving the auditing challenge message $Chal = \{c, k_1, k_2\}$ from the smart contract, instead of generating the correct response auditing proof information honestly, $\mathcal{A}$ falsifies a response auditing proof as $Proof' = \{\xi', \delta, \psi\}$ for the damaged/lost data block $m_{i_\theta}^{*\prime}$, where $\xi' = \sum_{\theta=1}^{\theta=c} \nu_\theta m_{i_\theta}^{*\prime}$, $\delta = \prod_{\theta=1}^{\theta=c} \delta_{i_\theta}^{\nu_\theta}$, and $\psi = \prod_{\theta=1}^{\theta=c} H(h_4(PSID, Name, n, \varpi))^{\nu_\theta}$. Hence, the forged response auditing proof $Proof' = \{\xi', \delta, \psi\}$ can pass the following integrity verification equation:

$$\tilde{e}(\delta, g) = \tilde{e}(\psi \cdot H(\xi') \cdot \prod_{\theta=1}^{\theta=c} H(\beta_{i_\theta})^{\nu_\theta}, R \cdot X^{h_3(PSID, X, R)}). \tag{7}$$

As a matter of fact, the CS can honestly generate a valid response auditing proof information $Proof = \{\xi, \delta, \psi\}$ as required. Thus, the correct response auditing proof $Proof$ can pass the following integrity verification equation:

$$\tilde{e}(\delta, g) = \tilde{e}(\psi \cdot H(\xi) \cdot \prod_{\theta=1}^{\theta=c} H(\beta_{i_\theta})^{\nu_\theta}, R \cdot X^{h_3(PSID, X, R)}). \tag{8}$$

According to the above Eq. (7) and Eq. (8), we can obtain that

$$\psi \cdot H(1 \cdot \xi') \cdot \prod_{\theta=1}^{\theta=c} H(\beta_{i_\theta})^{\nu_\theta} = \psi \cdot H(1 \cdot \xi) \cdot \prod_{\theta=1}^{\theta=c} H(\beta_{i_\theta})^{\nu_\theta}.$$

Since $\Delta\xi = \xi' - \xi \neq 0$, and assuming that the output of $\mathcal{O}_H$ in response to query "1" is g^a, we can further get $g^{a \cdot \Delta\xi} = 1_{\mathbb{G}}$. Thus, we can get $g^a = g^{-\Delta\xi}$ as the solution of the DL problem, where $-\Delta\xi$ is the inversion of $\Delta\xi$ such as $-\Delta\xi + \Delta\xi \equiv 0 \mod q$. Note that the probability of game failure is the same as the probability of $\Delta\xi = 0 \mod q$. The probability of $\Delta\xi = 0 \mod q$ is $1/q$. Therefore, we can address the DL problem with a probability of $\varepsilon' = 1 - 1/q$, which contradicts the hardness assumption of the DL problem.

Game 2: This game is same as **Game 1**, with the exception of one different. When some data block $m_{i_\theta}^*$ outsourced to the cloud is lost or damaged, $\mathcal{A}$ not only generates the combined message ξ' but also forges the corresponding aggregate signature information δ', thus trying to pass the integrity verification process. The details are described below.

Upon receiving an auditing challenge message $Chal = \{c, k_1, k_2\}$ from the smart contract, the well-trained $\mathcal{A}$ tries to forge the response auditing proof as $Proof' = \{\xi', \delta', \psi\}$ with a non-negligible probability ε, where $\xi' = \sum_{\theta=1}^{\theta=c} \nu_\theta m_{i_\theta}^{*\prime}$, $\delta' = \prod_{\theta=1}^{\theta=c} \delta_{i_\theta}^{\prime\nu_\theta}$, and $\psi = \prod_{\theta=1}^{\theta=c} H(h_4(PSID, Name, n, \varpi))^{\nu_\theta}$. Thus, the forged response auditing proof $Proof' = \{\xi', \delta', \psi\}$ can pass the following integrity verification equation:

$$\tilde{e}(\delta', g) = \tilde{e}(\psi \cdot H(\xi') \cdot \prod_{\theta=1}^{\theta=c} H(\beta_{i_\theta})^{\nu_\theta}, R \cdot X^{h_3(PSID,X,R)}). \tag{9}$$

Actually, when receiving an auditing challenge message $Chal = \{c, k_1, k_2\}$ from the smart contract, the honest cloud server can generate the corresponding response auditing proof $Proof = \{\xi, \delta, \psi\}$ as required. The proof satisfies the following integrity verification equation:

$$\tilde{e}(\delta, g) = \tilde{e}(\psi \cdot H(\xi) \cdot \prod_{\theta=1}^{\theta=c} H(\beta_{i_\theta})^{\nu_\theta}, R \cdot X^{h_3(PSID,X,R)}). \tag{10}$$

According to the Eq. (9) and Eq. (10), we can get that

$$\delta'/\delta = H(1 \cdot \xi')/H(1 \cdot \xi).$$

Similarly, since $\Delta\xi = \xi' - \xi \neq 0$, and assuming that the output of $\mathcal{O}_H$ in response to query "1" is g^a, we can obtain $\delta' \cdot \delta^{-1} = g^{a \cdot \Delta\xi}$. Without loss of generality, we can set $g^{ab} = \delta' \cdot \delta^{-1} \in \mathbb{G}$ as the solution of the CDH problem. As a result, if there exists an $\mathcal{A}$ successfully winning the **Game 2** by forging a valid response auditing proof with a non-negligible probability ε, then the challenger $\mathcal{C}$ will also have a non-negligible probability $\varepsilon' = \varepsilon$ to address the hardness assumption of CDH problem by running $\mathcal{A}$ as a subroutine.

Therefore, based on the above-detailed security analysis of **Game 1** and **Game 2**, we conclude that the proposed BPPA scheme achieves storage correctness guarantees. It is computationally infeasible for an $\mathcal{A}$ to falsify a valid response auditing proof information to pass the verification process of the smart contract, as long as the hardness assumptions of DL and CDH problems hold.

Theorem 2. *The proposed BPPA scheme can resist malicious miners in the blockchain, thus ensuring the randomness of the auditing challenge message.*

Proof. For challenge message generation, the smart contract computes $\{k_1, k_2\}$ which are locators to the subset of data file M^*, by concatenating the hash values of the latest ϑ blocks. Each hash value $\hbar(\nabla)_i, i \in \{Bl - \vartheta + 1, Bl - \lceil \vartheta/2 \rceil\}$ determines to the final concatenation, and further influences the generation of k_1 (similarly in k_2). We denote the block with BlockHeight $Bl - \lceil \vartheta/2 \rceil$ as $\mathbf{Bl_1}$, and the block with BlockHeight Bl as $\mathbf{Bl_2}$. Since the hash value of a block is unknown until the block is generated, the most effective way for an adversary $\mathcal{A}$ (including the malicious cloud server) is to ensure the subsets located by the hash values of $\mathbf{Bl_1}$ and $\mathbf{Bl_2}$ would not contain the damage storage data blocks.

We define a mapping function $\chi : \varepsilon \mapsto \{0, 1\}$, where ε is the set containing all the blocks in the blockchain. For a decisive block $\mathbf{Bl}$, if the hash value of it generates the corrupted data block index, then $\chi(\mathbf{Bl}) = 1$. Otherwise, $\chi(\mathbf{Bl}) = 0$. We assume there are games among honest miners and malicious miners tempted by an adversary $\mathcal{A}$ [11]. For one decisive block, a game starts when a fixed initial block indexed by 0 is broadcasted. Further, we denote the n-th block as the decisive block. Once the decisive block **Bl** is broadcasted by honest miners and $\chi(\mathbf{Bl}) = 0$, malicious miners can still find a satisfying block **Bl′** such that $\chi(\mathbf{Bl'}) = 1$. By constructing a longer chain, they publish **Bl′** and Δ additional blocks such that they could replace **Bl** with **Bl′**. Thus, the game ends when $n + \Delta$ blocks are appended after the initial block. Under this case, for a decisive block **Bl**, the winning probability of $\mathcal{A}$ can be $P_{\mathcal{A}} = F(\mu, \lambda, n, \Delta)$ [11], where μ is the probability of $\chi(\mathbf{Bl}) = 1$, λ is dishonest hashrate and Δ is the number of additional blocks. In the proposed model, $\mathcal{A}$ requires to bias the decisive blocks **Bl₁** and **Bl₂**. That is, only by winning the following *Game 1* and *Game 2*, can $\mathcal{A}$ control the final auditing results.

Game 1: In this game, the goal of $\mathcal{A}$ is to ensure **Bl₁** indexed by n is satisfying. Since the distance between **Bl₁** and **Bl₂** is $\lceil \vartheta/2 \rceil$ blocks, if honest miners publish a dissatisfying block **Bl₁**, $\mathcal{A}$ still has $\lceil \vartheta/2 \rceil - 1$ additional blocks to build a longer chain for replacing **Bl₁** with a satisfying block. Hence, the probability of $\mathcal{A}$ winning this game is at most $F(\mu, \lambda, n, \lceil \vartheta/2 \rceil - 1)$.

Game 2: Only $\mathcal{A}$ wins *Game 1*, could this game get started and then the end block of *Game 1* becomes the initial block of this game. In this game, the goal of $\mathcal{A}$ is to ensure **Bl₂** is satisfying. If $\chi(\mathbf{Bl_1}) = 1$, $\mathcal{A}$ requires no additional block to build a longer chain to replace **Bl₁** such that **Bl₁** is indexed by 0. Hence the index of **Bl₂** would be at most $\lceil \vartheta/2 \rceil$. With Δ additional blocks, the probability of $\mathcal{A}$ winning this game is at most $F(\mu, \lambda, \lceil \vartheta/2 \rceil, \Delta)$.

Thus, we can obtain an upper bound of winning probability of $\mathcal{A}$ which is denoted as $P_{\hat{\mathcal{A}}}$:

$$P_{\hat{\mathcal{A}}} = F(\mu, \lambda, n, \lceil \vartheta/2 \rceil - 1) \cdot F(\mu, \lambda, \lceil \vartheta/2 \rceil, \Delta) > P_{\mathcal{A}}.$$

To better evaluate the impact caused by malicious miners, we assume that 1% fraction of data is corrupted in CS, thus $\mu = 1 - (1 - 1\%)^c$. We assume $\vartheta = 20$, $n = 18$ and implement it in experiment. The experimental result is presented in Fig. 2. It demonstrates that when $\lambda < 51\%$, the adversary $\mathcal{A}$ wins with a negligible probability. Therefore, our auditing scheme can resist malicious miners and ensure the randomness of auditing challenge messages.

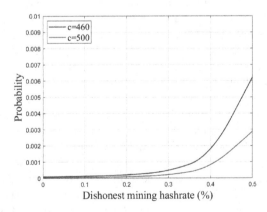

Fig. 2. The upper bound on winning probability of the adversary $\mathcal{A}$

6   Comprehensive Performance Evaluation

In this section, we compare our scheme with existing related auditing schemes IPANM [4], CIPPPA [21], and CPVPA [22] in terms of the computation costs (including the signature computation costs of a single data block, and the auditing computation costs on auditor side) and auditing communication overhead between auditor and CS. Meanwhile, we measure the gas cost of the proposed scheme's on-chain operation. Then we implement our scheme in experiments, which is based on a Windows 10 system with Intel(R) Core(TM) i5-10500 CPU 3.10 GHz and 16 GB of RAM. All the algorithms are programmed in Java language with JDK 1.7 version and JBPC library 2.0. A type-A bilinear pairing is employed, and the element size in $\mathbb{Z}_q^*$ is $|q| = 160$-bits accordingly. The smart contract runs on the Ethereum blockchain and is designed in the Solidity language. In addition, all the experiments are performed in 20 trials on average.

6.1   Comparison of the Computation Costs

To better evaluate the performance of our scheme and other related auditing schemes IPANM [4], CIPPPA [21], CPVPA [22] in aspects of the computation costs, we unify the cryptographic operations. The notations of execution time for cryptographic operations are listed in Table 1.

Table 1. Notations of Cryptographic Operations

Notation	Descriptions
T_{ha}	The running time of a general hash operation
T_{Ha}	The running time of a hash-to-point operation
T_{Ad}	The running time of a point addition operation
T_{Pa}	The running time of a bilinear pairing operation
T_{Mu}	The running time of a scalar multiplication operation
T_{mu}	The running time of a general multiplication operation
T_{Ex}	The running time of a modular exponentiation operation

Firstly, we evaluate the performance of the signature computation costs of a single data block between our scheme and existing related schemes IPANM [4], CIPPPA [21], CPVPA [22]. Here, the number of users in the ring signature of IPANM [4] is $u = 12$. The comparison of the implementation results is shown in Table 2, and the corresponding diagram is presented in Fig. 3. It demonstrates that our scheme is much less than the others in the signature computation cost. This is because the signature tag of a data block in BPPA is constructed by the homomorphic hash function, which can convert the modular exponentiation operation into a general multiplication without sacrificing the security of the signature.

Table 2. Computation Costs of Signature

Scheme	Signature computation costs	Executing time (μs)
IPANM [4]	$uT_{Mu} + uT_{mu} + T_{Ha} + 2T_{Ex}$	≈ 33826.2
CIPPPA [21]	$2T_{Mu} + T_{Ha} + T_{Ad}$	≈ 9836.6
CPVPA [22]	$4T_{mu} + 3T_{Ha} + 2T_{ha} + 6T_{Ex}$	≈ 23518.2
Our scheme	$2T_{mu} + T_{Ha}$	≈ 5494.8

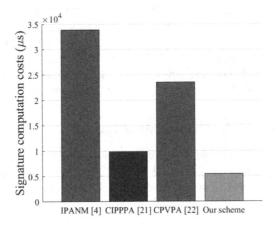

Fig. 3. Comparison of signature computation costs

Then, we further evaluate our scheme and existing related schemes IPANM [4], CIPPPA [21], CPVPA [22] in terms of the computation costs on the auditor side. The auditing verification computation costs are listed in Table 3, which are shown in Fig. 4. As we can see, with the increase in the number of challenged data blocks, the computation costs on the auditor side of our scheme are much lower than others. At the same time, only BPPA can achieve identity anonymity conditionally, and public cloud auditing, simultaneously.

Table 3. Computation Costs on Auditor Side

Scheme	Computation costs on auditor side	Executing time (μs)
IPANM [4]	$2T_{Pa} + (c+1)T_{mu} + cT_{Ha} + (c+3)T_{Ex}$	$\approx 6663.9c + 14364.9$
CIPPPA [21]	$3T_{Pa} + (c+1)T_{Mu} + T_{mu} + cT_{Ha} + T_{ha} + (c-1)T_{Ad}$	$\approx 7671.4c + 18447.7$
CPVPA [22]	$4T_{Pa} + (2c+1)T_{mu} + (c+4)T_{Ha} + (2c+3)T_{ha} + (3c+2)T_{Ex}$	$\approx 9023.1c + 46044.3$
Our scheme	$2T_{Pa} + (c+2)T_{mu} + T_{Ha} + T_{ha} + (c+1)T_{Ex}$	$\approx 1170.9c + 17526.6$

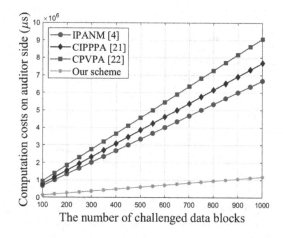

Fig. 4. Comparison of computation costs on auditor side

6.2   Comparison of the Communication Overhead

Now, we demonstrate our scheme has more advantages in communication over-head between auditor and CS compared with existing related schemes IPANM [4] CIPPPA [21], and CPVPA [22]. To provide performance comparison at the same security level, we set the size of the two multiplicative cyclic groups $\mathbb{G}$ and $\mathbb{G}_T$ to be 1024-bits and 2048-bits, respectively. We also set the total num-ber of data blocks $n = 2^{14}$, the size of user identity is $|ID| = 64$-bits, the size of a timestamp in Ethereum is $|T| = 32$-bits and the size of a hash value in Ethereum is $|Hash| = 256$-bits. According to the analysis of chain security in literature [12], we set the number of blocks extracted hash value each time in Ethereum as $\vartheta = 20$. In the proposed BPPA, the smart contract (acting as an auditor) conveys the challenge message $Chal = \{c, k_1, k_2\}$ to the cloud server, with overhead $\log_2 n + 2|q|$. Then the cloud server is demanded to respond with the proof information $Proof = \{\xi, \delta, \psi\}$, with overhead $|q| + 2|\mathbb{G}|$. Hence the total communication overhead between smart contract and CS in our scheme is $3|q| + 2|\mathbb{G}| + \log_2 n \approx 2542$-bits. The total communication overhead between auditor and CS in the IPANM [4] is $3c|q|+(u+3)|\mathbb{G}|+u|ID| \approx 480c+16128$-bits, CIPPPA [21] is $(c+1)|q|+3|\mathbb{G}|+c\log_2 n+|ID| \approx 174c+3296$-bits, and CPVPA [22] is $2|q| + 2|\mathbb{G}| + |T| + \vartheta|Hash| \approx 7520$-bits.

The comparison of communication overhead between auditor and CS between our scheme and others is depicted in Fig. 5, which demonstrates that our scheme is much less than IPANM [4], and CIPPPA [21] in the auditing communication overhead. Moreover, the communication overhead between the auditor and CS of CPVPA [22] and our scheme remains constant with the growth in the number of challenged data blocks, but our scheme requires less than CPVPA [22].

In conclusion, compared with existing related auditing schemes IPANM [4], CIPPPA [21], CPVPA [22] in terms of the computation costs and auditing com-munication overhead, the proposed BPPA scheme is efficient, which is more practical for cloud storage platform with capacity-limited devices.

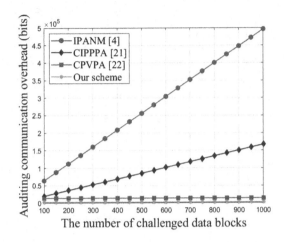

Fig. 5. Comparison of auditing communication overhead

6.3 On-Chain Consumption Evaluation

We evaluate the gas cost of the smart contract on the Ethereum virtual mechanism (EVM), including deploying the smart contract, generating challenges, and verifying proof. When we studied this research, the exchange rate of 1 ETH was about 1200 USD. The gasPrice is set to be 1 GWei, where 1 GWei is 10^{-9} ETH. The gas consumption of each operation is shown in Fig. 6. When $c = 460$, deploying the smart contract costs about 0.9 USD, generating challenge costs about 0.09 USD, and verifying proof costs about 1.81 USD. Thus, it only consumes about 1.9 USD to perform a public auditing task by using the smart contract, which is acceptable for DO.

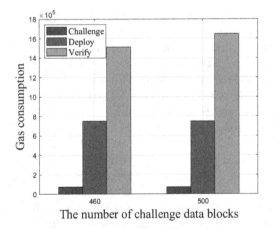

Fig. 6. Evaluation of gas costs

7   Conclusions and Future Work

In this paper, we have proposed a blockchain-assisted privacy-preserving public auditing scheme (BPPA). To get rid of an untrusted TPA, we employ the smart contract in Ethereum to verify the integrity of the user's remote data. We choose hash values from the nearest blocks on the blockchain as the factor to generate a challenge message that further locates to the subset of data blocks segmentally, thus eliminating the threat posed by malicious miners. Meanwhile, BPPA realizes conditional identity anonymity so that it can protect the identity privacy of cloud users. We provide in-depth security analysis and comprehensive performance evaluation to demonstrate our proposed BPPA is secure and deployable in the cloud storage platform. For the subsequent work, we will further focus on how to integrate other types of blockchain technology into public auditing mechanisms while ensuring its security, efficiency, and functionality.

Acknowledgements. This work is supported by the Shenzhen Science and Technology Program (Grant No. GXWD20220817124827001, JCYJ20210324132406016). This work is also supported by the National Key R&D Program of China (Grant No. 2021YFB2700900), the Shenzhen Key Technical Project (Grant No. 2022N009), the Fok Ying Tung Education Foundation of China (Grant No. 171058), Guangdong Provincial Key Laboratory of Novel Security Intelligence Technologies (Grant No. 2022B1212010005).

References

1. Ateniese, G., et al.: Provable data possession at untrusted stores. In: Proceedings of the 14th ACM conference on Computer and Communications Security, pp. 598–609 (2007)
2. Gennaro, R., Katz, J., Krawczyk, H., Rabin, T.: Secure network coding over the integers. In: Nguyen, P.Q., Pointcheval, D. (eds.) Public Key Cryptography – PKC 2010: 13th International Conference on Practice and Theory in Public Key Cryptography, Paris, France, May 26-28, 2010. Proceedings, pp. 142–160. Springer Berlin Heidelberg, Berlin, Heidelberg (2010). https://doi.org/10.1007/978-3-642-13013-7_9
3. Guo, J., Han, L., Yang, G., Liu, X., Tian, C.: An improved secure designated server public key searchable encryption scheme with multi-ciphertext indistinguishability. J. Cloud Comput. **11**(1), 1–12 (2022)
4. Huang, L., et al.: Ipanm: incentive public auditing scheme for non-manager groups in clouds. IEEE Trans. Dependable Secure Comput. **19**(2), 936–952 (2022)
5. Juels, A., Kaliski Jr, B.S.: Pors: proofs of retrievability for large files. In: Proceedings of the 14th ACM Conference on Computer and Communications Security, pp. 584–597 (2007)
6. Li, A., Chen, Y., Yan, Z., Zhou, X., Shimizu, S.: A survey on integrity auditing for data storage in the cloud: from single copy to multiple replicas. IEEE Trans. Big Data **8**(5), 1428–1442 (2020)
7. Li, J., Wu, J., Jiang, G., Srikanthan, T.: Blockchain-based public auditing for big data in cloud storage. Inform. Process. Manage. **57**(6), 102382 (2020)

8. Li, J., Yan, H., Zhang, Y.: Identity-based privacy preserving remote data integrity checking for cloud storage. IEEE Syst. J. **15**(1), 577–585 (2020)
9. Li, R., Qin, Y., Wang, C., Li, M., Chu, X.: A blockchain-enabled framework for enhancing scalability and security in iiot. IEEE Trans. Industr. Inf. **19**(6), 7389–7400 (2023)
10. Liu, X., Luo, Y., Yang, X., Wang, L., Zhang, X.: Lattice-based proxy-oriented public auditing scheme for electronic health record in cloud-assisted wbans. IEEE Syst. J. **16**(2), 2968–2978 (2022)
11. Pierrot, C., Wesolowski, B.: Malleability of the blockchain's entropy. Cryptogr. Commun. **10**(1), 211–233 (2018)
12. Rosenfeld, M.: Analysis of hashrate-based double spending. ArXiv abs/1402.2009 (2014)
13. Shacham, H., Waters, B.: Compact proofs of retrievability. In: Pieprzyk, J. (ed.) ASIACRYPT 2008. LNCS, vol. 5350, pp. 90–107. Springer, Heidelberg (2008). https://doi.org/10.1007/978-3-540-89255-7_7
14. Shu, J., Zou, X., Jia, X., Zhang, W., Xie, R.: Blockchain-based decentralized public auditing for cloud storage. IEEE Trans. Cloud Comput. **10**(4), 2366–2380 (2022)
15. Wang, C., Wang, Q., Ren, K., Lou, W.: Privacy-preserving public auditing for data storage security in cloud computing. In: 2010 proceedings IEEE infocom, pp. 1–9. IEEE (2010)
16. Wang, H., Wang, Q., He, D.: Blockchain-based private provable data possession. IEEE Trans. Dependable Secure Comput. **18**(5), 2379–2389 (2019)
17. Wood, G., et al.: Ethereum: a secure decentralised generalised transaction ledger. Ethereum project yellow paper **151**(2014), 1–32 (2014)
18. Xu, Y., Jin, C., Qin, W., Shan, J., Jin, Y.: Secure fuzzy identity-based public verification for cloud storage. J. Syst. Architect. **128**, 102558 (2022)
19. Xu, Z., He, D., Wang, H., Vijayakumar, P., Choo, K.K.R.: A novel proxy-oriented public auditing scheme for cloud-based medical cyber physical systems. J. Inform. Secur. Appl. **51**, 102453 (2020)
20. Zhang, C., Xu, Y., Hu, Y., Wu, J., Ren, J., Zhang, Y.: A blockchain-based multi-cloud storage data auditing scheme to locate faults. IEEE Trans. Cloud Comput. **10**(4), 2252–2263 (2021)
21. Zhang, X., Zhao, J., Xu, C., Li, H., Wang, H., Zhang, Y.: Cipppa: conditional identity privacy-preserving public auditing for cloud-based wbans against malicious auditors. IEEE Trans. Cloud Comput. **9**(4), 1362–1375 (2019)
22. Zhang, Y., Xu, C., Lin, X., Shen, X.: Blockchain-based public integrity verification for cloud storage against procrastinating auditors. IEEE Trans. Cloud Comput. **9**(3), 923–937 (2019)
23. Zhao, J., Huang, H., Gu, C., Hua, Z., Zhang, X.: Blockchain-assisted conditional anonymity privacy-preserving public auditing scheme with reward mechanism. IEEE Syst. J. **16**(3), 4477–4488 (2021)
24. Zhao, J., Zheng, Y., Huang, H., Wang, J., Zhang, X., He, D.: Lightweight certificateless privacy-preserving integrity verification with conditional anonymity for cloud-assisted medical cyber-physical systems. J. Syst. Architect. **138**, 102860 (2023)
25. Zhao, Y., Chang, J.: Certificateless public auditing scheme with designated verifier and privacy-preserving property in cloud storage. Comput. Netw. **216**, 109270 (2022)

MANet: An Architecture Adaptive Method for Sparse Matrix Format Selection

Zhenglun Sun[iD], Peng Qiao[✉], and Yong Dou

National University of Defense Technology, Changsha, China
{zhenglun_sun,pengqiao,yongdou}@nudt.edu.cn

Abstract. The proliferation of modern computer architectures brings a great challenge to sparse matrix-vector multiplication (SpMV), which is widely used in scientific computing and artificial intelligence. Providing a suitable sparse matrix format for SpMV is crucial to achieve high performance by enhance data locality and cache performance. However, for different architectures, the best sparse matrix format varies. In this paper, we propose a novel architecture adaptive sparse matrix format selection method, MANet, to select proper format to optimize performance of SpMV. This method transforms a sparse matrix into a high-dimensional image, with the matrix sparseness feature and architecture feature combined as inputs. To evaluate the effectiveness of this method, we generated a dataset that includes various scientific problems and architectures with augmentation. Results show that MANet improves sparse matrix selection accuracy by 6% compared to previous works and can achieve a speedup of up to 230% compared to methods with a fixed format. When adapting to an architecture that is not presented in the training, it can still provide 88% selection accuracy and 14% higher than the previous approaches, without further training.

Keywords: Sparse Matrix-Vector Multiplication · Sparse Matrix Format Selection · High-Performance Computing · Architecture adaptive · Deep Learning

1 Introduction

In high-performance computing, the Sparse Matrix-Vector Multiplication (SpMV) kernel plays an important role in scientific computing [15] and related numerical computing fields such as weather forecasting [8], computational fluid dynamics [10]. High computation efficiency of SpMV is hard to achieve due to the irregular memory access patterns introduced by sparse matrices.

Many researchers have sought to accelerate SpMV from both architectural and algorithmic perspectives. In terms of architecture, attempts have been made to increase CPU and memory frequencies for direct speedup, as well as to employ larger caches to reduce repetitive data I/O.

© The Author(s), under exclusive license to Springer Nature Singapore Pte Ltd. 2024
Z. Tari et al. (Eds.): ICA3PP 2023, LNCS 14488, pp. 311–326, 2024.
https://doi.org/10.1007/978-981-97-0801-7_18

From algorithmic perspective, numerous sparse matrix formats and algorithms have been proposed to improve data locality. Here we focus particularly on the sparse matrix format level. A number of storage formats have been developed to address the heterogeneity of nonzeros distribution, with impacts on data locality, cache performance, and thus overall performance. Examples of such formats include COO, CSR, and BSR, which are designed to meet the demands of modern architectures and scientific applications. With an appropriate selection of sparse matrix format, the performance of SpMV can be significantly enhanced [6]. In this work, we aim to select an appropriate format with different architecture settings.

Existing methods for sparse matrix selection rely on machine learning algorithms [3,12,19], which will extract nonzeros features and predict matrix format with the support of Convolutional Neural Networks (CNN) or the decision tree. However, the prediction results of previous studies may not be well-adapted to different architectures. The rapid progress in computer architecture has made it challenging to maintain a single optimal format that can be used across different CPUs or accelerators [2,9,13]. Therefore, it is necessary to develop a format selection method that is able to adapt to varying architectures.

To improve the generalizability of the model across various architectures, we introduce a sparse matrix selection approach with architecture generalization entitled MANet. Our proposed method utilizes matrix pooling-like normalization as preprocessing technique and employs a multiple-input CNN to select the most suitable storage format for the matrix based on the features of the architecture and the matrix.

Our contribution consists of three parts, and we will provide the corresponding code after finishing the organization process:

- We proposed a format selection network for sparse matrices called MANet, which is adaptive to architecture. In terms of adaptability, MANet improves prediction accuracy by 20% when adapting to other platforms with different architecture settings. When adapting to platforms with architectures that have not been previously encountered, MANet yields a 14% improvement compared to related works.
- After adaptation to the approximate platform, the SpMV speedup is 230% after format selection using MANet compared to the COO storage format. Moreover, 89.3% of the matrices can reach the minimum time after using MANet for format selection.
- Our research further probes into the effect of architecture settings on the format distribution in a CPU platform. We indicate that the computation speed of SpMV is determined by the hardware architecture setting, including processor frequency, memory bandwidth, cache size, etc.

2   Background

2.1   Sparse Matrix Storage Format

The main objective of sparse matrix storage formats is to compress matrices, thereby improving locality and reducing storage overhead. Various storage formats, custom-tailored to different distributions of nonzeros, have been developed. The selection of an appropriate format entails consideration of characteristics associated with the nonzeros distribution and architecture.

Our work focuses on three commonly used formats, COO, CSR, and BSR [5].

- The Coordinate Format (COO) storage structure organizes a matrix into three arrays: *col*, *row*, and *val*. The elements in each array specify the column indices of nonzeros entries, the row indices of nonzeros entries, and their values respectively.
- The Compressed Sparse Row (CSR) format stores the column coordinates of the nonzeros into an array *ind*, as well as containing the nonzeros values themselves in another array *val*. Additionally, CSR contains a pointer array *ptr* which helps to quickly identify the intervals of each row, as if listing it out as a list. The CSR format provides a practical approach to representing sparse matrices in three relatively small arrays.
- The Block Sparse Row (BSR) format divides a matrix into multiple dense blocks and stores them using the same CSR-style indexing used by the CSR format. The index array *ptr* gives an interval for each row of each block, the value array *val* contains the nonzeros entries, and the column coordinates of the nonzeros are given by *ind*.

2.2   Influence of Nonzeros Distribution

The standard SpMV computation paradigm is $y = Ax$, where y and x are vectors, and A is a sparse matrix. When performing SpMV computations, each non-zero element A_{ij} is accessed only once. Our work explores strategies to reuse both y and x, enabling the exploitation of data locality opportunities present in SpMV sessions.

In dense storage format, SpMV multiplication is known to be inefficient due to the traversal of zero values during updating the vector. Moreover, non-optimized memory and cache access can lead to significant drops in overall performance. A suitable storage format can alleviate these issues while simultaneously providing opportunities for parallel computing [13].

By taking the matrix stored in CSR format as an example and employing CSR as a storage structure, the memory access pattern is changed. The pseudocode of SpMV when dealing with a matrix stored in CSR is presented in Algorithm 1. This approach only requires $m + 3nnz$ memory accesses when without cache for a given matrix of size $m \times n$ and nonzeros nnz, resulting in significant speed improvements compared to the traditional dense format, which

Algorithm 1. SpMV using CSR format

Input: A: input matrix with $m \times n$ size
       val: nonzeros in A
       ptr: row offsets in CSR
       ind: column index in CSR
       x: input vector
Output: y: output vector
 1: **for** $i = 0, 1, 2, ..., m - 1$ **do**
 2:   $y[i] \leftarrow 0$
 3:   **for** $j = ptr[i], ..., ptr[i+1] - 1$ **do**
 4:     $y[i] \leftarrow y[i] + val[j] * x[ind[j]]$
 5:   **end for**
 6: **end for**
 7: return y

---

would require $m * n$ memory accesses. And for best case, it will only require $m + n + 2 \times nnz$.

From the above, it can be found that the access optimization due to the sparse matrix storage format is directly related to the distribution of nonzeros. Different storage formats will correspond to different SpMV algorithms, which will have an impact on the access pattern of x. To select proper format, CNN, decision tree and graph neural network (GNN) are used.

3 Methodology

3.1 Overview

In this section, we present a multi-input sparse matrix format selector that is able to adapt to various architectures while overcoming several challenges. Our proposed approach combines matrix pre-processing with architecture features that are associated with the distribution of non-zero data. This leads to improved performance when adapting to platforms with varying architecture settings. We provide a detailed overview of the construction process, from data pre-processing to classification.

The left part of Fig. 1 contains 4 steps to preprocess data and construct a muti-input CNN network. Assuming there is already a sparse matrix dataset D and target architectures P_i, where $i = 1, 2...n$.

(1) This work requires a matrix training label for the first step. To achieve the best performance in sparse matrix format, the training labels are acquired by running SpMV on P_i 100 times and measuring the SpMV computing time. By measuring the SpMV computing time, this step selects a format that makes SpMV execute at the fastest speed as a label and attaches the label to with matrix ID. (2) The second step will normalize the matrix into a fixed size by pooling the matrix. With a fixed size, the sample formed by the sparse matrix can be used for CNN training and classification. (3) The third step of the process

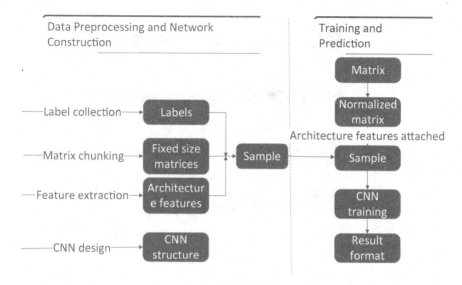

Fig. 1. Overview of the data preprocessing, network construction and training.

is focused on extracting architecture features from P_i that affect SpMV execution time. Data preprocessing allows us to obtain high-dimensional image data associated with both matrix nonzeros features and architecture features. (4) The final step of the left part in Fig. 1 forms a multi-input CNN network to extract features from the sample. (5) The right part of Fig. 1 performs the CNN training and prediction procedure.

The main challenges we face are in Steps 3 and 4. Since architectures have many features, it is necessary to conduct experiments to determine which of these features should be encoded into the sample. The high dimension of the samples makes it difficult for CNN to extract features for training. Designing an appropriate network structure to remain amenable to an architecture adaptive format selector remains a problem for our work.

3.2 Matrix Labeling

For the classification task, CNN requires labels to provide ground truth. In order to achieve this, 100 iterations of SpMV were executed on platform P_i, and the format with the shortest average time was selected as the label.

As previously discussed, data locality and other factors influencing the speed of SpMV can be highly variable, depending on the architecture configuration. In practical computing scenarios, the CPU retrieves data, including requested and surrounding data, thus creating a data locality. Variations in CPU frequency, cache size, etc. can have an effect on data locality and other factors that impact the speed of SpMV.

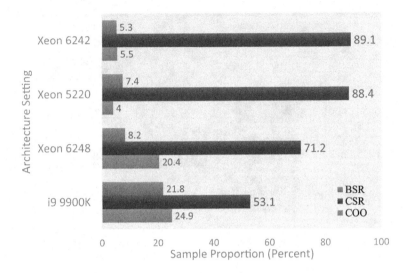

Fig. 2. The generated format label distribution on different architecture.

This experiment investigated the performance of SpMV across multiple platforms with different architectures, and also involved labeling matrices from the entire dataset. The results indicated that the proportions of samples varied depending on the architecture utilized, as demonstrated in Fig. 2. However, when selectors lack adaptability to these architecture changes, altering sample proportions could lead to poor performance results as shown in Fig 5. Notably, as depicted in Fig. 2, for certain parts of the dataset, COO format still outperforms other formats. Specifically, in the E5 platform, COO performed better than all other formats for the HB group in SuitSparse Matrix Collection [4].

3.3 Matrix Normalization and Matrix Feature Extraction

The size of matrices may vary, complicating their use in CNN training and inference. To address this issue, we propose a normalization method that extracts the matrix features and pools them into a fixed size suitable for CNN utilization. This method is based on the observation of the matrix non-zeros distribution features, as described in [2], and the BSR design proposed in [18]. In this part, we first describe the pooling method before detailing the feature extraction method.

Matrix Pooling. Algorithm 2 outlines a pooling-like Algorithm for normalizing a given matrix with a size-adjustable sliding window. Given a matrix, the sliding window slides through matrix A and maps its features to the corresponding pixel in the sample. After normalization, a three-channel sample is obtained, containing the matrix features. This sample is further processed to add the necessary architecture settings.

Algorithm 2. Matrix pooling with sliding window

Input: A: input matrix with $m \times n$ size

 val: nonzeros in A

 m_s: input sample row size

 n_s: input sample column size

 w: sliding window with $m_w \times n_w$ and nonzeros density $density$

Output: $Sample$: CNN input sample

1: set $m_w = \frac{m}{m_s}$

2: set $n_w = \frac{n}{n_s}$

3: **for** $i = 0, 1, 2, ..., \frac{m}{m_w}$ **do**

4:    **for** $j = 0, 1, 2, ..., \frac{n}{n_w}$ **do**

5:      moves the top left corner index of the w to the element $A_{(i*m_w, j*n_w)}$

6:      $density = \frac{nnz}{n_w \times m_w}$

7:      **if** $density \geq threshold$ **then**

8:        assigns matrix features to the position (i, j) of the sample

9:      **else**

10:        assigns zero to the position (i, j) of the sample

11:      **end if**

12:    **end for**

13: **end for**

14: **return** $Sample$

Taking a sparse matrix of size 100×100 as an example, and assuming that the input size of the CNN is 10×10, the sliding window size according to Algorithm 2 is also 10×10. With the *stride* shown in Eq. (1), the matrix is divided into 100 parts with pooling, each of which is mapped to a set of sample pixels. If the density of non-zero elements in a given part exceeds the threshold, the corresponding pixel is assigned with the matrix features; otherwise, it remains blank.

$$stride = \frac{m}{m_s} \tag{1}$$

Matrix Feature Extraction. We extract key features from the sparse matrix and store them in the relative dimensions of the sample, as demonstrated in previous work [2]. The extracted features include:

- matrix size: $m * n$ (require normalization)
- matrix density: $\frac{nonzeros}{matrix\ size}$
- nonzeros number of a row which contains most nonzeros: max_{row} (require normalization)

As these features can vary greatly, we normalize them to a range of $[0, 255]$ using Eq. (2). Limiting the feature values to the range of $[0, 255]$ enables a more accurate description of the density distribution of non-zero values covered by the sliding window across all color channels, thereby providing a more refined description of the non-zero value distribution for the entire sample. If normalization is not used, the significant differences in feature values may have an impact on the training process.

$$normalized\ feature = \frac{feature - min}{max - min} \times 255 \tag{2}$$

The density threshold is an important factor to consider when describing the nonzeros distribution of a sparse matrix. If the threshold is set too low, the detail of nonzeros distribution will not be described accurately, while setting the threshold too high will result in too many empty blocks without features. It is therefore necessary to carefully consider the optimal density threshold to ensure an accurate description of the sample.

3.4   Architecture Feature Extraction

Architecture has been shown to influence SpMV performance and format selection (Fig. 2). To ensure our model can adapt to other architectures without additional training, we incorporate architecture features into extra dimensions of the sample. The resulting sample has six dimensions: the first three dimensions correspond to matrix features, while the remaining three dimensions refer to architecture features.

By concatenating the architecture feature matrix with the matrix feature matrix, we can generate a sample with 6 dimensions. Initially, employing lscpu and mbw [1], we can retrieve the CPU architecture settings and memory bandwidth. Subsequently, according to Eq. (3), the architecture feature is normalized. Then, three architecture feature matrices are constructed based on corresponding architecture features. Finally, these matrices are stacked onto the sparse matrix feature matrix, resulting in a high-dimensional sample.

Architecture feature dimensions do not pay attention to the sparse matrix-related features. Setting the corresponding dimension to a solid value allows the CNN to focus on the value, eliminating the interference of sparse matrix features, and thereby improving accuracy.

In Sect. 4.4, our experiments revealed three significant architecture features that have an impact on the SpMV speed. Features of architecture we extract are shown below, those features require normalization following Eq. (3):

- The size of L3 cache size: $L3$.
- The basement frequency of CPU: $Frequency$.
- The bandwidth of memory: $Bandwidth$.

$$normalized\ feature = \frac{feature}{max} \times 255 \tag{3}$$

Normalization methods for matrix features cannot be used for architecture features. As can be observed by examining the hardware architecture settings in Table 1, applying the normalization method given in Eq. (2), a transformation in the distribution of the sample space can be observed. For instance, the Euclidean distance of the feature vector with Xeon 6248 and i9 9900K decreased from 2.2176 to 1.833 with the inclusion of the Xeon 6242 processor, indicating an overall 17% shrinkage and a corresponding shift in the position of the sample in the sample space. To this end, a scalable normalization approach is presented in Eq. (3), wherein max represents the maximum value of the matrix feature while $feature$ stands for the feature value of the matrix.

Table 1. Architecture features of each platform

Architecture Property	i9 9900K	Xeon 5220	Xeon 6248	Xeon 6242	E5 2640
Basement Frequency	3.6 GHz	2.2 GHz	2. 5 GHz	3.1 GHz	2.6 GHz
Cores	8	18	20	16	6
L3Cache	16 MB	24.75 MB	27.5 MB	22 MB	15MB
Memory	126 GB	119.2 GB	377.6 GB	119.2 GB	119.2 GB
Memory Bandwidth of Node	9137 MiB/s	4353 MiB/s	5024 MiB/s	5583 MiB/s	6069 MiB/s
Launch Date	Q4'18	Q2'19	Q2'19	Q2'19	Q1'12

3.5 Network Design

For traditional image classification tasks, images are typically input into a single-input network, with the variation in the structure of the network mainly in terms of the number of layers, the internal parameters, and the organization form. However, in the case of sparse matrix data, the feature distribution has some distinct differences from traditional image feature distributions, as the information present in each dimension may be different from other dimensions.

In order to facilitate the network's ability to extract information from each channel, we present MANet illustrated in Fig. 3 as an abridged general view. This network decouples the elements of the samples into six inputs at the beginning. Further, the tensor with six dimensions is fed into the CNN layer after which it is concatenated and fed into a fully-connected layer. By separating the samples at the start and merging the features at the end, this architecture enables the efficient extraction of relevant features while also reducing the complexity of training.

The MANet architecture consists of four CNN layers, with each layer containing six channels. Following each CNN layer, a max pooling layer is connected. The final CNN layer is followed by three fully connected layers that output the prediction results. The output channels for the CNN layers are 96, 256, 256, and 256, respectively. Additionally, the kernel size for each CNN layer is 11×11, 5×5, 3×3, and 1×1 with padding and stride, respectively. The input and output sizes for the fully connected layers are 13824×4096, 4096×4096, and 4096×3.

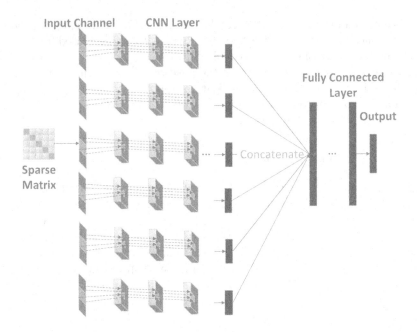

Fig. 3. Network structure of the matrix format selection method with architecture awareness.

4   Evaluation

4.1   Experiment Setting

Hardware Platform. We evaluated MANet on five platforms of varying architectures: i9 9900K (Coffee Lake), Xeon 5220 and 6248 (Cascade Lake), Xeon 6242R (Cascade Lake), and E5 2640 (Broadwell). Nvidia V100 GPU is used for training acceleration. The architecture features of the CPUs are shown in Table 1. These CPU platforms encompass a wide range of usage scenarios, from personal devices (like i9 9900k) to servers (like E5 6242), and span a considerable time period from Q1'12 to Q2'19.

Software Platform. PyTorch 1.7 [11] was utilized for constructing neural networks. To analyze the cache, the PAPI 6.0.0.1 [16] interface was employed to access the hardware counters during measuring the SpMV computation time. The memory bandwidth was then measured by the MBW [1] benchmark. To ensure stable labeling, all SpMV operations were performed with a single core using the SciPy library [17], which was also used to convert the matrix to the corresponding format.

Parameter Setting. The shape of the sample is 256×256 after preprocessing. The density used for normalization is 0.01. For training, the learning rate is

$1e-4$, we use Adam as the optimizer and use cross entropy as the loss function. The model was trained over 100 epochs with a batch size of 64. The weighting of each label's sample was set to 5.74, 1.41, and 8.56, respectively.

Dataset. This paper utilizes the sparse matrices from the Suit Sparse Matrix Collection [4] as training, testing, and validation data. The same 2150 matrices are preprocessed and used across five CPU platforms for training, testing, and validation. Matrix with size less than 256×256 is excluded resulting in 1702 original matrix for further argumentation. Furthermore, the data set is augmented by chunking into four equal parts as four new matrices, transposing, etc., resulting in 14168 sparse matrices. Matrix with size less than 256×256 is excluded. As indicated in Sect. 3, the feature to be extracted is as follows.

We employ an 80%–20% split for training and testing, respectively, with validation data directly sampled from the Suit Sparse Matrix Collection and excluded from the training and test sets.

Our work mainly focuses on three commonly used formats, COO, CSR, and BSR, as the storage format for sparse matrices. The three storage formats are implemented using the methods in Python Scipy Sparse V1.5.4 [17]. BSR is stored in blocks, and the block size of BSR in this experiment is selected by the heuristic Algorithm in Scipy Sparse library.

We have attempted to use other formats such as DIA, DOK, and LIL as labels, but we discovered that these formats only cover a mere 4% of the dataset. As a result, for SpMV in CPU, the COO, CSR, and BSR formats are capable of meeting most needs. Therefore, we have chosen to use COO, CSR, and BSR as labels for our dataset.

Due to the wide range of sparse matrices, the dataset used in this paper is more universal and challenging than earlier approaches. As matrices are utilized in various scientific computing applications, their feature distributions vary greatly and exhibit a strong long-tail effect when normalized.

4.2 Speedup

In this section, we use sparse matrix samples from several domains to produce their format prediction results and compare the time performance of SpMV using the corresponding formats of our method and two comparison methods.

In light of the potential for architecture adaptation and release time considerations, we compare the two methods proposed by Pichel et al. [12] and Zhao et al. [19]. The MANet is trained using the data set output from the i9 9900K and Xeon 6248 platforms, while both the Pichel et al. [12] and Zhao et al. [19] methods are trained using the data set output from the i9 9900K. Comparing datasets from two different architectures using traditional methods is not feasible since the labels for the samples may not match. However, MANet is designed to extract architecture-specific features, which eliminates this issue. We compare the results of these two methods on five different platforms and measure the number of best formats each method selects. The final results are shown

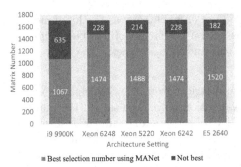
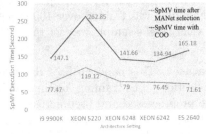

(a) Best selection number of MANet.

(b) SpMV time using MANet compared with COO format.

Fig. 4. SpMV speedup evaluation with MANet selector.

in Fig. 4. Figure 4a demonstrates that MANet is capable of selecting the best format in common scenarios with an accuracy of 89.3%.

When compared to the commonly used COO format, our proposed MANet shows a significant increase in SpMV computation speed of up to 230%, as demonstrated in Fig. 4b. As demonstrated in Sect. 3.2, the COO format remains an appropriate choice for enhancing the speed of SpMV.

4.3 Adaptation and Accuracy

In this section, we conduct an experiment to investigate the adaptation performance of our proposed method, MANet, in both approximate and non-approximate architecture adaptation scenarios. The comparison methods are the same as in the previous section.

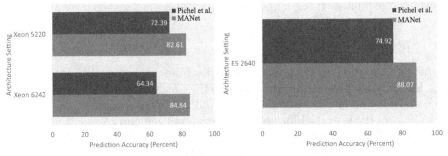

(a) Adapt to approximate architecture.   (b) Adapt to non-approximate architecture.

Fig. 5. Prediction accuracy when adapting to other architecture.

By utilizing a novel CNN structure and preprocessing technique, we demonstrate that MANet is capable of selecting the sparse matrix format in an environment that has not been previously encountered.

Approximate Adaptation. In approximate architecture adaptation, the test-set generated in Xeon 5220 and Xeon 6242 is used for testing. The architecture of the Xeon 5220 and Xeon 6242 is more similar to that of the Xeon 6248. Thus we choose Xeon 5220 and Xeon 6248 for the approximate architecture adaptation experiment.

The results obtained from the Intel Xeon 6242 and Xeon 5220 platforms indicate that the MANet architecture significantly outperforms the comparison works, with an improvement in prediction accuracy of up to 20% as shown in Fig 5a.

Non-approximate Adaptation. In non-approximate architecture adaptation, the data generated on an Intel Xeon E5 2640 Broadwell architecture platform was used to evaluate the performance of the MANet system. As shown in Fig. 5b, the system achieved a prediction accuracy of over 14%.

4.4   The Influence of Architecture

Architecture setting influence on the sample proportion is investigated in this section. CPU frequency, memory bandwidth, and cache size as architecture parameters are discussed in detail. Results indicate that those parameters have a significant impact on the sample distribution.

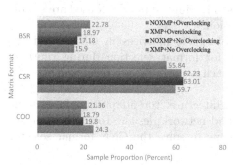

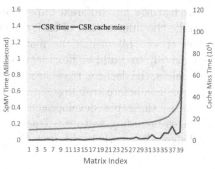

(a) The influence of CPU and memory overclocking .

(b) Relationship between cache miss and SpMV performance.

Fig. 6. The architecture setting influence towards SpMV and matrix labeling.

The effect of memory frequency and CPU frequency on the dataset from Intel i9 9900K was analyzed. Memory overclocking (XMP) and CPU overclocking were used to study the sample proportion change. The results of the sample proportion after using memory overclocking (XMP) and CPU overclocking, respectively, are presented in Fig. 6a. With lower memory frequency leading to a lower share of COO format and a greater bias towards CSR and BSR formats. Furthermore, higher CPU main frequencies have been found to have a greater bias towards BSR formats.

Experimental results demonstrate a direct correlation between the total number of L2 cache misses and the elapsed time for SpMV, as illustrated in Fig. 6b. This relationship is more evident, due to the relatively small size of the L2 cache, which helps to reduce the impact of noise.

It has been demonstrated that there is a direct correlation between the number of cache misses and the SpMV computation time, as illustrated in the above Figure. Utilizing the larger size of the L3 cache, this paper takes advantage of its more stable performance label during labeling, making it an important architecture parameter.

5   Related Work

Recent research has focused on utilizing machine learning methods, such as decision tree and deep neural networks (DNNs), to select the appropriate sparse matrix format for optimal performance of the sparse matrix-vector (SpMV) product. More recently, convolutional neural networks (CNNs) have been employed to address this problem.

Zhao et al. [19] introduced DNN into the task of format selection for sparse matrices. However, the feature extraction method only considers the spatial distribution of non-zeros without architecture features. Pichel et al. [12] propose an approach that directly turns the matrix into a fixed-size image and encodes relative features into different dimensions. The single input CNN used are unable to handle the information of each dimension.

Qiu et al. [14] utilize the features of sparse matrices in GNNs, and use XGBoost [3] to build a flexible model for format selection during runtime. Nonetheless, the decision tree-based approach requires the pre-definition of which feature to use before training.

Due to the rapid development of architecture, the adaptation faces some challenges including feature extraction and network design. To the best of our knowledge, this work is the first that aims to solve the CNN format selector adaptation problem.

6   Conclusion

Previous models for selecting sparse matrices did not consider the influence of architecture on their performance and typically required retraining or fine-tuning when adapting to different platforms. To address this limitation, we propose MANet, a sparse matrix selection model with architectural adaptability. Our approach incorporates architecture-specific data preprocessing and a multi-input neural network, enabling MANet to achieve superior accuracy without the need for retraining or fine-tuning during migration to diverse platforms. Consequently, it enhances the interoperability of sparse matrix selection models across architectures. In our future work, we will focus on adaptation across heterogeneous computation devices and addressing challenges in unbalanced sparse matrix classification that have not been adequately resolved.

Acknowledgements. We express our thanks to the anonymous reviewers for their insightful comments that improved the quality of the manuscript and Suit Sparse Matrix Collection for the dataset.

References

1. Mbw; Memory bandwidth benchmark (2010). http://manpages.ubuntu.com/manpages/lucid/man1/mbw.1.html
2. Chen, D., Fang, J., Chen, S., Xu, C., Wang, Z.: Optimizing sparse matrix-vector multiplications on an armv8-based many-core architecture. Int. J. Parallel Prog. **47**(3), 418–432 (2019). https://doi.org/10.1007/s10766-018-00625-8
3. Chen, T., et al.: Xgboost: extreme gradient boosting. R package version 0.4-2 **1**(4), 1–4 (2015)
4. Davis, T.A., Hu, Y.: The university of florida sparse matrix collection. ACM Trans. Math. Softw. **38**(1), 1–25 (2011)
5. Grossman, M., Thiele, C., Araya-Polo, M., Frank, F., Alpak, F.O., Sarkar, V.: A survey of sparse matrix-vector multiplication performance on large matrices. arXiv preprint arXiv:1608.00636 (2016)
6. Langr, D., Tvrdik, P.: Evaluation criteria for sparse matrix storage formats. IEEE Trans. Parall. Distrib. Syst.**27**(2), 428–440 (2016). https://doi.org/10.1109/tpds.2015.2401575, https://ieeexplore.ieee.org/document/7036061/
7. Li, M.L., Chen, S., Chen, J.: Adaptive learning: a new decentralized reinforcement learning approach for cooperative multiagent systems. IEEE Access **8**, 99404–99421 (2020). https://doi.org/10.1109/ACCESS.2020.2997899
8. Muhammed, T., Mehmood, R., Albeshri, A., Katib, I.: Suraa: A novel method and tool for loadbalanced and coalesced spmv computations on gpus. Appl. Sci. **9**(5), 947 (2019)
9. Nisa, I., Siegel, C., Rajam, A.S., Vishnu, A., Sadayappan, P.: Effective machine learning based format selection and performance modeling for spmv on gpus. In: 2018 IEEE International Parallel and Distributed Processing Symposium Workshops (IPDPSW). IEEE. https://doi.org/10.1109/ipdpsw.2018.00164, https://ieeexplore.ieee.org/document/8425531/
10. Oyarzun, G., Peyrolon, D., Alvarez, C., Martorell, X.: An fpga cached sparse matrix vector product (spmv) for unstructured computational fluid dynamics simulations. arXiv preprint arXiv:2107.12371 (2021)
11. Paszke, A., et al.: Pytorch: An imperative style, high-performance deep learning library. In: Advances in NeurIPS 32, pp. 8024–8035 (2019). http://papers.neurips.cc/paper/9015-pytorch-an-imperative-style-high-performance-deep-learning-library.pdf
12. Pichel, J.C., Pateiro-Lopez, B.: A new approach for sparse matrix classification based on deep learning techniques. In: 2018 IEEE International Conference on Cluster Computing (CLUSTER). IEEE. https://doi.org/10.1109/cluster.2018.00017, https://ieeexplore.ieee.org/document/8514858/
13. Pichel, J.C., Pateiro-Lopez, B.: Sparse matrix classification on imbalanced datasets using convolutional neural networks. IEEE Access **7**, 82377–82389 (2019). https://doi.org/10.1109/access.2019.2924060, https://ieeexplore.ieee.org/document/8742660/

14. Qiu, S., You, L., Wang, Z.: Optimizing sparse matrix multiplications for graph neural networks. In: Li, X., Chandrasekaran, S. (eds.) Languages and Compilers for Parallel Computing: 34th International Workshop, LCPC 2021, Newark, DE, USA, October 13–14, 2021, Revised Selected Papers, pp. 101–117. Springer International Publishing, Cham (2022). https://doi.org/10.1007/978-3-030-99372-6_7

15. Sun, X., Zhang, Y., Wang, T., Zhang, X., Yuan, L., Rao, L.: Optimizing spmv for diagonal sparse matrices on gpu. In: 2011 International Conference on Parallel Processing, pp. 492–501 (2011). https://doi.org/10.1109/ICPP.2011.53

16. Terpstra, D., Jagode, H., You, H., Dongarra, J.: Collecting performance data with PAPI-C. In: Müller, M.S., Resch, M.M., Schulz, A., Nagel, W.E. (eds.) Tools for High Performance Computing 2009: Proceedings of the 3rd International Workshop on Parallel Tools for High Performance Computing, September 2009, ZIH, Dresden, pp. 157–173. Springer Berlin Heidelberg, Berlin, Heidelberg (2010). https://doi.org/10.1007/978-3-642-11261-4_11

17. Virtanen, P., et al.: SciPy 1.0 Contributors: SciPy 1.0: fundamental algorithms for scientific computing in Python. Nature Methods **17**, 261–272 (2020). https://doi.org/10.1038/s41592-019-0686-2

18. Vuduc, R., Demmel, J.W., Yelick, K.A.: Oski: A library of automatically tuned sparse matrix kernels. In: Journal of Physics: Conference Series. vol. 16, p. 071 (2005)

19. Zhao, Y., Li, J., Liao, C., Shen, X.: Bridging the gap between deep learning and sparse matrix format selection. In: Proceedings of the 23rd ACM SIGPLAN Symposium on Principles and Practice of Parallel Programming. ACM. https://doi.org/10.1145/3178487.3178495, https://dl.acm.org/doi/pdf/10.1145/3178487.3178495

Service-Aware Cooperative Task Offloading and Scheduling in Multi-access Edge Computing Empowered IoT

Zhiyan Chen[1(✉)], Ming Tao[1], Xueqiang Li[1], and Ligang He[2]

[1] School of Computer Science and Technology, Dongguan University of Technology, Dongguan, China
z.chen.8@warwick.ac.uk, {taom,lixq}@dgut.edu.cn
[2] Department of Computer Science, University of Warwick, Coventry, UK
ligang.he@warwick.ac.uk

Abstract. Multi-access edge computing(MEC) enables computation task offloading and data processing at close proximity to provide rich end-users services with ultra-low latency in Internet of things(IoT). However, the high heterogeneity of the edge node configuration and the diversity of services pose challenges in fully utilizing the computing capacity in MEC. In this paper, we consider the problem of service-aware cooperative task offloading and scheduling in a three-tier MEC empowered IoT where the service requests from IoT devices can be distributed among edge nodes or further offloaded to remote cloud. As this problem is proven to be NP-hard, we proposed a two-layer Cooperative workload Initialization and Distribution Algorithm (CIDA) to solve the problem with low time complexity by decomposing it into two subproblems: 1) the optimization problem of offloading profile under dynamic resource allocation determined by the workload type, and 2) optimization problem of computation resources allocation under given offloading profile. Extensive experiments demonstrate that CIDA achieves superior performance compared to other approaches and scales well as the system size increases.

Keywords: Multi-access edge computing · Services-aware offloading · Task scheduling · Resource allocation

1 Introduction

The Internet of Things (IoT) has been introduced to connect and coordinate the rapidly growing number of smart devices over the past decade [1,2]. And cloud computing enables IoT devices to enhance their computational capabilities and extend battery life by offloading the computation requests to cloud servers for execution. However, the extensive connectivity and data exchange in the IoT have led to substantial communication latency, which poses challenges in guaranteeing the quality of IoT services [3]. As a result, edge computing has emerged as a promising paradigm to address these challenges. By deploying computing resources at the edge where data is generated, computation requests

© The Author(s), under exclusive license to Springer Nature Singapore Pte Ltd. 2024
Z. Tari et al. (Eds.): ICA3PP 2023, LNCS 14488, pp. 327–346, 2024.
https://doi.org/10.1007/978-981-97-0801-7_19

can be processed at the closer edge node instead of offloading to the remote cloud, which significantly reduce the latency and save energy [4].

Compared to cloud servers with abundant computational resources, edge server is generally resource-limited. Severe resource competition and data processing congestion could be caused by the offloading of computation intensive requests without cooperation. Furthermore, the absence of mutual collaboration among edge servers can result in certain servers being underutilized while popular edge servers are experiencing an overwhelming computational workload [5]. Additionally, given the diverse range of services in the IoT, the requirements on response time and computation resources can vary significantly. Both processing computation-intensive requests on relatively low-capacity edge servers and the improper allocation of resources on edge servers can lead to a degradation in system performance. Therefore, the establishment of collaborative offloading and scheduling mechanisms between edge servers and remote cloud servers is crucial to fully utilize the computational resources and achieve efficient operation of the system.

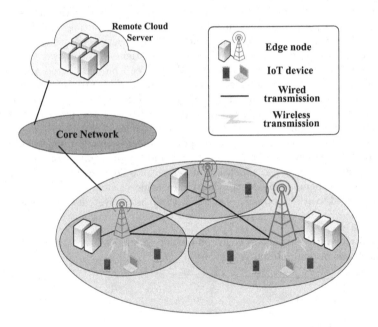

Fig. 1. Three-tier multi-access edge computing architecture

In this paper, we consider a multi-access edge computing empowered IoT system with N IoT devices, M base stations and a remote cloud server, which forms a three-tier architecture as shown in Fig. 1. Each base station is seamlessly integrated with an edge server, which is referred to as an edge node. IoT devices are connected to the edge nodes via wireless channels, while neighbouring edge nodes establish connectivity through local area networks or wired peer-to-peer links [6]. Furthermore, edge nodes possess the capability to further offload computation tasks to remote cloud server via the core network. In such system, individual IoT device may requests multiple types of services. And each specific service request made by an IoT device can be handled in one of the following ways: offloaded to its associated edge node for processing, transferred to another edge node within the network for processing, or further offloaded via the core network to remote cloud server with much higher computing capacity. Different types of services on an IoT device may offloaded to different destination for processing to improve the system performance, which can be achieved by a centralized offload controller such as SDN controller [7,8]. However, ensuring efficient system operation and resource utilization in such system presents a significant challenge as it requires an appropriate offloading and computing capacity allocation profiles to be obtained at low time cost. The close interplay of these profiles and the large searching space further complicate the problem.

To solve this problem, we discuss the design and implementation of CIDA, which is a two-layer heuristic Cooperative workload Initialization and Distribution Algorithm. The first layer of CIDA initialize the system offloading profile with the transmission delays for system requests are minimized regardless of the capacity constraint of edge nodes, then CIDA iteratively update the offloading profile by reallocating the requests to the execution platform with their response time are minimized. The second layer of CIDA obtains the optimal computing capacity allocation at each edge nodes by solving the KKT conditions based on the offloading profile in the first layer.

To summarize, we have made the following contributions in this paper:

(1) The problem of cooperative service-aware task offloading and scheduling in a three-tier multi-access edge computing system is investigated, which is formulated as an joint optimization problem of offloading profile and computing capacity allocation profile.
(2) The NP-hardness of the joint optimization problem is proven by demonstrating a reduction from the well-known NP-hard makespan scheduling problem to a specific case of the joint optimization problem.
(3) We design a two-layer heuristic algorithm CIDA to solve the joint optimization problem at a low time complexity by decompose it into two subproblems. In the first layer, an offloading profile is derived base on allocating the service requests to the execution platform with minimum response time iteratively. In the second layer, as the convexity of computing capacity allocation problem in each edge node is proved, the optimal computing capacity allocation profile could be obtained.

The rest of the paper is organized as follows. We review the related works in Sect. 2. Section 3 presents the overall system and the joint optimization problem formulation. In Sect. 5, the design and implementation of algorithm CIDA is presented. The results of the numerical simulations are provided in Sect. 6. Finally, this paper is concluded in Sect. 7.

2 Related Work

To fully utilize the potential of multi-access edge computing, the problem of tasks offloading under various scenarios attracted much attention of the research community lately. One major objective is to reduce the task completion latency and energy consumption [9,10]. In [9], an optimization algorithm based on genetic algorithm and particle swarm optimization is proposed to minimize the energy consumption of the system in a densely deployed distributed small cell network. In [10], a distributed association scheme is proposed to minimize the latency of data flows in a fog computing IoT system by iteratively associating IoT devices to suitable base station until convergence. Zhu et al. [11] addressed the triple-objective optimization problem of energy cost, task completion delay and network deployment cost. A metaheuristic algorithm based on whale optimization is adopted to produce a set of Pareto-optimal solutions for the problem. Besides the offloading profile optimization, the computation resource allocation optimization has been investigated in some research [12]. A cooperative offloading model based on software defined network is proposed in [12] for optimizing task scheduling in LTE-advanced networks. And [13] proposed an UAV-aided framework in federated-WSN-enabled IoT for trust-worthy data collection. However, these literature mainly emphasize the offloading between devices and the edge servers, without considering the further offloading from edge servers to the remote cloud.

There are also a few work done on the cooperation workload scheduling in the heterogeneity three-tier MEC network [14–17]. [16] studies the online workload scheduling among edge-clouds architecture and introduced an online job dispatching and scheduling algorithm to minimized the job response time. [18] proposed a joint optimization approach to allocate the computation resource to IoT devices in a three-tier fog IoT framework. In [17], the offloading decisions, communication resources and computation resources are taken into account to minimize the overhead in a three-tier MEC system. [19] focus on minimizing the energy cost and allocating the computation resource with the latency deadlines of mobile devices are satisfied. Three algorithms are presented to solve the problem under different time complexity performance. Nevertheless, all the above work have overlooked that an IoT device might request multiple type of services. Given the heterogeneous nature of service applications deployed on IoT devices, incorporating the consideration of various types of service requests within the same device becomes imperative when optimizing task scheduling in the IoT. Such an approach facilitates the provision of enhanced and finer-grained offloading strategies for MEC system.

3 System Model and Problem Formulation

For the sake of readability, we summarize the symbols used in Table 1. Denote $\mathcal{N} = \{1, 2, 3, ..., N\}$ as the set of IoT devices and $\mathcal{M} = \{1, 2, 3, ..., M\}$ as the set of edge nodes. Denote $\mathcal{J} = \{1, 2, 3, ..., J\}$ as the set of services provided in the system, and $\mathcal{J}_i$ as the set of services on IoT device i. The service request of type j is described as (L_j, C_j), where L_j represents the average input data length associated with the request, and C_j represents the average computation requirements of the corresponding task (in CPU cycles) for type j services. As in other paper [20], we assume the arrival of each type of service requests on IoT devices follows a Poisson process with λ_{ij} denotes the expect arrival rate of type j service requests on IoT device i.

Table 1. Summary of symbols

Symbol	Definition
$\mathcal{N}$	The set of IoT devices
$\mathcal{M}$	The set of edge nodes
$\mathcal{J}$	The set of services provided in the system
$\mathcal{J}_i$	The set of services on IoT device i
L_j	Average input data length associate with type j service request
C_j	Average computation requirements for type j service request
λ_{ij}	Expected arrival rate on type j service request on IoT device i
a_{ijk}^e	Indicator to denote if j request on device i is executed on edge node k
a_{ij}^c	Indicator to denote if j request on device i is executed on remote cloud
b_i	Associated edge node of IoT device i
T_{ijk}^e	Average response time for j request on device i executed on edge node k
T_{i,b_i}^{trans}	Transmission delay from IoT device i to its associated edge node b_i
$T_{b_i,k}^{trans}$	Transmission delay from edge node b_i to edge node k
T_{ijk}^{exe}	Execution delay for j request on device i to be executed on edge node k
R_{i,b_i}	Data rate achieved for device i to associated edge node b_i
B_{i,b_i}	Wireless channel bandwidth between device i to associated edge node b_i
P_i	Transmission power of IoT device i
G_{i,b_i}	Channel gain between device i to associated edge node b_i
σ_i^2	Noise power of device i
$\bar{L}_i$	Average input data length of the requests on device i
f_k	Computing capacity of edge node k
f_{jk}	Computing capacity allocated to type j request on edge node k
T_{ij}^c	Average response time for j request on device i executed on remote cloud
$T_{b_i,c}^{trans}$	Transmission delay from edge node b_i to remote cloud
$A(ij)$	Indicator of execution platform for type j service requests on IoT device i
$\boldsymbol{A}$	Offloading profile for all the requests on IoT devices in the system
$\boldsymbol{F}_k$	Computing capacity allocation profile on edge node k
$\boldsymbol{F}$	Computing capacity allocation profile for all edge nodes in the system
$T_{ij}(\boldsymbol{A},\boldsymbol{F})$	Average response time for type j request on device i under profile $\boldsymbol{A}$ and $\boldsymbol{F}$
ΔT_{ij}	Improvement on average response time for type j request on device i by moving it to its optimal edge node
$\mathcal{R}$	The set of remaining requests to be reallocated in the system
r_{ij}	The index of the type j request on device i in the set $\mathcal{R}$

Let a_{ijk}^e and a_{ij}^c be the binary indicator $(a_{ijk}^e, a_{ij}^c \in \{0,1\})$ which denote if type j service requests of IoT device i to be executed on edge node k or the remote cloud server, respectively. A service request is either be processed on an edge node or the remote cloud server, thus we have

$$\sum_{k \in \mathcal{M}} a_{ijk}^e + a_{ij}^c = 1, i \in \mathcal{N}, j \in \mathcal{J}_i \tag{1}$$

3.1   Edge Computing Model

For each IoT device in the system, it is associated to its closest edge node to offloading the service requests. Note that the associated edge node might not be the eventual execution platform for the requests on IoT devices. Denote b_i as the associated edge node of IoT device i. When type j requests on IoT device i are to be executed on edge node k, the total response time consist of the following four parts: the time it takes for the requests to be transferred from IoT device i to its associated edge node b_i (denoted by T_{i,b_i}^{trans}), the time it takes for the requests to be transferred from associated edge node b_i to the execution edge node k (denoted by $T_{b_i,k}^{trans}$), the time it takes for the requests to be executed on edge node k (denoted by T_{ijk}^{exe}), and the time it takes for sending the computation outcome back to IoT device i.

Similar to many studies such as [21–23], we omit the time overhead associated with the transmission of computation outcomes back to IoT device, which is based on the observation that the length of computation outcomes tends to be significantly smaller compared to the length of input data for numerous services such as image recognition and speech recognition. Hence, the total response time of type j requests on IoT device i to be executed on edge node k is given by

$$T_{ijk}^e = T_{i,b_i}^{trans} + T_{b_i,k}^{trans} + T_{ijk}^{exe} \tag{2}$$

During the wireless transmission between IoT device i and its associated edge node b_i, the data rate R_{i,b_i} can be generally expressed as a logarithmic function according to the Shannon theorem. Denote B_{i,b_i} as the channel bandwidth between IoT device i and edge node b_i, G_{i,b_i} as the channel gain between them, P_i as the transmission power and σ_i^2 as the noise power of IoT device i. The data rate of IoT device i can be expressed as

$$R_{i,b_i} = B_{i,b_i} \log(1 + \frac{P_i G_{i,b_i}}{\sigma_i^2}). \tag{3}$$

As mentioned earlier that the arrival of each type of service request on IoT devices is a Poisson process, the arrival of all the requests on IoT device i, which is a sum of multiple independent Poisson processes, follows a Poisson process with expected rate $\sum_{j \in \mathcal{J}_i} \lambda_{ij}$. Thus, the offloading from IoT device i to its associated edge node b_i can be modelled as an $M/M/1$ queue, and the average response time is given by

$$T_{i,b_i}^{trans} = \frac{1}{R_{i,b_i}/\bar{L}_i - \sum_{j \in \mathcal{J}_i} \lambda_{ij}} \tag{4}$$

where $\bar{L}_i$ denotes the average input data length for the service requests on IoT device i and $\bar{L}_i = \sum_{j \in \mathcal{J}_i} L_j \lambda_{ij} / \sum_{j \in \mathcal{J}_i} \lambda_{ij}$. To keep the queue stable we have $R_{i,b_i}/\bar{L}_i > \sum_{j \in \mathcal{J}_i} \lambda_{ij}$.

After the requests being offloaded to associated edge node b_i, they will be transferred to the execution edge node k. The transmission delay from b_i to k is denoted by $T^{trans}_{b_i,k}$, which can be measured and recorded by the SDN controller [24]. Note that $T^{trans}_{b_i,k} = 0$ when $b_i = k$. According to [25], the transmission delay between edge nodes is modelled as a linear function of the distance between them, which is given by

$$T^{trans}_{b_i,k} = \alpha \cdot d_{b_i,k} + \beta \tag{5}$$

where $d_{b_i,k}$ denotes the distance between edge node b_i and edge node k. And typically, $\alpha = 5$ and $\beta = 22.3$.

When type j requests on IoT device i are transferred to edge node k for execution, it shares the computing capacity allocated to type j requests on edge node k with same type of requests which are offloaded from other IoT devices. Given that the offloading of individual IoT device follows a Poisson process, it can be inferred that the arrival of each type of service request on the edge node also exhibits a Poisson process. The processing for each type of requests on the edge nodes can be modelled as an $M/M/1$ queue. Denote f_k as the computing capacity of edge node k and f_{jk} as the computing capacity allocated to type j request on edge node k. The computation delay for type j requests of IoT device i on edge node k is calculated as

$$T^{exe}_{ijk} = \frac{1}{f_{jk}/C_j - \sum\limits_{i \in \mathcal{N}} a^e_{ijk} \lambda_{ij}}. \tag{6}$$

To ensure the queue is stable, we have $\sum_{j \in \mathcal{J}} f_{jk} \leq f_k$ and $f_{jk}/C_j > \sum_{i \in \mathcal{N}} a^e_{ijk} \lambda_{ij}$.

3.2 Cloud Computing Model

For the service requests on IoT device i which are offloaded to remote cloud server for execution, the total response time consist of following four parts: the time it takes for the requests to be transferred from IoT device i to its associated edge node b_i, the time it takes for the requests to be transferred from edge node b_i to the remote cloud server via the core network, the time it takes for processing the requests on the remote cloud, and the time it takes for sending the computation results back to device i.

Similar to the edge computing model in Sect. 3.1, we disregarded the transmission delay of computation results from remote cloud to the IoT devices. Additionally, as the computing capacity of cloud servers exceeds that of edge nodes significantly, and the primary source of delay in cloud computing is attributed to data offloading from IoT devices to the remote cloud server, we have omitted the service execution delay on the cloud. Thus, denote $T^{trans}_{b_i,c}$ as the transmission delay between edge node b_i and the remote cloud, the total response time of type j requests on IoT device i to be executed on remote cloud server is given by

$$T_{ij}^c = T_{i,b_i}^{trans} + T_{b_i,c}^{trans}. \tag{7}$$

4   Problem Formulation

In this section, we present the problem formulation for service-aware task offloading and scheduling, focusing on the minimization of the average response time for all service requests. Denote $A(ij)$ as the indicator of execution platform for type j service requests on IoT device i, which is defined as $A(ij) = \{a_{ijk}^e, a_{ij}^c | k \in \mathcal{M}\}$. And A denotes the offloading profile for all the service requests on IoT devices in the system, which is defined as $A = \{A(ij) | i \in \mathcal{N}, j \in \mathcal{J}_i\}$. Furthermore, Denote F_k as the computing capacity allocated to each type of service on edge node k and $F_k = \{f_{jk} | j \in \mathcal{J}\}$. And F denotes the computing capacity allocation for all the edge nodes, which is given by $F = \{F_k | k \in \mathcal{M}\}$. Therefore, the service-aware task offloading and scheduling problem can be formulated as the problem of minimizing the average response time for all service requests with respect to A and F, which defined as:

$$\mathbf{P1}: \min_{A,F} \sum_{i \in \mathcal{N}, j \in \mathcal{J}_i} \frac{\lambda_{ij}}{\sum_{i \in \mathcal{N}, j \in \mathcal{J}_i} \lambda_{ij}} \left(\sum_{k \in \mathcal{M}} a_{ijk}^e T_{ijk}^e + a_{ij}^c T_{ij}^c \right) \tag{8}$$

$$s.t \quad \sum_{k \in \mathcal{M}} a_{ijk}^e + a_{ij}^c = 1, i \in \mathcal{N}, j \in \mathcal{J}_i \tag{9}$$

$$a_{ijk}^e, a_{ij}^c \in \{0, 1\}, i \in \mathcal{N}, j \in \mathcal{J}_i, k \in \mathcal{M} \tag{10}$$

$$R_{i,b_i} / \bar{L}_i > \sum_{j \in \mathcal{J}_i} \lambda_{ij}, i \in \mathcal{N} \tag{11}$$

$$f_{jk} / C_j > \sum_{i \in \mathcal{N}} a_{ijk}^e \lambda_{ij}, j \in \mathcal{J}, k \in \mathcal{M} \tag{12}$$

$$\sum_{j \in \mathcal{J}} f_{jk} \leq f_k, k \in \mathcal{M} \tag{13}$$

$$f_{jk} \in [0, f_k], j \in \mathcal{J}, k \in \mathcal{M} \tag{14}$$

The constraints (9) and (10) delineate the processing of each type of service requests on IoT devices are either on an edge node or on the remote cloud server. The constraint (11) ensures that the total traffic data on each IoT device remains within the wireless channel capacity between the device and its associated edge node. The constraint (12) guarantees the workload assigned to each type of requests on each edge node does not exceed the specific computing capacity allocated to that request type. The constraint (13) ensures that the collective computing capacity assigned to each type of request within an edge node should not exceed the overall computing capacity of the edge node. Lastly, constraint (14) enforces that the computing capacity allocated to each request type remains within the total computing capacity available of the edge node.

However, solving problem **P1** optimally with low searching cost could be quite challenging. The typical size of edge computing models tends to be substantial, which results in a large searching space that optimization algorithm need to explore. Additionally, the high heterogeneity on the configuration of edge nodes and the diversity of services on IoT devices further complicate the problem. Moreover, the interplay between the IoT devices offloading profile **A** and the edge node capacity allocation profile **F** must be considered during the optimization process, making problem **P1** hard to tackle. We will provide a proof of its NP-hardness to demonstrate the computational complexity of problem **P1** next.

Theorem 1. *Problem* **P1** *is at least NP-hard.*

Proof. Here, we present an analysis of a simplified case of problem **P1** to demonstrate its NP-hardness. In this particular scenario, only a single type of service is considered in the system. And we make the following assumptions: the transmission delay from IoT devices to their associated edge node are assumed to be 0, i.e., $T_{i,b_i}^{trans} = 0$; the transmission delay between edge nodes are assumed to be 0, i.e., $T_{b_i,k}^{trans} = 0$; the transmission delay from edge nodes to the remote cloud server are assumed to be infinite, i.e., $T_{b_i,c}^{trans} = \infty$, which indicates an infinite response time for requests executed on the remote cloud according to Equation (7). As a result, all the requests should be processed on edge nodes if capable.

By making these assumptions, problem **P1** is reduced to the classic makespan scheduling problem. In this transformation, the service requests on each IoT device are projected as jobs to be scheduled, while the edge nodes are projected as the processing machines in the makespan scheduling problem. As the classic makespan scheduling problem, which requires exploring all possible combinations of job allocation to find the optimal solution, has non-polynomial computation complexity for even there are only two identical processing machines, the NP-hardness of problem **P1** is established.

5   Algorithm Design

As stated above, even the simplified case of problem **P1** remain challenging to be solved in polynomial time complexity. As we can observed that problem **P1** involve a jointly optimization of the offloading profile of IoT devices **A** and the computing capacity allocation profile of edge nodes **F**. To solve problem **P1**, we carefully decompose the joint optimization involved into two sub problem:

(1) The optimization problem of offloading profile **A** under a certain computing capacity allocation policy.
(2) The optimization problem of computing capacity allocation profile **F** under a specific offloading profile **A**.

Accordingly, we present a two-layer Cooperative workload Initialization and Distribution Algorithm (CIDA) with the first layer minimizing the transmission delay and distributing the workload among edge nodes and the cloud server to obtain a suboptimal offloading profile, and the second layer allocate computing capacity to each type of services optimally on the edge nodes based on the offloading profile obtained in the first layer.

5.1 Offloading Profile Optimization Problem

To tackle the optimization problem of offloading profile, CIDA follows a two-step approach. Firstly, the service requests on IoT devices are assigned to their respective associated edge nodes with the transmission delay is minimized, establishing the initial offloading profile. Subsequently, an iterative process is performed to search for and reallocate requests which has the highest response time improvement by moving them from the current execution platform to the optimal execution platform. This process continues until no further improvement in response time can be achieved for any remaining requests in the system.

Offloading Profile Initialization. To initialize the offloading profile, we first focus on minimizing the average transmission delay from IoT devices to corresponding execution edge node for all the requests, which is given by

$$T_{ij}^{trans} = \sum_{k \in \mathcal{M}} a_{ijk}^{e}(T_{i,b_i}^{trans} + T_{b_i,k}^{trans}) + a_{ij}^{c}(T_{i,b_i}^{trans} + T_{b_i,c}^{trans}) \qquad (15)$$

Thus, the transmission delay minimization problem is formulated as

$$\textbf{P2} : \min_{A} \sum_{i \in \mathcal{N}, j \in \mathcal{J}_i} \frac{\lambda_{ij}}{\sum_{i \in \mathcal{N}, j \in \mathcal{J}_i} \lambda_{ij}} T_{ij}^{trans} \qquad (16)$$

$$s.t \sum_{k \in \mathcal{M}} a_{ijk}^{e} + a_{ij}^{c} = 1, i \in \mathcal{N}, j \in \mathcal{J}_i \qquad (17)$$

$$a_{ijk}^{e}, a_{ij}^{c} \in \{0,1\}, i \in \mathcal{N}, j \in \mathcal{J}_i, k \in \mathcal{M} \qquad (18)$$

$$R_{i,b_i}/\bar{L}_i > \sum_{j \in \mathcal{J}_i} \lambda_{ij}, i \in \mathcal{N} \qquad (19)$$

It is easy to observe that to minimize T_{ij}^{trans}, all the requests on IoT devices will be allocated to its associated edge node for execution, i.e., $A(ij) = b_i, i \in \mathcal{N}, j \in \mathcal{J}_i$, as the transmission delay of type j requests on IoT device i from associated edge node to the execution edge node $T_{b_i,k}^{trans} = 0$ when $b_i = k$, and $T_{b_i,b_i}^{trans} < T_{b_i,k}^{trans}$ holds when $b_i \neq k$ according to Equation (5)(15). Denote A_{ini} as the corresponding offloading profile, i.e., $A_{ini} = \{A(ij) = b_i | i \in \mathcal{N}, j \in \mathcal{J}_i\}$.

However, offloading profile A_{ini} might not be a feasible solution for problem **P1** since the following constraints might be violated:

$$\sum_{i \in \mathcal{N}, j \in \mathcal{J}_i} a_{ijk}^{e} \lambda_{ij} C_j < f_k, k \in \mathcal{M}. \qquad (20)$$

Constraint (20) can be derived from constraint (12) and (13). Specifically, allocating the requests on the IoT devices to its associated edge nodes could minimize the transmission delay in the system. Nevertheless, due to the variation in geographical distribution of IoT devices, the edge nodes with more IoT devices are associated to will be assigned a greater computational workload which might exceed the computing capacity of the edge nodes, while the workload on other edge nodes remain at a relatively low level. To address this issue, the workload need to be reallocated among edge nodes and remote cloud server to ensure efficient utilization of computing resources and minimize total response time for all requests.

Workload Reallocation. This subsection presents the main idea of algorithm design to reallocate the requests workload in the system. We first define the computing capacity allocation profile on the edge nodes during the reallocation process. Let the computing capacity allocated to each type of request to be determined based on the percentage of corresponding type of workload in the total workload on the edge node, which is given by

$$
f_{jk} = \frac{\sum\limits_{i\in\mathcal{N}} a^e_{ijk}\lambda_{ij}f_k}{\sum\limits_{i\in\mathcal{N},j\in\mathcal{J}_i} a^e_{ijk}\lambda_{ij}}, j \in \mathcal{J}, k \in \mathcal{M} \tag{21}
$$

In addition, denote $T_{ij}(\boldsymbol{A}, \boldsymbol{F})$ as the total response time of type j service of IoT device i under the current offloading profile $\boldsymbol{A}$ and computing capacity allocation profile $\boldsymbol{F}$. Note that $T_{ij}(\boldsymbol{A}, \boldsymbol{F}) = \infty$ for the requests on the edge node with constraint (20) is violated. Denote $\boldsymbol{A}_{-ij}$ as the offloading profile of the system other than type j service requests on IoT device i. If $\boldsymbol{A}_{-ij}$ is given, it can be determined that the optimal execution platform for type j service on device i to minimize its total response time. Denote $A(ij)^*$ as the indicator of optimal execution platform and $T_{ij}(A(ij)^*, \boldsymbol{A}_{-ij}, \boldsymbol{F})$ as the total response time of type j request on device i on execution platform $A(ij)^*$. Denote ΔT_{ij} as the response time improvement for reallocating type j requests on device i from current execution platform $A(ij)$ to optimal execution platform $A(ij)^*$. Thus we have

$$
T_{ij}(\boldsymbol{A}, \boldsymbol{F}) = \begin{cases} \infty & a^e_{ijk} = 1, \sum\limits_{i\in\mathcal{N},j\in\mathcal{J}_i} a^e_{ijk}\lambda_{ij}C_j < f_k \\ \sum\limits_{k\in\mathcal{M}} a^e_{ijk}T^e_{ijk} + a^c_{ij}T^c_{ij} & otherwise \end{cases} \tag{22}
$$

$$
A(ij)^* = \arg\min_{A(ij)} \{T_{ij}(A(ij), \boldsymbol{A}_{-ij}, \boldsymbol{F}) | i \in \mathcal{N}, j \in \mathcal{J}_i\} \tag{23}
$$

$$
\Delta T_{ij} = T_{ij}(\boldsymbol{A}, \boldsymbol{F}) - T_{ij}(A(ij)^*, \boldsymbol{A}_{-ij}, \boldsymbol{f}) \tag{24}
$$

Through a calculation of response time improvement for the requests which has not been reallocated in the system, the requests with highest response time improvement will be put into a set and wait for update. One of them will be selected randomly and reallocated to its optimal execution platform at each iteration. If there are multiple execution platforms that can achieve the optimal response time for a selected request, one of them is randomly chosen for real-location. This process repeats until all the requests in the system is reallocated or no further improvement on response time can be achieve for the remaining requests in the system.

Denote r_{ij} as the index of type j request on IoT device i. Denote $\mathcal{R}$ as the set of remaining service requests to be reallocated in the system, i.e., $\mathcal{R} = \{r_{ij}|i \in \mathcal{N}, j \in \mathcal{J}_i\}$. Denote R_{update} as the set for the requests having maximum response time improvement and waiting to be updated. The first layer of CIDA is shown in Algorithm(1).

Algorithm 1. The first layer: offloading profile optimization

---

Input: $\{L_j, C_j | j \in \mathcal{J}\}, \{\lambda_{ij}|i \in \mathcal{N}, j \in \mathcal{J}_i\}, \{B_{i,b_i}, P_i, G_{i,b_i}, \sigma_i^2, T_{b_i,c}^{trans}|i \in \mathcal{N}\}, \alpha, \beta, \{f_k|k \in \mathcal{M}\}$
Output: A solution of offloading profile $\boldsymbol{A} = \{A(ij)|i \in \mathcal{N}, j \in \mathcal{J}_i\}$, where $A(ij) = \{a_{ijk}^e, a_{ij}^c|k \in \mathcal{M}\}$.

1:   $A(ij) \leftarrow b_i, i \in \mathcal{N}, j \in \mathcal{J}_i$
2: **for** Iteration $j = 1, 2, 3, \ldots$ **do**
3:     $R_{update} \leftarrow \varnothing$
4:     $\Delta T_{max} = 0$
5:     **for** Each request $r_{ij} \in \mathcal{R}$ **do**
6:       Calculate $A(ij)^*$ and ΔT_{ij} by Equation $(1) \sim (7)$ and $(21) \sim (24)$
7:       **if** $\Delta T_{ij} > \Delta T_{max}$ **then**
8:         $R_{update} \leftarrow \varnothing \cup \{r_{ij}\}$
9:         $\Delta T_{max} \leftarrow \Delta T_{ij}$
10:      **else if** $\Delta T_{ij} = \Delta T_{max} \&\& \Delta T_{max} \neq 0$ **then**
11:        $R_{update} = R_{update} \cup \{r_{ij}\}$
12:      **end if**
13:     **end for**
14:     **if** $R_{update} \neq \varnothing$ **then**
15:       Select r_{ij} in R_{update} randomly for update
16:       $A(ij) \leftarrow A(ij)^*$
17:       $\mathcal{R} = \mathcal{R} - \{r_{ij}\}$
18:     **else**
19:       **return** the offloading profile $\boldsymbol{A}$
20:     **end if**
21: **end for**

We will now prove the finite termination property of Algorithm 1. Initially, the size of the set of remaining service requests to be reallocated is $|\mathcal{R}| = \sum_{i \in \mathcal{N}} |\mathcal{J}_i|$. At each iteration, $|\mathcal{R}|$ is decreased by one, as the request with the maximum response time improvement is successfully reallocated to its optimal execution platform and removed from the set $\mathcal{R}$. Consequently, it takes $|\mathcal{R}|$ iterations to reallocate all the requests in the system, resulting in Algorithm 1 terminating when $\mathcal{R} = \varnothing$. Alternatively, Algorithm 1 may terminate earlier if there are no requests in the system that can further improve their response times through reallocation ($\Delta T_{max} = 0$).

The time complexity of Algorithm 1 is analyzed as follows: Step 1 has a time complexity of $O(|\mathcal{R}|)$, as it performs a certain operation for each request in $\mathcal{R}$. In the worst case scenario, Step 2 is iterated $|\mathcal{R}|$ times. As step 5 and step 6 involve calculating the response time of requests on all execution platforms, which have the size of $|\mathcal{R}|$ and $|\mathcal{M}+1|$, the time complexity of these steps is $O(|\mathcal{R}||\mathcal{M}+1|)$. And step 7 to step 20 have a constant time complexity of $O(1)$. Taking all these steps into account, the overall time complexity of Algorithm 1 can be approximated as $O(|\mathcal{R}|^2(\mathcal{M}+1))$.

5.2 Computing Capacity Allocation Profile Optimization Problem

With the offloading profile $\boldsymbol{A}$ is determined, the transmission delay for all the requests can be calculated. And **P1** is transformed to a computing capacity allocation problem, with the objective of minimizing the execution delay for the requests in the system. Since the execution delay for requests on one edge node is independent of requests on other edge nodes, the computing capacity allocation problem for the system can be decomposed into multiple computing capacity allocation problems for each individual edge node. Thus the allocation problem on edge node k is given by:

$$\textbf{P3}: \min_{\boldsymbol{F}_k} \quad \sum_{i\in\mathcal{N}, j\in\mathcal{J}_i} \frac{a_{ijk}^e \lambda_{ij}}{\sum_{i\in\mathcal{N}, j\in\mathcal{J}_i} \lambda_{ij}} T_{ijk}^{exe} \tag{25}$$

$$f_{jk}/C_j > \sum_{i\in\mathcal{N}} a_{ijk}^e \lambda_{ij}, j \in \mathcal{J} \tag{26}$$

$$\sum_{j\in\mathcal{J}} f_{jk} \le f_k \tag{27}$$

$$f_{jk} \in [0, f_k], j \in \mathcal{J} \tag{28}$$

Theorem 2. *Problem* **P3** *is a convex optimization problem over the computing capacity allocation profile* $\boldsymbol{F}$.

Proof. An optimization problem is a convex optimization problem with following conditions are satisfied: the objective function is a convex function; the constraint functions are convex functions; the feasible region is a convex set. It is easy to observed that the constraints (26), (27), (28) are convex and the feasible region is a convex set. And we can prove the objective function of problem **P3** by showing the positive definiteness of its Hessian matrix.

Denote Z_k as the objective function of problem **P3**, thus we have $Z_k = \sum_{i\in\mathcal{N}, j\in\mathcal{J}_i} \frac{a_{ijk}^e \lambda_{ij}}{\sum_{i\in\mathcal{N}, j\in\mathcal{J}_i} \lambda_{ij}}(\frac{1}{f_{jk}/C_j - \sum_{i\in\mathcal{N}} a_{ijk}^e \lambda_{ij}})$. The Hessian matrix of Z is denoted by $H = \left[h_{jj'}\right]_{j\times j'}$. When $j = j'$, $h_{jj'} = \sum_{i\in\mathcal{N}} \frac{a_{ijk}^e \lambda_{ij}}{\sum_{i\in\mathcal{N}, j\in\mathcal{J}_i} \lambda_{ij}} \cdot \frac{2}{C_j^2(f_{jk}/C_j - \sum_{i\in\mathcal{N}} a_{ijk}^e \lambda_{ij})^3}$, otherwise $h_{jj'} = 0$, meaning that H is a positive definite matrix and Z is a convex function. Therefore, Problem **P3** is a convex optimization problem over the computing capacity allocation profile $\boldsymbol{F}_k$.

As problem **P3** is a convex optimization problem, it can be solved optimally via solving Karush-Kuhn-Tucker(KKT) conditions. We first construct the Lagrange function of Problem **P3** as

$$
\begin{aligned}
L_k(\boldsymbol{F}_k, \boldsymbol{\chi}_k, \gamma_k, \boldsymbol{\phi}_k, \boldsymbol{\omega}_k) = Z_k - \sum_{j \in \mathcal{J}} \chi_{jk}(f_{jk} - \sum_{i \in \mathcal{N}} a^e_{ijk} \lambda_{ij} C_j) \\
+ \gamma_k(\sum_{j \in \mathcal{J}} f_{jk} - f_k) + \sum_{j \in \mathcal{J}} \phi_{jk}(f_{jk} - f_k) - \sum_{j \in \mathcal{J}} \omega_{jk} f_{jk}
\end{aligned}
\tag{29}
$$

with $\boldsymbol{\chi}_k$, γ_k, $\boldsymbol{\phi}_k$ and $\boldsymbol{\omega}_k$ are the set of Lagrange multipliers, i.e., $\boldsymbol{\chi}_k = \{\chi_{jk} | j \in \mathcal{J}\}$, $\boldsymbol{\phi}_k = \{\phi_{jk} | j \in \mathcal{J}\}$ and $\boldsymbol{\omega}_k = \{\omega_{jk} | j \in \mathcal{J}\}$.

The KKT conditions of problem **P3** is given by constraints (26) $\sim$ (28) and equations (30) $\sim$ (35).

$$
\frac{\partial L_k(\boldsymbol{F}_k, \boldsymbol{\chi}_k, \gamma_k, \boldsymbol{\phi}_k, \boldsymbol{\omega}_k)}{\partial f_{jk}} = 0, \quad \forall j \in \mathcal{J}
\tag{30}
$$

$$
\chi_{jk}(f_{jk} - \sum_{i \in \mathcal{N}} a^e_{ijk} \lambda_{ij} C_j) = 0, \quad \forall j \in \mathcal{J}
\tag{31}
$$

$$
\gamma_k(\sum_{j \in \mathcal{J}} f_{jk} - f_k) = 0
\tag{32}
$$

$$
\phi_{jk}(f_{jk} - f_k) = 0, \quad \forall j \in \mathcal{J}
\tag{33}
$$

$$
\omega_{jk} f_{jk} = 0, \quad \forall j \in \mathcal{J}
\tag{34}
$$

$$
\chi_{jk}, \gamma_k, \phi_{jk}, \omega_{jk} \geq 0, \quad \forall j \in \mathcal{J}
\tag{35}
$$

As the KKT conditions of problem **P3** is solved, an optimal computing capacity allocation profile of edge node k can be derived. By aggregating the allocation profiles obtained for each individual edge node $\boldsymbol{F}_k$, the optimal allocation profile for the entire system $\boldsymbol{F}$ can be acquired.

6   Simulation Results

In this section, extensive simulations are conducted to evaluate the system performance of CIDA. The simulations are implemented with a MATLAB program on the executing host: 64-bit Windows 11 operating system, Intel Core i5-9600K CPU@3.70 GHz, and 32GB of RAM. In the simulations, 1000 IoT devices and 50 edge nodes are randomly distributed in a 200 km × 200 km area. The system offers 10 types of services in total, and each IoT device receives five random service requests out of the available 10. Unless otherwise stated, the default value of parameters are specified in Table 2.

We will start by evaluating the convergence of CIDA. Figure 2 illustrates the average response time of all service requests in the system over iterations. It can be observed that the system average response time prior to the 877th iteration remains infinite. This can be attributed to the initial allocation of requests to the associated edge nodes without considering the capacity of these

Table 2. Configuration of Simulation Parameters

Parameter	Value/Range	Parameter	Value/Range
λ_{ij}	$[2,3]$	G_{i,b_i}	$128.1 + 37.5 \log d_{i,b_i}$
L_j	$[80, 120]$KB	B_{i,b_i}	1MHz
C_j	$[10, 15]$ MegaCycles	P_i	23dBm
σ_i^2	-100dBm	$T_{b_i,c}^{trans}$	0.5s
f_k	$[5.4, 6.6]$GHz		

nodes. Consequently, requests on overwhelmed edge nodes experience an infinite execution delay, leading to an overall average response time of infinity for the system. However, after the 877th iteration, there is a significant decrease in the system average response time, and it eventually converges at the 2480th iteration as there are no requests left that can further improve the system's response time. Therefore, this observation verifies the convergence of CIDA.

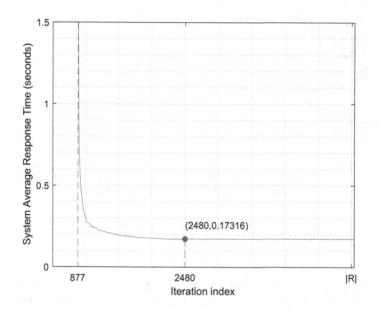

Fig. 2. The system average response time over iterations

We implemented the following algorithms as benchmark for comparison purpose:

(1) **Density Based Clustering Strategy (DBCS)**: All the requests will be offloaded to the edge node with minimum average response time until the workload on the edge node exceed the average workload among all the edge nodes, as in [26].

(2) **Distributed Offloading Decision Making (DODM)**: Service requests are the entities to make the offloading decision to minimized their own average response time, as in [27].

(3) **Application-aware Workload Allocation (AWA)**: As in [7], all the requests are assigned to the edge node with minimum transmission delay first. Then requests are reallocated to the edge node with minimum response time iteratively. However, there is no cooperation between edge nodes and the remote cloud server as service requests are not allowed to be further offloaded to remote cloud.

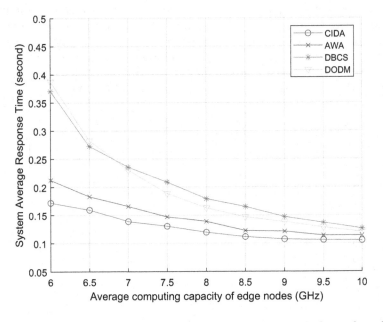

Fig. 3. Average response time over average computing capacity of edge nodes achieved by different algorithms

Figure 3 illustrates the system request average response time of the four algorithms mentioned above under different average computing capabilities of edge nodes (ranging from 6GHz to 10GHz). It can be observed that the system average response time of all algorithms exhibits a decreasing trend as the average computing capacity of edge nodes increases. When the average computing capability of edge nodes is 6GHz, CIDA reduces the system response time by 18.3%, 53.7%, and 55.8% compared to the AWA, DBCS, and DODM algorithms, respectively. As the capabilities of edge nodes increase, the performance gap between the other algorithms and CIDA gradually decreases. This is because the execution latency of system requests is no longer a performance bottleneck in the DBCS and DODM algorithms due to the enhanced computing capabilities of the edge nodes. Relatively speaking, there is no significant reduction on the system average response time of CIDA and AWA algorithms, indicating that the execution latency of system requests has been effectively optimized.

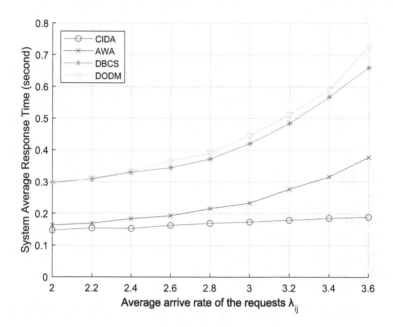

Fig. 4. Average response time over average request arrival rate λ_{ij} achieved by different algorithms

To investigate the performance of the above algorithms under different system workload level, we increased the average arrival rate of each type of request from 2 to 3.6. The experimental results are shown in Fig. 4. When $\lambda_{ij} = 2$, CIDA algorithm exhibits a significantly lower system average response time compared to the DBCS and DODM algorithms, while the system average response time of the AWA algorithm is relatively close to the CIDA algorithm. This result can be explained by the fact that in the DBCS algorithm, the assignment of system requests depends on the average workload level among edge nodes without considering the heterogeneity of them. In the DODM algorithm, individual offloading decisions are made by each system request based on its own average response time, resulting in the poor cooperation among edge nodes. These factors contribute to the inferior performance of the DBCS and DODM algorithms compared to the CIDA and AWA algorithms. However, as λ_{ij} increases, the performance gap between the other algorithms and the CIDA algorithm gradually widens. When λ equals to 3.6, CIDA algorithm reduces the system average response time by 49.7%, 71.4%, and 74.4% compared to AWA, DBCS and DODM algorithms, respectively. For AWA algorithm, since there is no cooperation between the cloud server and the edge nodes, the edge nodes are getting heavy-loaded as the average request arrival rate increases, which leads to much higher execution delay for the requests. In contrast, CIDA can further offload a portion of the requests with high computation requirement to the remote cloud server, ensuring efficient system operation even under high average requests arrival rate.

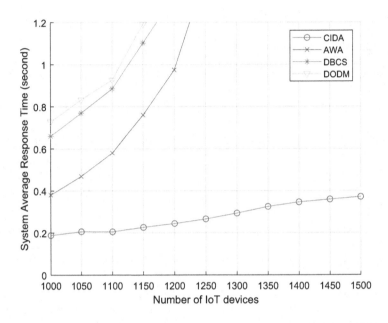

Fig. 5. Average response time over numbers of IoT devices achieved by different algorithms

To further validate the scalability of our algorithm, we increased the number of IoT devices from 1000 to 1500 while maintaining the average request arrival rate λ_{ij} at 3.6. The results are shown in Fig. 5. It can be observed that as the number of devices increases, the system average response time of AWA, DBCS, and DODM algorithms dramatically increases, and reach the system capacity limit as the number of devices approaches 1300. Relatively speaking, by leveraging the collaboration between edge nodes and the remote cloud server, CIDA exhibits only a slight increase in the system average response time, which clearly indicates the effectiveness and high scalability of CIDA.

7   Conclusions

In this paper, we systematically investigate the problem of service-aware cooperative task offloading and scheduling in a three-tier multi-access edge computing IoT environment. This problem involves the joint optimization of the system offloading profile and computing capacity allocation profile, which has been proven to be at least NP-hard. To address this problem, we propose a two-layer heuristic algorithm CIDA, where the first layer obtain a suboptimal offloading profile with minimized response time for the service requests at a low time complexity, and the second layer allocate the computing capacity on the edge nodes to each type of services optimally. Simulation results demonstrate that CIDA outperforms other approaches in terms of minimizing the average response time for service requests. Additionally, CIDA exhibits high scalability as the system size increases, making it suitable for large-scale deployment.

Acknowledgements. This work was supported in part by the Guangdong Key Construction Discipline Research Ability Enhancement Project (Grant No. 2021ZDJS086); in part by the Guangdong University Key Project (Grant No. 2019KZDXM012); in part by the Natural Science Foundation of Guangdong Province (Grant No. 2021A1515010656); in part by Guangdong Basic and Applied Basic Research Foundation (2022B1515120059); in part by the research team project of Dongguan University of Technology (Grant No. TDY-B2019009).

References

1. Laghari, A.A., Wu, K., Laghari, R.A., Ali, M., Khan, A.A.: A review and state of art of internet of things (iot). Archives of Computational Methods in Engineering, pp. 1–19 (2021)
2. Tao, M., Li, X., Wei, W., Yuan, H.: Jointly optimization for activity recognition in secure iot-enabled elderly care applications. Appl. Soft Comput. **99**, 106788 (2021)
3. Zhu, R., Liu, L., Song, H., Ma, M.: Multi-access edge computing enabled internet of things: advances and novel applications (2020)
4. Mach, P., Becvar, Z.: Mobile edge computing: a survey on architecture and computation offloading. IEEE Commun. Surv. Tutorials **19**(3), 1628–1656 (2017)
5. Xu, J., Chen, L., Zhou, P.: Joint service caching and task offloading for mobile edge computing in dense networks. In: IEEE INFOCOM 2018-IEEE Conference on Computer Communications, pp. 207–215. IEEE (2018)
6. Ma, X., Zhou, A., Zhang, S., Wang, S.: Cooperative service caching and workload scheduling in mobile edge computing. In: IEEE INFOCOM 2020-IEEE Conference on Computer Communications, pp. 2076–2085. IEEE (2020)
7. Fan, Q., Ansari, N.: Application aware workload allocation for edge computing-based iot. IEEE Internet Things J. **5**(3), 2146–2153 (2018)
8. Tao, M., Xueqiang, L., Kaoru, O., Mianxiong, D.: Single-cell multi-user computation offloading in dynamic pricing-aided mobile edge computing. IEEE Trans. Comput. Social Syst. (2023). https://doi.org/10.1109/TCSS.2023.3308563
9. Guo, F., Zhang, H., Ji, H., Li, X., Leung, V.C.: An efficient computation offloading management scheme in the densely deployed small cell networks with mobile edge computing. IEEE/ACM Trans. Network. **26**(6), 2651–2664 (2018)
10. Fan, Q., Ansari, N.: Towards workload balancing in fog computing empowered iot. IEEE Trans. Netw. Sci. Eng. **7**(1), 253–262 (2018)
11. Zhu, X., Zhou, M.: Multiobjective optimized cloudlet deployment and task offloading for mobile-edge computing. IEEE Internet Things J. **8**(20), 15582–15595 (2021)
12. Cui, Y., Song, J., Ren, K., Li, M., Li, Z., Ren, Q., Zhang, Y.: Software defined cooperative offloading for mobile cloudlets. IEEE/ACM Trans. Netw. **25**(3), 1746–1760 (2017)

13. Tao, M., Li, X., Yuan, H., Wei, W.: Uav-aided trustworthy data collection in federated-wsn-enabled iot applications. Inf. Sci. **532**, 155–169 (2020)
14. Tong, L., Li, Y., Gao, W.: A hierarchical edge cloud architecture for mobile computing. In: IEEE INFOCOM 2016-The 35th Annual IEEE International Conference on Computer Communications, pp. 1–9. IEEE (2016)
15. Tao, M., Ota, K., Dong, M.: Dsarp: dependable scheduling with active replica placement for workflow applications in cloud computing. IEEE Trans. Cloud Comput. **8**(4), 1069–1078 (2020)
16. Tan, H., Han, Z., Li, X.Y., Lau, F.C.: Online job dispatching and scheduling in edge-clouds. In: IEEE INFOCOM 2017-IEEE Conference on Computer Communications, pp. 1–9. IEEE (2017)
17. Chen, M.H., Dong, M., Liang, B.: Resource sharing of a computing access point for multi-user mobile cloud offloading with delay constraints. IEEE Trans. Mob. Comput. **17**(12), 2868–2881 (2018)
18. Zhang, H., Xiao, Y., Bu, S., Niyato, D., Yu, F.R., Han, Z.: Computing resource allocation in three-tier iot fog networks: a joint optimization approach combining stackelberg game and matching. IEEE Internet Things J. **4**(5), 1204–1215 (2017)
19. El Haber, E., Nguyen, T.M., Assi, C.: Joint optimization of computational cost and devices energy for task offloading in multi-tier edge-clouds. IEEE Trans. Commun. **67**(5), 3407–3421 (2019)
20. Chen, Z., He, L.: Modelling task offloading mobile edge computing. In: 2022 The 8th International Conference on Computing and Data Engineering, pp. 15–21 (2022)
21. Lyu, X., Tian, H., Sengul, C., Zhang, P.: Multiuser joint task offloading and resource optimization in proximate clouds. IEEE Trans. Veh. Technol. **66**(4), 3435–3447 (2016)
22. Chen, M., Hao, Y.: Task offloading for mobile edge computing in software defined ultra-dense network. IEEE J. Sel. Areas Commun. **36**(3), 587–597 (2018)
23. Tao, M., Ota, K., Dong, M., Yuan, H.: Stackelberg game-based pricing and offloading in mobile edge computing. IEEE Wireless Commun. Lett. **11**(5), 883–887 (2022)
24. Van Adrichem, N.L., Doerr, C., Kuipers, F.A.: Opennetmon: network monitoring in openflow software-defined networks. In: 2014 IEEE Network Operations and Management Symposium (NOMS), pp. 1–8. IEEE (2014)
25. Sun, X., Ansari, N.: Latency aware workload offloading in the cloudlet network. IEEE Commun. Lett. **21**(7), 1481–1484 (2017)
26. Jia, M., Cao, J., Liang, W.: Optimal cloudlet placement and user to cloudlet allocation in wireless metropolitan area networks. IEEE Trans. Cloud Comput. **5**(4), 725–737 (2015)
27. Gao, B., He, L., Jarvis, S.A.: Offload decision models and the price of anarchy in mobile cloud application ecosystems. IEEE Access **3**, 3125–3137 (2015)

Dynamic Multi-bit Parallel Computing Method Based on Reconfigurable Structure

Lin Jiang[1], Shuai Liu[1], Jiayang Zhu[1(✉)], Rui Shan[2], and Yuancheng Li[1]

[1] Xi'an University of Science and Technology, Xi'an 710600, Shaanxi, China
zhujiayoung@163.com

[2] Xi'an University of Posts and Telecommunications, Xi'an 710100, Shaanxi, China

Abstract. Reconfigurable architecture has great potential in computation-intensive and memory-intensive applications due to its flexible information configuration. Aiming at the problem of low computing efficiency caused by the inconsistency between different granularity data and the underlying hardware structure in applications such as communication baseband signal processing, a parallel computing method supporting multi-bit data is proposed, and a dynamic granularity configuration structure used this method is designed based on reconfigurable array processors. The structure divides the calculation granularity into 8 bits, 16 bits, and 32 bits, and realizes four functions: data-combination, data-splitting, parallel-addition, and parallel-multiplication. These features increase the parallelism and flexibility of array structures. The experimental results show that the speedup ratio can reach 1.5 within a certain error range, the running time is reduced by about 20%, and the code complexity is also significantly reduced. In addition, the maximum operating frequency of the dynamic configuration circuit is 133.5 MHz by FPGA comprehensive implementation, which can realize the dynamic configuration of different granularity data in the calculation and achieve parallel computing of multi-bit data.

Keywords: Reconfigurable Architecture · Parallel Computing · Computing Granularity · Array Processor

1 Introduction

As an architecture that combines the flexibility of software and the efficiency of hardware, reconfigurable computing is more balanced between key indicators such as performance, power consumption, and flexibility, filling the gap between general computing architecture and dedicated computing architecture, and now it has gained more and more extensive research and application [1]. Due to the difference in the granularity of different communication baseband algorithms, the dynamic configuration of granularity can greatly improve the performance of reconfigurable array processors. At present, the most used technology is the

Supported by National Key R&D Program of China

subword parallelism [2], which is a data unit with lower precision than a word. By compressing multiple subwords into a word, the whole word is processed in parallel, and finally, the word decompression completes the restoration of bit width. Subword parallelism is equivalent to a Single Instruction Multiple Data (SIMD) operation, which can fully exploit data-level parallelism and improve memory efficiency. The existing hardware platform solves the problems of flexibility, parallelism, and efficiency that are common in most applications to a certain extent, but it does not consider the characteristics of data to be processed, and the lack of coordination between data and the underlying hardware architecture reduces the overall performance of the processor. Some dynamic configuration methods of computing granularity are proposed to improve computing performance.

In ref. [3], a granularity configuration method for bit fusion is proposed, which is a bit-level dynamic combination architecture for Deep Neural Network (DNN). This design provides dynamic bit-level fusion and decomposition functions and can realize the combination and flexible mapping of bit-level micro-architectures. Although no dedicated hardware unit is designed to support the computing granularity configuration, it is realized by combining PE method, which complicates the design of array processor control unit and data path. Ref. [4] proposes a customization matrix multiplication architecture based on the Intel HARPv2 platform, which completes dynamic precision switching with highly customization hardware templates and supports universal matrix multiplication calculation of single-precision floating point numbers and 16-bit, 8-bit, 4-bit, and 2-bit fixed-point numbers. However, the architecture is only designed for matrix multiplication, which cannot meet the computing requirements of various algorithms with insufficient expansivity. A Reconfigurable Constant Coefficient Multipliers (RCCMs) for speeding up the application of deep learning in ref. [5] is proposed. RCCM only uses adder, subtracter, shifter, and multiplexer to obtain the result of multiplication by selecting a finite number of coefficients multiplied by the input value. RCCM reduces the use of resources compared with general multipliers and has the optimal structure under specific coefficients, but it can only design the multiplication circuit by the predefined coefficient set, and its calculation accuracy is low. DNN can also be accelerated by PMU (Parallel Multiplication Unit) [6] and FILM-QNN (FPGA for Intra-Layer, Mixed-Precision Quantized DNNs) [7] architectures, which parallelized low-precision operations to speed up the inference performance of neural networks.

In the neural network, different layers have different requirements for data bit width. The fixed data bit width will cause bit width redundancy [8], which is not conducive to model compression, ref. [3–5] have made in-depth research on this topic. Same as the neural network algorithm, the communication baseband algorithm also focuses on multiplication and addition [9]. Different algorithms have different computing granularity and lack coordination between data characteristics and the underlying hardware architecture, which reduces the processor's overall performance. This paper analyzes the computing granularity requirements of different algorithms in baseband signal processing, designs instructions to control the operation of data under different granularity and proposes a computing granularity dynamic configuration structure based on reconfigurable array processor, thereby achieving parallel computing of multi-bit data, which has high flexibility and versatility.

2   Dynamic Configurable Scheme Design

2.1   Requirements Analysis of Computational Granularity

At present, the operation bit width of baseband processing platform is uneven and there is no unified standard. For example, in matrix multiplication operation, the operation bit width includes 8 bits and 32 bits [10]. In the triangular approximation semi-definite relaxation algorithm, the fixed-point word length of the operator is 14 bits [11]. In the large-scale MIMO algorithm, the bit width of 16 bits is enough to meet the accuracy requirements of multiplication and addition operation, while the bit width of linear signal detection algorithm is basically concentrated in 32 bits [12]. In this section, the digital signal processing-oriented test set in ref. [13] is selected as the target program for computing granularity analysis. Table 1 shows the application scope and data bit width of the target program.

Table 1. Introduction to object program

Object Program	Applied Range	Data Width
FIR-8	Voice signal processing, channel balancing	8–16 bit
IIR-8	Audio processing	8–16 bit
Vector Dot Product	Convolution, matrix multiplication	8–32 bit
Vector Add	Matrix calculation	16 bit
Vector Maximum	Error control coding	16 bit
FFT-64	Radar, spectrum analysis	16–32 bit

To sum up, although communication baseband signal processing algorithms have different requirements for calculation accuracy, most of the calculation granularity is concentrated in 8-bit, 16-bit, and 32-bit. To make full use of the computing power of hardware resources, it is necessary to design a dynamic configuration structure of computing granularity that supports different bit widths and operation formats.

2.2   Multi-bit Data Parallel Processing Method Design

Based on the four-stage pipeline structure of reconfigurable array processor, the dynamic configuration of computing granularity is implemented by top-down method. Firstly, the granularity of computation is divided into 8, 16 and 32 bits. Secondly, corresponding parallel processing methods are proposed for different computing granularity. Then, granularity dynamic configuration instructions

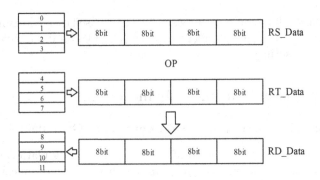

Fig. 1. Method of data parallel processing

are designed, including data-combination, data-splitting, parallel-addition, and parallel-multiplication. Finally, according to the instruction function, we achieve the corresponding hardware circuit design and complete the function verification and performance analysis. Figure 1 takes the parallel processing method of 8-bit data as an example to show the execution flow of the computation granularity dynamic configuration structure. As can be seen from the data parallel processing method in Fig. 1, the acquisition of operands in "RS" and "RT" registers requires data to be read from DM and splices according to the operation granularity, data-combination instructions are required. Parallel computing instructions are required to perform operations on the data. Before the operands are written to memory, the data needs to be split and then stored by location, so the data-splitting instruction is required. For parallel data processing, the data-combination and data-splitting are named "COMB" and "SPL", parallel-addition and parallel-multiplication are named "PADD" and "PMUL". The instructions are shown in Table 2.

2.3   Hardware Structure of Reconfigurable Array

The reconfigurable array processor based on H-type transmission network, which is shown in Fig. 2 [14], was jointly developed by Xi'an University of Science and Technology and Xi'an University of Posts and Telecommunications. It consists of host interface, global controller, reconfigurable processing unit, input memory and output memory.

The reconfigurable processing unit is the core of the array processor, which is composed of 1024 PEs, with each group of 4 × 4. Every group forms a Process Element Group (PEG), Fig. 2 shows only four PEGs, and the rest can be extended on this architecture. Each PE contains a data storage unit and an instruction storage unit. The former has 512 rows, and the bit width of each row is 32 bits; The latter also has 512 lines with the 30 bits instruction. Adjacent interconnection can be used to transfer data between adjacent PEs, and routing can be used to transfer data between adjacent PEGs [15].

Table 2. Granularity dynamic configuration instructions

Type	Mnemonic	Configuration method
Data-Combination	COMB4	RD = {memory[RS+3], memory [RS+2], memory [RS+1], memory [RS]}
	COMB2	RD= memory [RS+1], memory [RS]
	COMB1	RD= memory [RS]
Data-Splitting	SPL4	memory [RD]= {24'd0, RS [7:0]}
		memory [RD+1] = {24'd0, RS [15:8]}
		memory [RD+2] = {24'd0, RS [23:16]}
		memory [RD+3] = {24'd0, RS [31:24]}
	SPL2	memory [RD]= {16'd0, RS [15:0]}
		memory [RD+1] = {16'd0, RS [31:16]}
	SPL1	memory [RD] = RS
Parallel-Addition	PADD4	RD= {RS [31:24] +RT [31:24], RS [23:16] +RT [23:16], RS [15:8] +RT [15:8], RS [7:0] +RT [7:0]}
	PADD2	RD={RS [31:16] +RT [31:16], RS[15:0] +RT[15:0]}
	PADD1	RD = RS + RT
Parallel-Multiplication	PMUL4	RD= {RS [31:24] *RT [31:24], RS [23:16] *RT [23:16], RS [15:8] *RT [15:8], RS [7:0] *RT [7:0]}
	PMUL2	RD= {RS [31:16] *RT [31:16], RS [15:0] *RT [15:0]}
	PMUL1	RD = RS * RT

3 Design of Dynamic Configuration Circuit

The design of corresponding hardware circuits should be based on the four-stage pipeline (fetching, decoding, execution and write-back) for the computation granularity dynamic configuration structure. The overall circuit shown in Fig. 3 is obtained with the requirements of different computing granularity. Firstly, the fetching module will get the instruction signal "cmu_instr" from the instruction memory and send the output signal "im_instr" to the decoding module. Next, the decoding module will decode the high 6 bits of the instruction code and send decoding result to the execution module through the control signal. Data-combination instructions "COMB2/COMB4" need 2/4 data respectively, so the signals "is_comb2/is_comb4" need to be sent back to the fetching module. Lastly, the signals of data-combination and data-splitting are sent to the write-back module, data is written into registers or data storage.

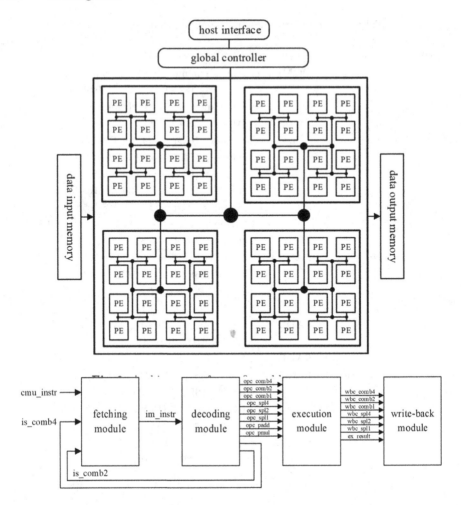

Fig. 3. The overall circuit of dynamic configuration structure

3.1 Design and Implementation of Fetching Module

Fetching module has the function of reading instructions from instruction storage and transferring them to the decoding module. The instructions continuously output from instruction memory with the update of Program Counter (PC). The update of PC value depends on the Finite State Machine (FSM), which is mainly used to calculate the storage address of the next instruction. When the circuit ends in the "RESET" state, the state machine is in the "INIT" state and then jumps unconditionally to the "NORM" state. In "NORM" state, FSM jumps to the next state according to different control signals. In the reconfigurable array processor, the reading delay is one clock cycle. Since the number of "COMB2/COMB4" reading from the data store is 2/4 respectively. "is_COMB2/is_COMB4" are designed because of reading delay, the former has higher priority than the latter. According to the jump order of the state machine, the circuit structure of the fetching module is shown in Fig. 4, which completes three functions: loading configuration instructions from the instruction memory, updating PC values through FSM, and sending configuration instructions to the decoding module.

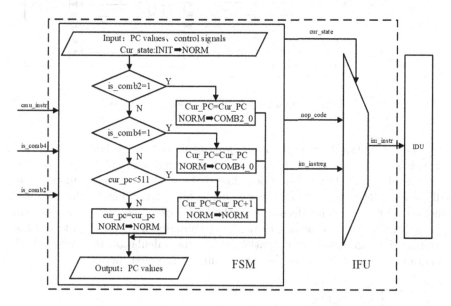

Fig. 4. Circuit diagram of instruction fetching module

3.2 Design and Implementation of Decoding Module

The decoding module decodes the 30-bit instructions and outputs the parsed information to the fetching or executing module. When "COMB2/COMB4" instruction needs to be suspended, the decoding module will transmit the control signal to the fetching module, and the fetching module updates the PC value according to the FSM and transmits the configuration instruction after the NOP instruction is inserted to the decoding level. The circuit structure of the decoding module is shown in Fig. 5, which is divided into decoding unit and register unit and has two decoding methods, one is directly reading the high 6 bits of the input signal "im_instr" and determines the instruction type according to the defined instruction code; the other is combining the operations of the same kind and analyzes them in groups. All signals through the decoding unit will use registers to delay one cycle, which will be used as the control signal of the execution module to complete the preparation work of the instruction execution.

3.3 Design and Implementation of Execution Module

The execution module is responsible for the execution of instructions, completes operations by the operation types and operands, and sends the calculation results

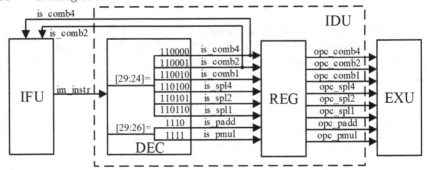

Fig. 5. Circuit diagram of decoding module

back. The execution level is the most important part of the four-level pipeline. Figure 6 is the circuit structure diagram of the execution module. For the operation signal obtained by the first decoding method, the corresponding operation will be completed directly at the execution level according to the command function. For the second decoding method, the executor determines the specific instruction type according to the last two digits of opcode input, and then performs the corresponding operation. After the calculation is completed, the execution module will output the "ex_result" to the next module.

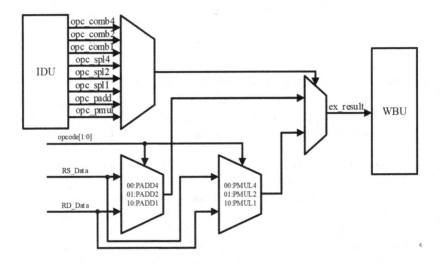

Fig. 6. Circuit diagram of execution module

The functions of data-combination and data-splitting need read and write from the data storage, and the reading and writing are closely related to the storage control unit, so data-combination and data-splitting complete the signal assignment related to the storage control unit only at the execution level.

3.4   Design and Implementation of Write-Back Module

The write-back module is responsible for sending the calculation results to the destination register or PE's data storage DM. This module includes the storage control unit CMU and the register read/write unit REG_RW. The circuit structure of the write-back module is shown in Fig. 7. The storage control unit interacts with the data storage to complete data reading and writing. The register read/write unit receives information from the decoding and the execution level, writes the execution result of the instruction to the register, or reads data from the register for calculation.

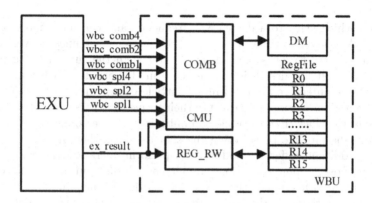

Fig. 7. Circuit diagram of write-back module

The data-splitting function uses control signals "wbc_spl1, wbc_spl2 and wbc_spl4" in the storage control unit as the mark of data writing to the memory. Through different control signals, the number of writing signals is assigned first, and data is divided by the operand granularity, the number of writing addresses is calculated, and data is stored in the write number address in turn.

4   FPGA Experimental Results and Analysis

Verilog is used to design and implement the dynamic configuration structure of computing granularity. After the circuit of each module passes the simulation verification, the ZC706 development board with Zynq-7000 XC7Z045-2ffg900c is used in Xilinx Vivado 2018.3 development tool. The synthesis and implementation results of fetching module, decoding module, execution module, write-back module and the overall circuit are shown in Table 3.

As can be seen from Table 3, the working frequency of the whole circuit is 133.5 MHz, occupying 2299 LUT resources and 1021 FF resources, with a minimum delay of 7.490 ns. The execution module has the lowest operating frequency among the four modules and occupies the highest sum of LUT and Flip Flop (FF) resources. Because the execution module undertakes the computational tasks of parallel addition and multiplication, the corresponding calculation delay

Table 3. Comprehensive realization result

	Frequency/MHz	LUT	FF	Minimum Delay/ns
fetching module	283.4	160	200	3.259
decoding module	650.6	284	100	1.537
execution module	172.2	477	58	5.806
write-back module	468.6	164	174	2.134
whole circuit	133.5	2299	1021	7.490

is greater, which leads to a decrease in frequency and consumes more on-chip resources. At the same time, factors such as interconnect lines, driver delays, often result in an overall system delay greater than the sum of the delays of each module, but the design is not the case in this article. On the one hand, the pipeline architecture designed in this article can divide tasks into multiple modules, and different modules execute their respective tasks sequentially under the drive of the clock. Each module can optimize the interconnection of various modules in the pipeline by selecting the shortest path, thereby allocating task delays. In addition, different modules can perform operations in parallel, and the design of overlapping operations can increase throughput and reduce overall latency. On the other hand, the Vivado2018.3 development tool will perform various optimizations on the circuit, such as logic optimization, timing optimization, and layout optimization. These optimizations will reduce the delay of certain modules, making the overall delay less than the sum of the delays of each module.

The design in this paper is compared with the dynamic configuration structure of calculation granularity based on FPGA at home and abroad, as shown in Table 4. The design of ref. [16] uses 1024 parallel 1-bit processors to realize from 1 to 16 bits multiplication and addition calculation. Although this design reduces the delay caused by combination logic through pipeline strategy, the working frequency can only be maintained at 100 MHz due to too many interconnected multiplication and addition units. Ref. [17] proposed a reconfigurable micro-processing unit, which realized parallelmultiplication and parallel-addition operations in the 2-8 bit interval. Although the working frequency of the reconfigurable micro-processing unit was slightly higher than that designed in this paper, the maximum bit width of the design in this paper was 32 bits, which

Table 4. Comparison of dynamic configuration structures

	Device Type	Frequency/MHz	Bit Width/bit	LUT/K	FF/K	BRAM	DSP	Reconfigurable
Ref. [16]	XCZU9EG	100	1–16	57		900	0	N
Ref. [17]	Ultra96-V2	150	2–8	44.8	60.2	204	336	Y
Ref. [18]	Artix-7	133.3	8/10/12/13	3.48	3.55	16	30	N
Our Work	XC7Z045	133.5	8/16/32	69.1	27.4	136	256	Y

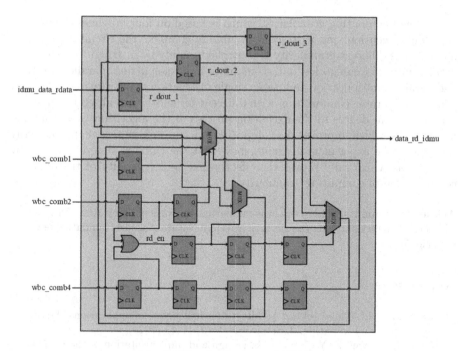

Fig. 8. Circuit diagram of COMB

has higher computational accuracy than the design of ref. [17]. The working frequency of the hardware circuit in ref. [18] is close to the design of this paper, but the design of ref. [18] only carries out the dynamic configuration of the calculation granularity for FFT algorithm, resulting in its lack of generality. The design in this paper is more granular and flexible, so it can meet the computing needs of different algorithms.

This multi-bit parallel computing method implements flexible configurations of three different data bit widths based on array processors to demonstrate good reconfigurability. Bit width configuration provides a flexible way to adapt to different computing needs and performance requirements by adjusting the bit width of data and signals. Reconfigurability provides the ability to dynamically configure and reconfigure, making bit width configuration an important means of configuration. This article combines bit width configuration and reconfigurability to meet specific design requirements and performance goals under different application requirements and algorithm characteristics, with high flexibility and resource utilization.

5 Summarize

Aiming at the problem that different communication baseband algorithms have different requirements for computing granularity, this paper researches on parallel computing methods for multi-bit data and designs a dynamic configuration structure of computing granularity, which is based on four pipelines of fetching, decoding, execution, and write-back module to complete the circuit design. The operation of different granularity is controlled mainly by instructions, which are divided into four categories: data-combination, data-splitting, parallel-addition, and parallel-multiplication and uses a single instruction to operate multiple data. The reconfigurable computation with different bit widths is realized by changing the parallelism degree in the execution process of the algorithm by assembling instructions, which improves the computing efficiency of the reconfigurable array processor. The experimental results show that the running time is reduced by about 20%, which fully proves the effectiveness and reliability of the dynamic configuration of computing granularity.

Acknowledgements. This work was supported by National Key R&D Program of China (2022ZD0119001); Key projects of National Natural Science Foundation of China (61834005).

References

1. Lu, Y., Liu, L., Zhu, J., et al.: Architecture, challenges and applications of dynamic reconfigurable computing. J. Semicond. **41**(2), 4–13 (2020)
2. Chiu, J.-C., Yan, Z.-Y., Liu, Y.-C.: Design and implementation of the CNN accelator based on multi-streaming SIMD mechanisms. In: Hsieh, S.-Y., Hung, L.-J., Klasing, R., Lee, C.-W., Peng, S.-L. (eds.) New Trends in Computer Technologies and Applications: 25th International Computer Symposium, ICS 2022, Taoyuan, Taiwan, December 15–17, 2022, Proceedings, pp. 460–473. Springer Nature Singapore, Singapore (2022). https://doi.org/10.1007/978-981-19-9582-8_40
3. Sharma, H., Park, J., Suda, N.: Bit Fusion: Bit-Level Dynamically Composable Architecture for Accelerating Deep Neural Networks. In: ACM/IEEE 45th annual international symposium on computer architecture (ISCA). IEEE **2018**, 764–775 (2018)
4. Moss, D.J., Krishnan, S., Nurvitadhi, E., et al.: A customizable matrix multiplication framework for the intel harpv2 xeon+fpga platform: a deep learning case study. In: 2018 ACM/SIGDA International Symposium on Field-Programmable Gate Arrays. ACM, pp. 107–116 (2018)
5. Faraone, J. Kumm, M., Hardieck, M., et al.: AddNet: deep neural networks using fpga-optimized multipliers. IEEE Trans. Very Large Scale Integr. (VLSI) Syst. **28**(1), 115–128 (2020)
6. Tang, S.N.: Area-efficient parallel multiplication units for CNN accelerators with output channel parallelization. IEEE Trans. Very Large Scale Integr. (VLSI) Systems. **31**(3), 406–410 (2023)
7. Sun, M., Li, Z., Lu, A., et al.: FILM-QNN: efficient FPGA acceleration of deep neural networks with intra-layer, mixed-precision quantization. In: Proceedings of the 28th ACM/SIGDA International Symposium on Field-Programmable Gate Arrays (FPGA), pp. 134–145 (2022)
8. Wang, N., Nia, J., Li, J., et al.: A compression strategy to accelerate LSTM meta-learning on FPGA. ICT Express **8**(3), 322–327 (2022)

9. Nataraj Urs, H.D., Venkata Siva Reddy, R., Gudodagi, R., et al.: A novel algorithm for reconfigurable architecture for software-defined radio receiver on baseband processor for demodulation. Sustainable Computing. Springer, Cham, pp. 187–206 (2023). https://doi.org/10.1007/978-3-031-13577-4_11

10. Umuroglu, Y., Conficconi, D., Rasnayake, L., et al.: Optimizing bit-serial matrix multiplication for reconfigurable computing. ACM Trans. Reconfigurable Technol. Syst. (TRETS) 12(3), 1–24 (2019)

11. Liu, K., Tian, Z., Li, Z., et al.: RfLoc: a reflector-assisted indoor localization system using a single-antenna AP. IEEE Trans. Instrum. Meas. 70(3), 1–16 (2021)

12. Wang, A., Xu, W., Sun, H., et al.: Arrhythmia classifier using binarized convolutional neural network for resource-constrained devices. In: 2022 4th International Conference on Communications, Information System and Computer Engineering (CISCE), Shenzhen, China, 2022, pp. 213–220 (2022)

13. Stepchenkov, Y.A., Khilko, D.V., Shikunov, Y.I.: Filter kernels preliminary benchmarking, DSP, for recurrent data-flow architecture. In: IEEE Conference of Russian Young Researchers in Electrical and Electronic Engineering (ElConRus). IEEE 2021, pp. 2040–2044 (2021)

14. Deng, J., Jiang, L., Zhu, Y., et al.: HRM: H-tree based reconfiguration mechanism in reconfigurable homogeneous PE array. J. Semiconductors. 41(2), 1–9 (2020)

15. Shan, R., Jiang, L., Wu, H., He, F., Liu, X.: Dynamical self-reconfigurable mechanism for data-driven cell array. J. Shanghai Jiaotong Univ. (Science) 26(4), 511–521 (2021). https://doi.org/10.1007/s12204-021-2319-z

16. Maki, A., Miyashita, D., Nakata, K., et al.: FPGA-based CNN processor with filterwise-optimized bit precision. In: 2018 IEEE Asian Solid-State Circuits Conference (A-SSCC). IEEE, pp. 47–50 (2018)

17. Chen, Y., Du, H., Chang, L.: A reconfigurable micro-processing element for mixed precision CNNs. In: 2022 14th International Conference on Measuring Technology and Mechatronics Automation (ICMTMA). IEEE, pp. 1–5 (2022)

18. Liu, W., Liao, Q., Qiao, F., et al.: Approximate designs for fast Fourier transform (FFT) with application to speech recognition. IEEE Trans. Circuits Syst. I Regul. Pap. 66(12), 4727–4739 (2019)

A Heuristic Method for Data Allocation and Task Scheduling on Heterogeneous Multiprocessor Systems Under Memory Constraints

Junwen Ding[1], Liangcai Song[1], Siyuan Li[1], Chen Wu[2],
Ronghua He[2], Zhouxing Su[1], and Zhipeng Lü[1]($\boxtimes$)

[1] Huazhong University of Science and Technology, Wuhan, China
{junwending,zhipeng.lv}@hust.edu.cn
[2] 2012 Lab, Huawei Technologies Co., Ltd., Shenzhen, China

Abstract. Computing workflows in heterogeneous multiprocessor systems are often depicted as directed acyclic graphs (DAGs) comprising tasks and data blocks. These elements represent computational modules and their interdependencies, where the output data generated by one task serves as input for other tasks. However, in certain workflows, such as task schedules in digital signal processors, it is essential to account for the memory capacity limitations of data blocks when scheduling the tasks. The main objective of this paper is to address the challenge of data allocation and task scheduling under memory constraints, particularly on shared memory platforms. We propose an integer linear programming model to formulate the problem, which can be viewed as an extension of the flexible job shop scheduling problem. The goal is to minimize the length of the critical path. We propose a tabu search algorithm (TS) to tackle the data allocation and task scheduling problem under memory constraints. The TS algorithm incorporates several distinguishing features, such as a greedy initial solution construction method and a mixed neighborhood evaluation strategy that combines exact and approximate evaluation methods. Experimental results on randomly generated instances demonstrate that the TS algorithm can obtain high-quality solutions within a reasonable computational time. On average, the TS algorithm improves the makespan by 5–25% compared to the widely used classical load balancing algorithms in the literature. Furthermore, we analyze some key features of the TS algorithm to identify the contributing factors to its success.

Keywords: Task scheduling · Data allocation · Heterogeneous multiprocessor · Tabu search

Supported by by the Special Project for Knowledge Innovation of Hubei Province under Grant No. 2022013301015175 and the National Natural Science Foundation of China under Grants No. 62202192 and 72101094.

1 Introduction

A digital signal processor (DSP) is a specialized microprocessor chip designed specifically for the computational needs of digital signal processing tasks [7]. DSPs are fabricated on metal oxide semiconductor (MOS) integrated circuit chips. They find extensive application in various fields, including audio signal processing, digital image processing, speech recognition systems, high performance computing centers, and everyday consumer electronic devices like mobile phones, notebook computers, smart watches, and intelligent wearable device [4].

The DSP chip consists of different types of cores and memories, with the cores being responsible for performing computations and memory serving as the storage unit. Based on the description provided in [8], the DSP chip incorporates various core types, including general-purpose cores and synergistic processor cores. In terms of memory, there are high-speed memory modules as well as low-speed memory modules such as DDR (Double Data Rate) memory. The cores are organized into clusters and groups, with each group associated with a local high-speed memory. On the other hand, other high-speed memory and low-speed memory modules are shared globally, accessible to all the cores on the DSP chip.

In the domain of parallel computing on multiprocessor systems, tasks are commonly represented using Directed Acyclic Task Graphs (DAGs), where nodes represent individual tasks and edges represent their dependencies [11]. When dealing with a series of tasks that need to be executed on DSP processors, along with the associated data blocks generated by these tasks (i.e., the task dependencies), the task scheduling problem arises. This problem involves assigning each task to specific cores, determining the storage location for the data blocks, and establishing the execution order of the tasks on each core. The primary objective is to minimize the overall completion time of all tasks while improving the utilization of the cores and memories. By optimizing the task assignment and execution order, it is possible to achieve better performance and efficiency in the execution of tasks on the DSP processor.

The job shop scheduling problem is a fundamental problem in the domains of intelligent manufacturing and high-performance computing. Its primary objective is to schedule priority resources in order to sequentially execute multiple tasks, with the aim of minimizing the maximum completion time of all tasks. For instance, let's consider a chip foundry where the production of chips involves a series of sequential processes, such as photolithography and etching, which are carried out on different machines [36]. We also encounter similar scenarios in parallel computing environments, where there exist dependencies between computing tasks, and the output of a predecessor task serves as the input for a successor task [17].

Scheduling problems encountered in real execution processes tend to be more complex due to the presence of multiple constraints from various dimensions. For example, in a cloud computing data center where multiple tasks are executed on shared multi-core processors, the allocation of cores to tasks and their concurrent scheduling becomes critical, taking into account energy and performance constraints [18]. In parallel computing scenarios, it is essential to consider not only the utilization of computing resources but also the memory resources occupied by concurrent tasks, ensuring that they do not exceed the maximum capacity limit. Moreover, heterogeneous chips integrate diverse computing units that are capable of handling compatible tasks, which further increases the complexity of task scheduling due to the intricate constraints among different memory types [19].

With the increasing prevalence of heterogeneous processors, it is common for operations of the same type to be processed by different cores, each with its own processing time and data capacity. Additionally, various components of a distributed shared-memory system exhibit significant heterogeneity in data access time [22,34]. As a result, several critical issues arise, including task-to-processor assignment, datum-to-memory allocation, and operation sequencing for both processing tasks and data retrieval, aiming to satisfy specific constraints and minimize the overall task completion time. This problem is formally referred to as the Heterogeneous Data Allocation and Task Scheduling problem (HDATS).

2   Literature Review

Processors and memories have always been valuable in large-scale computations, as highlighted in [30]. The problem of scheduling large-scale scientific workflows using distributed resources has been identified by [26]. Their work was further expanded upon in [24], which proposed two genetic algorithms to handle computing tasks. Chen et al. [8] introduced an online heterogeneous dual-core scheduling algorithm for dynamic workloads with real-time constraints, and carried out a series of extensive experiments to compare different workloads and scheduling algorithms. This problem also arises in the domain of sparse direct solvers, as investigated by [1], who analyzed the impact of processor mapping on memory consumption in multi-frontal methods. Building on the research conducted on sparse direct solvers in [20], Aupy et al. [3] proposed a heuristic approach that utilizes problem-specific knowledge to minimize peak memory usage.

Sbîrlea et al. [28] introduced a bounded memory scheduling algorithm for parallel workloads represented by dynamic task graphs. The algorithm sets an upper limit on the peak memory usage within the computing environment. Sergent et al. [29] explored the integration of a task-based distributed application with a run-time system that manages memory subscription levels during the processing period. Ergu et al. [12] presented a model for task-oriented resource allocation in a cloud computing environment. The model ranks resource allocation task using techniques such as pairwise comparison matrix and the analytic hierarchy process, taking into account both available resources and user preferences. Praveenchandar and Tamilarasi [25] proposed an enhanced approach for task scheduling and power minimization to achieve efficient dynamic resource allocation. Their approach combines a prediction mechanism with a dynamic resource table updating algorithm.

In the domain of digital signal processing (DSP), there have been studies on modeling the task scheduling problem with specific constraints as the flexible job shop scheduling problem. The FJSP, introduced in [6] as an extension of the job shop scheduling problem, is a well-studied combinatorial optimization problem. For the FJSP with the objective of minimizing the makespan, exact approaches were proposed by [23,27], who developed mixed-integer linear programming (MILP) models. Another MILP was presented in [5] for the FJSP, considering an extension that incorporates precedence relations between operations of a job, specified by an arbitrary directed acyclic graph. Hansmann et al. [15] suggested a combination of MILP and branch and bound algorithm to address the FJSP with restricted machine accessibility. In the context of task scheduling in virtual controllers and multiple clusters of remote radio heads, Xia et al. [35] transformed the problem into a matroid constrained submodular maximization problem. They proposed heuristic algorithms to obtain approximate optimal solutions. Fu et al. [13] introduced a unified graph to model both map task scheduling and reduce task scheduling. They transformed the problem into a well-known graph problem called minimum weighted bipartite matching.

In the context of cloud computing based on containers for smart manufacturing, Yin et al. [36] developed a task scheduling model that takes into account the role of containers. They designed a task scheduling algorithm and a reallocation mechanism specifically considering container characteristics, with the objective of minimizing task delays. Yuan et al. [37] presented a spatial task scheduling and resource optimization method for distributed green cloud data centers. Their approach aims to minimize the total cost for providers by effectively scheduling heterogeneous application tasks while meeting delay-bound constraints and optimizing resource utilization. Hu et al. [16] investigated the task scheduling problem in heterogeneous distributed systems. Their objective was to minimize the schedule length for parallel applications while considering energy constraints. Zhuge et al. [38] proposed a polynomial-time algorithm based on dynamic programming. Additionally, they introduced a global data allocation algorithm and a heuristic maximal similarity scheduling approach. The main focus of their methods is to reduce memory traffic and minimize memory access costs.

For data allocation and task scheduling on heterogeneous multiprocessor systems, the primary objective is to obtain a schedule ensuring that memory usage never exceeds its maximum capacity during execution. To address this challenge, we propose a tabu search (TS) algorithm that combines distinguishing features, including a hybrid initial solution construction method using both greedy and random approaches, and a mixed neighborhood evaluation strategy combining exact and approximate evaluations. Experimental results on randomly generated instances demonstrate that the proposed algorithm can achieve high quality solutions within feasible computational time. Additionally, we analyze key features of the TS algorithm to the success of its performance.

The rest of the paper is organized as follows: Sect. 3 provides a description of the problem and its mathematical formulation. Section 4 details the proposed tabu search algorithm and its components. Section 5 presents the computational results and analyzes the key features, and Sect. 6 concludes the paper and pro-. vides suggestions for future research directions.

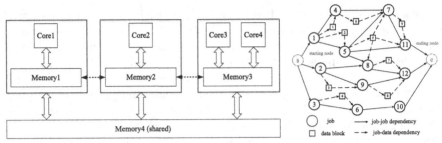

(a) The architecture of the heterogeneous distributed shared-memory multiprocessor system

(b) An example with 12 tasks and 9 data blocks

Fig. 1. A typical DSP architecture and an illustrative example

3   Problem Definition and Formulation

3.1   Problem Description

The architecture of the DSP is a heterogeneous distributed shared-memory multiprocessor system as illustrated in Fig. 1a. The architectural scheme encompasses a set P of n connected heterogeneous processors, i.e., $P = \{P_1, P_2, \ldots, P_n\}$. Each processor P_i is associated with its own local memory M_i, while the collective set of local memories forms a distributed physical memory that is shared globally. For instance, M_1 represents the local virtual memory of processor P_1, whereas M_2 and M_3 represent remote physical memories. For processor P_2, M_2 serves as its local memory, whereas M_1 and M_3 are remote memories. As all the distributed memories are merged into a global shared space, every processor can access the global shared memory. It is noteworthy that the access time by different processors on the same memory may be different due to the non-uniform structure of the memory paradigm.

Given a direct acyclic task graph DAG $G = (V, E)$, V is the node set and E is the edge set, nodes $s, e \in V$ represent the starting and ending nodes, respectively. By formulating a memory access operation as a node, the traditional DAG can be extended to a memory-access data flow graph (MDFG). Figure 1b gives an illustrative example of the HDATS problem, where cycle blocks represent tasks and square blocks represent data blocks depending on them or being depended.

A MDFG is a node-weighted directed graph extended from a DAG which is represented as $G' = (V_1, V_2, \ldots, E, D, var, P, M, AT, ET)$, where the notations are explained as follows:

- $V_1 = \{v_1, v_2, \ldots, v_{N_1}\}$ represents a set of N_1 task nodes.
- $V_2 = \{u_1, u_2, \ldots, u_{N_2}\}$ represents a set of N_2 memory access operation nodes.
- E is a set of edges, where $E \subset V \times V$, $V = V_1 \cup V_2$. An edge $(i, j) \in E$ denotes the dependency between node i and node j, expressing that task or operation i has to be executed before task or operation j.
- D is a set of initial input data.

- $var : V_1 \times V_2 \times D \rightarrow \{0,1\}$ is a binary mapping relationship, in which $var(v, u, w)$ represents whether memory access operation $u \in V_2$ is delivering data $w \in D$ for task $v \in V_1$.
- $P = \{P_1, P_2, \ldots, P_n\}$ represents a set of n heterogeneous processors.
- $M = \{M_1, M_2, \ldots, M_m\}$ represents a set of m local memories.
- AT is the memory access time functions.
- $PT(v_i, P_j) = et_j(i)$ is the processing time of task v_i when it is processed on processor P_j.

Therefore, the heterogeneous data allocation and task scheduling problem is formally defined as MDFG, aiming to find a solution represented by a triple (Mem, AS, SC). Here, Mem is a data allocation $Mem : D \rightarrow M$, where $Mem(h) \in M$ indicates the memory assigned to $h \in D$; AS is the task assignment function $AS : V_1 \rightarrow P$, where $A(v)$ denotes the processor assigned to task $v \in V_1$; SC represents the schedule, $SC : V_1 \cup V_2 \rightarrow \mathbb{R}$, specifying the starting time for each task in V_1 and each memory access operation in V_2. The objective is to minimize the overall completion time $T(G')$, ensuring that the total amount of data blocks assigned to memory M_i, denoted as T_i, does not exceed its capacity $S(M_i)$, i.e., $T_i <= S(M_i)$. The HDATS problem has been proven to be NP-hard [31].

3.2   The Integer Linear Programming Formulation of HDATS

This section presents the integer linear programming (ILP) formulation for the HDATS problem, which consists of task assignment with processor constraints, data allocation with memory size and concurrency constraints, precedence constraints, and a time constraint. Given an MDFG, the ILP model for the HDATS problem consists of two main parts: processor assignment and memory allocation. The processor assignment aims to assign tasks in the given MDFG to appropriate processors, while the memory allocation aims to allocate the required data for task processing. The objective is to minimize the maximum completion time of all tasks, i.e.,

$$\min \max\{RT(i,j) + PT(v_i, P_j)\}, \forall i \in [1, N_1], j \in [1, n] \tag{1}$$

Task assignment and processor constraints:

$$\sum_{j=1}^{n} \sum_{k=1}^{|P_j|} x_{ijk} = 1, \quad \forall i \in [1, N_1] \tag{2}$$

$$\sum_{i=1}^{N_1} \sum_{j=1}^{n} x'_{ijm} \leq n, \quad \forall m \in [1, S] \tag{4}$$

$$\sum_{i=1}^{N_1} x'_{ijm} \leq 1, \quad \forall m \in [1, S] \tag{3}$$

In the processor part, two binary variables, x_{ijk} and x'_{ijm}, are introduced to indicate whether task v_i in G' starts processing at step k and is processed in stage m on processor P_j, respectively. Constraint (2) ensures that each task node can perform execution on only one stage and one processor. Constraint (3) guarantees that at most one task is scheduled in any stage on any processor. Constraint (4) ensures that the number of tasks processed in each stage does not exceed the number of processors. Task v_i is assigned to processor $P(i)$.

Data Allocation and Memory Constraints:

$$\sum_{j=1}^{n} d_{hj} = 1, \ \forall h \in [1, N_d] \quad (5)$$

$$\sum_{j=1}^{n} \sum_{k=1}^{S} y_{ljk} = 1, \ \forall l \in [1, N_2] \quad (8)$$

$$\sum_{h=1}^{N_d} d(h) \times d_{hj} \leq S_j, \ \forall j \in [1, n] \quad (6)$$

$$Mem(h) = \sum_{j=1}^{m} j \times d_{hj}, \ \forall h \in [1, N_d] \quad (7)$$

$$\sum_{l=1}^{N_2} y'_{ljm} \leq MA, \quad (9)$$
$$\forall j \in [1, n], \forall m \in [1, S]$$

In the memory part, binary variable d_{ij} represents whether data i is allocated to memory M_j. Binary variables y_{ljk} and y'_{ljk} indicate whether memory access operation node u_l starts processing and is scheduled in stage k on memory M_j, respectively. The capacity of memory M_j is denoted by Sj. The dependency between data allocation and memory access operations is represented by $M(l)$.

Constraint (5) ensures that each data block is allocated to exactly one local memory. Constraint (6) guarantees that the total size of all data allocated in M_j does not exceed S_j. Constraint (7) denotes the local memory $Mem(h)$ to store data h. Constraint (8) ensures that each memory access operation node can start processing in only one stage and one local memory. Constraint (9) guarantees that the number of memory access operation nodes in each stage does not exceed the maximum access number of a local memory.

Precedence Constraints:

$$\sum_{j=1}^{n} \sum_{k=1}^{S} (k + RT(u, j)) \times x_{ujk} \leq \sum_{j=1}^{n} \sum_{k=1}^{S} k \times x_{vjk}, \quad (10)$$
$$\forall e(u, v) \in E, \forall u \in [1, N_1], \forall v \in [1, N_1]$$

$$\sum_{j=1}^{n} \sum_{k=1}^{S} (k + RT(u, j)) \times x_{ujk} \leq \sum_{j=1}^{n} \sum_{k=1}^{S} k \times y_{vjk}, \quad (11)$$
$$\forall e(u, v) \in E, \forall u \in [1, N_1], \forall v \in [1, N_2]$$

$$\sum_{j=1}^{n}\sum_{k=1}^{S}(k + RA\_t(u,j)) \times y_{ujk} \leq \sum_{j=1}^{n}\sum_{k=1}^{S} k \times y_{vjk}, \tag{12}$$

$$\forall e(u,v) \in E, \forall u \in [1, N_2], \forall v \in [1, N_2]$$

$$\sum_{j=1}^{n}\sum_{k=1}^{S}(k + RA\_t(u,j)) \times y_{ujk} \leq \sum_{j=1}^{n}\sum_{k=1}^{S} k \times x_{vjk}, \tag{13}$$

$$\forall e(u,v) \in E, \forall u \in [1, N_2], \forall v \in [1, N_1]$$

In a given MDFG, edge $e(u,v) \in E$ represents the precedence relation from node u to node v. Equations (10)-(13) ensure that each task and memory access operation accurately respects the precedence constraints. Equations (10) and Equations (12) respectively formulates the precedence relation among tasks and memory access operations. Equations (11) and (13) define the precedence constraints between tasks and memory access operations. Generally, these equations state that u must be completed before v starts.

Execution and Memory Access Time Constraints:

$$RT(i,j) = \sum_{k=1}^{S} x_{ijk} \times PT(v_i, P_j), \quad \forall i \in [1, N_1], \forall j \in [1, n] \tag{14}$$

$$\sum_{m=k}^{k+RT(i,j)-1} x'_{ijm} = RT(i,j), \quad \forall i \in [1, N_1], \forall j \in [1, n] \tag{15}$$

$$RT\_t(l,j) = \sum_{i=1}^{N_1}\sum_{h=1}^{N_d}\sum_{k=1}^{S} y_{ljk} var(v_i, u_l, h) AT(P(i), M_jd(h)), \tag{16}$$

$$\forall l \in [1, N_2], \forall j \in [1, n]$$

$$\sum_{m=k}^{k+RT(i,j)-1} y'_{ljm} = RA\_t(l,j), \quad \forall l \in [1, N_2], \forall j \in [1, n]. \tag{17}$$

In addition, the real processing time of task v_i on processor P_j is denoted as $RT(i,j)$, as defined in Eq. (14). If $x_{ijk} = 1$, variable x'_{ijm} must satisfy constraint (15), indicating that the processing of a task should not be interrupted.

The real memory access time of a memory access operation u_l on memory M_j is denoted as RA_t, as expressed in constraint (16), If $y_{ijk} = 1$, variable y'_{ijm} must satisfy Eq. (17).

4 Algorithm Description

The proposed tabu search algorithm comprises a greedy initial solution construction procedure, a tabu search procedure, and a memory update procedure, which are described in detail in the subsequent sections.

Algorithm 1 Greedy construction procedure for initial solution

---

1: **Input**: Problem instance
2: **Output**: A feasible initial solution S_{init}
3: $S_{init} \leftarrow InitS()$, $taskSet, R, Q, Slack \leftarrow Init()$, $t \leftarrow -1$
4: **while** $taskSet$ is not empty **do**
5:     $t \leftarrow selectTaskAccodingToRQSlack()$
6:     $availCores \leftarrow getAvailableCores(t)$;
7:     $endTime \leftarrow InitET(availCores)$
8:     **for** each core c of in $availCores$ **do**
9:         $N \leftarrow getPredecessorsSet(t)$
10:        $startTime \leftarrow \max\{getFinishTime(p)|p \in N\}$
11:        **for** each data d of task t **do**
12:            **if** memory of highType2 is enough at startTime **then**
13:                $tryAssignMemory(d, highType2)$
14:            **else if** memory highType1 is enough at startTime **then**
15:                $tryAssignMemory(d, highType1)$
16:            **else**
17:                $tryAssignMemory(d, lowType)$
18:            **end if**
19:        **end for**
20:        $endTime[c] \leftarrow calcuEndTime(t, c)$
21:    **end for**
22:    $C \leftarrow \arg\min\{getEndTime(c)|c \in availCores\}$
23:    $assignToCore(t, C), updateSolution(S_{init})$
24:    $freshRQSlack(t, C), freshMemory()$
25:    $taskSet \leftarrow tastSet \setminus \{t\}$
26: **end while**
27: **return** S_{init}

4.1 Greedy Construction Procedure for Initial Solution

The construction of an initial solution involves assigning each task to a specific core and each data block to a particular memory block. However, not all assignments are valid due to the difficulty of satisfying the precedence relationship between tasks and the capacity constraints of each memory.

Prior to the construction procedure, a preprocessing step is necessary to generate a profitable job sequence. First, we use topological sorting to obtain a valid sequence considering only job-job constraints and job-data constraints. Subsequently, we apply a dynamic programming procedure to the topological sequence and compute the R, Q, makespan, and $Slack$ values. Here, $taskSet$ represents the list of candidate tasks that have not yet been assigned.

Algorithm 1 presents the pseudo code for the greedy construction procedure used to generate the initial solution. The main idea can be summarized as iteratively selecting the most significant task from the set of currently unallocated tasks, and then assigning it to the optimal core and selecting the suitable memory for the produced data.

Then, on line 5, the task at the front of the candidate list is selected. If multiple eligible tasks are present, we prioritize them based on the following lexicographical order: 1) R value; 2) *Slack* value; 3) the minimum *Slack* value of the successor jobs.

After selecting a task, the start time of task t on each available core in *avail-Cores* must be calculated based on the constraints imposed by its predecessors, and the data generated by the task must be allocated to memory. Specifically, the data blocks produced by the task are sorted based on the minimum slack value of the task dependent on each data block. For each data block generated by the task, we prioritize global high-speed memory first and then local high-speed memory within the same core group to minimize move-out time. Notably, the data can be released from memory once all dependent tasks have been executed.

The greedy construction procedure typically selects the core and memory assignments that enable the earliest completion of the task as the task's assignment result (line 22). Subsequently, the solution S_{init} and the R, Q, *Slack* values of the node are updated (lines 23–24). Once the update is finished, the currently selected task is removed from the candidate set (line 25), and the next round of assignment starts to select the next task and allocate memory in a similar manner. When *taskSet* becomes empty, a complete solution is generated.

4.2 The Proposed Tabu Search Procedure

After an initial solution is obtained by the greedy construction procedure, the solution undergoes further optimization through the tabu search procedure.

Considering both task scheduling and data allocation simultaneously is highly complex. Hence, this paper introduces a two-layered local search procedure. The outer layer focuses on scheduling the task sequence on the machine, while the inner layer addresses memory allocation. If the memory constraints are ignored, the problem can be formulated as the flexible job shop scheduling problem (FJSP).

Neighborhood Structures. The optimization of the outer layer can utilize the classic neighborhood structures proposed in [10]. The machine re-assignment is performed on the k-insertion neighborhood (called N^{α} here [14]) for all the critical tasks, and the sequence change is performed on the neighborhood (called N^{π} here [21]) for all the critical blocks. Specifically, the execution time of tasks within the same critical block may not be continuous due to the presence of transfer time.

Tabu Structure. To prevent revisiting previously searched areas within a short period, we employ a tabu table in the local search process. This ensures that a job will not be moved to the same machine within a specified tabu period.

Algorithm 2 The proposed tabu search procedure for HDATS
___
1: **Input**: Greedy Solution S_{init}, λ, $\bar{T}$, $\bar{K}$
2: **Output**: The best found solution S^*
3: $S_c \leftarrow S_{init}$, $S^* \leftarrow S_{init}$, $N \leftarrow \emptyset$, $Iter \leftarrow 0$, $Duration \leftarrow 0$
4: **while** $Iter < \lambda$ and $Duration < T$ **do**
5:     **for** each critical task t in S_{init} **do**
6:         $N^\pi \leftarrow constructN7(S_{init}, t)$
7:         $N^\alpha \leftarrow constructChangeCore(S_{init}, t)$
8:         $N \leftarrow N \cup N^\pi \cup N^\alpha$
9:         $N \leftarrow checkTabuList(N)$
10:        **if** N is empty **then**
11:            $S' \leftarrow randomPerturbation(S_c)$
12:        **else**
13:            $topkSet \leftarrow selectApproximateTopK(N)$
14:            $S' \leftarrow \arg\min\{getMakespan(S)|S \in topkSet\}$
15:        **end if**
16:        add $Move(S_c, S')$ to tabu list
17:        $S_c \leftarrow S'$, $S' \leftarrow memoryReassign(S')$
18:        **if** $getMakespan(S') < getMakespan(S^*)$ **then**
19:            $S^* \leftarrow S'$, $Iter \leftarrow 0$
20:        **end if**
21:    **end for**
22:    $N \leftarrow \emptyset$; $Iter \leftarrow Iter + 1$
23:    $Duration \leftarrow getDuration()$
24: **end while**
25: **return** S^*
___

Neighborhood Evaluation Method. We adopt the same approximate evaluation method used in [10], the approximate makespan f^{appr} of a neighboring solutions is experssed as follows:

$$f^{appr} = \max_{i \in W}\left(R\left(JP\left[i\right]\right), R\left(MP\left[i\right]\right)\right) + T[i] + \max_{i \in W}\left(Q\left(JS\left[i\right]\right), Q\left(MS\left[i\right]\right)\right), \quad (18)$$

where W is the set of affected operations when applying the k-insertion move. Similar evaluation method can be adopted for the sequence change move.

Algorithm 2 describes the input requirements, including the initial solution S_{init} constructed by the greedy strategy, the maximum number of unimproved iterations λ, the maximum number K of accurately evaluated solutions at each iteration, and the search duration T. The output of the tabu search is the best solution S^* found so far.

First, we construct the neighborhood N^π and N^α (lines 6–7), respectively. Let N denote the union of the two neighborhoods, and remove the solutions in the tabu state (line 9). If N is empty, indicating that all neighborhood moves are in tabu state, a random perturbation operation is performed on the current solution S_c (line 11).

If N is not empty, we approximately evaluate each neighborhood solution. Then, we select the first K solutions and store them in $topkSet$. Since the approximate makespan are often inaccurate, it is necessary to accurately evaluate each solution in $topkSet$, calculate its actual makespan, and select the solution S' with the smallest makespan to replace the current solution. Subsequently, this neighborhood move is added to the tabu table (line 16) and the current solution is replaced with the neighborhood solution S' (line 17).

Since both N^π and N^α neighborhood moves alter the job sequences on the machines, it is also necessary to re-allocate memory and update the memory allocation status for each data block (line 17). To fulfill this purpose, we propose a memory update algorithm described in detail in Sect. 4.3.

4.3 Memory Update Procedure

The overall time of each task consists of the transfer time and the execution time. The transfer rates of high-speed memory and low-speed memory are not the same. Additionally, high-speed memory has a limited capacity, thus the memory allocation strategy of data blocks will affect the final outcome. Throughout the algorithm, the memory update procedure is repeatedly invoked during each iteration. The memory update strategy is mainly based on two basic greedy criteria:

1. Assign as many data blocks as possible to the fast memory without violating capacity constraints.
2. Prioritize "important" data blocks for placement in high-speed memory. Since shortening the length of the critical path is crucial for optimizing the makespan, we measure the importance of each data block by the number of transfers occurring on the critical path.

Once the memory is updated, the local search determines the task sequence, and when all memory is assigned to low-speed, a complete solution is generated. Consequently, we can determine the start time and duration of each stage, the entry and exit time of each data block in the memory, and also identify the critical path and critical tasks. The number of times that a data block appears on the critical path corresponds to the number of critical tasks having dependence relationships with it. The data blocks are sorted in a descending order based on their occurrences.

When attempting to place important data blocks in high-speed memory, it is possible for the peak memory usage to exceed the memory capacity. Therefore, a judgment strategy is required to ensure compliance with memory capacity constraints. This strategy is designed as follows: first, calculate the lifespan of all data blocks, then determine the memory usage per second using the differential array, and finally judge whether it exceeds the capacity limit. However, acquiring per-second information is time-consuming due to the huge size of makespan compared to the number of data block nodes. Since the peak memory usage occurs when data blocks are placed in memory, we can discretize all the time

---

Algorithm 3 The memory updating procedure

1: **Input**: The temp solution S' in tabu search, problem instance p
2: **Output**: The true solution S^t
3: $S' \leftarrow InitMemory(S')$, $R, Q, Slack \leftarrow Init()$
4: $dataSet \leftarrow getAllData()$, $taskSeq \leftarrow getAllTask()$
5: **while** $dataSet is not empty$ **do**
6:     $topoSeq \leftarrow TopoSort(taskSeq, p)$
7:     $calcuRQSlack(topoSeq, p, S')$
8:     $D \leftarrow -1$, $maxUseT \leftarrow 0$
9:     **for** each data d in $dataSet$ **do**
10:        $criticalUse \leftarrow countCriIn(d) + countCriOut(d)$
11:        **if** $criticalUse > maxUseT$ **then**
12:            $maxUseT \leftarrow criticalUse$
13:            $D \leftarrow d$
14:        **end if**
15:    **end for**
16:    **if** memory of highType2 is enough **then**
17:        $AssignToMemory(d, highType2)$
18:    **else if** memory highType1 is enough **then**
19:        $AssignToMemory(d, highType1)$
20:    **else**
21:        $AssignToMemory(d, lowType)$
22:    **end if**
23:    $updataSolution(S^t)$
24:    $dataSet \leftarrow dataSet \setminus \{d\}$
25: **end while**
26: **return** S^t

---

nodes that induce memory usage changes and differentiate them into an array. This enables us to determine if the peak memory usage surpasses the capacity limit.

Algorithm 3 presents the pseudocode for the memory updating procedure. Initially, it allocates all data blocks to low-speed memory and initializes $dataSet$. At each iteration, it performs topological sorting on all tasks and calculates R, Q, and $Slack$ values. It then sequentially attempts to allocate the most important data block that has not yet been assigned to memory. If the memory usage does not exceed the capacity limit, the data block is allocated to memory.

Subsequently, the critical path may be changed because the allocation of a data block is determined. Therefore, in the next iteration, it is necessary to recalculate the R, Q, and $Slack$ values and reevaluate the importance of the remaining data based on this information. The memory updating procedure is completed when $dataSet$ is empty because the data block is deleted from $dataSet$ each time once it is allocated to memory.

5   Experiment Design and Analysis

5.1   Parameter Settings and Experimental Protocol

In this section, we conduct extensive experiments to evaluate the performance of the proposed TS algorithm. Synthetic task graphs were generated using the TGFF tool [9]. Each benchmark was compiled and extracted as a directed acyclic graph (DAG) with different numbers of tasks and different in/out degree limits. The data-DAG is constructed by adding a node for each edge in the corresponding task-DAG. The TS algorithm was implemented in C++ and executed on an AMD Ryzen 7 5800H CPU with Radeon Graphics 3.20 GHz. Table 1 provides the descriptions and settings of the parameters used in TS. The last column indicates the settings for the entire set of instances. These parameter values were determined through extensive preliminary experiments.

Table 1. Parameter settings in TS

Para	Description	Value
K_{max}	maximum accurate evaluation	100
p	memory update round	100
θ_1	tabu tenure for N^k	$m + rand()\%(2*m)$
θ_2	tabu tenure for N^π	$n + rand()\%n$
λ	depth of tabu search	100000
T_{max}	maximum run time of TS	600 s

Table 2. The basic information of the benchmarks

Item	Description	Value
DAG	Num. of tasks	20/100/500
	Num. of cores	2 high-speed + 2/4/6/8 general
time	$T_{in} : T_{proc} : T_{out}$	1:1:1/1:10:1
	$S_{high} : S_{low}$	1.2:1
data	data size	[90,110]/[900,1100]

To comprehensively compare the performance of the algorithms, five groups of random instances were generated for different conditions, including varying numbers of tasks, ratios of move time to processing time, and numbers of cores. Table 2 presents the basic information of the instances. Columns T_{in}, T_{proc}, and T_{out} represent the move-in time from memory before task execution, the processing time, and the move-out time from memory after task execution, respectively. S_{high} and S_{low} represent the data transfer rate of high-speed and low-speed memories, respectively. It should be noted that all instances have the same memory sizes, and the size of low-speed memory is infinite.

5.2   Implementation of Reference Algorithms and Comparison

We compare TS with three well-known scheduling algorithms: HEFT [32], PEFT [2], and HSIP [33]. However, these algorithms do not consider memory constraints. In order to make a relatively fair comparison, we implemented these algorithms and used the same memory allocation strategy as our proposed TS algorithm. This allowed us to compare the performance of their algorithms with our proposed approach under the same computing environment.

Table 3 presents the results of the proposed tabu search algorithm and other scheduling algorithms on randomly generated instances with 20, 100, and 500 tasks. The second row represents different ratios between average processing time and average memory access time, while the third row represents different numbers of cores. The superiority of the proposed TS algorithm is evident in various instance sizes and parameter settings.

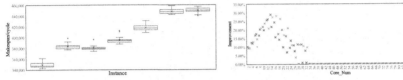

(a) The boxplot of makespan obtained by   (b) The change of makespan with respect
TS on 10 instances   to the number of high speed cores

Fig. 2. The results of TS on different instances

In this section, we analyze the stability of the tabu search algorithm by executing TS on 10 instances. To achieve this goal, TS is applied to each instance for 20 independent runs, each with a distinct initial solution. The objective is to determine variations in the quality of the best-found solutions. The computational results are depicted in Fig. 2a. It indicates that the range of makespan and the disparity between the minimum and maximum makespan are relatively small for all instances, thus confirming the stability of the proposed tabu search algorithm.

In this section, we analyze the impact of the number of cores on the performance of the tabu search procedure. To accomplish this point, we apply TS and the greedy initial solution construction procedure to three randomly generated instances and display the results in Fig. 2b. The x-axis represents the number of cores, and the y-axis represents the makespan improvement rate achieved by TS compared to the greedy initial solution construction procedure. It is worth noting that, for architectural heterogeneity to be ensured, a minimum of two synergistic high-speed cores is required.

The results in Fig. 2b demonstrate an increase in the improvement rate from 10% to 30% as the number of DSP cores increases from 2 to 12. However, this rate declines to 0 as the number of DSP cores continues to increase beyond 12, reaching around 28 or more. This trend indicates that with a small number

Table 3. The computational results of four algorithms under different conditions

| Seed | Alg. | $N_1 = 20$ | | | | | | | | $N_1 = 100$ | | | | | | | | $N_1 = 500$ | | | | | | | |
| | | 1:1 | | | | 10:1 | | | | 1:1 | | | | 10:1 | | | | 1:1 | | | | 10:1 | | | |
		2	4	6	8	2	4	6	8	2	4	6	8	2	4	6	8	2	4	6	8	2	4	6	8
1	HEFT	44327	40506	40506	40506	25413	21956	20834	18566	142838	125209	121255	117545	86200	67094	58974	55981	531514	408633	355633	326992	374705	267036	221685	199169
	PEFT	42158	39012	39012	37093	22410	21847	19872	17850	138843	117349	116242	107117	87185	64286	58107	51358	537119	390220	317894	271964	378171	259278	230174	193827
	HSIP	38214	37093	37093	37093	21491	19557	19554	17560	140717	112394	106272	104639	80089	61542	53552	50385	504722	383069	308476	272578	380380	260418	209679	177912
	TS	38214	37093	37093	37093	21461	17794	17595	17125	109116	93924	92242	91691	74990	57898	47732	41366	451650	335906	266270	230430	357981	254248	205101	171970
2	HEFT	35472	31254	31254	31254	21894	17144	14179	13555	175838	142877	132199	125228	86795	65090	56973	51952	683606	514133	441661	407742	371125	263496	217368	188053
	PEFT	35328	31254	31254	31254	21894	17288	14028	13566	158420	143772	128865	124771	86273	64088	54031	52176	708835	501795	442187	391977	370291	265549	211768	180799
	HSIP	35139	31254	31254	31254	22038	17654	13830	13770	157526	138277	125270	123976	84740	63873	52688	49036	705381	523769	441550	388522	368066	261222	212892	181086
	TS	34831	31254	31254	31254	21775	14447	13830	13236	138121	124426	122159	120323	78801	58795	49733	44265	632482	470421	384499	330176	363739	253758	205181	173116
3	HEFT	40536	40536	40536	40536	22834	19060	16310	16310	218763	202092	202092	202092	80579	64276	60685	58518	789272	631328	582678	582381	366987	267996	226653	209475
	PEFT	40536	40536	40536	40536	19980	16802	15714	15714	213557	208329	203274	194167	81624	67121	56372	56103	812663	671557	593465	579634	364583	268464	227483	203976
	HSIP	40569	40569	40569	40569	19964	16391	15615	15615	218725	218005	197233	197233	76279	58551	55107	55231	881100	692472	578347	577114	365854	262534	219162	188998
	TS	39526	39526	39526	39526	19060	16306	14677	14123	181713	174705	172243	170520	71511	55380	50173	50134	689694	515847	442179	431712	351126	255135	205987	177347
4	HEFT	58790	55119	55119	55119	30702	28894	27031	27031	215793	190214	185081	185081	104212	74551	65474	60518	816131	625849	588229	566346	384862	276731	231496	205836
	PEFT	54271	53486	53486	53486	29574	28512	27031	27031	204938	189557	184377	184261	104581	70625	65138	59302	814377	619869	585849	573881	380675	278494	228072	203920
	HSIP	53486	53486	53486	53486	28409	26902	25401	25401	205817	181561	187531	187378	102341	71298	61659	58284	842445	662378	579880	574375	388936	276951	225700	193574
	TS	53486	53486	53486	53486	28032	25401	25401	25401	180313	171093	170078	164582	98605	64946	56774	51433	756300	556861	464385	461273	366579	268317	219257	186801
5	HEFT	60614	59415	59415	59415	27745	27816	25683	25683	201191	165057	160791	160791	100213	74471	67176	64990	767750	591527	552643	554056	359263	279351	231408	209674
	PEFT	62638	60149	59627	59627	27716	25439	25439	25439	198744	163828	147171	145329	98971	65166	60531	58705	825499	641847	572067	534886	370644	278157	228169	210082
	HSIP	64019	58511	58511	58511	27302	24003	24003	24003	192259	147623	140788	140788	93642	67044	60373	58479	838369	637518	608188	523021	369224	277763	224231	206780
	TS	58308	58308	58308	58308	26447	23747	23747	23747	153481	144182	135408	130065	85110	65258	56817	53798	721304	528344	467106	461068	348140	266763	219154	195602

of DSP cores, multiple unrelated tasks are assigned to the same cores, causing dependencies among them. Conversely, when a large number of cores are available, the predecessors of a task can be assigned to different cores, enabling parallel processing.

5.3 Effect of Mixed Evaluation Strategy

It is well-known that neighborhood evaluation is one of the most important ingredients in the local search procedure. In order to alleviate the computational burden in tabu search, we propose a mixed evaluation strategy in this paper. Specifically, at each iteration of TS, we first apply an approximate evaluation method to evaluate all neighboring solutions. Although this method may not provide highly accurate makespan calculations, it significantly improves computational efficiency. Next, we sort these neighboring solutions in an ascending order of makespan. Subsequently, we apply an exact evaluation method to the top k solutions and select the one with the minimum makespan to replace the current solution before proceeding to the next round of local search.

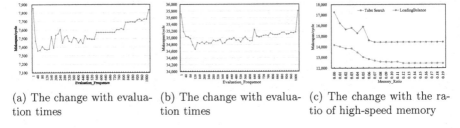

(a) The change with evaluation times

(b) The change with evaluation times

(c) The change with the ratio of high-speed memory

Fig. 3. The change of makespan with different evaluation frequency and memory ratios

Figure 3a and Fig. 3b illustrate the results of TS for different ratios of exact evaluation on a random instance and a larger instance, which is five times the size of the former. It can be observed that when $k = 1$ (the leftmost point on the x-axis), the best solution is evaluated only using the approximate method. Both curves show a decreasing trend for $k \in [1, 30]$ and $k \in [1, 120]$, followed by a slight increase with further increasing k. This is primarily due to the fact that, with the same cutoff time, an excessive number of runs using the exact evaluation strategy leads to fewer iterations of the tabu search, thereby diminishing its effectiveness. These findings indicate that the mixed evaluation strategy is a compromise between exact and approximate methods, enabling the quick selection of relatively high-quality solutions.

5.4 Effect of High Speed Memory Ratio on Makespan

In this section, we intend to investigate the impact of the high-speed memory ratio on makespan. We perform TS and the greedy initial solution construction procedure separately for 20 independent runs, and present the results in Fig. 3c. It can be observed that TS consistently outperforms the greedy initial solution construction procedure across the entire range of memory ratios from 0 to 0.19, with a minimum makespan difference of 2000. This may be attributed to the fact that, due to the greedy strategy employed by the memory allocation approach, both algorithms exhibit an unusual scenario in which a slight increase in the proportion of high-speed memory leads to an increase in makespan.

Besides, when the high-speed memory is insufficient, the makespan obtained by tabu search exhibits a slight increase, suggesting that tabu search is more effective in mitigating the impact of insufficient high-speed memory. Moreover, the makespan obtained by tabu search at low speed remains lower than that of load balancing at high speed. Consequently, by restricting the utilization of high-speed memory, a superior scheduling scheme can be obtained that optimizes both makespan and high-speed memory.

6 Conclusions

This paper addresses the problem of data allocation and task scheduling under memory constraints in heterogeneous multiprocessor systems, where an integer linear programming model is introduced and a tabu search algorithm with distinct features is proposed to minimize the critical path of the graph. Experimental results show that the TS algorithm produces high-quality solutions within a reasonable computational time, outperforming the classical load balancing algorithm by 5-25%.

In particular, several conclusions are drawn as follows: First, the proposed TS algorithm outperforms other scheduling algorithms on different scale instances, and its stability is demonstrated through experiments on representative cases. Second, the tabu search procedure can efficiently optimize the initial solution constructed with the greedy strategy, and the improvement ratio varies with the number of cores, which means that the number of cores with the maximum promotion rate under different conditions is not necessarily the same. Third, hybrid neighborhood evaluation technique balances evaluation accuracy and evaluation time, and ultimately enables the local search to obtain high quality solutions.

In the future, it is an interesting research direction to combine population-based metaheuristic methods and problem-specific knowledge to enhance the performance of the current algorithm. Besides, solution-based tabu strategy is also worthy to attempt in order to improve the search intensification. Furthermore, another extension to this study could involve energy-aware information allocation and task scheduling aimed at minimizing both the total workload execution and overall energy consumption.

References

1. Agullo, E., Amestoy, P.R., Buttari, A., Guermouche, A., Excellent, J.Y., Rouet, F.H.: Robust memory-aware mappings for parallel multifrontal factorizations. SIAM J. Sci. Comput. **38**(3), 256–279 (2016)
2. Arabnejad, H., Barbosa, J.G.: List scheduling algorithm for heterogeneous systems by an optimistic cost table. IEEE Trans. Parallel Distrib. Syst. **25**(3), 682–694 (2014)
3. Aupy, G., Brasseur, C., Marchal, L.: Dynamic memory-aware task-tree scheduling. In: 2017 IEEE International Parallel and Distributed Processing Symposium (IPDPS), pp. 758–767 (2017)
4. Baruah, S., Fisher, N.: The partitioned multiprocessor scheduling of deadline-constrained sporadic task systems. IEEE Trans. Comput. **55**(7), 918–923 (2006)
5. Birgin, E.G., Feofiloff, P., Fernandes, C.G., De Melo, E.L., Oshiro, M.T., Ronconi, D.P.: A MILP model for an extended version of the flexible job shop problem. Optimization Lett. **8**(4), 1417–1431 (2014)
6. Brucker, P., Schlie, R.: Job-shop scheduling with multi-purpose machines. Computing **45**(4), 369–375 (1990)
7. Chantem, T., Hu, X.S., Dick, R.P.: Temperature-aware scheduling and assignment for hard real-time applications on mpsocs. IEEE Trans. Very Large Scale Integration Syst. **19**(10), 1884–1897 (2010)
8. Chen, Y.S., Liao, H.C., Tsai, T.H.: Online real-time task scheduling in heterogeneous multicore system-on-a-chip. IEEE Trans. Parallel Distrib. Syst. **24**(1), 118–130 (2012)
9. Dick, R.P., Rhodes, D.L., Wolf, W.: Tgff: task graphs for free. In: Proceedings of the Sixth International Workshop on Hardware/Software Codesign, pp. 97–101. IEEE (1998)
10. Ding, J., Lü, Z., Li, C.M., Shen, L., Xu, L., Glover, F.: A two-individual based evolutionary algorithm for the flexible job shop scheduling problem. In: Proceedings of the AAAI Conference on Artificial Intelligence, vol. 33, pp. 2262–2271 (2019)
11. Du, J., Wang, Y., Zhuge, Q., Hu, J., Sha, E.H.M.: Efficient loop scheduling for chip multiprocessors with non-volatile main memory. J. Signal Process. Syst. **71**(3), 261–273 (2013)
12. Ergu, D., Kou, G., Peng, Y., Shi, Y., Shi, Y.: The analytic hierarchy process: task scheduling and resource allocation in cloud computing environment. J. Supercomput. **64**(3), 835–848 (2013)
13. Fu, Z., Tang, Z., Yang, L., Liu, C.: An optimal locality-aware task scheduling algorithm based on bipartite graph modelling for spark applications. IEEE Trans. Parallel Distrib. Syst. **31**(10), 2406–2420 (2020)
14. González, M.A., Vela, C.R., Varela, R.: Scatter search with path relinking for the flexible job shop scheduling problem. Eur. J. Oper. Res. **245**(1), 35–45 (2015)
15. Hansmann, R.S., Rieger, T., Zimmermann, U.T.: Flexible job shop scheduling with blockages. Math. Methods Oper. Res. **79**(2), 135–161 (2014)
16. Hu, Y., Li, J., He, L.: A reformed task scheduling algorithm for heterogeneous distributed systems with energy consumption constraints. Neural Comput. Appl. **32**(10), 5681–5693 (2020)
17. Ilavarasan, E., Thambidurai, P.: Low complexity performance effective task scheduling algorithm for heterogeneous computing environments. J. Comput. Sci. **3**(2), 94–103 (2007)

18. Kang, Q., He, H., Song, H.: Task assignment in heterogeneous computing systems using an effective iterated greedy algorithm. J. Syst. Softw. **84**(6), 985–992 (2011)
19. Kang, S., Dean, A.G.: Darts: techniques and tools for predictably fast memory using integrated data allocation and real-time task scheduling. In: The 16th IEEE Real-Time and Embedded Technology and Applications Symposium, pp. 333–342. IEEE (2010)
20. Liu, J.: An application of generalized tree pebbling to sparse matrix factorization. SIAM J. Algebraic Discrete Methods (1987)
21. Mastrolilli, M., Gambardella, L.M.: Effective neighbourhood functions for the flexible job shop problem. J. Sched. **3**(1), 3–20 (2000)
22. Ouni, B., Ayadi, R., Mtibaa, A.: Partitioning and scheduling technique for run time reconfigured systems. Intern. J. Comput. Aided Eng. Technol. **3**(1), 77–91 (2011)
23. Özgüven, C., Özbakır, L., Yavuz, Y.: Mathematical models for job-shop scheduling problems with routing and process plan flexibility. Appl. Math. Model. **34**(6), 1539–1548 (2010)
24. Peris, A.D., Hernández, J., Huedo, E.: Distributed late-binding scheduling and cooperative data caching. J. Grid Comput. **15**, 235–256 (2017)
25. Praveenchandar, J., Tamilarasi, A.: Dynamic resource allocation with optimized task scheduling and improved power management in cloud computing. J. Ambient. Intell. Humaniz. Comput. **12**(3), 4147–4159 (2021)
26. Ramakrishnan, A., et al.: Scheduling data-intensiveworkflows onto storage-constrained distributed resources. In: Seventh IEEE International Symposium on Cluster Computing and the Grid, vol. 1, pp. 401–409 (2007)
27. Roshanaei, V., Azab, A., ElMaraghy, H.: Mathematical modelling and a meta-heuristic for flexible job shop scheduling. Int. J. Prod. Res. **51**(20), 6247–6274 (2013)
28. Sbîrlea, D., Budimlić, Z., Sarkar, V.: Bounded memory scheduling of dynamic task graphs. In: Proceedings of the 23rd International Conference on Parallel Architectures and Compilation, pp. 343–356 (2014)
29. Sergent, M., Goudin, D., Thibault, S., Aumage, O.: Controlling the memory subscription of distributed applications with a task-based runtime system. In: 2016 IEEE International Parallel and Distributed Processing Symposium Workshops (IPDPSW), pp. 318–327 (2016)
30. Sethi, R., Ullman, J.D.: The generation of optimal code for arithmetic expressions. J. ACM **17**(4), 715–728 (1970)
31. Shao, Z., Zhuge, Q., Xue, C., Sha, E.M.: Efficient assignment and scheduling for heterogeneous dsp systems. IEEE Trans. Parallel Distrib. Syst. **16**(6), 516–525 (2005)
32. Topcuoglu, H., Hariri, S., Wu, M.Y.: Performance-effective and low-complexity task scheduling for heterogeneous computing. IEEE Trans. Parallel Distrib. Syst. **13**(3), 260–274 (2002)
33. Wang, G., Wang, Y., Liu, H., Guo, H.: Hsip: a novel task scheduling algorithm for heterogeneous computing. Sci. Program. **2016**(2), 1–11 (2016)
34. Wang, Y., Li, K., Chen, H., He, L., Li, K.: Energy-aware data allocation and task scheduling on heterogeneous multiprocessor systems with time constraints. IEEE Trans. Emerg. Top. Comput. **2**(2), 134–148 (2014)
35. Xia, W., Quek, T.Q., Zhang, J., Jin, S., Zhu, H.: Programmable hierarchical c-ran: from task scheduling to resource allocation. IEEE Trans. Wireless Commun. **18**(3), 2003–2016 (2019)

36. Yin, L., Luo, J., Luo, H.: Tasks scheduling and resource allocation in fog computing based on containers for smart manufacturing. IEEE Trans. Industr. Inf. **14**(10), 4712–4721 (2018)
37. Yuan, H., Bi, J., Zhou, M.: Spatial task scheduling for cost minimization in distributed green cloud data centers. IEEE Trans. Autom. Sci. Eng. **16**(2), 729–740 (2018)
38. Zhuge, Q., Guo, Y., Hu, J., Tseng, W.C., Xue, C.J., Sha, E.H.M.: Minimizing access cost for multiple types of memory units in embedded systems through data allocation and scheduling. IEEE Trans. Signal Process. **60**(6), 3253–3263 (2012)

ACDP-Floc: An Adaptive Clipping Differential Privacy Federation Learning Method for Wireless Indoor Localization

Xuejun Zhang$^{(\boxtimes)}$, Xiaowen Sun, Bin Zhang, Fenghe Zhang, Xiao Zhang, and Haiyan Huang

School of Electronic and Information Engineering, Lanzhou Jiaotong University, Lanzhou 730070, China
`xuejunzhang@mail.lzjtu.cn`

Abstract. With the increasing demand for location services, the fingerprint recognition technology based on the received signal strength (RSS) has been paid more and more attention and applied due to its advantages of mature infrastructure and easy implementation, Federated Learning (FL) has been applied to indoor localization to solve data silos and privacy security problems in recent research work. To prevent eavesdroppers from inferring private information and model features of the client by analyzing parameter information. Some researchers introduce differential privacy (DP) technology into FL for privacy protection, but the addition of noise seriously affects the availability of data and models. We investigate the privacy loss measurement and tracking methods of DP and propose ACDP-Floc, an adaptive clipping differential private federated learning method for indoor location, the usability of data and model is improved by adaptive clipping of model gradient. Experimental results show that: when the privacy budget $\varepsilon = 1.0$, which indicates that the algorithm adds a large noise, ACDP-Floc achieves 92.53%, 93.61% and 96.54% classification accuracy for the Mall Area, Mall-Wi-Fi and UIJIIndoorLoc three real datasets, respectively.

Keywords: Privacy protection · Federal learning · Indoor localization · Differential privacy · Adaptive clipping

1 Introduction

Indoor location technology makes up for the defect that GPS/BDS is prone to failure in indoor environment, its importance and commercial value have become increasingly apparent, which has been widely concerned by the society [1]. Recent progress in indoor location has led a number of technologies, including those based on infrared positioning technology, ultrasonic positioning technology, Bluetooth positioning technology, Wi-Fi location technology, ultra-bandwidth location technology, radio frequency identification location technology, and geomagnetic location technology [2], as well as hybrid positioning systems that combine

Z. Tari et al. (Eds.): ICA3PP 2023, LNCS 14488, pp. 381–393, 2024.
https://doi.org/10.1007/978-981-97-0801-7_22

these technologies. Fingerprint positioning technology based on received signal strength (RSS) has become the mainstream trend of indoor positioning due to its advantages such as simple implementation, low cost, low power consumption and mature infrastructure [3]. The distributed learning framework represented by FL solves the problems of the centralized learning framework, such as excessive network transmission load, insufficient computing resources, response delay and network congestion [4], it enhances the performance of location services. Recent research has proposed deploying DL-based indoor fingerprint positioning systems into FL to provide low-latency, real-time responsiveness, and high-precision indoor positioning services to the external users. Bo Gao et al. [5] propose an FL framework (FedLoc3D) for both building-floor classification (BFC) and latitude-longitude regression (LLR), which based on a convolutional neural network with depthwise separable convolutions. However, some studies has shown that attackers can use differential attacks and model inversion attacks to extract user information through model parameters, which poses a serious threat to user privacy [6]. Existing methods mostly employ techniques such as encryption, k anonymous mechanism or DP to ensure privacy protection in the FL scenario [7]. The use of encryption-based techniques require significant computational resources and time for encryption and decryption operations, it also leads to increased communication overhead during data transmission, its effectiveness remains to be further validated. As a result, there is not suitable for FL scenarios with resource-constrained end device and high-frequency data updates. Therefore, researchers tend to prefer combining federated learning with DP techniques. Shen X et al. [8] develop a performance-enhanced DP-based FL (PEDPFL) algorithm, where a classifier-perturbation regularization method improve the robustness of the trained model against DP-injected noise. Nevertheless, the added noise significantly impairs data availability and the learning process, resulting in a substantial degradation in model performance [9], thereby preventing users from obtaining satisfactory indoor localization services. In response to the issue of decreased model accuracy due to differential privacy [10], Xu Z et al. [11] develop an adaptive and fast convergent learning algorithm with a provable privacy guarantee, which mitigates the negative effect of differential privacy upon the model accuracy by introducing adaptive noise. [12]improve the image classification ability of CNN models under differential privacy protection that improve the classification utility. These methods have made some advancements in terms of model performance and data availability, but there are still significant limitations, particularly when applied to real-world location datasets.

In order to solve the above problems and challenges, and provides high precision, security, and real-time indoor location services for users, we study the ways and methods of disclosing user privacy and model parameter privacy during indoor location model training through DP and FL. This paper design a dynamic privacy protection method for indoor localization that safeguards the privacy of user data and model parameters by implementing a finer-grained gradient clipping mechanism. Finally, we set up a comprehensive experiment on three real data sets to evaluate the performance of our method in the real environment, and found that our method is superior to other methods mentioned in this paper at the comprehensive evaluation of positioning accuracy and response time.

- To the best of our knowledge, the ACDP-Floc is the first work to address the substantial decline of FL model accuracy caused by privacy protection while using DP. Our ACDP-Floc succeeds in utilizing the privacy tracking and adaptive gradient clipping technique to mitigate the impact of DP mechanisms on the precision of FL models.
- We design an adaptive privacy protection method for wireless indoor localization to protect the privacy of users' data and model parameters by implementing a finer-grained gradient clipping mechanism.
- we set up a comprehensive experiment on three practical datasets to evaluate the performance of our method in the practical environment, and find that our method is superior to other methods mentioned in this paper at the comprehensive evaluation of localization accuracy and response time.

The rest of the paper is organized as follows. Section 2 is the details of ACDP-Floc. In Sect. 3, we present the experiments and performance evaluation. Section 4 shows the conclusion.

2 Our Proposed Method

In this section, we present FL localization framework, propose adaptive DP deep learning localization technique.

2.1 System Framework

Our group conducted in-depth research on the privacy protection of localization systems with FL, and advanced both a differential private indoor localization federated learning model and a framework for edge computing in our previous work [13]. We propose a novel method to optimize the approach of adding noise to the model parameters. The framework of ACDP-Floc is shown in Fig. 1.

The framework of ACDP-Floc consists of three entities: end devices, edge servers, and cloud server. The end devices collect and stores local indoor RSS data and sends it to nearby edge servers, while independently preprocessing and adding noise to the collected RSS datasets; The edge servers first receive the disturbed RSS fingerprint data uploaded by nearby end devices, and aggregate these fingerprint data into RSS fingerprint data containing multiple user information. At the same time, it utilize these aggregated data for trusted training of local localization sub models, and upload the trained local sub model parameters to the cloud server. The cloud server use the FedAvg algorithm to update the global model parameters, and distribute the updated model parameters to each edge servers for the next round of iterative training until the optimal training model is obtained.

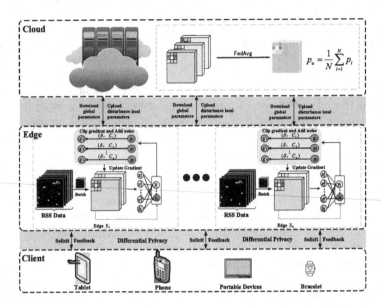

Fig. 1. The framework of ACDP-Floc

The privacy threats to the FL framework are listed as follows.

Threat 1: Potential privacy leakage. Although the FL training process only exchanges model parameters without sharing original data, the latest research shows attackers can still use model inversion attacks and gradient backward inference attacks to obtain user's privacy through model parameters.

Threat 2: Data privacy leakage. The servers, as the data collector, are not trustworthy for the users yet, so privacy threaten are primary concerns for users before submitting data.

To counter the **Threat 1** and **Threat 2**, our group have introduced DP techniques to protect user's privacy by adding appropriate Laplace noise to the model parameters of the original data and each participant under the FL protocol. The specific methods for addressing **Threat 1** and **Threat 2** faced by the framework are detailed in Sect. 2.3 of the article We addressed the requirements of privacy protection partially, but the method performed poorly in model localization accuracy and data availability.

In this paper, we propose ACDP-Floc to optimize the approach of adding noise to the model parameters, so as to improve the data availability and model accuracy.

2.2 Adaptive Privacy Protection

To address the security threats to FL, all edge servers are required to protect the privacy of the model parameters obtained from the training of local datasets when they are uploaded. Zheng et al. and Yang et al. [14, 15] proposed to utilize the schemes based on Paillier encryption and Secret sharing algorithm to protect privacy respectively, yet both suffer from excessive computational overhead. In contrast, DP techniques are less computationally intensive and more suitable for resource-constrained edge computing devices. To better characterize the differences in model parameters of the neural network on any two datasets, we utilize DP-SGD to control the effect of data on the model. Zhang et al. [13] used DP techniques to add noise matching the Laplace distribution on model parameters directly to protect parameters privacy, but it causes a large loss of model accuracy. Abadi. et al. [16] employed moment account to track the privacy loss during training, clipped the gradient with a global threshold C and added Gaussian noise to protect the parameters privacy. But in this method the threshold C is a fixed value. DP-SGD clips the L_2 norms of the gradient vector by a threshold C, which limiting the gradient of each example during each iteration $\nabla_\theta \mathcal{L}(\theta, x)$. So the selection of C value is crucial for DP deep learning, which will add too much noise if it is too large and will overclip the gradient if it is too small.

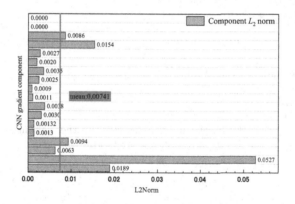

Fig. 2. L_2 norms of the gradient component of the model

Xu et al. [11] presented gradient norms decreases as the model converges in CNN, but the L_2 norms of each gradient component is different, as shown in Fig. 2. Therefore, gradient clipping with a fixed value of C is not reasonable. Here come the reasons. First, the fixed value cannot adapt itself to the decreasing gradient as the model converges. Second, it cannot adapt to different gradient magnitudes.

In this paper, we propose an adaptive DP mechanism for model gradient, which can adaptively adjust the clipping threshold according to the training process to reduce the negative impact of noise on model accuracy. For the continuity and progressivity of the model optimization process, the historical gradient can estimate the value of the current gradient. We use the average historical gradient as a priori knowledge to predict the gradient in the current round.

Algorithm 1 presents ACDP-Floc. In each iteration, a batch of examples are drawn from the training datasets and the algorithm calculates the loss, training accuracy and gradient of the step. The gradient clipping bounds per-example gradients by L_2 norms clipping with a threshold C. Gaussian mechanism adds random noise $\xi^{\sim}_{gausssian}(0, \sigma, C_t)$ to the clipped gradient g', which is shown in Fig. 3. Ultimately, the next round of gradient clipping threshold is calculated from the current gradient and the calculation process of historical gradient is implemented as follows.

Algorithm 1 The algorithm of ACDP-Floc
___

Input: Training examples $\{x_1, x_2, x_3, \ldots, x_N\}$; Learning rate lr; Gradient norms bound C; Noise scale σ; Batch size L; Client numbers N; Communication numbers T; Total privacy budget ε_{total} ; Model training times $Epoch$.
Output: Model parameters w_{i+1}.

1: **for** i in T **do**
2:     Initialise $w_0, E[\tilde{g}^2]_0$
3:     **for** per client in system **do**
4:         $w_i^e \leftarrow w_i$; // Receive parameters
5:         **for** t in $Epoch$ **do**
6:             $rdp \leftarrow rdp\_account(L/N, orders, \sigma, Epoch * T)$; // Calculate rdp value
7:             $\varepsilon_{curr} \leftarrow AdpBudgetAlloc(\varepsilon_{total}, i, T, schedual)$; // Calculating Privacy Budget Losses
8:             **if** $\varepsilon_{curr} > \varepsilon_{total}$ **then**
9:                 break;
10:            **end if**
11:            **for** s in $step$ **do**
12:                Take a batch of data samples B_s from the training dataset;
13:                $g_t^{e,i} \leftarrow \nabla w_t \mathcal{L}(w_t^e, x)$; // Compute gradient
14:                $(g_t^{e,i})' \leftarrow g_t^{e,i}/max(1, \frac{\|g_t^{e,i}\|_2}{C})$; // Clip gradient
15:                $(\hat{g}_t^{e,i})' \leftarrow \frac{1}{B}\left\{(g_t^{e,i})' + g_{gausssian}(0, \sigma, C_t)\right\}$; // Add noise
16:                $E[(\tilde{g}_t^{e,i})'^2]_{t+1} = \gamma E[(\hat{g}_t^{e,i})'^2]_t + (1-\gamma)E[(\hat{g}_t^{e,i})'^2]_{t-1}$; // Estimate the next gradient
17:                **if** $E[(\tilde{g}_t^{e,i})'^2]_{t+1} > G$ **then**
18:                    $C_{t+1} = \beta\sqrt{E[(\tilde{g}_t^{e,i})'^2]_{t+1}}$;
19:                **end if**
20:                $w_t^e \leftarrow w_t^e - lr * \hat{g}_t^{e,i}$;
21:            **end for**
22:        **end for**
23:    **end for**
24:    $w_{i+1} = \frac{1}{N}\sum_{e=1}^{N} w_i^e$; //Parameters aggregation;
25:    Parameters Broadcast: Broadcast the aggregated parameters to the client;
26: **end for**

We forecast the next round of gradients $E[\tilde{g}^2]_{t+1}$ using historical gradients $E[\hat{g}^2]_{t-1}$ and current gradients $E[\hat{g}^2]_t$, which is used to calculate the clipping

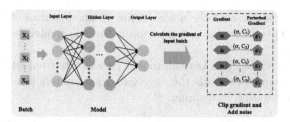

Fig. 3. Local sub-models add noise

threshold C_{t+1} for the next round, defined as $C_{t+1} = \beta\sqrt{E[\tilde{g}^2]}_{t+1}$, where β is local clipping factor. The priori knowledge $E[\hat{g}^2]_{t-1}$ and $E[\hat{g}^2]_t$ are given by:

$$E[\hat{g}^2]_0 = \vec{0} \tag{1}$$

$$E[\tilde{g}^2]_{t+1} = \gamma E[\hat{g}^2]_t + (1-\gamma)E[\hat{g}^2]_{t-1}, (t > 0) \tag{2}$$

Here, prior knowledge $E[\hat{g}^2]_0 = 0^{\rightarrow}$ at the first training round, which will lead to $C_{t+1} = \beta\sqrt{E[\tilde{g}^2]}_{t+1} = 0$, cannot be used for gradient cropping. So we set a threshold value G, let the gradient clipping threshold take a fixed value C when the priori knowledge of the gradient is not enough at the early stage of training. To conclude, the process of noise added for edge server e in round t is as shown follows:

$$(\hat{g}_t^{e,i})' \leftarrow \frac{1}{B}\left\{(g_t^{e,i})' + g_{gausssian}(0,\sigma,C_t)\right\} \tag{3}$$

where

$$C_{t+1} = \begin{cases} C & , E[\tilde{g}^2]_{t+1} < G \\ \beta\sqrt{E[\tilde{g}^2]}_{t+1} & , E[\tilde{g}^2]_{t+1} > G \end{cases} \tag{4}$$

From the aforementioned, as the model converging, the local clipping threshold C_{t+1} decreases with $E[\tilde{g}^2]_{t+1}$, which makes the added noise $\xi^\sim_{gausssian}(0,\sigma,C_t)$ also smaller and smaller, and contributes to the convergence of the model.

2.3 Framework Detail

The algorithmic framework involves three main steps.

a) In order to avoid users directly leaking their privacy to untrustworthy edge servers, to address the second type of privacy security threat, end devices first use PCC to eliminate the weakly correlated RSS fingerprint data, convert it to grayscale images, and use (ε, δ)-DP techniques to differentially scramble the converted grayscale images for protecting their own data privacy.

b) To address the first type of privacy security threat, during the model training process, the edge servers use the perturbed data for model training. They calculate the privacy budget loss, adjusts the degree of perturbation of the parameters gradient with the achieved fine-grained control of the noisy data, adaptively. Then the edge servers submit the sub-model parameters to the cloud server.

c) The cloud server receives the sub-model parameters from each edge servers, aggregates them, then sends the aggregated parameters to each edge servers for the next iteration of training.

In summary, the ACDP-Floc as a whole satisfies (ε, δ)-DP, and it is difficult for any attacker to obtain privacy of user's data and model parameters using malicious means such as differential analysis, model inversion attacks, and inference attacks.

3   Experimental Evaluation

In this section, we conduct comprehensive experiments on three pratical datasets to qualify the performance of ACDP-Floc, and set a FL method NO-DP based on SGD without privacy protection, DPSGD [16], DP-FLocEC [13], ICGC-DP [12] as the baselines. In the following, we initially describe the experimental settings which includes the experimental setup, FL framework, datasets and parameters settings. Additionally, we introduce the evaluation metrics. Finally, we compare the accuracy and time overhead of ACDP-Floc with the others.

3.1   Experimental Settings

Experimental Setup. The configuration used for the experimental procedure is as follows: the operating system is Windows 10, the RAM size is 32G, the GPU is GeForce RTX 3060, and the code is mainly based on the TensorFlow 2.0 framework.

FL Framework. In order to simulate the FL protocol of indoor positioning in edge computing environment, we construct an indoor positioning model using TensorFlow 2.0, and simulate two edge servers with the same data volume. Socket protocol is used to realize the communication between edge servers and cloud server. And the optimizer uses SGD.

Datasets. We verifiy the validity of the method on 3 publicly practical location datasets: Mall Area [17], Mall_Wi-Fi [17] and UIJIIndoorLoc [13].

Parameters Settings. The fixed cropping threshold C, the privacy budget $\varepsilon = 1.0$, and $\beta = 1.2$ in the adaptive optimization algorithm, the number of training rounds on dataset Mall Area, Mall_Wi-Fi and UIJIIndoorLoc are 1000, 500 and 500, the learning rate on methods NO-DP, ICGC-DP and ACDP-Floc are 0.015 while on DPSGD is 0.001.

3.2 Evaluation Metrics

Cosine Similarity (CS), CS measures how similar the two time series data are in terms of direction, and takes values in the range $[-1, 1]$, with values closer to 1 indicating that two time series data are more similar. The formulas of CS as follows:

$$CS(A, B) = \frac{A \bullet B}{||A|| * ||B||} \tag{5}$$

Here, A and B are two time series data. $\bullet$ represents the dot product (inner product) of the vectors. $||A|| * ||B||$ represent the L_2 norms of vectors A and B, respectively.

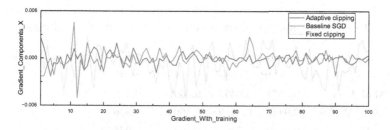

Fig. 4. Variation of random one gradient of the model with training

3.3 Gradient Descent

In purpose of measuring the utility of the proposed adaptive gradient clipping algorithm in model training, we compare the method adaptive clipping with the method fixed clipping. On the UIJIIndoorLoc dataset, we observe the historical gradient that is under the adaptive gradient clipping and the gradient clipping with a fixed threshold compared to the baseline SGD. Then we find ACDP-Floc to be better at retaining some information as shown in the Fig. 4.

Figure 4 illustrates the historical value that is the average value of a random gradient component in the CNN model with the above models of clipping. It is clear to gain that our method satisfies DP while retaining more gradient knowledge during model training.

Here, we use CS to measure the evolution in the gradient of the model by different clipping methods in Table 1.

As shown in Table 1, the CS between adaptive clipping, fixed clipping and N0-DP are 0.2882 and 0.0235. We know that compared with the fixed threshold method, the clipped gradient values and gradient changes of adaptive clipping method are more consistent with the original gradient, which makes the gradient retain more of the original information. The performance and effectiveness of the model are maintained while data privacy is protected.

Table 1. Cosine Similarity between NO-DP and other methods

Metrics	Between NO-DP and different method	
	DPSGD	DP-FLocEC
CS	0.2882	0.0235

3.4 Accuracy

To validate the performance of our method on practical datasets, we conduct experiments on three practical location datasets, and the accuracy of different methods on practical datasets is shown in Table 2.

Table 2. Accuracy comparisons on practical localization datasets

Dataset	Accuracy comparisons on practical localization datasets			
	NO-DP	DPSGD	ICGC-DP	ACDP-Floc
Mall Area	0.9371	0.6355	0.5697	0.9253
Mall_Wi-Fi	0.9442	0.7219	0.5694	0.9361
UIJIIndoorLoc	0.9689	0.7567	0.7037	0.9654

As shown in Table 2, on the Mall Area datasets, the accuracy of NO-DP and ACDP-Floc reach 0.9371 and 0.9253, respectively. But the DPSGD and ICGC-DP can only reach 0.6355 and 0.5697. On the Mall_Wi-Fi datasets, the accuracy of NO-DP and ACDP-Floc reach 0.9442 and 0.9461, respectively, but the DPSGD and ICGC-DP can only reach 0.7219 and 0.5694. On the UIJIIndoorLoc datasets, the accuracy of NO-DP and ACDP-Floc reach 0.9689 and 0.9654, respectively. But the classical DPSGD and ICGC-DP can only reach 0.7037. It should be noted that the accuracy significantly decreases and the model is unusable when we directly add Laplace noise to the weights of our model using the DP-FLocEC for privacy protection.

Table 2 illustrates that compared with NO-DP, which cannot ensure privacy security, our ACDP-Floc accuracy only loses 0.0118, 0.0041, and 0.0018 on the Mall Area, Mall_Wi-Fi, and UIJIIndoorLoc datasets, respectively. In addition, our method achieves significant improvements of 0.2893, 0.2142, 0.2087 and 0.3556, 0.3667, 0.2617 on the three practical datasets, respectively, when compared to the DPSGD and ICGC-DP, which also ensure privacy preservation. This is because our ACDP-Floc performs a finer-grained clip of the gradient and adds less noise in the same steps.

In addition, we evaluate the average time consumption of the ACDP-Floc, DPSGD, ICGC and NO-DP methods on the Mall Area, Mall_Wi-Fi and UIJIIndoorLoc datasets, respectively. As shown in Fig. 5, we observe the average time consumption of these methods over 500 training rounds on different datasets.

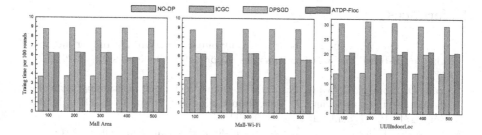

Fig. 5. Training time of different methods on three practical datasets

The time consumption of model training is inconsistent across different datasets. For example, in a 500-round training of the UIJIIndoorLoc dataset, the total time and time per round for training the NO-DP model are 114.10 min and 13.69 s. Due to gradient clip and noise addition, the total time and each round time for training the DPSGD and ICGC are 166.92, 253.4 min and 20.03, 30.41 s, respectively. Meanwhile, the total time and time per round for training the ACDP-Floc model are 171.95 min and 20.63 s. We conclude from Fig. 5 and Tab. 2 that the extra cost of time is negligible compared to the progress we have made in terms of accuracy.

4 Conclusion

In this paper, we propose an adaptive clipping differential private FL method which improves our previous work for wireless indoor location. First, we present an adaptive privacy protection framework for indoor localization. Specifically, we differentiate the gradient tensor of the model in training and obtain clipping factors for each gradient component, and then add adaptive noise based on these clipping factors. Finally, our ACDP-Floc pays more attention on the gradient clipping, resulting in more time spent in experiments. But we achieve an accuracy improvement of 37.2% compared to the ICGC-DP while spending 32.5% less time. Even compared with DPSGD, we achieve a 27.6% accuracy improvement while incurring only an additional 3.01% time cost. Therefore, we demonstrate that, ACDP-Floc has better performance in practical wireless indoor localization scenarios compared with other baselines in this paper. In the future, we will study the asynchronous FL to solve the communication problems caused by synchronous FL, and strive to improve the real-time performance and effectiveness of wireless indoor localization services.

Acknowledgements. This work was supported by the National Natural Science Foundation of China under grant number 61762058, Education Industry Support Plan of Gansu Provincial Department under grant number 2022CYZC-38 and the Natural Science Foundation of Gansu Province under grant number 21JR7RA282.

References

1. Zhu, X., Qu, W., Qiu, T., Zhao, L., Atiquzzaman, M., Wu, D.: Indoor intelligent fingerprint-based localization: principles. Approaches Challenges **22**(4), 2634–2657 (2020)
2. Yan, D., Song, W., Wang, X., Hu, Z.: Domestic indoor positioning technology development status review. J. Navigation Positioning **7**(04), 5–12 (2019). https://doi.org/10.16547/j.cnki.10-1096.20190402
3. Nagia, N., Rahman, M.T., Valaee, S.: Federated learning for wifi fingerprinting. In: ICC 2022 - IEEE International Conference on Communications, pp. 4968–4973 (2022). https://doi.org/10.1109/ICC45855.2022.9838945
4. Cheng, X., Liu, T., Shu, F., Ma, C., Li, J., Wang, J.: Providing location information at edge networks: a federated learning-based approach. IEEE Netw. **36**(5), 114–120 (2022). https://doi.org/10.1109/MNET.001.2200212
5. Gao, B., Yang, F., Cui, N., Xiong, K., Lu, Y., Wang, Y.: A federated learning framework for fingerprinting-based indoor localization in multibuilding and multifloor environments. IEEE Internet Things J. **10**(3), 2615–2629 (2023). https://doi.org/10.1109/JIOT.2022.3214211
6. Liu, Y., et al.: ML-Doctor: holistic risk assessment of inference attacks against machine learning models. In: 31st USENIX Security Symposium (USENIX Security 2022), pp. 4525–4542. USENIX Association, Boston, MA (2022)
7. Nieminen, R., Järvinen, K.: Practical privacy-preserving indoor localization based on secure two-party computation. IEEE Trans. Mob. Comput. **20**(9), 2877–2890 (2021). https://doi.org/10.1109/TMC.2020.2990871
8. Shen, X., Liu, Y., Zhang, Z.: Performance-enhanced federated learning with differential privacy for internet of things. IEEE Internet Things J. **9**(23), 24079–24094 (2022). https://doi.org/10.1109/JIOT.2022.3189361
9. Jiang, B., Li, J., Wang, H., Song, H.: Privacy-preserving federated learning for industrial edge computing via hybrid differential privacy and adaptive compression. IEEE Trans. Industr. Inf. **19**(2), 1136–1144 (2023). https://doi.org/10.1109/TII.2021.3131175
10. Koskela, A., Honkela, A.: Learning rate adaptation for differentially private learning. In: Proceedings of the Twenty Third International Conference on Artificial Intelligence and Statistics, vol. 108, pp. 2465–2475. PMLR (2020)
11. Xu, Z., Shi, S., Liu, A.X., Zhao, J., Chen, L.: An adaptive and fast convergent approach to differentially private deep learning. In: 2020 IEEE Conference on Computer Communications, pp. 1867–1876 (2020). https://doi.org/10.1109/INFOCOM41043.2020.9155359
12. Ma, C., Kong, X., Huang, B.: Image classification based on layered gradient clipping under differential privacy. IEEE Access **11**, 20150–20158 (2023). https://doi.org/10.1109/ACCESS.2023.3249575
13. Xuejun, Z., Fucun, H., Jiyang, G., Junda, B., Haiyan, H., Xiaogang, D.: A differentially private federated learning model for fingerprinting indoor localization in edge computing. J. Comput. Res. Dev. **59**(12), 2667–2688 (2022)
14. Yang, Z., Järvinen, K.: The death and rebirth of privacy-preserving wifi fingerprint localization with paillier encryption. In: IEEE INFOCOM 2018 - IEEE Conference on Computer Communications, pp. 1223–1231 (2018)
15. Yang, X., Luo, Y., Xu, M., fu, S., Chen, Y.: Privacy-preserving wifi fingerprint localization based on spatial linear correlation. In: Wireless Algorithms, Systems,

and Applications: 17th International Conference, WASA 2022, Dalian, China, 24–26 November 2022, Proceedings, Part I, pp. 401–412 (2022). https://doi.org/10.1007/978-3-031-19208-1_33

16. Abadi, M., Chu, A., Goodfellow, I., McMahan, H.B., Mironov, I., Talwar, K.: Deep learning with differential privacy. In: Proceedings of the 2016 ACM SIGSAC Conference on Computer and Communications Security, CCS 2016, pp. 308–318. Association for Computing Machinery, New York (2016). https://doi.org/10.1145/2976749.2978318

17. Zhang, X., et al.: A differentially private indoor localization scheme with fusion of wifi and bluetooth fingerprints in edge computing. Neural Comput. Appl. **34**(6), 4111–4132 (2022). https://doi.org/10.1007/s00521-021-06815-9

Label-Only Membership Inference Attack Against Federated Distillation

Xi Wang[1], Yanchao Zhao[1(✉)], Jiale Zhang[2], and Bing Chen[1]

[1] College of Computer Science and Technology, Nanjing University of Aeronautics and Astronautics, Nanjing 211106, China
yczhao@nuaa.edu.cn
[2] School of Information Engineering, Yangzhou University, Yangzhou 225009, China

Abstract. Federated learning is a prevailing distributed machine learning paradigm that aims to protect data privacy by training models locally. However, it is still vulnerable to various attacks, such as federated membership inference attacks, which can reveal the data or model information of the participants. To prevent these attacks, some protection measures have been proposed, such as data encryption/distortion or federated distillation (FD). It is a more sophisticated framework that communicates logits instead of model parameters, which can enhance the resistance to attacks. Nevertheless, in this paper, we investigate membership inference attacks in FD and demonstrate that malicious users can still infer the membership status and even reconstruct the data of the clients in FD. Moreover, we design two black-box membership inference attacks against FD and improve the attack accuracy by using data reconstruction as a pre-attack. Our experimental results show that even without access to model gradients in FD, our method can achieve over 80% attack accuracy on the server side for the EMNIST and CIFAR-100 datasets. We also show that our method can boost attack effectiveness by incorporating data reconstruction as a pre-attack.

Keywords: Federated distillation · membership inference attack · data reconstruction

1 Introduction

Nowadays, data privacy protection is becoming increasingly critical. In traditional machine learning, large volumes of data are required for model training. However, direct transmission of such sensitive data leads to serious privacy breaches. In contrast, federated learning (FL) [23], a distributed model training framework, offers a more secure solution where each data participant trains their model locally and uploads their gradient parameters to the central server in each communication round while holding native data locally. This distributed learning approach can almost achieve the accuracy of centralized machine learning, while effectively avoiding direct data transfer.

Nonetheless, the transmission of gradient parameters during federated learning training may still put privacy at risk. For example, adversaries can obtain the target gradient and carry out data reconstruction attacks based on the gradient, such as using generative adversarial networks (GANs) [15] for data reconstruction based on the gradient passed by the model [2,29]. More seriously, the adversary can launch a membership inference attack as a training participant, which seriously leaks the user's private data [16].

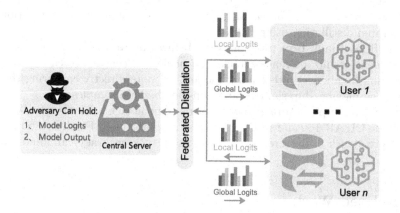

Fig. 1. Passive and curious servers in federated distillation.

Correspondingly, federated distillation (FD) [10,11], combined with knowledge distillation technology [8], avoids direct transmission of gradients containing sensitive information, thus playing a critical role in defense against attacks. As shown in Fig. 1, in FD, the malicious server can only obtain the uploaded logits and output labels of the target model. Thus, so other forms of federated learning architectures struggle to defend against membership inference attacks if such attacks were to occur during federated distillation. However, to the best of our knowledge, few works have investigated privacy and security issues in federated distillation.

To this end, we present two server-side initiated membership inference attacks, each capable of achieving user-level inference granularity, which initiates membership inference from the server side.

The two attack methods are briefly described below.

(A1.) Shadow Model Attack: Assuming that the adversary server possesses a shadow dataset and has trained a shadow model locally, with a data distribution that is identical to that of the target client, the server can leverage the shadow model to perform a membership inference attack on the target client.

(A2.) Adversarial Attack: Compared to shadow model attacks, the challenge of malicious server-launching attacks is more severe. Specifically, attackers no longer rely on shadow models to launch membership inference attacks. Instead, the attacker directly exploits the gap between member and non-member samples that fit in the same model. By adding noise to the samples using an adversarial approach [3], the malicious server can capture the gap between member samples and non-member samples.

Our contributions can be summarized as follows.

- **Data reconstruction against the federated distillation:** We find that although logit parameters cannot be directly used for data reconstruction, the shadow model established by logit parameters has a close fit to the data of the target model. Based on this, we implement data reconstruction for federated distillation using GANs.
- **Excellent black-box membership inference attacks:** We propose that two membership inference attacks can achieve excellent attack accuracy by building highly accurate shadow models, or more directly capturing the gap between membership samples and non-membership samples.
- **Extensive experimental evaluations:** Our attack achieves over 80% and 85% on CIFAR-100 and EMNIST datasets, respectively. Further, after adding the pre-attack of data reconstruction, the attack accuracy can almost reach 90%.

2   Related Work

In this section, we expound on membership inference attacks, separately in centralized and distributed scenarios, in addition to data reconstruction attacks predicated on gradient implementations. Meanwhile, we also briefly consider certain defense mechanisms.

2.1   Membership Inference Attacks

When a target model is trained on a sensitive dataset, membership inference attacks (MIA) can be employed by adversaries to determine whether a target sample belongs to the said private dataset. In centralized machine learning, Shokri et al. [18] were the pioneers of black-box membership inference attacks. They employed multiple shadow models to train an attack model and utilized it to identify member samples of the target model. Li et al. [13] proposed two variants of membership inference attacks, which take advantage of the model's overfitting on member samples to achieve favorable MIA outcomes in the black-box scenario. In federated learning, Shokri et al. [16] extended the prior research and proposed the white-box MIA in both federated and centralized scenarios. They introduced a novel MIA method based on gradient parameters, where each training data is associated with a unique trace on the gradient of the loss function, and the attacker can actively exploit this information by manipulating the gradient of a dataset to infer participant data. Yang et al. [24] proposed a membership inference attack on federated distillation, whereby the attacker conducts membership inference on other participants as a local participant.

At present, the extant membership inference attacks primarily focus on conventional federated learning models [2,9,14,16], as well as centralized machine learning models [2,12,13,19,25].

2.2   Data Reconstruction Attack

In the federated learning architecture, Zhu et al. [29] discovered that an adversarial server could reconstruct data through the gradient of the initial target model, resulting in entirely random reconstructed data. To address this issue, improved DLG (iDLG) was developed by Zhao et al. [27] based on the properties of the activation function used in DLG. Wang et al. [21] improved GANs to reconstruct data while also discerning the properties of target clients and their own properties. Boenisch et al. [1] proposed a method where the adversarial server restores the client's data by manipulating the gradient received from the client. The client uses the tampered gradient to train with only a small number or even a single sample in each batch in the most extreme cases. The next time gradient upload only contains the training information of these actual samples, making it susceptible to data reconstruction attacks by an adversary server. Geiping et al. [4] developed a method to improve the loss function used for data reconstruction by gradients through cosine similarity. Lastly, Yin et al. [26] introduced an algorithm for gradient inversion using the last fully connected layer and a regular term to ensure image quality based on multi-seed optimization.

2.3   Defense Method Against Privacy Leakage

Federated learning is faced with so many attack methods, defense is particularly important. For example, Xu et al. [22] utilized homomorphic encryption to avoid potential privacy breaches due to data reconstruction attacks from adversaries. However, in order to achieve an optimal level of privacy protection, a larger amount of encrypted data would be necessary, increasing the computational burden significantly. On the other hand, Geyer et al. [5] have introduced differential privacy to inject noise into parameters, while the degree of privacy security could diminish the performance of the model. Finding the proper balance between these two factors is challenging. In light of this, Sun et al. [20] proposed a novel approach to federated learning that adds a defensive layer to the network structure in order to address the issue of data entanglement which could drastically reduce data quality. Shejwalkar et al. [17] utilizes reference data either generated or selected to train the model, and ultimately utilizes knowledge distillation to obtain a protected model. The key idea of [17] is to shorten the gap between member samples and non-member samples.

Based on the above research work, there is almost no work discussing membership inference against federated distillation, and there is also no attempt to perform data reconstruction against federated distillation. In normal federated learning, these two kinds of attacks are often achieved with the help of model gradients. However, in federated distillation, the attacker does not have access to additional information beyond the model output labels and logits parameters. Moreover, there is also a research gap for the defense work in FD.

3 Preliminaries

In this section, we present the background, including the federated distillation pipeline, generative adversarial networks, and the threat model in attack methods.

3.1 Federated Distillation

Knowledge distillation [7] serves as an efficient approach to compress large-scale models by facilitating knowledge transfer from a complex model to a simpler one. This process entails the introduction of novel parameters into the widely-used softmax layer in neural networks, as follows:

$$q_i = \frac{\exp\left(z_i/T\right)}{\sum_j \exp\left(z_j/T\right)} \tag{1}$$

Here, the probability of the i-th class is represented by q_i, z_i denotes the i-th logits parameter (i.e., input to the softmax layer) and T represents the temperature parameter. A higher value of T results in a smoother probability distribution of the softmax layer output.

Further, federated distillation (FD) [10], which combines federated learning with knowledge distillation, can transfer information between federated models. Essentially, it is a form of federated learning that employs knowledge distillation to transfer parameters. Following pre-training on a public dataset, participants train their models locally using local data and cross-entropy, as depicted:

$$L_{CE} = \frac{1}{n}\sum_{i=1}^{n}L_i = -\frac{1}{n}\sum_{i=1}^{n}\sum_{c=1}^{C} y_{ic}\log\left(p_{ic}\right) \tag{2}$$

where n is the number of samples, C is the number of categories, y_{ic} is the classification result of sample i on category c, for category c then go to 1, and vice versa 0, p_{ic} is the probability that sample i is belongs to category c. After local training is completed, participants upload logit values normalized by the softmax function for each training round to the server. Next, the server receives all parameters, aggregates, and averages before sending them to each client. The aggregation formula can be expressed as:

$$Z_t = \frac{1}{K}\sum_{k=1}^{K} Z_{t-1}^k \tag{3}$$

where Z_t is the global logits trained at the t-th iteration, Z_{t-1}^k is the local logits uploaded by user k at the iteration of $t-1$, and K is the number of participant models. Each client can be considered as a student model independently, while the other participants are teacher models, and the average of other teacher models is obtained for learning in each iteration of training. The model uses cross entropy as a loss regularizer for students to measure the difference between the teacher and student models. FD does not train the global model on the server side, and each participant ends up with its local model as the final model.

3.2 Generative Adversarial Networks

Generative adversarial networks (GANs) [6] are generative models trained in an adversarial manner and contain two competing neural network models: a discriminator $\mathcal{D}$ and a generator $\mathcal{G}$. The $\mathcal{G}$ generates random samples (with random noise, e.g., gaussian or uniform distribution) as input from z. In contrast, $\mathcal{D}$ discriminates between actual samples and samples generated by $\mathcal{G}$. The loss of $\mathcal{D}$ on different samples will feed back to $\mathcal{G}$. The objective function of GANs can describe as follows:

$$
\min_{\mathcal{G}} \max_{\mathcal{D}} V(\mathcal{D}, \mathcal{G}) = E_{x \sim P_{data}(x)}[\log(\mathcal{D}(x))]
$$
$$
+E_{z \sim P_z(x)}[\log(1 - \mathcal{D}(\mathcal{G}(z)))]
$$
(4)

where P_{data} and P_z are the data distributions, x represents real data, while z represents random data with the same structure as x, $\mathcal{D}$ is the discriminator that utilizes the parameters of the shadow model, and $\mathcal{D}(x)$ represents the discriminant result of the discriminator on x, when $\mathcal{D}$ and $\mathcal{G}$ reach Nash equilibrium, GNAs can generate samples that approximate the actual samples.

3.3 Threat Model

Here, we elaborate on the conditions of the adversary.

Adversary's Objectives. FD diverges from the traditional practice of training the global model on the central server. In this method, the server is merely responsible for parameter aggregation and distribution. By taking advantage of a curious and passive server, the attacker succeeds in securing the parameters from the target client. The final objective is to ascertain whether the sample under review falls within the private training data of the target client. We assess the effectiveness of our attack model using two essential metrics: 1) the efficacy of the local model and 2) the accuracy of inference performed on the server side.

Adversary's Observations. The adversary has access to the logits parameter uploaded by the local model and can make black-box queries to each local model in every training iteration. Specifically, the attacker can navigate the target model and retrieve the output results of the model. In federated learning's training setup, the client model is usually delivered by the server, and therefore, it is natural for the server to exercise black-box access to the client model in such an environment.

Adversary's Capabilities. The adversarial server has the capability to construct a shadow model on its own side and maintain a corresponding shadow dataset, given that the shadow data is characterized by the same data distribution as the local data of the target model, but not sharing any overlapping instances. The adversary's server-generated shadow model bears a resemblance in structure to that of the client-local model.

4   Proposed Membership Inference Attack in FD

In this section, we describe the detail of the membership inference attacks in federated distillation. Specifically, we focus on the attacker as a curious and passive server-side launching the attack.

4.1   Pre-attack Based on Data Reconstruction

In the reconstruction method, adversary server $\mathcal{A}$ uses shadow model M_s to build a GANs model to achieve the accurate reconstruction of the local data of the target model. GANs consist of a discriminative network $f(x; \theta_D)$ and a generation network $g(z; \theta_G)$. The generator $\mathcal{G}$ is initialized with random noise, and the discriminant $\mathcal{D}$ is initialized with the parameters of the shadow model. After model initialization, $\mathcal{G}$ sends the generated samples and training samples (taken from the shadow dataset) to $\mathcal{D}$, and $\mathcal{D}$ feeds back the discrimination results to $\mathcal{G}$. Let x_i be the original image in the training set, x_{gen} are generated images. The optimization algorithm can be described as:

$$\min_{\theta_{\mathcal{G}}} \max_{\theta_{\mathcal{D}}} \sum_{i=1}^{n_+} \log(f(x_i; \theta_{\mathcal{D}})) + \sum_{i=1}^{n_-} \log(1 - f(g(x_{gen}; \theta_{\mathcal{G}}); \theta_{\mathcal{D}})) \tag{5}$$

$$\mathcal{L}_{\mathcal{G}}(\theta_g) = \mathbb{E}_{z \sim p(z)}[\log(\mathcal{D}(\mathcal{G}(z)))] \tag{6}$$

where x_i is the training sample and x_{gen} is the generated sample. After iterating N rounds to complete data reconstruction, we will add the reconstructed data to the dataset to be tested and mark the reconstructed samples as member samples. Compared with starting the membership inference attack directly after the training is completed, performing data reconstruction as a pre-attack is necessary to improve inference accuracy.

Shadow Model. The majority of our methods are heavily dependent on the shadow model implemented by the server. This entails that the server tends to adopt a structure akin to the intended target model. Conversely, Shokri et al. [18] employed several shadow models to emulate the intended target model, leveraging the data obtained from these models to train the final attack model once the shadow model training process concluded. Our strategy, on the other hand, only necessitates one shadow model for the attack to be successful.

In this regard, we utilize the shadow model in both the data reconstruction and membership inference operations. By conducting numerous experiments, we observed that:

- The larger the shadow dataset is, the better the shadow model imitates, but it requires more frequent access to the target model.
- We use the logits of the target model to participate in the training of the shadow model, and we found that this step is necessary to achieve the current experimental results.

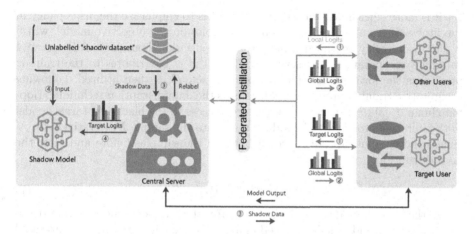

Fig. 2. Overview of shadow model training process in FD. In each iteration of the training round: ① all local clients upload model logits (server saves logits of attack targets); ② server sends down global logits; ③ server queries target model using samples from shadow dataset to obtain model output and marks shadow dataset; ④ server trains shadow model using saved logits and shadow dataset.

4.2   Attack Overview

Shadow Model Attack. As depicted in Fig. 2, we illustrate the training process of the shadow model in a federated distillation environment. In this setting, there are K participants, one of which is the victim $\mathcal{V}$, while the adversary $\mathcal{A}$ is a curious but passive server. $\mathcal{A}$ maintains a shadow dataset $\mathcal{S}$ that shares the same data distribution as $\mathcal{V}$'s local data and a corresponding shadow model. During training's k-th iteration, $\mathcal{V}$ uploads the logits parameter x of the local model to $\mathcal{A}$. Then, $\mathcal{A}$ utilizes the data in the shadow dataset to query the target model in this iteration, then leverages the obtained query results to label $\mathcal{S}$. The k-th iteration concludes after utilizing the labeled $\mathcal{S}$ and x to train the shadow model. Eventually, the iterative training results in the simultaneous training of the shadow model locally trained by $\mathcal{A}$.

After the successful training of the shadow model, $\mathcal{A}$ applies the model to reconstruct the training examples of the target model. Notably, this process is non-parametric and shall be further elaborated on later. The reconstructed samples are subsequently included in the test dataset and processed by the shadow model alongside other samples. Here in, samples with a cross-entropy loss exceeding the predetermined threshold t are categorized as member samples, while those below t are deemed non-member samples.

Adversarial Attack. This attack is more strenuous compared to the prior one as the adversary is unable to retain both the shadow dataset and shadow model, yet it is more efficacious. The central concept is to emphasize the discrepancy in the fitness of member and non-member samples on the objective model, which is the discrepancy in the distance from the decision boundary. Adversary $\mathcal{A}$ continues to operate as the server, while victim $\mathcal{V}$ contributes to the training process as the local model. The test sample set, D, is then utilized. Following the iterative training procedure, $\mathcal{A}$ leverages the data disruption technique, Hop-SkipJump $\mathcal{F}$ [3] ($\mathcal{F}$ necessitates unfettered access to the black-box target model in the iterative process), to introduce noise into the sample D_i iteratively for $\mathcal{N}$ times, and compute the distances d between the samples before and after the data disruption:

$$d = \|\mathcal{W}' - \mathcal{W}\|^2 \tag{7}$$

where the $\mathcal{W}$ and $\mathcal{W}'$ disorder before and after the samples respectively. Furthermore, the threshold y is determined empirically, we compare d to the threshold and label the samples with d greater than y as member samples, and vice versa as non-member samples.

4.3 Attack Algorithm

The level of data correspondence achieved by the shadow model is already in close proximity to that of the target model. Subsequently, the adversarial server shall submit the specimens, which require testing, to the shadow model. In the event of earlier data reconstruction, the reconstructed specimens will be appended to the testing specimen group. After acquiring the output of the model, we shall categorize the specimens that display cross-entropy loss exceeding the threshold value as "OUT" i.e., non-member specimens; alternatively, specimens with less loss shall be categorized as "IN" i.e., member specimens. Algorithm 1 outlines the attack detail.

After the training phase is complete, the adversarial server initiates the attack. Leveraging the disparity between the member and non-member specimens, we can employ [3] to disturb the samples in a manner that brings them close to the decision boundary. This, in turn, enables us to determine the distance prior to and post perturbation. To enhance the efficiency of the attack technique, we determine a suitable iteration count through rigorous experimentation. The addition of noise is not contingent upon it resulting in a change in the data outcome of the target model. In fact, following N iterations, the majority of sample labels fail to alter. Following the assessment of the variation in distance before and after perturbation, we classify samples with less than

---

Algorithm 1 Shadow Model Attack.

Require: Test dataset D (size $= m$), shadow model M_s, threshold y.
Ensure: The inference result 'IN' or 'OUT'.
 1: **for** $(i = 1; i <= m; i++)$ **do**
 2:    Input test sample D_i into M_s
 3:    M_s output cross-entropy loss L_i
 4:    **if** $L_i < y$ **then**
 5:       Mark D_i as 'IN'
 6:    **else**
 7:       Mark D_i as 'OUT'
 8:    **end if**
 9: **end for**
10: **Output:** Mark every record as 'IN' or 'OUT', where 'IN'represents the Victim's training sample.

---

the stipulated threshold distance loss as "OUT" i.e., non-member samples, and those exceeding the threshold as "IN" i.e., member samples. In the experimental phase of the adversarial attack, we set the query count at 40 without testing for the alteration of the sample label, thus launching a membership inference attack. The adversarial attack we generalize to Algorithm 2.

Algorithm 2 Adversarial Attack.

Require: Test dataset D (size $= m$), $HopSkipJump$ $\mathcal{F}$, iterations N, threshold y.
Ensure: The inference result 'IN' or 'OUT'.
 1: **for** $(i = 1; i <= m; i++)$ **do**
 2:    Record sample D_i parameters as W
 3:    **for** $(j = 1; j <= N; j++)$ **do**
 4:       Use $\mathcal{F}$ to disrupt D_i (add noise)
 5:    **end for**
 6:    Record sample D_i parameters as W'
 7:    $d = \|W' - W\|^2$
 8:    **if** $d > y$ **then**
 9:       Mark D_i as 'IN'
10:    **else**
11:       Mark D_i as 'OUT'
12:    **end if**
13: **end for**
14: **Output:** Mark every record as 'IN' or 'OUT', where 'IN' represents the Victim's training sample.

---

5   Performance Evaluation

In this section, we evaluate our proposed methods, including data reconstruction and membership inference attacks.

5.1   Datasets and Evaluation Goals

We conducted federated distillation on four distinct datasets. Our experimentation involved utilizing CIFAR-10 and MNIST as public datasets, and CIFAR-100 and EMNIST as the corresponding private datasets. Additionally, we carried out experiments in both the I.I.D and NonI.I.D [28] contexts for each dataset.

Public Datasets

- **MNIST**: The aforementioned dataset is a collection of digital images consisting of ten distinct character categories (i.e., ten different numbers). The training set comprises 60,000 images with corresponding labels, while the test set consists of 10,000 images with labels. Both sets possess a resolution of 28 × 28 pixels.
- **CIFAR-10**: The dataset referred to consists of a training data set and a test data set comprising 60,000 and 10,000 images, respectively. Each image is 32 × 32 pixels and comprises three channels, with ten classes in total. The images primarily feature animals such as cats, dogs, and horses, among others.

Private Datasets

- **EMNIST**: The aforementioned dataset comprises a total of 62 character categories, including ten digits, 26 lowercase letters, and 26 uppercase letters. All images in the dataset possess a resolution of 28 × 28 pixels, resulting in a total of 814,255 samples.
- **CIFAR-100**: The dataset consists of 100 categories of colored images, such as animals, flowers, planes, and people, each of which is 32 × 32 in size. Each category contains 600 images.

To comprehensively assess our attack methodology, we establish the following benchmarks: 1) local model accuracy: the accuracy of the indigenous model in FD; 2) membership inference accuracy: the precision of the membership inference attack under the federated distillation circumstance; 3) the improvement of inference accuracy by reconstructing the samples.

5.2   Experimental Settings

We introduce a certain degree of model heterogeneity, and the shadow model also chooses a model structure similar to the target model for the reasons we have mentioned before. To further validate the inference, we conducted experiments with 3 to 10 participants for different scenarios and 10 participants for different data distributions. The structure of the local model contains three convolutional layers, each followed by a pooling layer, and finally a fully connected and aggregated layer, where we chose to give a 3×3 convolutional kernel with ReLU as the activation function. To ensure the accuracy of the local model, we use only 8 and 4 samples per local class as private datasets in the case of using the EMNIST dataset and the CIFAR-100 dataset, respectively. Since the sample is not large, only 20 iterations of training are needed.

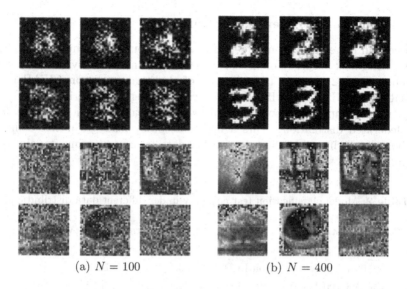

(a) $N = 100$ (b) $N = 400$

Fig. 3. Reconstruction based on GANs.

Table 1. The enhancement effect of data reconstruction on membership inference

Data Reconstruction		Shadow Model Attack	Adversarial Attack
EMNIST & MNIST	Before	82.0%	85.1%
	After	**85.9%**	**89.7%**
CIFAR100 & CIFAR10	Before	74.9%	79.7%
	After	**80.1%**	**84.6%**

5.3 Performance of Data Reconstruction

In order to showcase the viability of our data reconstruction methodology, we reconstructed the data in the case of five participants. We constructed a GANs model on the server side, utilizing the parameters of the shadow model to construct the discriminator $\mathcal{D}$, and random parameters for constructing the generator $\mathcal{G}$ in the GANs. As depicted in Fig. 3, we present the reconstructed samples generated from continuous iterations, on the left after 100 iterations, and on the right after 400 iterations. It is observable that the reconstructed images on the right can be roughly distinguished.

After integrating the reconstructed samples into the sample set for testing purposes, we evaluated the performance of both attack methods, as presented in Table 1. Our experimental results indicate that the accuracy of the membership inference attack can be further enhanced by reconstructed samples. Based on these findings, we infer that the inference accuracy can be further augmented by implementing GANs on the server side.

5.4   Performance of Membership Inference

To ascertain the efficacy of membership inference, our main emphasis is on two aspects: the task accuracy of the local model and the accuracy of membership inference. Table 2 presents the accuracy of the local model on CIFAR-100 & CIFAR-10, EMNIST & MNIST datasets with varying data distributions. Specifically, when compared to conventional federated learning, our results indicate that federated distillation incurs a higher accuracy loss. We utilize a limited number of experimental samples to gather data for our experiments, aiming to maximize model accuracy.

Table 2. Model classification test accuracy under different data distributions

Data Distribution	Top Accuracy	Mean Accuracy
EMNIST & MNIST I.I.D	85.3%	83.7%
EMNIST & MNIST Non I.I.D	80.1%	78.5%
CIFAR-100 & CIFAR-10 I.I.D	75.2%	74.2%
CIFAR-100 & CIFAR-10 Non I.I.D	72.1%	71.6%

We evaluate the adversary server's inference accuracy against target participants with varying participant numbers, ranging from 3 to 10, and conduct our experiments using two datasets with disparate data distributions. For this subsection's experiments, we select a test dataset that does not contain reconstructed samples, i.e., the current inference accuracy of the adversary server is obtained without recourse to data reconstruction. As depicted in Fig. 4, the accuracy of the adversarial attack is higher than that of the shadow model attack in most cases since the latter tends to lose some information during the training process of the shadow model. We also observe that a smaller participant number leads to higher inference accuracy. This outcome is likely due to the broader global influence of the target model, resulting in higher information exposure risk with fewer participants. Regarding datasets, we note that attacks perform better on EMNIST & MNIST datasets in most cases. Nevertheless, the more simple MNIST dataset does not predispose the model towards overfitting in the NonI.I.D case of adversarial attack, resulting in lower inference accuracy.

To establish the superiority of our attack, we employ a comparative analysis with the work of FD-Leaks [24], whereby we subject the adversarial attack to identical conditions. As illustrated in Fig. 5, the results evince that the shadow model attack surpasses FD-Leaks when exposed to the independent and identically distributed condition. Conversely, our adversarial attack outmatches FD-Leaks across diverse data distribution and dataset settings. FD-Leaks initiate attacks from the client side, while our work conducts attacks from the server side. The amalgamation of both strategies yields conspicuous evidence of the federated distillation's seemingly impregnable defense architecture being vulnerable to significant privacy risks.

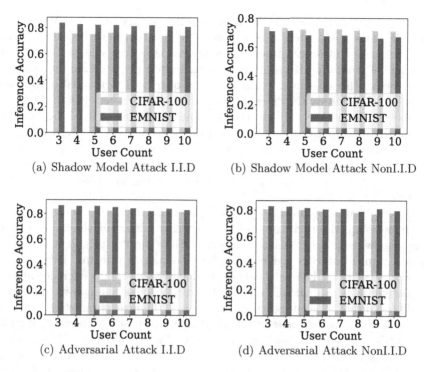

Fig. 4. Inference accuracy of the two attacks, where the dataset setting and data distribution are different.

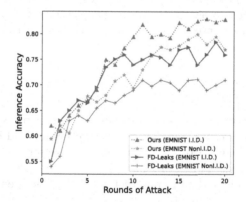

Fig. 5. Comparison of inference accuracy between adversarial attack and FD-Leaks [24].

6   Discussion

Federated distillation leverages parameter transfer of logit parameters to facilitate model training, and this parameter transfer is deemed as an inherent safeguard measure against privacy leaks. However, our work brings to light the inadequacy of logit parameter transfer in preventing privacy breaches. We demonstrate that logit parameters remain susceptible to data reconstruction attacks and membership inference attacks. The adversarial server side can bolster the target model's imitation by utilizing logit parameters derived from the uploaded parameters of the local model and continual queries. The foundation of these tactics lies in exploiting the divergence in sample fitting across identical models, whereby the incomplete data distribution and overfitting of local data exacerbate the fitting disparity in federated learning. Therefore, the adversary server side can execute membership inference attacks under both passive and adversary conditions. Worse still, the server can isolate the target model comprehensively under the active and adversary settings, culminating in catastrophic privacy breaches.

To forestall privacy breaches in our experiments, local clients can constrain the number of queries to a bare minimum, thereby mitigating the establishment of the shadow model to a considerable extent. However, this technique also poses practical implementation challenges. The fundamental principle behind using shadow models and adversarial attacks to devise membership inference attacks lies in exploiting the fitting discrepancy between member and non-member samples. In the future, we can incorporate noise into the logit parameters, albeit at the cost of wider model accuracy. Given that the accuracy of federated distillation models is already low, such an approach may prove untenable. Nevertheless, one plausible solution to reducing the fitting gap between member samples and non-member samples is to train using generated data to complement the data distribution. For instance, DMP [17] partially supplements the data distribution of the training data by selecting or creating eligible reference data and leveraging knowledge distillation to obtain the ultimate defense model. This approach not only delivers robust defenses against privacy breaches but can also enhance the model's accuracy to a certain extent.

7 Conclusion

This paper proposes two kinds of membership inference attacks against federated distillation, and a corresponding data reconstruction method leveraging GANs. The attackers are assumed to be curious and passive servers, operating within a black-box access framework. To further refine our approach, we introduce the concept of employing data reconstruction within federated distillation using GANs, with the reconstructed data serving as a pre-attack mechanism to improve the overall accuracy of the membership inference attacks. Experimental evaluation of our approach utilized CIFAR-10/MNIST and CIFAR-100/EMNIST as public and private datasets respectively. By examining the impact of varying data distributions and participant numbers, we offer insight into factors influencing experimental outcomes. Empirical results indicate that sample reconstruction improves inference accuracy.

Acknowledgement. This work was supported in part by the National Natural Science Foundation of China under Grant(No. 62172215) and in part by the Natural Science Foundation of Jiangsu Province(No. BK20200067), in part by the A3 Foresight Program of NSFC (Grant No. 62061146002).

References

1. Boenisch, F., Dziedzic, A., Schuster, R., Shamsabadi, A.S., Shumailov, I., Papernot, N.: When the curious abandon honesty: federated learning is not private. arXiv preprint arXiv:2112.02918 (2021)
2. Chen, J., Zhang, J., Zhao, Y., Han, H., Zhu, K., Chen, B.: Beyond model-level membership privacy leakage: an adversarial approach in federated learning. In: Proceedings of ICCCN, pp. 1–9 (2020)
3. Chen, J., Jordan, M.I., Wainwright, M.J.: Hopskipjumpattack: a query-efficient decision-based attack. In: Proceedings of SP, pp. 1277–1294 (2020)
4. Geiping, J., Bauermeister, H., Dröge, H., Moeller, M.: Inverting gradients-how easy is it to break privacy in federated learning? In: Proceedings of NeurIPS, pp. 16937–16947 (2020)
5. Geyer, R.C., Klein, T., Nabi, M.: Differentially private federated learning: a client level perspective. arXiv preprint arXiv:1712.07557 (2017)
6. Goodfellow, I., et al.: Generative adversarial networks. Commun. ACM **63**(11), 139–144 (2020)
7. Gou, J., Yu, B., Maybank, S.J., Tao, D.: Knowledge distillation: a survey. Int. J. Comput. Vision **129**, 1789–1819 (2021)
8. Hinton, G., Vinyals, O., Dean, J., et al.: Distilling the knowledge in a neural network. arXiv preprint arXiv:1503.02531 (2015)
9. Hitaj, B., Ateniese, G., Perez-Cruz, F.: Deep models under the gan: information leakage from collaborative deep learning. In: Proceedings of the 2017 ACM SIGSAC Conference on Computer and Communications Security, pp. 603–618 (2017)
10. Jeong, E., Oh, S., Kim, H., Park, J., Bennis, M., Kim, S.L.: Communication-efficient on-device machine learning: Federated distillation and augmentation under non-iid private data. arXiv preprint arXiv:1811.11479 (2018)
11. Li, D., Wang, J.: Fedmd: Heterogenous federated learning via model distillation. arXiv preprint arXiv:1910.03581 (2019)
12. Li, J., Li, N., Ribeiro, B.: Membership inference attacks and defenses in supervised learning via generalization gap. arXiv preprint arXiv:2002.12062 (2020)
13. Li, Z., Zhang, Y.: Membership leakage in label-only exposures. In: Proceedings of the 2021 ACM SIGSAC Conference on Computer and Communications Security, pp. 880–895 (2021)
14. Melis, L., Song, C., De Cristofaro, E., Shmatikov, V.: Inference attacks against collaborative learning. arXiv preprint arXiv:1805.04049 (2018)
15. Mirza, M., Osindero, S.: Conditional generative adversarial nets. arXiv preprint arXiv:1411.1784 (2014)
16. Nasr, M., Shokri, R., Houmansadr, A.: Comprehensive privacy analysis of deep learning: Passive and active white-box inference attacks against centralized and federated learning. In: Proceedings of SP, pp. 739–753 (2019)
17. Shejwalkar, V., Houmansadr, A.: Membership privacy for machine learning models through knowledge transfer. In: Proceedings of AAAI, pp. 9549–9557 (2021)
18. Shokri, R., Stronati, M., Song, C., Shmatikov, V.: Membership inference attacks against machine learning models. In: Proceedings of SP, pp. 3–18 (2017)

19. Song, L., Shokri, R., Mittal, P.: Privacy risks of securing machine learning models against adversarial examples. In: Proceedings of the 2019 ACM SIGSAC Conference on Computer and Communications Security, pp. 241–257 (2019)

20. Sun, J., Li, A., Wang, B., Yang, H., Li, H., Chen, Y.: Provable defense against privacy leakage in federated learning from representation perspective. arXiv preprint arXiv:2012.06043 (2020)

21. Wang, Z., Song, M., Zhang, Z., Song, Y., Wang, Q., Qi, H.: Beyond inferring class representatives: user-level privacy leakage from federated learning. In: Proceedings of INFOCOM, pp. 2512–2520 (2019)

22. Xu, W., Fan, H., Li, K., Yang, K.: Efficient batch homomorphic encryption for vertically federated xgboost. arXiv preprint arXiv:2112.04261 (2021)

23. Yang, Q., Liu, Y., Chen, T., Tong, Y.: Federated machine learning: concept and applications. ACM Trans. Intell. Syst. Technol. (TIST) 10(2), 1–19 (2019)

24. Yang, Z., Zhao, Y., Zhang, J.: Fd-leaks: Membership inference attacks against federated distillation learning. In: Proceedings of Web and Big Dat, pp. 364–378 (2023)

25. Yeom, S., Giacomelli, I., Fredrikson, M., Jha, S.: Privacy risk in machine learning: Analyzing the connection to overfitting. In: Proceedings of CSF, pp. 268–282 (2018)

26. Yin, H., Mallya, A., Vahdat, A., Alvarez, J.M., Kautz, J., Molchanov, P.: See through gradients: Image batch recovery via gradinversion. In: Proceedings of CVPR, pp. 16337–16346 (2021)

27. Zhao, B., Mopuri, K.R., Bilen, H.: idlg: improved deep leakage from gradients. arXiv preprint arXiv:2001.02610 (2020)

28. Zhao, Y., Li, M., Lai, L., Suda, N., Civin, D., Chandra, V.: Federated learning with non-iid data. arXiv preprint arXiv:1806.00582 (2018)

29. Zhu, L., Liu, Z., Han, S.: Deep leakage from gradients. In: Proceedings of NeurIPS, pp. 14774–14784 (2019)

Efficient Proactive Resource Allocation for Multi-stage Cloud-Native Microservices

Pengfei Liao[1], Guanyan Pan[2], Bei Wang[3], Xingzhen He[1], Wenbing Peng[2], Minhui Fang[2], Fanding Huang[4], Yifei Chen[4], and Yuxia Cheng[1(✉)]

[1] Hangzhou Dianzi University, Hangzhou, China
yxcheng@hdu.edu.cn
[2] Taizhou Urban and Rural Planning and Design Institute, Taizhou, China
[3] School of Computer Science, Zhejiang University, Hangzhou, China
[4] HDU-ITMO Joint Institute, Hangzhou Dianzi University, Hangzhou, China

Abstract. Microservices deployment in the cloud often faces a prevalent challenge: how to maximize resource utilization while maintaining high quality-of-service (QoS). Existing automatic scaling tools frequently exhibit limited adaptability, particularly when handling frequent request load fluctuations, which exacerbates the challenge. To address this issue, we introduce a proactive runtime deployment optimization method for multi-stage microservices, aiming to ensure both resource efficiency and QoS.

Our proposed method encompasses four interrelated modules–forecasting, constraint planning, judgment selection, and execution–which collaboratively work towards optimizing runtime resource allocation, generating viable deployment plans, and identifying cost-efficient solutions without compromising QoS. Through a set of experiments, we demonstrate that the proposed proactive deployment optimization method can potentially reduce computational resource usage by 35% while maintaining the desired quality of service.

Keywords: Microservice · QoS · resource utilization · runtime deployment optimization

1 Introduction

Software-as-a-Service (SaaS) has emerged as a prominent business model in which software applications are offered as on-demand cloud services, allowing users to access them anytime and anywhere via the Internet [7,27]. In this model, SaaS providers can swiftly launch services with minimal deployment and maintenance costs. To enhance economic efficiency, these providers must deliver efficient software application services while minimizing production cost investments [15,21].

Dynamic request frequency and fluctuating resource demands make it challenging to estimate resource requirements for SaaS services. Providers often

Z. Tari et al. (Eds.): ICA3PP 2023, LNCS 14488, pp. 411–432, 2024.
https://doi.org/10.1007/978-981-97-0801-7_24

reserve excess computing resources to ensure high availability [23,29]. However, this approach can lead to resource wastage, service performance degradation, and even service instance failures, ultimately causing economic losses. Existing research has explored dynamic resource management for optimization targets such as service performance, server cost, and service migration cost [4,8,9,16,28,36].

Many contemporary cloud computing applications have transitioned from monolithic application architectures to microservice architectures, as exemplified by companies like Google and Amazon [31]. Microservice architecture allows applications to be composed of loosely coupled program modules, enhancing scalability, elasticity, and responsiveness to business changes. However, the complex topology of dependencies among microservices increases deployment and management difficulties [11,37]. Existing management tools, such as Kubernetes, offer automatic scaling features based on machine or custom metrics. Nonetheless, reactive heuristic adjustments often fail to address service level degradation in a timely manner [10]. This approach usually results in partial adjustments and may lead to resource wastage. Consequently, data-driven approaches have been proposed to capture the impact of dependencies between microservices on performance [9,28,36].

In this paper, we propose a queuing network model to identify key microservices contributing to performance degradation in SaaS services. By predicting service request frequency and incorporating constraint planning, we provide an overall optimization scheme for SaaS services. Our approach evaluates the optimization scheme's ability to guarantee SaaS service quality while considering the cost of computing resource usage and potential revenue loss to maximize economic benefits.

The key contributions of this paper are as follows:

1. We construct a queuing network model to pinpoint the critical microservices causing performance degradation in SaaS services.
2. We incorporate service request frequency prediction and constraint planning to offer an overall optimization scheme for SaaS services.
3. .Our approach evaluates the optimization scheme by considering both computing resource usage costs and potential revenue loss, aiming to maximize economic benefits while ensuring service quality.

These contributions demonstrate a novel approach to optimizing SaaS service resource allocation and management, addressing the challenges posed by microservice architectures and dynamic request frequencies.

2   Related Works

Current research on the deployment optimization problem of SaaS services primarily falls into two categories: initial deployment optimization and runtime deployment optimization of services.

In the context of initial deployment optimization, the work by Bhardwaj S [4] formulates the SaaS service deployment problem as a combinatorial optimization problem, aiming to minimize service deployment cost. A particle swarm approach is employed to solve the problem, and the algorithm's solution quality is shown to be superior compared to a greed-based heuristic algorithm. The work by Fu K [8] maps microservice instances with frequent interactions to the same worker node, reducing time overhead due to data transfer between microservices and subsequently minimizing completion time. Nautilus demonstrates a reduction in data interaction between microservices, ranging from 39.2% to 82.4%, in comparison to Kubernetes' automatic deployment.

While initial deployment optimization can accommodate various request loads, it lacks adaptability to load changes. Persisting with the initial deployment scheme during changes in request frequency could lead to resource wastage, increased costs, and service performance degradation. In such cases, runtime deployment optimization becomes crucial.

Runtime deployment optimization is explored in [16], which defines the problem for SaaS services with complex dependencies and multiple instances as a fractional polynomial problem (FPP). The objective is to minimize the average response time. The problem is transformed into a quadratic sum-fractional problem (QSRFP) to reduce the high computational complexity of solving the FPP problem. The work by Zhang Y [36] introduces Sinan, a data-driven cluster manager for interactive cloud microservices. It utilizes convolutional neural networks (CNN) to predict the response time of a microservice under varying computing resource allocation conditions and employs a boosted tree (BT) model to evaluate the performance impact of delay queuing effects on overall SaaS service. The work by Qiu H [28] dynamically adjusts the computing resources occupied by the microservice. Additionally, they present FIRM, an intelligent fine-grained resource management framework that reduces shared resource contention across microservices, improves computing resource reuse rate, and employs a two-level machine learning (ML) framework to identify and mitigate performance bottlenecks in key microservices.

Data-driven machine learning approaches have demonstrated the ability to analyze, diagnose, and predict performance issues in microservice architecture applications accurately. However, these methods necessitate extensive historical data for training and real-time monitoring of system metrics during operation. This can impact performance, increase production costs, and may require retraining when the microservice architecture changes. Considering the scalability of SaaS environments and the complexity of microservice architectures, traditional performance analysis methods are not fully applicable. Accurately assessing the performance of SaaS services and maintaining their efficient operation at a lower cost remains a challenging task.

3   Adaptive Deployment Optimization for Microservice-Based SaaS Environments

This work focuses on the runtime deployment optimization of microservice architecture applications in a SaaS environment. The primary challenges associated with runtime deployment optimization of microservice architecture applications in a SaaS environment include: adjusting the allocation of computational resources to ensure service quality and reasonable resource utilization in response to dynamic fluctuations in request loads [5]; evaluating the impact of varying resource allocation schemes on the performance of microservice modules [38]; and identifying key microservice modules that cause performance bottlenecks in SaaS services in order to allocate limited reserved computing resources effectively to mitigate performance degradation [35].

To address these challenges, we construct a queuing network model [20, 32] based on the topological order of microservices within the SaaS service. This model evaluates the impact of different deployment schemes on the overall performance of the SaaS service under predicted request loads and selects the scheme that guarantees quality of service and maximizes overall economic efficiency for deployment.

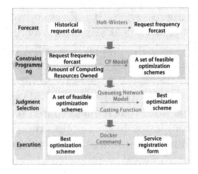

Fig. 1. Adaptive deployment optimization process.

Our proposed adaptive deployment optimization method comprises four modules: a forecast module, a constrained programming module, a judgment selection module, and an execution module (see Fig. 1). The forecast module predicts the number of service requests in the next period based on historical service request data, thereby determining the average arrival rate of service requests (λ). The constraint planning module generates feasible deployment solutions by planning the number of instances of microservices in the SaaS service, considering computing resources (CPU and memory) and the average arrival rate of service requests (λ) as constraints. The judgment selection module employs the queuing network model constructed from SaaS services to calculate the average queuing time of feasible solutions, discarding those that exceed the set delay.

Subsequently, an economic efficiency calculation function is utilized to select the deployment solution with the highest comprehensive benefit. Finally, the execution module updates the service registry according to the deployment solution generated by the judgment selection module.

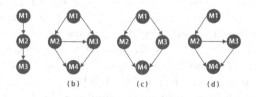

Fig. 2. Service flow of microservice applications.

3.1   Multi-stage Applications Composed of Cascading Microservices

Typically, a SaaS service request necessitates the collaboration of two or more microservices with distinct business capabilities. These microservices operate concurrently or sequentially, ultimately providing feedback on the results. Consequently, the service flow can be represented as a directed acyclic graph (DAG) [19] (see Fig. 2). Our proposed adaptive deployment optimization approach is applicable to SaaS applications of forms a and c, as these two forms can be transformed into multi-stage applications composed of cascading microservices. In this configuration, application services are executed sequentially by one or more microservice combinations across one or more stages. Each microservice combination within a stage consists of one or more microservices with no execution dependencies among them.

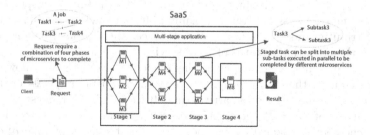

Fig. 3. Multi-stage application composed of cascading microservices.

Figure 3 presents an example of a multi-stage application composed of cascading microservices. The example application comprises four stages. The request first enters the initial stage, where three microservices execute tasks in parallel. Upon completion, the results are passed to the second and third stages, each consisting of two microservices. Finally, the fourth stage microservice executes to produce the ultimate outcome.

The total response time of a multi-stage application is the sum of the response times for each stage. As each stage contains at least one microservice, the response time of each stage is determined by the microservice with the longest response time within the stage, also known as the critical microservice. This is in accordance with the short board effect. The response time of a microservice is the sum of its computing time and the queuing delay of the request in the microservice's request queue. Generally, the computing time of the microservice itself is variable and influenced by factors such as programming, hardware configuration, and request type. Given a fixed hardware configuration, the average computing time of a microservice can be predicted by testing various request types. In multi-stage applications composed of cascading microservices, there are no execution dependencies between microservices. According to queuing theory [3], the average queuing delay of a microservice's request queue is determined by the number of instances of the microservice and the arrival frequency of the service requests it receives. As a result, SaaS service providers must implement an appropriate deployment scheme to guarantee the quality of service (QoS) for SaaS services.

Our proposed adaptive deployment optimization approach aids SaaS service providers in identifying microservices that cause performance bottlenecks during periods of high request frequency and increasing their parallelism accordingly to ensure SaaS service quality. When request frequency decreases, the approach assists SaaS service providers in reducing the appropriate number of microservice instances, thereby lowering production expenses while maintaining efficient SaaS services.

3.2   Request Scheduler

To implement this adaptive deployment optimization approach, we designed a request scheduler for multi-stage applications. The scheduler is responsible for providing unified access to requests, maintaining a request queue for microservices, distributing tasks to microservice instances through load balancing, and recording the operational status of SaaS services.

Figure 4 presents an example of the request scheduler's workflow. Upon receiving requests, the scheduler updates the total number of requests for the current time period and the request arrival time. It then adds the request ID_i to the request queues of $MicroService_1$, $MicroService_2$, and $MicroService_3$ in stage 1. The scheduler periodically examines the request queues of $MicroService_i$ and queries the service registry for an idle instance of $MicroService_i$. If an idle instance is available, the scheduler assigns $Task_i$ to the idle instance and updates the service registry. When the instance of $MicroService_i$ completes $Task_i$ and provides feedback to the scheduler, the scheduler updates the instance's status and checks whether all Microservices in the current stage have completed $Task_i$. If all tasks are completed, the scheduler adds $Task_i$ to the request queue of $MicroService_4$ and $MicroService_5$ in stage 2. This process continues until

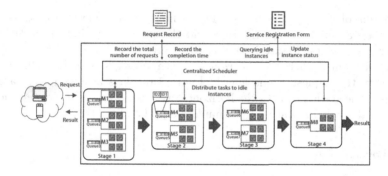

Fig. 4. Workflow of the request scheduler.

$MicroService_8$ of stage 4 completes $Task_i$. The scheduler records the request completion time, marking the request as completed.

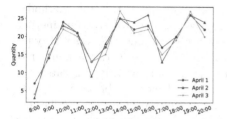

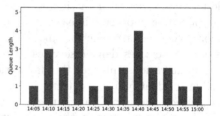

Fig. 5. Hourly service request statistics from 8:00 am to 8:00 pm on April 1, April 2 and April 3.

Fig. 6. Change in request queue length for the microservice between 2pm and 3pm (recorded every 5 min).

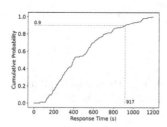

Fig. 7. Cumulative distribution of response time statistics.

By analyzing the data recorded by the scheduler, it is possible to examine the changes in request frequency, the operational status of each microservice within the application, and the average response time of requests. Figure 5 provides

an example of the dynamic change in the number of requests over time, with
the horizontal axis representing time and the vertical axis indicating the total
number of requests. Figure 6 demonstrates an example of the request queue
length over time, where the horizontal axis represents time and the vertical
axis shows the request queue length. Figure 7 displays an example of average
response time statistics, with the horizontal axis denoting the response time and
the vertical axis representing the cumulative probability.

3.3   Forecasting

To estimate future request numbers, we opt to utilize the sequence of total
request numbers from historical periods. The scheduler will calculate the total
number of requests in each historical period, which is uniformly recorded at the
end of each time period. Regarding the prediction of this time series, we have
experimented with the following methods.

Exponential smoothing methods [12,17] are frequently used in production to
analyze time series trends. The simple exponential smoothing method is suitable for time series without significant trend changes, while the Holt method is
appropriate for capturing linear trends in time series. The Holt-Winters method
is more effective for predicting time series with linear trends and seasonality [6].
Since SaaS services deliver services to users over the Internet, the number of
requests at various times of the day varies dynamically with user habits. It is
possible to identify a linear trend in the number of requests between adjacent
time periods and a cyclical change in the number of requests within a 24-hour
cycle. Therefore, we employ the triple exponential smoothing method to predict
the number of requests in future time periods.

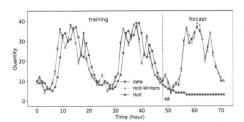

Fig. 8. Forecasting the number of third-day requests for intelligent typeset service
using different exponential smoothing methods.

Figure 8 presents an example of forecasting by different methods, where the
horizontal axis represents time and the vertical axis denotes the total number of
requests. The blue line illustrates the true historical total number of requests,
while the green line displays the total number of requests fitted and predicted
by Holt-Winters. The graph includes the true total number of requests for three
days, demonstrating that the historical total number of requests exhibits both

linear and seasonal trends. Using the true request totals for two days as the training set. The Holt method and Holt-Winters method are used to forecast the number of requests for the third day. The maximum error rate for the number of requests forecasted using the Holt-Winters method is 11.1% compared to the true value. The error using Holt method is larger. Forecasting using Holt-Winters method is more reliable.

3.4 Constrained Programming

The constrained programming module utilizes the request frequency λ generated by the forecast module and the amount of computing resources (CPU, memory) set by the SaaS service provider to plan and solve for the number of instances of microservices in the SaaS service, deriving a feasible set of deployment solutions.

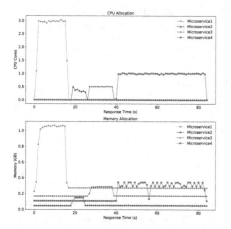
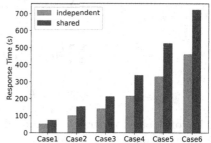

Fig. 9. Records of CPU and memory used by the four microservices in the intelligent typeset service to complete a service request.

Fig. 10. Comparison of the response time of requesting intelligent typeset service using six sets of test cases with different number of headings and different number of images.

Currently, container technology [24] is widely adopted for deploying microservice architecture applications, allowing for independent allocation of computational resources for each container and monitoring the running state of the containers using commands. Thus, the basic computational resource requirements for the runtime of different microservices can be analyzed. Figure 9 presents an example of the dynamics of computing resource usage for microservices over time, with the horizontal axis representing time and the vertical axis denoting usage. To determine the specific allocation of microservice container resources (CPU_i, MEM_i), we tested whether the microservice can run normally and whether the service completion time is within the normal range under different resource allocation scenarios by inputting numerous test cases. Subsequently, we selected the computational resource allocation scheme that ensures the normal and efficient operation of the microservice.

We transformed the problem of deploying microservice containers in a server into a constraint planning problem [2, 25]. A server with a total CPU capacity of CPU_{total} and a total memory capacity of MEM_{total} can deploy any combination of microservice instances $[k_1, k_2, ..., k_i]$ whose sum of CPU is less than or equal to CPU_{total} and whose sum of memory is less than or equal to MEM_{total}. (k refers to the number of microservice instances.)

Considering the range of feasible solutions, the number of computational resources (CPU_{total}, MEM_{total}) of the server can only provide the upper limit of feasible solutions (Eqs. 1 and 2). According to the requirements of the queuing network model established in the judgment selection module, the service intensity μ_i of each microservice must be less than 1; otherwise, the average queue length calculated by the model will be infinite. Therefore, we used the request frequency λ and the average service completion time μ of each microservice to limit the minimum value of the number of microservice instances deployed (Eq. 3).

$$\sum_{i=1}^{n} k_i CPU_i \leq CPU_{total} \tag{1}$$

$$\sum_{i=1}^{n} k_i MEM_i \leq MEM_{total} \tag{2}$$

$$k_i \mu_i > \lambda \tag{3}$$

The constrained programming module calculates the computational resource occupation of microservice instances based on the allocation of independently used computational resources. This approach may lead to a decrease in the overall utilization of server computational resources (Sharing computational resources among microservice instances can improve the overall utilization of computational resources, but additional management of shared resources is required to solve the resource contention problem [13, 14]). However, it ensures the stable and efficient operation of microservices, preventing competition for CPU and memory usage, which could result in operational errors or significant degradation of service performance. Figure 10 provides an example of microservice execution task time, illustrating the comparison of computing time between two microservice instances sharing computing resources (without implementing resource contention management) and using resources independently. The computing time increases by 50% to 60% when sharing computing resources, compared to when using resources independently. A higher degree of sharing may lead to worse microservice performance.

3.5   Judgment Selection

The judgment selection module uses the queuing network model to evaluate whether the average queuing delay of the deployment solution generated by the constraint planning module exceeds the set delay. If it does, the solution is removed from the set of feasible solutions. The deployment solution with the highest overall benefit among the deployment solutions that pass the evaluation is then selected using the economic efficiency function.

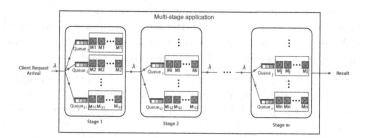

Fig. 11. Analytical model of a multi-stage application cascaded by microservice compositions.

To evaluate the performance of multi-stage applications cascaded by microservice compositions, we constructed a queuing network model [1,18,30, 33]. Figure 11 presents the analytical model of a multi-stage application cascaded by microservice compositions. Suppose a multi-stage application consists of n microservices, ($MicroService_1$, $MicroService_2$, $MicroService_3$, ..., $MicroService_n$), which belong to m stages ($0 < m <= n$). The n microservices have n request queues. The queuing network model consists of these n queuing models: $queue_1$, $queue_2$, $queue_3$, ..., $queue_n$.

We assumed that request arrivals follow a Poisson process with rate λ, and the average request arrival interval is $1/\lambda$. We also assumed that the computing times for each type of microservice are exponentially distributed. A queuing model in which request arrivals follow a Poisson process and service times are exponentially distributed can provide approximately pessimistic predictions for the time distribution (normal or random distribution) of other kinds of request arrivals [26,34].

Among the m phases, the microservice with the longest response time within the first phase has an impact on the arrival and performance of requests in later phases. When the request arrives in a Poisson process with rate λ, all microservices $MicroService_1$, $MicroService_2$, $MicroService_3$, ..., $MicroService_{s1}$ within the first stage receive the task simultaneously. The task arrival of all microservices in the first stage follows a Poisson process with rate λ. According to the assumption that the operation times of all microservices in the SaaS application follow an exponential distribution μ_1, μ_2, μ_3, ..., μ_n, the request queues of all microservices within the first stage can be modeled as the M/M/k queuing model (k $>=$ 1). When the microservice with the longest response time in the first stage completes its task, the requests immediately move to the second stage. According to Burke's theorem [22], the output process of the M/M/k queue is a Poisson process that has the same rate λ as its input process. Therefore, the arrivals of all microservices in the second stage also follow a Poisson process with rate λ. The request queues of all microservices within the second stage can also be modeled as an M/M/k queuing model (k $>=$ 1). This reasoning can be extended to all m stages, where the request queues of all microservices can be modeled as an M/M/k queuing model (k $>=$ 1). The performance metrics of all microservices can then be calculated, and the performance metrics of the microservice with the longest response time in each phase will be combined to evaluate the overall performance of the SaaS application.

Performance indicators of the M/M/1 queuing system:

1. The service intensity of the system, which indicates the workload of a microservice with only one instance. The closer the request frequency λ is to μ, the higher the probability that an instance of that microservice is in a busy state.

$$\rho = \frac{\lambda}{\mu} \tag{4}$$

2. The average number of requests in the waiting state in the system indicates the average queue length of the request queue for a microservice with only one instance per unit time. The higher the request frequency λ, the higher the probability of request queuing.

$$L_q = \frac{\lambda^2}{\mu(\mu - \lambda)} \tag{5}$$

3. The average number of requests in the system indicates the average capacity of the request queue for a microservice with only one instance per unit time. The higher the request frequency λ, the longer the average length of the request queue.

$$L_s = \frac{\lambda}{\mu - \lambda} \tag{6}$$

4. The average wait time of a request in the system indicates the average queuing time of a request in the request queue of a microservice with only one instance. The higher the request frequency λ, the longer the average queuing time of the request.

$$W_q = \frac{\lambda}{\mu(\mu - \lambda)} \tag{7}$$

5. The average sojourn time of requests in the system indicates the average response time per unit time for microservices with only one instance. The higher the request frequency λ, the longer the average response time.

$$W_S = \frac{1}{\mu - \lambda} \tag{8}$$

Performance indicators of the M/M/k (k >= 2) queuing system:

1. The service intensity of the system, which indicates how busy a microservice with k instances is. The closer the request frequency λ is to $k\mu$, the higher the probability that an instance of that microservice is in a busy state.

$$\rho = \frac{\lambda}{k\mu} \tag{9}$$

2. The probability of having c requests in the system represents the steady-state probability of having c requests in the request queue of a microservice with k instances.

$$P_0 = \left[\sum_{c=0}^{k-1} \frac{1}{c!} \left(\frac{\lambda}{\mu}\right)^c + \frac{1}{k!} \frac{1}{1-\rho} \left(\frac{\lambda}{\mu}\right)^k \right]^{-1} \tag{10}$$

$$P_c = \frac{1}{c!} \left(\frac{\lambda}{\mu}\right)^c \qquad (0 < c < k) \tag{11}$$

$$P_c = \frac{1}{k! k^{c-k}} \left(\frac{\lambda}{\mu}\right)^c P_0 \qquad (c \geq k) \tag{12}$$

3. The average number of requests in the waiting state in the system indicates the average queue length of the request queue for a microservice with k instances per unit time. The higher the request frequency λ, the higher the probability of request queuing.

$$L_q = \frac{(k\rho)^k \rho}{k! (1-\rho)^2} P_0 \tag{13}$$

4. The average number of requests in the system indicates the average capacity of the request queue for a microservice with k instances per unit time. The higher the request frequency λ, the longer the average length of the request queue.

$$L_s = Lq + k\rho \tag{14}$$

5. The average waiting time of a request in the system indicates the average queuing time of a request in the request queue of a microservice with k instances per unit time. The higher the request frequency λ, the longer the average queuing time of the requests.

$$W_q = \frac{L_q}{\lambda} \tag{15}$$

6. The average sojourn time of requests in the system indicates the average response time of microservices with k instances per unit time. The higher the request frequency λ, the longer the average response time.

$$W_s = Wq + \frac{1}{\mu} \tag{16}$$

Integrated average queuing delay:

$$Tq = \sum_{i=1}^{m} L_{qi} W_{qi} \tag{17}$$

L_{qi} and W_{qi} are determined by the microservice with the longest response time (longest average sojourn time) in phase i. If the integrated average queuing delay exceeds the threshold set by the SaaS service provider, then the current deployment option is discarded.

Economic efficiency function:

$$V = \sum_{i=1}^{n} k_i CPU_i W_{CPU} + \sum_{i=1}^{n} k_i MEM_i W_{MEM} + Tq W_T \lambda \tag{18}$$

CPU_i and MEM_i are the compute resource allocations of the corresponding microservices, which are determined by the SaaS service provider after testing the microservices it develops. W_{CPU} and W_{MEM} are the cost weights of the corresponding compute resources, which can be determined based on the purchase or rental price of the server. W_T is the time cost weight, which is determined by the general pricing of the SaaS service. This function calculates the expected compute resource usage cost invested by the SaaS service provider in the next time period if the current solution is deployed, as well as the potential loss of request revenue due to queuing delays. The judgment selection module selects the solution with the smallest calculation result among all deployment solutions as the optimal deployment solution, which maximizes the expected combined revenue of the SaaS service provider.

3.6   Execution Module Implementation

The execution module, which is a custom Kubernetes controller, assumes the responsibility of implementing the optimal deployment plan generated by the judgment selection module (Kubernetes service). Its primary function is to maintain regular communication with the judgment selection module's service, adjusting the number of instances for each microservice as necessary to ensure the deployment plan's successful execution.

Periodically, the execution module will retrieve the current count of registered microservice instances (with a user-configurable time interval), comparing it against the optimal deployment plan provided by the judgment selection module's service. If there is a disparity between the two, the module will initiate the creation or termination of instances as required, giving precedence to creating new instances over terminating existing ones to minimize service disruption.

In summary, the execution module plays a pivotal role in achieving optimal deployment solutions, dynamically adjusting the instance count for each microservice, and guaranteeing the efficient operation of SaaS applications in accordance with expected performance standards.

4 Evaluation

To validate the effectiveness of our adaptive deployment optimization approach, we conducted real deployment tests using an intelligent typeset service and compared it to a commonly used automatic scaling approach (with the request queue's capacity as the threshold and a 5-minute expansion and contraction trigger interval) in four environments with limited computational resources. Additionally, we designed a SaaS service (Table 2) comprising nine microservices for simulation experiments (Simulation experiments differ from real experiments in that they utilize generated service times instead of actual service times, while all other aspects remain consistent).

The intelligent typeset service is a composite of four stages of microservices, each containing one microservice: $MicroService_1$, $MicroService_2$, $MicroService_3$, $MicroService_4$. We obtained the base computing resources (CPU, memory, and disk space) required by the four microservices and their average computing time after extensive testing using 21 different types of currently supported test cases (Table 1).

Table 1. Base computing resources and average computing time of the intelligent typeset service.

	CPU(cores)	Memory (GB)	Average computing time (s)	Hard Disk Drive (GB)
$MicroService_1$	3	1.2	163	3.61
$MicroService_2$	0.5	0.39	18	1.15
$MicroService_3$	0.5	0.39	19	1.15
$MicroService_4$	1	0.78	90	9.19

Table 2. Base computing resources and average computing time of the designed SaaS service.

	CPU(cores)	Memory (GB)	Average computing time (s)
$MicroService_1$	0.4	0.1	41
$MicroService_2$	0.4	0.6	26
$MicroService_3$	2.0	1.2	14
$MicroService_4$	1.2	0.4	15
$MicroService_5$	1.4	1.2	17
$MicroService_6$	2.2	1.0	17
$MicroService_7$	0.8	0.3	30
$MicroService_8$	1.4	0.7	24
$MicroService_9$	0.4	0.8	23

4.1 Experiment Setup

We utilized six cloud servers (OS Ubuntu 20.04.5, 4*Intel(R) Xeon(R) Platinum 8163CPU@2.50 GHz CPU, 8 GB RAM, 128G HDD, Docker version: 20.10.17) for the experiments. The scheduler was deployed on a separate server, while instances of the four microservices were deployed on the other five cloud servers. We employed Docker to allocate and limit the CPU and memory usage of the containers individually. The containers communicated via an intranet and shared data using a network file system.

By employing the constraint planning module and the queuing network model, the intelligent typeset service deployed on five cloud servers could accept request input streams with an average arrival rate of up to 89 requests per hour (with a total average queuing delay of no more than 290 s, i.e., no more than the average completion time of a request). No changes to the initial deployment were required when the average arrival rate of request input streams did not exceed 8 per hour (with one container for each microservice). Consequently, we conducted experiments using Poisson flow requests with a frequency ranging from 9 to 89 requests per hour.

The experimental setup ensured a robust comparison between the adaptive deployment optimization approach and the commonly used automatic scaling approach. The results from the experiments can provide insights into the performance improvements and resource utilization efficiency offered by the adaptive deployment optimization approach. By analyzing these results, we can determine the effectiveness and feasibility of our proposed approach in real-world scenarios, particularly in environments with limited computational resources.

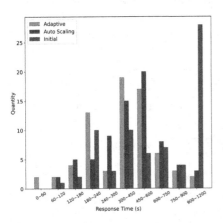
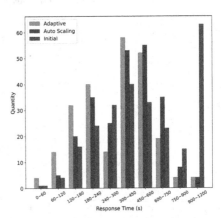

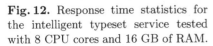

Fig. 12. Response time statistics for the intelligent typeset service tested with 8 CPU cores and 16 GB of RAM.

Fig. 13. Response time statistics for the intelligent typeset service tested with 12 CPU cores and 24 GB of RAM.

Figure 12 presents the response time statistics for the intelligent typeset service using different deployment optimization methods in an environment with 8 cores and 16 GB of CPU and memory. The Poisson flow request frequency ranges from 9 to 25 requests per hour. The adaptive method has an average response time of 413 s, while the auto-scaling method's average response time is 472 s. Both methods achieve approximately a 50% performance improvement compared to the initial deployment scheme. The adaptive method reduces the average response time by 13% compared to the auto-scaling method. Although both methods result in similar deployment optimization, the auto-scaling method's overall performance is slightly worse due to multiple expansions and contractions during runtime, which increases the average response time.

Figure 13 displays the response time statistics for the intelligent typeset service with different deployment optimization methods in an environment with 12 cores and 24 GB of CPU and memory, with the Poisson flow request frequency ranging from 9 to 49 requests per hour. The adaptive method's average response time is 373 s, whereas the auto-scaling method's average response time is 421 s. The adaptive method improves service performance by 11% compared to the auto-scaling method. The adaptive method allocates reserved resources to $MicroService_1$ and $MicroService_4$, while the auto-scaling method prioritizes allocation to $MicroService_1$, $MicroService_2$, and $MicroService_3$. This difference leads to increased average queue length and overall average response time for the auto-scaling method.

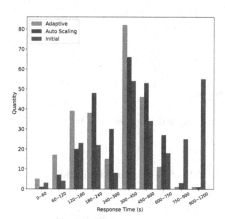

Fig. 14. Response time statistics for the intelligent typeset service tested with 16 CPU cores and 32 GB of RAM.

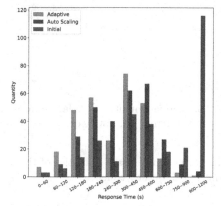

Fig. 15. Response time statistics for the intelligent typeset service tested with 20 CPU cores and 40 GB of RAM.

Figure 14 shows the response time statistics for the intelligent typeset service using different deployment optimization methods in an environment with 16 cores and 32 GB of CPU and memory, with the Poisson flow request frequency ranging from 20 to 69 requests per hour. The adaptive approach's average response time is 331 s, while the auto-scaling approach's average response time is 442 s. This results in a 25% improvement in response latency for the adaptive approach compared to the auto-scaling approach. The adaptive method consistently allocates reserved resources to $MicroService_1$ and $MicroService_4$, while the auto-scaling method's progressive tuning affects the overall service performance.

Figure 15 provides the response time statistics for the intelligent typeset service with different deployment optimization methods in an environment with 20 cores and 40 GB of CPU and memory, with the Poisson flow request frequency ranging from 33 to 88 requests per hour. The average response time for the adaptive approach is 322 s, and the average response time for the auto-scaling approach is 385 s. The adaptive approach improves the service performance by 16% compared to the auto-scaling approach. In this environment, the adaptive method continues to allocate reserved resources to $MicroService_1$ and $MicroService_4$, prioritizing $MicroService_1$. The auto-scaling method also prioritizes resource allocation for $MicroService_1$, but the allocation of remaining resources among $MicroService_2$, $MicroService_3$, and $MicroService_4$ varies, affecting the overall average response time.

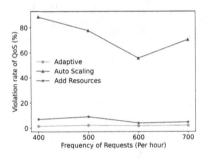

Fig. 16. Comparison of QoS violation rates of the designed SaaS services by simulation testing at different request frequencies.

Table 3. Resource usage statistics of the designed SaaS service by simulation testing at different request frequencies.

Request frequency (per hour)	Initial resources		After adding resources	
	CPU(cores)	Memory (GB)	CPU(cores)	Memory (GB)
400	36	24	56	40
500	40	28	64	40
600	48	32	72	44
700	56	36	84	56

Figure 16 and Table 3 present the quality of service (QoS) violation statistics and comparison of computational resource usage for different deployment optimization methods. The tests are conducted on the designed SaaS service with Poisson flow request frequencies of 400, 500, 600, and 700 requests per hour. The adaptive approach's average QoS violation rate is 1.9%, while the auto-scaling approach's average QoS violation rate is 73%. The adaptive approach avoids

71.1% of QoS violations compared to the auto-scaling approach. Although the auto-scaling approach reduces the QoS violation rate after increasing computational resource usage, it increases CPU resource usage by 53% and memory resource usage by 50% on average compared to the adaptive approach.

In conclusion, the proposed adaptive deployment optimization method effectively regulates the number of instances of microservice modules within SaaS applications under limited computing resources. It successfully addresses the service performance degradation problem and reduces computational resource usage by 35% compared to the traditional auto-scaling approach.

5   Conclusion

We proposed a queue analysis-based adaptive deployment optimization method for multi-stage applications cascaded by microservice compositions. By adjusting the number of instances of microservice modules within SaaS applications, our method effectively addresses the service performance degradation problem under limited computing resources while maintaining cost efficiency. The effectiveness of this method was validated through real-world tests on intelligent typeset services and simulation tests.

However, the current method primarily focuses on tandem microservice compositions and does not consider other topologies of microservice invocation chains. Future work will explore the extension of the method to various microservice chain topologies and investigate more efficient optimization techniques to address scalability issues in larger-scale microservice deployments, ensuring effectiveness across a wide range of application scenarios.

Acknowledgements. This research was funded by the Basic Public Welfare Research Project of Zhejiang Province grant number LY20F020014 and the National Science Foundation for Young Scientists of China grant number 61802096.

References

1. Alenizi, A., Ammar, R., Elfouly, R., Alsulami, M.: Queue analysis for probabilistic cloud workflows. In: 2020 IEEE International Symposium on Signal Processing and Information Technology (ISSPIT), pp. 1–6 (2020). https://doi.org/10.1109/ISSPIT51521.2020.9408967
2. Alsarhan, A., Itradat, A., Al-Dubai, A.Y., Zomaya, A.Y., Min, G.: Adaptive resource allocation and provisioning in multi-service cloud environments. IEEE Trans. Parallel Distrib. Syst. 29(1), 31–42 (2018)
3. Baccelli, F., Bremaud, P.: Elements of Queueing Theory. Elements of Queueing Theory (1961)
4. Bhardwaj, S., Sahoo, B.: A particle swarm optimization approach for cost effective saas placement on cloud. In: International Conference on Computing, Communication and Automation, ICCCA 2015, pp. 686–690 (2015). https://doi.org/10.1109/CCAA.2015.7148462

5. Chainbi, W., Sassi, E.: A multiswarm for composite saas placement optimization based on pso. Softw.- Pract. Exp. **48**(10), 1847–1864 (2018). https://doi.org/10.1002/spe.2600

6. Chatfield, C.: The holt-winters forecasting procedure. J. Roy. Stat. Soc. **27**(3), 264–279 (1978). https://doi.org/10.2307/2347162

7. Dikaiakos, M.D., Katsaros, D., Mehra, P., Pallis, G., Vakali, A.: Cloud computing: distributed internet computing for it and scientific research. IEEE Internet Comput. **13**(5), 10–11 (2009). https://doi.org/10.1109/MIC.2009.103

8. Fu, K., Zhang, W., Chen, Q., Zeng, D., Guo, M.: Adaptive resource efficient microservice deployment in cloud-edge continuum. IEEE Trans. Parallel Distrib. Syst. **33**(8), 1825–1840 (2022). https://doi.org/10.1109/TPDS.2021.3128037

9. Gan, Y., Liang, M., Dev, S., Lo, D., Delimitrou, C.: Sage: practical and scalable ml-driven performance debugging in microservices. In: International Conference on Architectural Support for Programming Languages and Operating Systems - ASPLOS, pp. 135–151 (2021). https://doi.org/10.1145/3445814.3446700

10. Gan, Y., et al.: An open-source benchmark suite for cloud and iot microservices. arXiv: 1905.11055 (2019)

11. Gan, Y., et al.: Seer: leveraging big data to navigate the complexity of performance debugging in cloud microservices. In: International Conference on Architectural Support for Programming Languages and Operating Systems - ASPLOS, pp. 19–33 (2019). https://doi.org/10.1145/3297858.3304004

12. Gardner, E.S.: Exponential smoothing: The State of the Art-part ii. Int. J. Forecast. **22**(4), 637–666 (2006)

13. Gevros, P., Crowcroft, J.: Distributed resource management with heterogeneous linear controls. Comput. Netw. **45**(6), 835–858 (2004)

14. Gias, A.U., Casale, G., Woodside, M.: Atom: model-driven autoscaling for microservices. In: Proceedings International Conference on Distributed Computing Systems 2019, pp. 1994–2004 (2019). https://doi.org/10.1109/ICDCS.2019.00197

15. Hajji, M.A., Mezni, H.: A composite particle swarm optimization approach for the composite saas placement in cloud environment. Soft. Comput. **22**(12), 4025–4045 (2018). https://doi.org/10.1007/s00500-017-2613-8

16. He, X., Tu, Z., Wagner, M., Xu, X., Wang, Z.: Online deployment algorithms for microservice systems with complex dependencies. IEEE Trans. Cloud Comput. (2022). https://doi.org/10.1109/TCC.2022.3161684

17. Hyndman, R.J., Koehler, A.B., Snyder, R.D., Grose, S.: A state space framework for automatic forecasting using exponential smoothing methods. Int. J. Forecast. **18**(3), 439–454 (2002)

18. Jia, R., Yang, Y., Grundy, J., Keung, J., Li, H.: A deadline constrained preemptive scheduler using queuing systems for multi-tenancy clouds. In: IEEE International Conference on Cloud Computing, CLOUD 2019, pp. 63–67 (2019)

19. Kannan, R.S., Subramanian, L., Raju, A., Ahn, J., Mars, J., Tang, L.: Grandslam: guaranteeing slas for jobs in microservices execution frameworks. In: Proceedings of the 14th EuroSys Conference 2019 pp. ACM Special Interest Group on Operating Systems (SIGOPS) (2019). https://doi.org/10.1145/3302424.3303958

20. Khazaei, H., Mii, J., Mii, V.B.: Modelling of cloud computing centers using m/g/m queues. In: Proceedings - International Conference on Distributed Computing Systems, pp. 87–92 (2011). https://doi.org/10.1109/ICDCSW.2011.13

21. Khazaei, H., Misic, J., Misic, V.B.: Performance analysis of cloud computing centers using m/g/m/m+r queuing systems. IEEE Trans. Parallel Distrib. Syst. **23**(5), 936–943 (2012)

22. Klevans, R.L., Stewart, W.J.: From queueing networks to markov chains: the xmarca interface. Perform. Eval. **24**(1–2), 23–45 (1995)
23. Liao, W.H., Chen, P.W., Kuai, S.C.: A resource provision strategy for software-as-a-service in cloud computing. Proc. Comput. Sci. **110**, 94–101 (2017). https://doi.org/10.1016/j.procs.2017.06.123
24. Merkel, D.: Docker: lightweight linux containers for consistent development and deployment. Linux J. **2014**(239) (2014). https://doi.org/10.5555/2600239.2600241
25. Moens, H., Truyen, E., Walraven, S., Joosen, W., Dhoedt, B., De Turck, F.: Cost-effective feature placement of customizable multi-tenant applications in the cloud. J. Netw. Syst. Manage. **22**(4), 517–558 (2014)
26. Mohammadi, M., Jolai, F., Rostami, H.: An m/m/c queue model for hub covering location problem. Math. Comput. Model. **54**(11–12), 2623–2638 (2011). https://doi.org/10.1016/j.mcm.2011.06.038
27. Pallis, G.: Cloud computing: the new frontier of internet computing. IEEE Internet Comput. **14**(5), 70–73 (2010). https://doi.org/10.1109/MIC.2010.113
28. Qiu, H., Banerjee, S.S., Jha, S., Kalbarczyk, Z.T., Iyer, R.K.: Firm: an intelligent fine-grained resource management framework for slo-oriented microservices. Proceedings of the 14th USENIX Symposium on Operating Systems Design and Implementation, OSDI 2020, pp. 805–825 (2020). https://doi.org/10.48550/arXiv.2008.08509
29. Reiss, C., Tumanov, A., Ganger, G.R., Katz, R.H., Kozuch, M.A.: Heterogeneity and dynamicity of clouds at scale: Google trace analysis. In: Proceedings of the 3rd ACM Symposium on Cloud Computing, SoCC 2012. ACM Special Interest Group on Management of Data (SIGMOD) (2012)
30. Tournaire, T., Castel-Taleb, H., Hyon, E., Hoche, T.: Generating optimal thresholds in a hysteresis queue: application to a cloud model. In: Proceedings - IEEE Computer Society's Annual International Symposium on Modeling, Analysis, and Simulation of Computer and Telecommunications Systems, MASCOTS 2019, 283–294 (2019). https://doi.org/10.1109/MASCOTS.2019.00040
31. Villamizar, M., et al.: Infrastructure cost comparison of running web applications in the cloud using aws lambda and monolithic and microservice architectures. In: Proceedings - 2016 16th IEEE/ACM International Symposium on Cluster, Cloud, and Grid Computing, CCGrid 2016, pp. 179–182 (2016). https://doi.org/10.1109/CCGrid.2016.37
32. Wada, H., Suzuki, J., Oba, K.: Queuing theoretic and evolutionary deployment optimization with probabilistic slas for service oriented clouds. In: SERVICES 2009–5th 2009 World Congress on Services (PART 1), pp. 661–669 (2009). https://doi.org/10.1109/SERVICES-I.2009.59
33. Wang, S., Li, X., Ruiz, R.: Performance analysis for heterogeneous cloud servers using queueing theory. IEEE Trans. Comput. **69**(4), 563–576 (2020). https://doi.org/10.1109/TC.2019.2956505
34. Wu, H., Sun, Y., Wolter, K.: Analysis of the energy-response time tradeoff for delayed mobile cloud offloading. Perform. Eval. Rev. **43**(2), 33–35 (2015). https://doi.org/10.1145/2825236.2825251
35. Yang, H., Chen, Q., Riaz, M., Luan, Z., Tang, L., Mars, J.: Powerchief: Intelligent power allocation for multi-stage applications to improve responsiveness on power constrained cmp. In: Proceedings - International Symposium on Computer Architecture Part **F128643**, pp. 133–146 (2017). https://doi.org/10.1145/3079856.3080224

36. Zhang, Y., Hua, W., Zhou, Z., Suh, G.E., Delimitrou, C.: Sinan: Ml-based and qos-aware resource management for cloud microservices. In: International Conference on Architectural Support for Programming Languages and Operating Systems - ASPLOS, pp. 167–181 (2021). https://doi.org/10.1145/3445814.3446693

37. Zhou, H., et al.: Overload control for scaling wechat microservices. In: SoCC 2018 - Proceedings of the 2018 ACM Symposium on Cloud Computing, pp. 149–161 (2018). https://doi.org/10.1145/3267809.3267823

38. Zhou, X., et al.: Fault analysis and debugging of microservice systems: Industrial survey, benchmark system, and empirical study. IEEE Trans. Software Eng. **47**(2), 243–260 (2021). https://doi.org/10.1109/TSE.2018.2887384

Reliable Function Computation Offloading in Cloud-Edge Collaborative Network

Shaonan Li, Yongqiang Xie$^{(\boxtimes)}$, Zhongbo Li, Jin Qi, and Yumeng Tian

Institute of System Engineering, Academy of Military Science, Beijing 100141, China
61ssec_xyq@sina.com

Abstract. In this paper, we focus on the cloud-edge collaborative network, where a task is decomposed into a set of functions and could be offloaded to different computing nodes, which is referred to as Function Computation Offloading (FCO). One of the most important problems in FCO is to schedule the functions in computing nodes to achieve low latency and high reliability. We formulate FCO scheduling in the Cloud-edge Collaborative Network as mixed-integer nonlinear programming. The objective is to minimise the end-to-end delay of a task while satisfying the latency and reliability constraints. To solve the problem, we propose an efficient mechanism to decide the redundancy of functions according to the reliability requirements. Then, we deploy the non-redundant functions on the computing nodes. Finally, we present a Reinforcement Learning (RL) to learn the scheduling policy of the redundant functions to further reduce the end-to-end delay of the task. Simulation results show that our proposed algorithm can significantly reduce tasks' completion time by about 13–26% with fewer iterations compared with other alternatives.

Keywords: Computation Offloading · Reliability · Task Decomposition · Reinforcement learning · Cloud-edge Collaboration

1 Introduction

The emergence of the Internet of Things (IoT) has led to a rapid increase in computing-sensitive terminal devices [17]. These devices can reduce latency and energy consumption by offloading computing to edge computing servers (EC) and cloud computing servers (CC) [27]. However, terminal devices such as mobile phones and IoT devices lack the resources to efficiently run intelligent applications [5]. To address this issue, deploying computing resources near the network edge is considered a promising solution. EC can provide low-latency services for UEs and guarantee their data security [16]. However, they may not have the same ability as the cloud to handle computation-intensive tasks [10]. While CC has the sufficient resources to handle computation-intensive tasks [14], it is unable to address the problem of long delays caused by data transmission [23].

Z. Tari et al. (Eds.): ICA3PP 2023, LNCS 14488, pp. 433–451, 2024.
https://doi.org/10.1007/978-981-97-0801-7_25

EC, on the other hand, is capable of reducing data transmission delays due to its shorter transmission distance and higher transmission rate between UEs and ECs. This makes it well-suited for handling tasks that require low latency. By working together, EC and CC can provide a better solution for meeting the diverse needs of users [5].

Computation offloading has been widely studied in many directions, such as Internet of Vehicles (IoV) [25], Industrial Internet of Things(IIoT) [1], smart city [9], Mobile Edge Computing(MEC) [8]. Recent works has shown that the dynamic characteristics of the environment [34], resource allocation [25], caching [4] and energy consumption [33] have been well addressed by complete and partial task offloading. However, task-based computational offloading cannot take advantage of resources in a cloud-edge collaborative environment. A computational task being offloaded to an EC node or CC node can only utilize limited resources. Meanwhile, the reliability of computation offloading is a gap in current research.

To address the problem, we propose splitting the computational task into functions and offloading them to different computing nodes using Function as a Service (FaaS) platform [13]. This approach provides a more code-fragmented software architecture paradigm and allows for a more granular program unit than services. FaaS infrastructure enables FCO. However, to enhance the reliability of a task, functions execution needs to have additional redundant instances, which occupy more computing resource and result in a higher waiting time for other network services. Therefore, an intelligent scheduler is essential to balance between latency and reliability [11].

To construct the function computation offloading system and make it work normally, there are mainly three challenges.

1) Previous studies have focused on task-level computation offloading and do not address function-level computation offloading.
2) FCO is a strategy that not only takes into account the optimal deployment node for a function, but also considers the data transfer between functions. This approach differs from traditional computation offloading strategies.
3) FCO is currently reliant on the FaaS platform, however, there is a need to improve the reliability of function triggers and wake-ups as any failure in function execution can lead to service failure. The redundancy of functions and the level of redundancy employed play a key role in determining the overall reliability of the task. The problem of function computation offloading is a composite problem that involves minimizing both data transmission and computation delays.

We proposed an algorithm based on RL and divide-and-conquer is formulated to deal with the mixed integer non-linear programing problem under the cloud-edge collaborative scenario. The contributions of this work are summarized as follows:

- We have developed a reliable FCO model and formulated the reliability-aware FCO problem as a Mixed Integer Non-Linear Programing (MINLP) problem, which is NP-hard and shows the complexity of the FCO scheduling problem and the difficulty to find a globally optimal solution.

- Based on the divide-and-conquer idea, we decompose the MINLP problem into three sub-problems (Redundancy Determining, Node Selection, Redundancy Node Selection). The complexity of the algorithm can be greatly reduced.
- We propose a reinforcement learning algorithm to provide a offloading strategy with lower latency. The effectiveness of our proposed approach is revealed through the simulations. The results show that the proposed approach outperforms other algorithms in term of end-to-end delay.

The rest of this paper is organized as follows. Section 2 introduces the related work. Section 3 presents the network model in this paper and mathematically formulates the reliability-aware FCO scheduling problem definition. In Sect. 4, we propose a redundancy determining algorithm and develop a reinforcement learning algorithm to schedule the FCO scheduling problem. Section 5 demonstrates the simulation results and analysis. Finally, we conclude our work in Sect. 6.

2   Related Work

The computation offloading strategy optimization problem has consistently been a hot research topic in industry and academia, and has been investigated extensively. Based on optimization objective, existing computation offloading approaches can be roughly divided into five categories: 1) latency; 2) energy consumption; 3) tradeoff between latency and energy consumption; 4) resource allocation; 5) computing architectures.

Latency. Guo et al. [8] proposed two online learning-based offloading algorithms to offload tasks from a device to the access points with full or partial current network information. Zhu et al. [34] proposed an MADRL based computation offloading algorithm to make offloading decisions for multiple mobile vehicles. In this algorithm, each agent selects its nearby edge server for offloading the task generated by a vehicle to minimize completion latency. Cao et al. [2] studied the problem of IIoT computation offloading using a software-defined architecture. They developed an optimization algorithm based on multiagent reinforcement learning to improve the success rate of multiuser channel access and reduce the user computing delay. To minimize the total latency, Zhang et al. [30] proposed a multistage stochastic programming-based scheme to jointly make decision on offloading, resource allocation, and migration.

Energy Consumption. Zhao et al. [31] decompose the computing offloading problem into three subproblems named as offloading ratio selection, transmission power optimization, and subcarrier and computing resource allocation.

Then, they proposed an iterative algorithm to deal with them in a sequence. Simulation results demonstrate that the proposed algorithm can save 20%–40% energy compared with the reference schemes, and can converge to local optimal solutions.

Tradeoff Between Latency and Energy Consumption. Sun et al. [22] proposed a heuristic algorithm and a Cauchy-Schwards inequality-based closed-form method to jointly optimize the server selection and resource allocation to minimize the weighted sum of latency and energy consumption. Peng et al. [15] designed an online resource coordinating and allocating scheme to simultaneously minimize latency and energy consumption in a D2D-assisted edge computing system. Zhang et al. [29] presented two DRL-based computation offloading methods for multiuser multiserver edge computing systems with energy harvesting. Yang et al. [26] developed a game theory-based offloading algorithm to minimize the weighted sum of latency and energy consumption in the D2D-enable edge computing system. Zhan et al. [28] modeled the offloading problem with information sharing as a decentralized computation offloading game and then formulated the problem without information sharing as a partially observable Markov decision process (MDP). Liang et al. [12] introduced a submodular theory-based computation offloading algorithm to minimize a weighted sum of the latency and energy consumption in an edge-cloud computing system.

Resource Allocation. Wang et al. [24] formulate the computation offloading decision, physical resource block (PRB) allocation, and MEC computation resource allocation as optimization problems and proposed a graph coloring method to proform the PRB allocation. Liang et al. [32] proposed a contract-based framework to deal with the joint task offloading, resource sharing and computation incentive (TORSCI) issue under the asymmetric information scenario. Simulation results demonstrate the efficiency of our proposed contract-based incentive approach.

Computing Architectures. Ding et al. [5] classify EECC into two computing architectures types according to the visibility and accessibility of the cloud to UEs, i.e., hierarchical end-edge-cloud computing (Hi-EECC) and horizontal end-edge-cloud computing (Ho-EECC) and construct a potential game for the EECC environment, in which each UE selfishly minimizes its payoff, study the computation offloading strategy optimization problems, and develop two potential game-based algorithms in Hi-EECC and Ho-EECC. Ren et al. [19] investigated the computation offloading optimization problem in the hierarchical edge-cloud computing architecture and developed a convex-based algorithm to decide the task slipping strategy. Shah-Mansouri et al. [21] developed a potential game-based algorithm to optimize the strategy in the horizontal edge-cloud computing architecture. Du et al. [6] considered the communication cost between co-resident and non-co-resident tasks, and designed an algorithm to obtain a sub-optimal strategy. Fantacci et al. [7] formulated the problem as a queueing system model and determined the strategy by maximizing the rate of UEs whose QoS can be satisfied. Chen et al. [3] proposed a two-level alternation method framework based on reinforcement learning (RL) and sequential quadratic programming (SQP) to solve the mixed-integer nonlinear optimization problem which constrained by computing resource and cache capacity of each access point.

3 System Model and Problem Formulation

3.1 Network Model

As illustrated in Fig. 1, we consider a cloud-edge network integrated a CC and n ECs. Each node is deployed with the Faas platform and contains all the functions required by the terminal user. The terminal initiates a task which consists of several different functions. These functions will be executed by different computing nodes. The terminal will send the data to the first function and then the calculated results will be passed sequentially in the order of the execution order to finally obtain the results. In this case, the end-to-end delay can be defined as the sum of processing time and data transmission time. The processing time of a function is calculated as $t_{calc} = d_f w_f / \varpi_j$, where d_f is the amount of data input to the function f, w_f is the frequency required by the function f to process a unit of data (i.e., CPU frequency, which is quantified by the number of cycles per second) and ϖ_j is the computation speed of computing node j. The data transmission time of a function is calculated as $t_{txm} = d_f \varrho_f / \gamma_j$, where d_f is the amount of data input to the function f, ϱ_f is a coefficient characterizing the relationship between the output data of the function and the input data and γ_j is the transmission speed of computing node j.

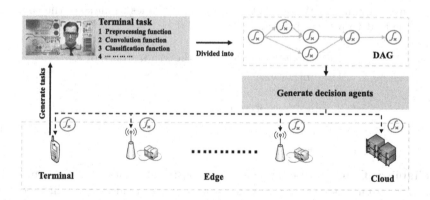

Fig. 1. FCO in Cloud-edge Collaborative Network.

According to the literature [18], the execution of any function may be interrupted since the trigger of the function failures and the hardware of software failures (e.g., unexpected restart/shutdown of a physical machine, network disconnection, software bugs, etc.). Therefore, the reliability of the FCO is the most important consideration besides performance. Effective ways to improve reliability include adding redundancy, fast recovery, offsite migration, etc. In this paper, we deploys multiple instances of the same function running at the same time to increase reliability. Because the redundancy function has the potential to further reduce the end-to-end delay of task. We define θ as the reliability of a computing node, which specifies the probability of successful completion of a function on a computing node. Then, the reliability of a function is defined as:

$$r_f = 1 - \prod_k^{R_f}(1 - \theta) \tag{1}$$

where r_f is the reliability of function f and R_f is the redundancy of f. k is the kth instance of the function. With respect to a task, the end-to-end reliability is calculated as the product of the reliability of the functions comprising the task. Thus, the reliability of a task is:

$$R_s = \prod_{f \in s} r_f = \prod_{f \in s} \left[1 - \prod_k^{R_f}(1 - \theta) \right] \tag{2}$$

where R_s is the reliability of the task. f specifies one of the functions s. k is the kth instance of the function f. Therefore, the more redundant instances are instantiated, the higher reliability task is.

3.2   Problem Formulation

Decision Variables. We define the following variables to indicate the computing node that to process function:

$$x_{ij}^k = \begin{cases} 1, & \text{node } j \text{ is processing the } k\text{th instance of } s_i \\ 0, & \text{otherwise.} \end{cases} \tag{3}$$

$$\hat{x}_{ij}^k = \begin{cases} 1, & \text{output of the } k\text{th instance of } s_i \text{ are transmitted} \\ & \text{to node } j \\ 0, & \text{otherwise.} \end{cases} \tag{4}$$

Reliability Constraints. The following constraint shows that the number of function redundant instances should ensure the task request reliability requirement.

$$R_s = \prod_{i \in s} r_i = \prod_{i \in s} \left[1 - \prod_j^N \prod_k^{R_i}(1 - \theta_j)x_{ij}^k \right] \geq \Phi_s, \tag{5}$$

$$\forall i \in s, j \in N, k \in R_i$$

where i, j, N, k denote ith function, jth computing node, the number of computing nodes, and the kth redundancy of the function. R_i denotes the number of redundancies of the ith function, respectively. Φ_s is the minimum reliability that the task needs to meet. Based on the above formula, it is possible to determine how many redundant functions are needed. However it is not possible to determine which function needs to be redundant. Therefore, we redundant functions by order. Moreover, the difference between two number in a redundancy list is less than 1, which can be proved to be the maximum reliability when a fixed amount of redundancies is given [11].

Lemma 1: Given two redundancy a, b, where $a, b \geq 2$ and $a + b \equiv R_f$, if $0 \leq a - b \leq 1$ than $Rel(a, b)$ is maximized. $Rel(a, b)$ is the reliability of the task composed of a and b.

Proof: To prove it by contradiction try and assume that the statement is false. Let $p = 1 - \theta$,where $0 < \theta < 1$, then

$$[1 - (1 - \theta)^a] \times [1 - (1 - \theta)^a] = (1 - p^a) \times (1 - p^b) \tag{6}$$

Therefore,

$$(1 - p^a) \times (1 - p^b) < (1 - p^{(a+1)}) \times (1 - p^{(b-1)})$$
$$p^a(1 - p) < p^{(b-1)}(1 - p)$$
$$p^a < p^{(b-1)} \tag{7}$$
$$a < b - 1$$

Here, it arrives to a contradiction with $a - b \geq 0$.

Calculation Delay and Constraint. The calculation delay of function can be formulated as:

$$t_{calc} = \sum_{i}^{|s|} \sum_{j}^{N} \sum_{k}^{R_i} \frac{d_i^k w_i}{\varpi_j} x_{ij}^k, \tag{8}$$
$$\forall i \in s, j \in N, k \in R_i$$

s.t.

$$d_i^k = \varrho_{i-1} d_{i-1}^k, \quad \forall i \in s, k \in R_i \tag{9}$$

$$\sum_{k}^{R_i} d_i^k = \varrho_{i-1} d_{i-1}^k, \quad \forall i \in s \tag{10}$$

where t_{calc} is calculation delay of all functions. i, j, N, k denote ith function, jth computing node, the number of computing nodes, and the kth redundancy of the function. R_i denotes the number of redundancies of the ith function, respectively. d_i^k is the amount of data input to the kth redundant instance of function i, w_i is the frequency required by the function i to process a unit of data. Constraint 9 indicates that the amount of input data of function i is the amount of output data of function $i - 1$; Constraint 10 indicates that redundant functions are not involved in the execution.

Transmission Delay and Constraint. The transmission delay of data between functions can be formulated as:

$$t_{txm} = \sum_i^{|s|} \sum_j^N \sum_k^{R_i} \frac{d_i^k \varrho_i}{\gamma_j} \hat{x}_{ij}^k, \tag{11}$$

$$\forall i \in s, j \in N, k \in R_i$$

s.t.

$$\hat{x}_{ij}^k = x_{i+1j}^k, \quad \forall i \in s, j \in N, k \in R_i \tag{12}$$

where t_{txm} is the transmission delay of task. i, j, N, k denote ith function, jth computing node, the number of computing nodes, and the kth redundancy of the function. R_i denotes the number of redundancies of the ith function, respectively. d_i^k is the amount of data input to the kth redundant instance of function i, ϱ_i is a coefficient characterizing the relationship between the output data of the function and the input data and γ_j is the uplink transmission speed of computing node j. Constraint 12 indicates that data can only be transferred along the order of the functions.

End-to-End Delay. The following equation guarantees the transmission delay of the task.

$$T_s = t_{calc} + t_{txm} \tag{13}$$

In this paper, we aim to minimize the end-to-end delay of a task request with given network resources. The objective is defined as follows:

$$\arg\min \sum_i^{|s|} \sum_j^N \sum_k^{R_v} T_s \tag{14}$$

The reliable function computation offloading problem is modeled as a Mixed Integer Non-Linear Problem (MINLP) with a series of constraints, which is very complex to solve. We decompose the reliable FCO problem into three subproblems and solve them one by one. In the next section, we present the specific solutions to those problem.

4   Solutions to Reliability-Aware FCO Scheduling

According to [20], the basic function scheduling problem can be regarded as an extended Flexible Job Shop Problem (FJSP) which is proved to be NP-hard and too difficult to find an optimal solution in polynomial time. Therefore, the problem defined in Sect. 3.2 is an NP-hard problem. In order to simplify the solution of the problem, we decompose the reliable FCO into three subproblems to reduce the complexity. The first subproblem is determining the optimal number of function redundancies to guarantee the reliability requirement. For this problem, we propose an algorithm to determine the optimal number of redundancies. The second subproblem is to determine where to deploy the functions based on the order of the functions. Redundant functions are not deployed for the time being.

We propose a heuristic algorithm to optimize the end-to-end delay of task. The third subproblem is the redundant function deployment problem. We develop an intelligent RL-based algorithm to handle this problem. Increase reliability by deploying redundancy functions while being able to find data forwarding paths with smaller end-to-end latency.

4.1 Overview of Reliable Function Computation Offloading

We will refer to the strategy proposed in this paper as the reliable function computation offloading (RLFCO). It includes three critical steps. When there comes an terminal task which has been decomposed into functions. **Redundancy Determining** is used to calculate the optimal redundancy of functions according to the required task reliability. This process determines which functions need redundancy and how many redundant functions should be deployed. **Heuristic node selection for non-redundant functions** provides a strategy that selects an appropriate computing node for the functions and calculate the end-to-end delay of task. The redundancy function is not in the decision. This procedure identifies a suboptimal functions deployment strategy and does not meet the reliability requirements. **Reinforcement Learning for Node Selection of Redundant Functions** is designed to provide a deployment strategy for redundant functions. If the redundant function is deployed to reduce the end-to-end delay of task, then the redundant function will replace the original function and the data forwarding path will be changed. The original function becomes a redundant function.

4.2 Redundancy Determining

Adding redundancy functions on cloud and edge computing nodes is the most effective paradigm to improve reliability. The approach improves reliability by boosting the overhead of hardware resources. Therefore a tradeoff between resource consumption and reliability is required. In order to obtain an optimal reliability-aware scheduling without excessive resource consumption, it is required to determine the minimum number of redundancies and decide which functions should be replicated. Our strategy is to start from the first function and add a redundancy of first function. Then evaluate whether the reliability requirements are satisfied. If adding a redundancy of last function still fails to meet the reliability requirement, then it will go back to the first function and repeat the above process.

Suppose a task s is composed of 5 functions. The initial number of all functions at the beginning is 1. Assume that the reliability probability of any function deployed on any node is $\theta = 0.96$. Task has a minimum reliability requirement of $\Phi_s \geq 0.95$. Hence the initialized task reliability is $Rel(s) = \theta^5 = 0.815 \leq 0.95$. Then we add a redundancy to the first function and the reliability of task is $Rel(s) = \theta^3(1 - (1 - \theta)^2) = 0.848$. The reliability of task reaches 0.95 until the fourth function adds a redundancy. Figure 2 shows the impact of redundant functions on reliability.

We propose a redundancy determining Algorithm 1 to obtain the number of redundant functions. We assume that the reliability of any function deployed on any node is the same, $\theta_{j \in N} = \theta$. Given the service function chain $s = f_i$, the reliability of a computing node θ, and the reliability requirement of the task Φ_s, the algorithm outputs a redundancy list Redundancy(f) that indicates the number of redundancies for every function and the length of Redundancy(f) $\mathcal{H}$.

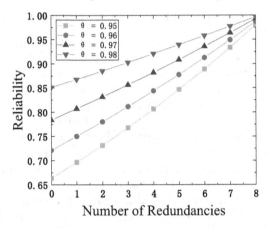

Fig. 2. The impact of redundant functions on reliability.

Algorithm 1: Redundancy Determining

Input: $s = \{f_i\}$, θ, Φ_s
Output: Redundancy(f) and the redundancy number of f_i
1 $\mathcal{L} \leftarrow$ The length of s;
2 Initialize the function as 1;
3 Calculating task reliability Φ_s;
4 **if** $Rel(s) \geq \Phi_s$ **then**
5 $\quad$ **return** *Task satisfy the given reliability;*
6 **else**
7 $\quad$ Set $i \leftarrow 1$;
8 $\quad$ **while** $i \leq \mathcal{L}$ **do**
9 $\quad\quad$ Add a redundancy to the function f_i;
10 $\quad\quad$ Calculating task reliability Φ_s according to Eq. 5;
11 $\quad\quad$ **if** $Rel_s \geq \Phi_s$ **then**
12 $\quad\quad\quad$ Redundancy(f) $\leftarrow f_i$;
13 $\quad\quad\quad$ $\mathcal{H} \leftarrow$ The length of Redundancy(f);
14 $\quad\quad\quad$ **return** *Redundancy(f) and $\mathcal{H}$;*
15 $\quad\quad$ **else**
16 $\quad\quad\quad$ Redundancy(f) $\leftarrow f_i$;
17 $\quad\quad\quad$ $i+ = 1$;

18 **end**

4.3   Heuristic Node Selection for Non-redundant Functions

This section presents a heuristic node selection algorithm, i.e., Algorithm 2, for non-redundant functions. Given the service function chain $s = f_i$, the properties of the function f_i, and the scenario information, i.e., the number of computing nodes, the properties of computing nodes, the algorithm outputs the computing node where the function f_i should be developed. The algorithm starts by selecting the node of the first function with the shortest end-to-end delay as the deployment node. Subsequent functions take the same procedure when it is deployed (line4-16). Note that different functions can not be deployed on the same nodes, but the resources are not shared.

Algorithm 2: Heuristic node selection for non-redundant functions

Input: $s = \{f_i \triangleq (d_i, \omega_i, \varrho_i)\}$,
$\quad\quad CN = \{\mathcal{C}_j \triangleq (\varpi_j, \gamma_j)\}$

Output: Functions offloading strategy Γ_{f_i}

1   $\mathcal{L} \leftarrow$ The length of s;
2   $\mathcal{N} \leftarrow$ The number of computing node (CN);
3   Initialize the function as 1;
4   **while** $i \leq \mathcal{L}$ **do**
5   Set an empty list $\tau_{i,j}$;
6   **if** $i == 1$ *or* $i == \mathcal{L}$ **then**
7   **while** $j \leq \mathcal{N}$ **do**
8            $\tau_{i,j} \leftarrow sum(\tau_i) + d_i\omega_i/\varpi_j$;
9            $j+ = 1$;
10        $\Gamma_{f_i} \leftarrow \arg\min \tau_i$;
11   **else**
12   **while** $j \leq \mathcal{N}$ **do**
13            $\tau_{i,j} \leftarrow sum(\tau_i) + d_i/\gamma_{\Gamma_{f_{i-1}}} + d_i\omega_i/\varpi_j$;
14            $j+ = 1$;
15        $\Gamma_{f_i} \leftarrow \arg\min \tau_i$;
16      $i+ = 1$;
17 **return** *Functions offloading strategy* Γ_{f_i}

Heuristic algorithms may not always produce the best results, but they can still provide decent results in a short amount of time. In situations where reliability is crucial, redundancy is necessary for certain functions. By implementing redundant functions, the end-to-end delay of tasks can be further reduced and a result that is closer to the optimal strategy can be achieved. As a result, the deployment strategy of redundant functions plays a significant role.

4.4 Reinforcement Learning for Node Selection of Redundant Functions

We develop a Reinforcement Learning (RL) Algorithm 3 that used to select the optimal node for the redundancy function. The RL agent takes as input the current state of the computing node and a redundancy function and outputs a offloading action. The agent obtains a reward after performing an action according to the reward function, and updates the agent parameters.

State. State is the input of the RL agent. The state consists of the redundancy function type, computing resources required for redundant functions, deployed functions and their locations, the residual computing resources of the computing node, the communication resources of computing nodes.

Algorithm 3: Reinforcement Learning for Node Selection of Redundant Functions

---

Input: $Redundancy(f) = \{f_i \triangleq (d_i, \omega_i, \varrho_i)\}$,
$\quad\quad CN = \{\mathcal{C}_j \triangleq (\varpi_j, \gamma_j), \Gamma_{f_i}\}$

Initialization: Greedy police EPSILON = 0.9
$\quad\quad\quad\quad\quad$ Learning rate ALPHA = 0.1
$\quad\quad\quad\quad\quad$ Discount factor GAMMA = 0.9
$\quad\quad\quad\quad\quad$ MAX_EPISODES = 500

1 $\mathcal{L}, \mathcal{N} \leftarrow$ the number of $Redundancy(f)$ and Computing node;
2 MEMORY $\leftarrow$ build_memory($\mathcal{L}, \mathcal{N}$);
3 **while** $episode \leq MAX\_EPISODES$ **do**
4 $\quad$ $CN \leftarrow$ Initialize env;
5 $\quad$ **for** $i \leq \mathcal{L}$ **do**
6 $\quad\quad$ A = choose an action based on minimum MEMORY;
7 $\quad\quad$ R $\leftarrow$ Calculation of task end-to-end delay based on decision A;
8 $\quad\quad$ predict $\leftarrow$ Prediction from MEMORY of the task end-to-end time delay after taking an A action;
9 $\quad\quad$ **if** $i \neq \mathcal{L}$ **then**
10 $\quad\quad\quad$ target = R + GAMMA * MEMORY.iloc[i+1, :].min();
11 $\quad\quad$ **else**
12 $\quad\quad\quad$ target = R;
13 $\quad\quad\quad$ MEMORY.loc[i, A] += target;
14 $\quad\quad\quad$ MEMORY.loc[i, A] += ALPHA * (target - predict);
15 $\quad\quad$ $CN \leftarrow$ update_env(A, f_i);
16 $\quad$ $episode += 1$;
17 **for** $i \leq \mathcal{L}$ **do**
18 $\quad$ $\Gamma'_{f_i} \leftarrow \arg\min$ MEMORY.iloc[1,:];
19 **return** $Redundant\ functions\ offloading\ strategy\ \Gamma'_{f_i}$

Action. Action is the output of the RL agent, which means which node the redundancy function should be deployed. Hence the action is a set of cloud and edge computing nodes.

Reward. Reward is the feedback from the environment after an action is performed. To guide the agent, we design a reward R based on the task end-to-end delay (Eq. 13). The rewards are stored in memory and progressively updated over many iterations. The update function is shown in line 14 of the Algorithm 3.

4.5 Complexity Analysis

Time Complexity. To analyse the time complexity of RLFCO, we need to analyse time complexity of the three algorithms separately. Algorithm 1 has 16 lines of code operations, 3 lines of constant complexity, an If conditional judgement and a While loop. The number of While loops is $\mathcal{L}$. The time complexity is therefore $O(\mathcal{L})$. Algorithm 2 contains a two-level nested while loop. The number of loops in the first while is $\mathcal{L}$ and the number of loops in the second while is $\mathcal{K}$. Thus the time complexity is $O(\mathcal{L} * \mathcal{K})$. Algorithm 3 has three loops, one of which is a For loop nested within a While loop. The number of executions of both For-loops is $\mathcal{L}$. The number of While loops is $\mathcal{L}$. Therefore the time complexity of Algorithm 3 is $O(MAX\_EPISON * \mathcal{L})$. In summary, the total time complexity of the RLFCO algorithm proposed in this paper is $O(MAX\_EPISON * \mathcal{L} + \mathcal{L} * \mathcal{K} + \mathcal{L})$.

5 Evaluation

In this section, we first present the setup in our simulation. Then we conduct simulations to evaluate the performance of our proposed RLFCO for solving the function scheduling problem.

5.1 Simulation Setup

Simulation Environment Parameters. We implement a Python-based FCO simulator to build and train the agent. All the simulations are executed on a machine with Intel Core i7 2.9 GHz and 48 GB RAM.

Baseline Algorithms. To prove the effectiveness of our algorithm, we implement a Q-Learning model and a Heuristic model as comparison method. We evaluate the following algorithms.

- Heuristic is a greedy algorithm that selects the one with the lowest end-to-end delay as the deployment node.
- The Q-Learning implementation of RL agent, which only use the value function to make actions. The Q-learning approach to directly decide which computing node to deploy functions and redundant functions.

5.2 Impact of the Number of Functions

We evaluate the performance with different number of functions of a task as shown in Fig. 3. Since the number of different functions has an impact on the reliability and redundant nodes of task, we let the number of nodes of redundant functions remain the same and set to 2 for a fair comparison. The result shows that the end-to-end delay of all increases with the increase of functions. The Q-learning (700) and Q-learning (7000) denote the results of the Q-learning method after 700 iterations and 7000 iterations, respectively. Note that, with the same computing capacity of computing nodes, the end-to-end delay is affected by the inter-node communication capabilities. The inter-node communication capabilities can be influenced by both computing nodes and scheduling policies. Our strategy can achieve better end-to-end delay with a shorter number of iterations. Q-learning performance is significantly worse than ours under the same conditions. After 7000 iterations the delay of the Q-learning approach decreases significantly and is still much higher than ours. After sufficient learning, Q-learning may achieve better performance than us, but the time complexity will be considerably higher than ours.

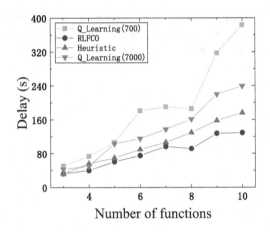

Fig. 3. Impact of the number of functions.

5.3   Impact of the Number of Computing Nodes

We evaluate the performance with different number of computing nodes in the system as shown in Fig. 4. The result shows that the end-to-end delay of all methods maintain stability with the increase of computing nodes. Although the number of nodes increases, the number of computing nodes to deploy the function is equal to the number of functions, and the impact of the number of computing nodes on the delay should be smaller while the capacity of the nodes remains stable. Our proposed method obtains the lowest end-to-end delay, which is slightly lower than that of the heuristic. Q-learning has the highest delay. Obviously, this solution meets our expectation, since our algorithm finds a better solution on the heuristic solution with the help of redundant functions. For the same number of iterations, the result of Q-learning is not ideal and is higher than the heuristic solution.

5.4   Impact of the Reliability

We evaluate the performance with different number of redundant functions as shown in Fig. 5. The redundancy functions have an overall similar impact on the end-to-end delay of task for different number of functions. However, it also shows that the reliability requirements have an impact on the algorithm proposed in this paper. Note that, when the redundant nodes are less than 4, the delay variation is small and proving that reliability requirements have little impact on the algorithm. When the redundant nodes are greater than 4, the delay of task all produces a significant increase. Since our algorithm is based on a heuristic solution for quadratic optimization. As the number of redundant nodes increases, the complexity of the algorithm increases, and the speed of convergence slows down as the number of functions increases for the same iteration count. When it is not feasible to converge quickly to the optimal value a sudden increase on the graph is observed.

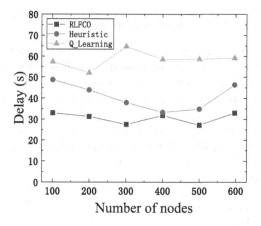

Fig. 4. Impact of the number of computing nodes.

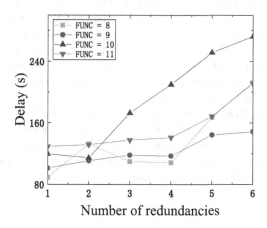

Fig. 5. FCO in Cloud-edge Collaborative Network.

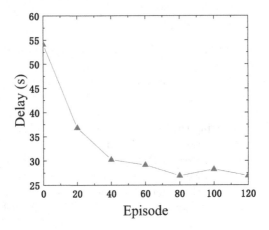

Fig. 6. Convergence of proposed method.

5.5 Convergence of Proposed Method

Figure 6 shows the learning curves of the proposed model, testing the models every 20 iterations. The experimental results show that our algorithm converges very quickly. When after 20 iterations, the end-to-end delay given by the algorithm approximate optimal value. This proves that our algorithm is indeed converging and also proves validity. The reliability requirements of task provide the opportunity to further optimize the latency. Since the number of redundant nodes is in most cases smaller than the number of functions, the complexity of our proposed algorithm is much smaller than that of other methods that take all functions as input.

6 Conclusion

In this paper, we study the reliability-aware FCO scheduling problem in a cloud-edge collaborative network. Our approach consists of three steps, namely redundancy determining, heuristic node selection for non-redundant functions and reinforcement learning for node selection of redundant functions. For the first two steps, we develop two mechanisms to determine the minimum number of redundancies to guarantee the reliability and generate a complete non-redundant function path. In the last step, we propose a scheduling algorithm based on reinforcement learning. Finally, the simulation results show the effectiveness of proposed approach in accelerating the deployment of functions between cloud-edge nodes. As further future work, we would like to consider a multi-resource network environment (i.e., bandwidth, CPU and memory), and to solve the FCO scheduling problem within real-world model.

References

1. Cai, J., Fu, H., Liu, Y.: Multitask multiobjective deep reinforcement learning-based computation offloading method for industrial internet of things. IEEE Internet Things J. **10**(2), 1848–1859 (2023). https://doi.org/10.1109/JIOT.2022.3209987
2. Cao, Z., Zhou, P., Li, R., Huang, S., Wu, D.O.: Multiagent deep reinforcement learning for joint multichannel access and task offloading of mobile-edge computing in industry 4.0. IEEE Internet Things J. **7**(7), 6201–6213 (2020). https://doi.org/10.1109/JIOT.2020.2968951
3. Chen, Q., Kuang, Z., Zhao, L.: Multiuser computation offloading and resource allocation for cloud-edge heterogeneous network. IEEE Internet Things J. **9**(5), 3799–3811 (2022). https://doi.org/10.1109/JIOT.2021.3100117
4. Chen, Z., Yi, W., Alam, A.S., Nallanathan, A.: Dynamic task software caching-assisted computation offloading for multi-access edge computing. IEEE Trans. Commun. **70**(10), 6950–6965 (2022). https://doi.org/10.1109/TCOMM.2022.3200109
5. Ding, Y., Li, K., Liu, C., Li, K.: A potential game theoretic approach to computation offloading strategy optimization in end-edge-cloud computing. IEEE Trans. Parallel Distrib. Syst. **33**(6), 1503–1519 (2022). https://doi.org/10.1109/TPDS.2021.3112604
6. Du, M., Wang, Y., Ye, K., Xu, C.: Algorithmics of cost-driven computation offloading in the edge-cloud environment. IEEE Trans. Comput. **69**(10), 1519–1532 (2020). https://doi.org/10.1109/TC.2020.2976996
7. Fantacci, R., Picano, B.: Performance analysis of a delay constrained data offloading scheme in an integrated cloud-fog-edge computing system. IEEE Trans. Veh. Technol. **69**(10), 12004–12014 (2020). https://doi.org/10.1109/TVT.2020.3008926
8. Guo, K., Gao, R., Xia, W., Quek, T.Q.S.: Online learning based computation offloading in MEC systems with communication and computation dynamics. IEEE Trans. Commun. **69**(2), 1147–1162 (2021). https://doi.org/10.1109/TCOMM.2020.3038875
9. Haber, E.E., Alameddine, H.A., Assi, C., Sharafeddine, S.: UAV-aided ultra-reliable low-latency computation offloading in future IoT networks. IEEE Trans. Commun. **69**(10), 6838–6851 (2021). https://doi.org/10.1109/TCOMM.2021.3096559
10. Hu, J., Li, K., Liu, C., Chen, J., Li, K.: Coalition formation for deadline-constrained resource procurement in cloud computing. J. Parallel Distrib. Comput. **149**, 1–12 (2021). https://doi.org/10.1016/j.jpdc.2020.10.004
11. Jia, J., Yang, L., Cao, J.: Reliability-aware dynamic service chain scheduling in 5G networks based on reinforcement learning. In: 40th IEEE Conference on Computer Communications, INFOCOM 2021, Vancouver, BC, Canada, 10–13 May 2021, pp. 1–10. IEEE (2021). https://doi.org/10.1109/INFOCOM42981.2021.9488707
12. Liang, B., Ji, W.: Multiuser computation offloading for edge-cloud collaboration using submodular optimization. Tongxin Xuebao/J. Commun. **41**(10), 25–36 (2020). communication resources;Computation offloading;Computing-task;Edge clouds;Greedy algorithms;Mode selection;Stable systems;Submodular optimizations. https://doi.org/10.11959/j.issn.1000-436x.2020205
13. Lin, C., Mahmoudi, N., Fan, C., Khazaei, H.: Fine-grained performance and cost modeling and optimization for Faas applications. IEEE Trans. Parallel Distrib. Syst. **34**(1), 180–194 (2023). https://doi.org/10.1109/TPDS.2022.3214783

14. Liu, G., Xiao, Z., Tan, G., Li, K., Chronopoulos, A.T.: Game theory-based optimization of distributed idle computing resources in cloud environments. Theor. Comput. Sci. **806**, 468–488 (2020). https://doi.org/10.1016/j.tcs.2019.08.019

15. Peng, J., Qiu, H., Cai, J., Xu, W., Wang, J.: D2d-assisted multi-user cooperative partial offloading, transmission scheduling and computation allocating for MEC. IEEE Trans. Wirel. Commun. **20**(8), 4858–4873 (2021). https://doi.org/10.1109/TWC.2021.3062616

16. Qiu, C., Wang, X., Yao, H., Du, J., Yu, F.R., Guo, S.: Networking integrated cloud-edge-end in IoT: a blockchain-assisted collective Q-learning approach. IEEE Internet Things J. **8**(16), 12694–12704 (2021). https://doi.org/10.1109/JIOT.2020.3007650

17. Qiu, T., Chi, J., Zhou, X., Ning, Z., Atiquzzaman, M., Wu, D.O.: Edge computing in industrial internet of things: architecture, advances and challenges. IEEE Commun. Surv. Tut. **22**(4), 2462–2488 (2020). https://doi.org/10.1109/COMST.2020.3009103

18. Qu, L., Assi, C., Shaban, K.B., Khabbaz, M.J.: A reliability-aware network service chain provisioning with delay guarantees in NFV-enabled enterprise datacenter networks. IEEE Trans. Netw. Serv. Manag. **14**(3), 554–568 (2017). https://doi.org/10.1109/TNSM.2017.2723090

19. Ren, J., Yu, G., He, Y., Li, G.Y.: Collaborative cloud and edge computing for latency minimization. IEEE Trans. Veh. Technol. **68**(5), 5031–5044 (2019). https://doi.org/10.1109/TVT.2019.2904244

20. Riera, J.F., Escalona, E., Batalle, J., Grasa, E., Garcia-Espin, J.A.: Virtual network function scheduling: concept and challenges, Vilanova i la Geltru, Spain (2014). complex scheduling;Network functions;Network services;Proof of concept;Routing function;Scheduling problem;State of the art;Virtual networks. https://doi.org/10.1109/SaCoNeT.2014.6867768

21. Shah-Mansouri, H., Wong, V.W.S.: Hierarchical fog-cloud computing for IoT systems: a computation offloading game. IEEE Internet Things J. **5**(4), 3246–3257 (2018). https://doi.org/10.1109/JIOT.2018.2838022

22. Sun, C., et al.: Task offloading for end-edge-cloud orchestrated computing in mobile networks. In: 2020 IEEE Wireless Communications and Networking Conference, WCNC 2020, Seoul, South Korea, 25–28 May 2020, pp. 1–6. IEEE (2020). https://doi.org/10.1109/WCNC45663.2020.9120496

23. Wang, C., Zhang, S., Chen, Y., Qian, Z., Wu, J., Xiao, M.: Joint configuration adaptation and bandwidth allocation for edge-based real-time video analytics. In: 39th IEEE Conference on Computer Communications, INFOCOM 2020, Toronto, ON, Canada, 6–9 July 2020, pp. 257–266. IEEE (2020). https://doi.org/10.1109/INFOCOM41043.2020.9155524

24. Wang, C., Yu, F.R., Liang, C., Chen, Q., Tang, L.: Joint computation offloading and interference management in wireless cellular networks with mobile edge computing. IEEE Trans. Veh. Technol. **66**(8), 7432–7445 (2017). https://doi.org/10.1109/TVT.2017.2672701

25. Yang, H., Xie, X., Kadoch, M.: Intelligent resource management based on reinforcement learning for ultra-reliable and low-latency IoV communication networks. IEEE Trans. Veh. Technol. **68**(5), 4157–4169 (2019). https://doi.org/10.1109/TVT.2018.2890686

26. Yang, Y., Long, C., Wu, J., Peng, S., Li, B.: D2D-enabled mobile-edge computation offloading for multiuser IoT network. IEEE Internet Things J. **8**(16), 12490–12504 (2021). https://doi.org/10.1109/JIOT.2021.3068722

27. You, C., Huang, K., Chae, H., Kim, B.: Energy-efficient resource allocation for mobile-edge computation offloading. IEEE Trans. Wirel. Commun. **16**(3), 1397–1411 (2017). https://doi.org/10.1109/TWC.2016.2633522

28. Zhan, Y., Guo, S., Li, P., Zhang, J.: A deep reinforcement learning based offloading game in edge computing. IEEE Trans. Comput. **69**(6), 883–893 (2020). https://doi.org/10.1109/TC.2020.2969148

29. Zhang, J., Du, J., Shen, Y., Wang, J.: Dynamic computation offloading with energy harvesting devices: a hybrid-decision-based deep reinforcement learning approach. IEEE Internet Things J. **7**(10), 9303–9317 (2020). https://doi.org/10.1109/JIOT.2020.3000527

30. Zhang, L., Cao, B., Li, Y., Peng, M., Feng, G.: A multi-stage stochastic programming-based offloading policy for fog enabled IoT-ehealth. IEEE J. Sel. Areas Commun. **39**(2), 411–425 (2021). https://doi.org/10.1109/JSAC.2020.3020659

31. Zhao, M., et al.: Energy-aware task offloading and resource allocation for time-sensitive services in mobile edge computing systems. IEEE Trans. Veh. Technol. **70**(10), 10925–10940 (2021). https://doi.org/10.1109/TVT.2021.3108508

32. Zhao, N., Du, W., Ren, F., Pei, Y., Liang, Y., Niyato, D.: Joint task offloading, resource sharing and computation incentive for edge computing networks. IEEE Commun. Lett. **27**(1), 258–262 (2023). https://doi.org/10.1109/LCOMM.2022.3220233

33. Zhou, H., Jiang, K., Liu, X., Li, X., Leung, V.C.M.: Deep reinforcement learning for energy-efficient computation offloading in mobile-edge computing. IEEE Internet Things J. **9**(2), 1517–1530 (2022). https://doi.org/10.1109/JIOT.2021.3091142

34. Zhu, X., Luo, Y., Liu, A., Bhuiyan, M.Z.A., Zhang, S.: Multiagent deep reinforcement learning for vehicular computation offloading in IoT. IEEE Internet Things J. **8**(12), 9763–9773 (2021). https://doi.org/10.1109/JIOT.2020.3040768

A Fast, Reliable, Adaptive Multi-hop Broadcast Scheme for Vehicular Ad Hoc Networks

Ping Liu[1], Xingfu Wang[1(✉)], Ammar Hawbani[1(✉)], Bei Hua[1], and Liang Zhao[2]

[1] School of Computer Science and Technology, University of Science and Technology of China, Hefei 230000, China
iacmy@mail.ustc.edu.cn, {wangxingfu,anmande,bhua}@ustc.edu.cn
[2] Department of Computer Science, Shenyang Aerospace University, Shenyang 110136, China
lzhao@sau.edu.cn

Abstract. Multi-hop broadcasting serves as an effective countermeasure against the negative impacts of rapid vehicular movement and internal interference on data transmission in Vehicular Ad Hoc Networks (VANETs), thereby enhancing support for safety-critical applications. The selection of candidate forwarding vehicles within this multi-hop broadcasting paradigm significantly impacts network performance. This paper proposes a Fast, Reliable, and Adaptive (FRA) multi-hop broadcast scheme, specifically designed to lower collision probability and reduce communication overhead. Initially, we examine the correlation between transmission distance and packet reception ratio under different channel conditions, leading to the development of an adaptive forwarding priority function. This function prioritizes data forwarding by preferentially selecting vehicles that exhibit higher packet reception ratios, depending on the channel's collision rates. Following this, we propose a contention-based scheme to assign waiting times to candidate vehicles based on the adaptive priority function. The proposed scheme can adaptively adjust the contention window size according to the forwarding priority and vehicular density, effectively differentiating waiting times among candidate vehicles and reducing the collision probability. Significantly, it guarantees that high-priority vehicles obtain shorter waiting times than their low-priority counterparts, aiming to minimize communication overhead. Simulation results indicate that FRA excels in terms of packet delivery ratio, average delivery time, and network overhead.

Keywords: Vehicular Networks · Multi-hop communication · Opportunistic Routing

1 Introduction

Underscoring the progression towards a more interconnected and intelligent transportation system, Vehicular Ad Hoc Networks (VANETs) have surfaced as a pivotal technology [14]. VANETs are capable of enhancing traffic flow efficiency and improving the passenger experience by enabling real-time collaboration among vehicles [10]. More importantly, VANETs play a crucial role in bolstering traffic safety [4,15]. Within VANETs, vehicles can communicate directly with each other using On-Board Units (OBUs), providing more efficient and direct support for safety-critical applications compared to other communication technologies like mobile cellular networks.

The fast and reliable transmission of emergency messages is one of the basic requirements for safety-critical applications [5]. However, the rapid changes in network topology caused by the high mobility of vehicles, coupled with the internal interference in wireless medium, can lead to instability in the communication links between vehicles. Opportunistic routing presents an effective solution for addressing the problem of link instability in VANETs [7]. Many opportunistic multi-hop broadcast schemes have been proposed to improve VANETs network performance [1]. In these schemes, the sender initially broadcasts emergency messages to neighboring vehicles. However, not all vehicles that receive the message are allowed to forward it. Allowing all vehicles to do so would result in excessive redundant transmissions, potentially causing severe collisions. Instead, only one vehicle is selected to re-forward the message. This process repeats until the message reaches the target Area of Interest (AoI). The selection of which vehicle will re-forward the message can be effectively managed by assigning different timers to the vehicles [9]. If a vehicle that has received the message does not detect that another vehicle has already re-forwarded the message after the specified waiting time, it will immediately re-forward the message itself. A well-designed policy for assigning waiting times can effectively reduce both the probability of collisions and the delivery delay required for message propagation. Opportunistic multi-hop broadcast schemes can be classified into two types based on how the waiting times are determined: sender-based and receiver-based [20]. In the sender-based scheme, the sender calculates the waiting times for the potential forwarding vehicles and communicates this information to them through wireless transmission. While effective, this scheme introduces additional control overhead. Furthermore, it is more prone to the hidden terminal problem, particularly when the sender is at a significant distance from the candidate vehicles. In the receiver-based scheme, vehicles that receive the message independently calculate their own waiting times in a distributed manner, without requiring communication from the sender. Moreover, as these potential forwarding vehicles are closer to the target AoI than the sender, they have a more accurate perception of channel conditions. This enables them to make more informed decisions about forwarding the message. In this paper, our primary focus is on the receiver-based multi-hop broadcast scheme.

Nevertheless, the receiver-based scheme presents some challenges that need to be addressed. While its distributed nature eliminates the need for additional communication overhead, it also results in a lack of coordination among vehicles. This inevitably leads to an increased likelihood of transmission collisions. Balancing the minimization of collision probability with reducing forwarding waiting times for vehicles presents a significant challenge. Additionally, most existing receiver-based schemes have not fully leveraged the channel conditions perceived by the candidate forwarding vehicles. This oversight makes it difficult for them to adapt to fluctuating network loads and vehicular densities. This paper begins by examining the relationship between transmission distance and packet reception ratio under varying channel conditions. Leveraging this relationship, we design an adaptive priority function tailored to specific channel conditions for candidate vehicles, thereby reducing communication overhead. Moreover, we propose an adaptive contention-based scheme to calculate the waiting time for potential forwarding vehicles, based on their forwarding priorities and the vehicular density. This scheme effectively differentiates waiting times between candidate vehicles. More importantly, it ensures that high-priority vehicles are assigned shorter waiting times than low-priority ones, thereby minimizing the collision probability and communication overhead for relaying messages.

2   Related Work

In the opportunistic multi-hop broadcasting schemes, simultaneous message forwarding by multiple candidate vehicles can cause transmission collisions. These collisions can severely degrade network performance. Various protocols have been proposed to decrease collision probability while minimizing communication overhead [8]. These protocols fall into three categories: waiting-based, probability-based, and contention-based.

Waiting-based schemes aim to assign deterministic waiting times to candidate forwarding vehicles based on their priorities. Both Briesemeister et al. [2] and Viriyasitavat et al. [16] employ a continuous function that maps the distance between the forwarding vehicle and the sender to a waiting time. According to these functions, a candidate vehicle farther from the sender will have a shorter forwarding wait time. In addition to distance, other factors have been introduced to more effectively distinguish the forwarding wait times of different candidate vehicles. Yang et al. [18] take both the position and velocity of the vehicles into account when calculating their waiting times, while Zemouri et al. [19] consider link quality. Naderi et al. [11] design an analytic hierarchy process to assign waiting times to candidate vehicles based on multiple factors, including congestion, link loss, limited bandwidth, and interference. However, the deterministic nature of waiting times may limit the adaptability of these protocols to the complex and variable vehicle distribution in real-world scenarios. Contrary to waiting-based schemes, probability-based schemes assign specific forwarding probabilities to each candidate vehicle. Wisitpongphan et al. [17] propose a weighted p-Persistence broadcasting scheme that assigns higher probabilities to candidate vehicles further away from the sender, aiming to maximize the hop progress. However, this scheme does not consider the impact of vehicular density. As vehicular density increases, the probability of collisions also inevitably rises. Pei et al. [13] address this limitation by taking the channel busy ratio into account, thereby making the vehicles' forwarding probability adaptive to vehicular density. Contention-based schemes are a hybrid of waiting-based and probability-based schemes. They assign different sizes of contention windows to candidate vehicles based on their priorities. Vehicles then uniformly sample a random number within the specified contention window to determine their waiting times. Palazzi et al. [12] adjust the size of vehicles' contention windows based on transmission distance. Vehicles further from the sender are assigned a smaller contention window to reduce the number of relay hops for messages. Francesco et al. [6] build upon [12] by adaptively adjusting the size of the contention window in response to vehicular density, thereby reducing collision probability in high-density scenarios. Bujari et al. [3] provide an analytical model for contention-based schemes, theoretically examining the number of relay hops and propagation delay. A significant drawback of these contention-based schemes is that the candidate forwarding vehicles with the highest priority do not always achieve the shortest waiting times. This inconsistency could potentially impede the propagation process of the messages.

3   Adaptive Contention-Based Multi-hop Broadcast Scheme

In this paper, we propose an adaptive contention-based opportunistic multi-hop broadcast scheme. This scheme aims to resolve the inconsistency found in traditional contention-based schemes and reduce the communication overhead. Our proposed scheme consists of two main components. First, we devise a novel forwarding priority function that adapts to channel conditions. Second, we propose an adaptive contention-based scheme to calculate the waiting time for forwarding vehicles, based on the designed forwarding priority function.

3.1   Adaptive Forwarding Priority Functions

In VANETs, vehicles equipped with on-board units communicate with each other over a wireless channel. Given that all participants in a wireless channel share the same transmission medium, these transmissions are highly susceptible to collisions. Technologies such as CSMA/CA of IEEE 802.11p have been widely adopted to mitigate the occurrence of collisions, but completely avoiding collisions remains challenging, especially under heavy network loads. Additionally, the hidden terminal problem further increases the probability of collisions. When a collision occurs, the packet delivery ratio inevitably decreases. Hence, conducting experiments to explore the impact of channel collision rate on vehicular transmission becomes essential.

In this section, we employ the network simulator NS-3 to calculate the packet reception ratios for all neighboring vehicles linked to a common sender on a specific road segment, under a variety of channel collision rates. The parameters used for the simulation are detailed in Sect. 4. We measure the channel collision rate P by analyzing the beacons transmitted and received by vehicles within the road segment each second, as formulated in Eq. (1). Here, N_t represents the number of beacons that vehicles are expected to receive each second, while N_r denotes the number of actual beacon packets received by the vehicles. A larger value of P indicates a higher degree of collision within the channel. During the simulation, we categorize the sender's neighboring vehicles into five groups based on their respective distances from the sender. These distances fall within the ranges of $[0\,\text{m}, 50\,\text{m})$, $[50\,\text{m}, 100\,\text{m})$, $[100\,\text{m}, 150\,\text{m})$, $[150\,\text{m}, 200\,\text{m})$, and $[200\,\text{m}, 250\,\text{m}]$ respectively.

$$P = 1 - \frac{N_r}{N_t} \qquad (1)$$

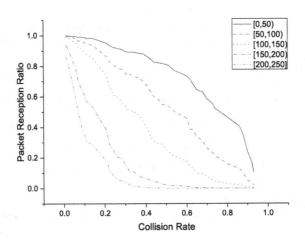

Fig. 1. The packet reception ratios for the five vehicle groups under different channel collision rates.

Figure 1 illustrates the packet reception ratios for the five vehicle groups across different channel collision rates. As depicted in Fig. 1, the influence of the channel collision rate on the packet reception ratio varies for receivers located at different distances from the sender. Generally, the packet reception ratio for vehicles starts to decline gradually as the channel collision rate escalates. However, the collision rate's impact on the packet reception ratio varies depending on the vehicle's distance from the sender. When the collision rate is low, vehicles at all distances can receive broadcast packets from the sender with relatively high probability. Conversely, when the collision rate in the channel is high, vehicles located farther from the sender scarcely receive packets, while those closer to the sender can still receive packets with a high likelihood. We hypothesize that this observation is because vehicles situated further away from the sender are more susceptible to internal interference and issues related to hidden terminal.

The design of the forwarding priority function in prior opportunistic multi-hop schemes generally follows the form shown in Eq. (2). In this equation, γ_i represents the forwarding priority of vehicle v_i, and d_i represents the distance between the forwarding vehicle v_i and the sender vehicle v_s. R_t denotes the communication range of the vehicles. According to Eq. (2), vehicles located farther from the sender are assigned higher priority, which helps to reduce the delivery delay required for message transmission.

$$\gamma_i = \frac{d_i}{R_t} \tag{2}$$

However, our simulation results presented in Fig. 1 reveal that, under high channel collision rates, vehicles located farther away encounter difficulties in receiving messages from the sender. In such scenarios, candidate vehicles positioned at a greater distance from the high-priority vehicles are likely to miss their forwarded messages. This situation leads to an increase in redundant transmissions, subsequently raising the channel collision rate. Therefore, we propose a modification to the forwarding priority function to assign lower priorities to vehicles situated farther from the sender under high channel collision rates. Specifically, forwarding vehicles assess the current channel collision rate P based on their own receipt of neighboring beacons. If P exceeds a predefined threshold σ, we use the modified priority function as formulated in Eq. (3) to calculate forwarding priorities. The ρ in Eq. (3) is a parameter used to define the valid range for candidate forwarding vehicles. In this paper, we select $\sigma = 0.5$ and $\rho = 0.6$ as these values correspond to the point at which the packet reception ratio for vehicles located more than 150 m away approaches zero, refer to Fig. 1.

$$\gamma_i = \begin{cases} \frac{d_i}{R_t} & \text{if } d_i \leq \rho \cdot R_t, \\ 0 & \text{otherwise.} \end{cases} \tag{3}$$

3.2 Adaptive Contention-Based Scheme

In this section, we introduce an adaptive contention-based scheme that calculates waiting times for forwarding vehicles based on their modified priority function. In previous work FMA [6], a vehicle, denoted as v_i, would randomly sample a waiting time from the contention window, defined as $[1, CWS_i]$. Here, CWS_i is determined by the forwarding priority of v_i, with higher priority values resulting in smaller CWS_i, and therefore potentially shorter forwarding waiting times. However, these traditional schemes set the lower bound of the contention window to 1, which implies a high likelihood that vehicles with differing forwarding priorities may end up with similar waiting times. Moreover, this design doesn't ensure that high-priority forwarding vehicles consistently obtain shorter waiting times compared to low-priority vehicles. In response to these limitations, we propose a novel contention window-based scheme. This new scheme aims to produce more differentiated forwarding waiting times for vehicles with different priorities and guarantees that high-priority vehicles consistently obtain shorter forwarding waiting times than those with lower priorities.

We denote the set of the candidate forwarding vehicles of the sender as V. In VANETs, all vehicles are required to periodically broadcast beacons that disseminate information such as speed, location, and other essential data to foster cooperative awareness and enable safety applications. Given this, vehicles can not only calculate their own forwarding priorities but also estimate the forwarding priorities of neighboring vehicles based on the beacon information they receive. From these calculated priorities, potential forwarding vehicles can be classified into different priority groups. Vehicles belonging to the same priority group will be assigned the same contention window. Assuming that the value range of the forwarding priority is $[0, P]$, this can be divided into m distinct groups. Each group will then have a length denoted by $l = P/m$. The quantity of forwarding vehicles within each priority group is symbolized as η, where $\eta = (\eta_1, \eta_2, ..., \eta_m)$. η can be computed using equation Eq. (4), where k ranges from 0 to $m - 1$.

$$\eta_k = \sum_{i \in |V|} [\gamma_i \in (l \cdot k, l \cdot k + l)] \tag{4}$$

The symbol $[...]$ appearing in Eq. (4) represents the Iverson bracket, the functionality of which is described in Eq. (5).

$$[P] = \begin{cases} 1 & \text{if } P \text{ is true,} \\ 0 & \text{otherwise.} \end{cases} \tag{5}$$

Subsequently, for a vehicle v_i belonging to the kth priority group, we assign its contention window $(\underline{CWS_k}, \overline{CWS_k})$ according to Eq. (6), where $\eta_m = 0$.

$$\begin{cases} \underline{CWS_k} = \sum_{j=k+1}^{m} \eta_j \cdot \alpha^{m-1-j} \\ \overline{CWS_k} = \sum_{j=k}^{m} \eta_j \cdot \alpha^{m-1-j} \end{cases} \tag{6}$$

As demonstrated in Eq. (6), contention windows with distinct start and end times are assigned to various priority groups. These contention windows do not overlap, which ensures that vehicles in higher-priority groups consistently experience shorter forwarding wait times than those in lower-priority groups. Furthermore, as deduced from Eq. (6), the length of the contention window corresponding to the kth priority group, denoted as $n_k \cdot \alpha^{m-1-k}$, hinges on two factors. The first is the number of candidate forwarding vehicles within the kth priority group, and the second is the priority level of the group itself. As the number of vehicles in a group increases, the contention window lengthens, reducing the probability of transmission collisions within the group. Conversely, as the number of vehicles decreases, the contention window shortens, which in turn reduces wait times. This dynamic design allows the contention window's size to adapt to vehicular density. Moreover, to differentiate waiting times effectively among vehicles from various priority groups, the size of the contention window is designed to increase exponentially as the priority level decreases. As a result, vehicles in lower-priority groups have significantly longer waiting times than those in higher-priority groups. The constant α in Eq. (6) can take on values such as 2, 3, or 4. Ultimately, vehicle v_i randomly selects a number uniformly within the corresponding contention window to be its waiting time.

Algorithm 1 outlines the procedure for calculating the waiting time for candidate forwarding vehicles.

Algorithm 1. The procedures for candidate forwarding vehicle v_i to compute its waiting time.

Require:
    The forwarding priority value: γ_i;
    The candidate forwarding vehicles set: V;
    The length of the priority group: l;
Ensure:
    v_i's forwarding waiting time: T;
1: $k = m - 1$;
2: **for** $k \geq 0$ **do**
3:    **if** $\gamma_i \in (l \cdot k, l \cdot k + l)$ **then**
4:      $\underline{CWS_k} = \sum_{j=k+1}^{m} \eta_j \cdot \alpha^{m-1-j}$;
5:      $\overline{CWS_k} = \sum_{j=k}^{m} \eta_j \cdot \alpha^{m-1-j}$;
6:      Sampling a random number t uniformly within $(\underline{CWS_k}, \overline{CWS_k})$;
7:      $T = t \cdot 100\mu s$;
8:    **end if**
9:    $k = k - 1$;
10: **end for**
11: **return** T;

---

4 Performance Evaluation

In VANETs, vehicular density and speed are two critical factors that significantly influence overall network performance. Vehicular density impacts the number of candidate forwarding vehicles, affecting both the collision probability and delivery delay of message transmission. Vehicular speed can influence the stability of network topology, thus affecting the robustness of the opportunistic multi-hop broadcast schemes. Therefore, this section evaluates the performance of the proposed protocol across various scenarios, altering vehicular densities and speeds. We first present detailed simulation settings in Sect. 4.1. Subsequently, in Sects. 4.2 and 4.3, we compare the performance of different protocols under various vehicular densities and speeds, respectively.

4.1 Simulation Settings

To validate the effectiveness of the proposed protocol FRA, we compare FRA with two counterparts, namely FMA and FLOOD. FMA [6] represents an efficient and representative contention-based opportunistic broadcast protocol. Within FMA, vehicles have the capacity to dynamically adjust their contention window size in line with vehicular density, thereby striving to minimize both collision probability and delivery delay. FLOOD is a naive opportunistic broadcast protocol, which can intuitively reflect network conditions and the effectiveness of other broadcast protocols. Therefore, it is adopted as the baseline protocol in this paper. Our simulations are conducted on a linear road segment that is 1 km in length and consists of four lanes. Vehicles enter the segment from both ends, with maximum speeds capped at 60 kilometers per hour. We employ the Simulation of Urban Mobility (SUMO, Version 0.32.0) to generate the requisite simulation scenarios and vehicular traffic.

Table 1. SIMULATION SETTINGS

Parameter	Value
Experimental Environment	Ubuntu 20.04 (64 bit)
Network Simulator	NS-3.30
Vehicular Mobility Simulator	SUMO 0.32.0
Vehicular Density	20, 40, 60, ..., 160 veh/km
Maximum Vehicular Speed	60 km/h
Vehicular Distribution	Poisson Distribution
Car Following Model	Krauss model
Number of Lanes	4 per direction
Wireless Radio Range	250 m
Channel Bandwidth	6 Mbps
MAC Layer Protocol	IEEE 802.11p
Radio Propagation Model	*Two Ray Ground Reflection*
Emergency Packet Size	1024 bytes
Total Simulation Time	350 s
Beacon Interval	100 ms

All protocols are implemented using the Network Simulator NS-3 (version 3.30) on an Ubuntu 20.04 (64-bit) platform. We employ IEEE 802.11p as the MAC layer protocol. The Two Ray Ground Reflection Model is also utilized in the simulations, with the channel bandwidth set at 6 Mbps. Vehicles are configured to broadcast beacons at intervals of 100 ms to enable cooperative awareness. To facilitate the evaluation of the protocol's performance, two vehicles, marked as A and B, are parked at each end of the segment. Every second, vehicle A broadcasts an emergency message intended for vehicle B at the other end. The following evaluation metrics are considered in our simulations: (1) Packet Delivery Ratio: Defined as the ratio of the emergency messages received by B to the total number of messages broadcast by A, this metric can evaluate the broadcast reliability of the opportunistic multi-hop protocol. (2) Average Delivery Time: This metric represents the average end-to-end transmission time of emergency messages. (3) Number of Redundant Packets: This metric can reflect the collision probability of different protocols. A larger number of redundant packets indicates that more messages are being re-forwarded by candidate vehicles. The simulation parameters are listed in Table 1.

4.2   Varying Vehicular Densities

In this section, we estimate the influence of vehicular density on different protocols' network performance by varying the number of vehicles per kilometer from 20 to 160. The maximum speed of vehicles is set to 60 km/h. The other simulation settings are the same as the default settings.

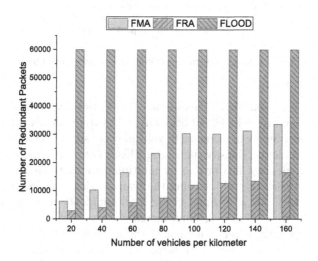

Fig. 2. Number of redundant packets varying the number of vehicles.

Figure 2 and Fig. 3 depict the influence of vehicular density on the number of redundant packets and the packet delivery ratio for each protocol, respectively. During wireless transmission, the simultaneous forwarding of messages by multiple candidate vehicles can cause channel collisions, leading to transmission failures. As vehicular density increases, the likelihood of simultaneous forwarding by multiple candidate vehicles also grows. In scenarios where an opportunistic multi-hop broadcast scheme, such as FLOOD, fails to effectively mitigate transmission collisions, the number of concurrently forwarding candidate vehicles can quickly rise. This increase is evidenced by a swift rise in redundant packets and a rapid decrease in the packet delivery ratio, as depicted in Fig. 2 and Fig. 3. Notably, FLOOD's number of redundant packets across all vehicular densities exceeds 60,000. For visualization purposes, we have marked its maximum value as 60,000 in Fig. 2. In contrast to FLOOD, the FMA protocol differentiates the forwarding times of candidate vehicles by assigning contention

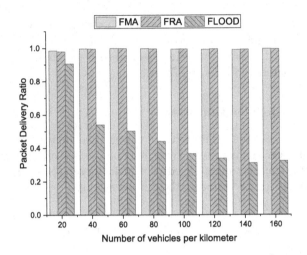

Fig. 3. Packet delivery ratio varying the number of vehicles.

windows with a common start time but varying lengths, determined by their distance from the sending vehicle. Additionally, FMA dynamically adjusts the size of the contention window in response to vehicular density, effectively lengthening the window as vehicular density increases. This, in turn, reduces the likelihood of simultaneous message forwarding by multiple candidate vehicles, leading to fewer redundant packets and lessening the transmission collision impact on the packet delivery ratio. As illustrated in Fig. 2, although the number of redundant packets in FMA does show a gradual increase with growing vehicular density, it remains significantly lower than that of the FLOOD protocol. FMA sustains a packet delivery ratio of around 1 across all vehicular densities, as shown in Fig. 3. In comparison to FMA, FRA implements a more efficient contention-based scheme. It employs an adaptive forwarding priority function to categorize candidate vehicles into different priority groups and assigns each group contention windows with distinct start and end times. This differentiation ensures that the contention windows for different priority groups do not overlap, thereby preventing transmission collisions between vehicles from different groups. Furthermore, as vehicular density increases and forwarding priority decreases, the contention window length expands exponentially, further reducing the collision likelihood among vehicles within the same priority group. In scenarios with high vehicular density, the adaptive priority function employed by the FRA protocol also mitigates the likelihood of redundant transmissions, as explained in Sect. 3.1. Figure 2 demonstrates that FRA generates significantly fewer redundant packets than both the FLOOD and FMA protocols in each scenario. Notably, the number of redundant packets generated by FRA is less than half that of the FMA protocol. Owing to FRA's remarkable capacity to reduce collision probability, it consistently upholds a high packet delivery ratio, regardless of vehicular density scenarios.

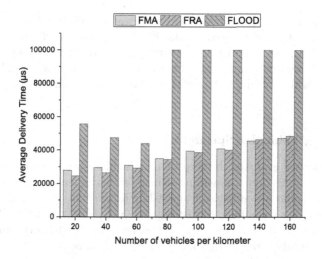

Fig. 4. Average delivery time varying the number of vehicles.

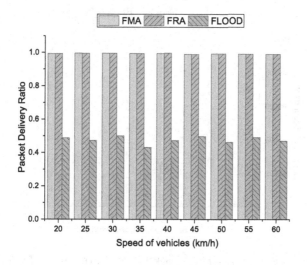

Fig. 5. Packet delivery ratio varying the speed of vehicles.

Figure 4 presents the variation in the average delivery time of different protocols in response to changes in the number of vehicles. In the FLOOD protocol, the maximum time is capped at 100,000 ms for better visualization. As illustrated in Fig. 4, both the FRA and FMA protocols are effective in reducing the average delivery time of emergency messages, even though their delivery times display a gradual increase in correspondence with a rise in vehicular density. This is largely due to the increased likelihood of collisions under conditions of higher vehicular density. Under conditions of low vehicular densities, FRA ensures that vehicles with higher priority are assigned shorter waiting times compared to those with lower priority. This strategy reduces vehicle wait times and results in a lower average delivery time compared to the FMA protocol. During high vehicular density scenarios, the adaptive forwarding priority function employed by FRA intentionally lowers the priority of vehicles located too far away to reduce redundant packets, which results in a slightly higher average delivery time compared to FMA.

4.3   Varying Vehicular Speeds

In this section, we evaluate the impact of vehicular speed on the performance of different protocols under a medium-density scenario, featuring 80 vehicles per kilometer. Vehicular speed varies from 20 km/h to 60 km/h, with all other simulation parameters remaining consistent with the default settings. The packet delivery ratio, number of redundant packets, and average delivery time of different protocols are depicted in Fig. 5, Fig. 6, and Fig. 7, respectively.

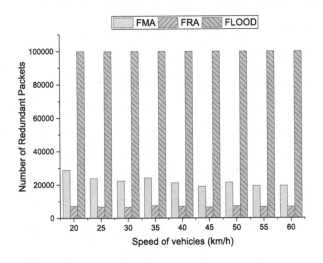

Fig. 6. Number of redundant packets varying the speed of vehicles.

Vehicular speed directly impacts the stability of the network topology in VANETs. The higher the speed, the more frequent the network topology changes. In urban environments, where vehicular speeds on roads fluctuate frequently, opportunistic broadcast protocols should exhibit robustness against these changes to ensure stable network performance. The simulation results depicted in Fig. 5, Fig. 6, and Fig. 7 demonstrate that alterations in speed do not significantly impact the performance of the three protocols. They exhibit insensitivity to changes in vehicle speed.

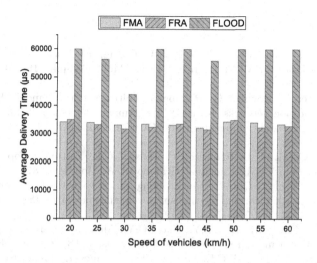

Fig. 7. Average delivery time varying the speed of vehicles.

5   Conclusion

This paper introduces an adaptive contention-based opportunistic multi-hop broadcast protocol, FRA, designed to enhance the reliability of vehicular ad hoc networks in complex environments while reducing communication overhead. We observe that when the channel collision rate is high, the packet reception ratio for receiving vehicles situated far from the transmitting vehicle is considerably lower than for those closer. In response, we develop an adaptive forwarding priority function that diminishes the priority of vehicles at greater distances under conditions of high collision rates. We also propose an adaptive contention-based scheme for calculating vehicle waiting times. This scheme dynamically adjusts the window size according to the vehicle's priority and the number of vehicles within the same priority group. More importantly, it ensures that high-priority vehicles consistently have shorter waiting times than their lower-priority counterparts. All these designs effectively reduce the collision probability and communication overhead. Simulation results demonstrate the effectiveness of the proposed protocol. However, the current approach to adjusting the contention window length could be refined further. In future work, we aim to design a method that allows for more finely-tuned window lengths, with the aim of further enhancing network performance.

Acknowledgements. This paper is supported by the Innovation Team and Talents Cultivation Program of the National Administration of Traditional Chinese Medicine (No. ZYYCXTD-D-202208).

References

1. Abbasi, H.I., Voicu, R.C., Copeland, J.A., Chang, Y.: Towards fast and reliable multihop routing in VANETs. IEEE Trans. Mob. Comput. **19**(10), 2461–2474 (2019)
2. Briesemeister, L., Hommel, G.: Role-based multicast in highly mobile but sparsely connected ad hoc networks. In: 2000 First Annual Workshop on Mobile and Ad Hoc Networking and Computing, MobiHOC (Cat. No. 00EX444), pp. 45–50. IEEE (2000)
3. Bujari, A., Conti, M., De Francesco, C., Palazzi, C.E.: Fast multi-hop broadcast of alert messages in VANETs: an analytical model. Ad Hoc Netw. **82**, 126–133 (2019)
4. Chen, C., Wang, C., Qiu, T., Xu, Z., Song, H.: A robust active safety enhancement strategy with learning mechanism in vehicular networks. IEEE Trans. Intell. Transp. Syst. **21**(12), 5160–5176 (2019)
5. Chou, Y.H., Chu, T.H., Kuo, S.Y., Chen, C.Y.: An adaptive emergency broadcast strategy for vehicular ad hoc networks. IEEE Sens. J. **18**(12), 4814–4821 (2017)
6. De Francesco, C., Palazzi, C.E., Ronzani, D.: Fast message broadcasting in vehicular networks: model analysis and performance evaluation. IEEE Commun. Lett. **24**(8), 1669–1672 (2020)
7. Guesmia, S., Semchedine, F., Djahel, S.: A scalable time-division-based emergency messages broadcast scheme for connected and autonomous vehicles in urban environment. Veh. Commun. **38**, 100544 (2022)
8. Li, P., Zeng, Y., Li, C., Chen, L., Wang, H., Chen, C.: A probabilistic broadcasting scheme for emergent message dissemination in urban internet of vehicles. IEEE Access **9**, 113187–113198 (2021)
9. Liu, B., et al.: A novel V2V-based temporary warning network for safety message dissemination in urban environments. IEEE Internet Things J. **9**(24), 25136–25149 (2022)
10. Mahi, M.J.N., et al.: A review on VANET research: perspective of recent emerging technologies. IEEE Access **10**, 65760–65783 (2022)
11. Naderi, M., Ghanbari, M.: Adaptively prioritizing candidate forwarding set in opportunistic routing in VANETs. Ad Hoc Netw. **140**, 103048 (2023)
12. Palazzi, C.E., Ferretti, S., Roccetti, M., Pau, G., Gerla, M., et al.: How do you quickly choreograph inter-vehicular communications? A fast vehicle-to-vehicle multi-hop broadcast algorithm, explained. In: CCNC, pp. 960–964 (2007)
13. Pei, Z., Chen, W., Li, C., Du, L., Liu, H., Wang, X.: Analysis and optimization of multihop broadcast communication in the internet of vehicles based on C-V2X Mode 4. IEEE Sens. J. **22**(12), 12428–12443 (2022)
14. Rani, P., Sharma, R.: Intelligent transportation system for internet of vehicles based vehicular networks for smart cities. Comput. Electr. Eng. **105**, 108543 (2023)
15. Taha, M.B., Alrabaee, S., Choo, K.K.R.: Efficient resource management of microservices in VANETs. IEEE Trans. Intell. Transp. Syst. **24**, 6820–6835 (2023)
16. Viriyasitavat, W., Tonguz, O.K., Bai, F.: UV-CAST: an urban vehicular broadcast protocol. IEEE Commun. Mag. **49**(11), 116–124 (2011)
17. Wisitpongphan, N., Tonguz, O.K., Parikh, J.S., Mudalige, P., Bai, F., Sadekar, V.: Broadcast storm mitigation techniques in vehicular ad hoc networks. IEEE Wirel. Commun. **14**(6), 84–94 (2007)
18. Yang, Y.T., Chou, L.D.: Position-based adaptive broadcast for inter-vehicle communications. In: 2008 IEEE International Conference on Communications Workshops, ICC Workshops, pp. 410–414. IEEE (2008)

19. Zemouri, S., Djahel, S., Murphy, J.: A fast, reliable and lightweight distributed dissemination protocol for safety messages in urban vehicular networks. Ad Hoc Netw. **27**, 26–43 (2015)
20. Zhang, H., Zhang, X., Sung, D.K.: A fast, reliable, opportunistic broadcast scheme with mitigation of internal interference in VANETs. IEEE Trans. Mob. Comput. **22**, 1880–1893 (2021)

A Grouping-Based Multi-task Scheduling Strategy with Deadline Constraint on Heterogeneous Edge Computing

Xiaoyong Tang, Wenbiao Cao, Tan Deng$^{(\boxtimes)}$, Chao Xu, and Zhihong Zhu

School of Computer and Communication Engineering,
Changsha University of Science and Technology, Changsha, Hunan 410114, China
dengtan0510@csust.edu.cn

Abstract. In heterogeneous edge computing, multiple tasks often compete for limited computing resources on the same edge server. These tasks request different edge computing services and usually have a deadline. Efficiently scheduling them is a complex and challenging problem. In this paper, we first develop a model for grouping and mapping limited edge computing resources. Then, we mathematically describe the multi-task scheduling problem with deadline constraints. Third, we propose a grouping-based multi-task scheduling strategy called GMTSS, which includes task regrouping and priority sorting, a resource-aware greedy scheduling algorithm, and a task adjusting method. Task regrouping and priority sorting are designed to balance the efficiency and fairness of scheduling multiple tasks. The greedy scheduling algorithm assigns tasks to an optimal node based on the status of resource groups. Additionally, task adjusting aims to achieve a better scheduling scheme that will meet the maximum number of deadlines or higher long-term satisfaction of system service, called LTSS. We conduct large-scale simulations, and the experimental results clearly show that our proposed GMTSS outperforms the current state-of-the-art benchmark strategy in terms of task completion rate within deadlines and LTSS. Furthermore, GMTSS performs well in terms of task completion time.

Keywords: Deadline constraint · Heterogeneous edge computing system · System service satisfaction · Task scheduling

1 Introduction

In recent years, the rapid development of IoT technology has driven the prosperity of cloud computing [7]. Unfortunately, deploying applications to public clouds leads to network congestion due to the high volume of data access and computing requests [13]. This approach will also result in a lower quality of experience (QoE) for delay-sensitive applications [9]. Therefore, edge computing has emerged as a promising computing paradigm to alleviate this problem [5]. However, limited and heterogeneous computing resources pose challenges for efficient task scheduling in edge computing [1]. Besides, the diversity of tasks leads

© The Author(s), under exclusive license to Springer Nature Singapore Pte Ltd. 2024
Z. Tari et al. (Eds.): ICA3PP 2023, LNCS 14488, pp. 468–483, 2024.
https://doi.org/10.1007/978-981-97-0801-7_27

to resource competition among multiple tasks on edge servers, which is also an urgent problem to be solved.

Table 1. Performance of Heterogeneous Edge Nodes.

Computing node	Cores & Frequency	Computing resource
Edge Gateway	4, 1.2–1.5 GHz	CPUs/GPUs
Local Server	4–24, 2.2–3.5 GHz	CPUs/GPUs
Small Data Centers	32–64, 2.6–3.5 GHz	CPUs/GPUs/FPGAs

Generally, computing resources in edge networks are often heterogeneous, as shown in Table 1. Meanwhile, multiple APs receive tasks offloaded from edge devices. These tasks are usually delay-sensitive and have deadlines [2]. Dispatching a deadline-constrained task to different heterogeneous nodes may result in different results in deadlines, missed or non-missed. Furthermore, edge servers deployed with limited services that cannot respond to all types of computing requests. So, multiple deadline-constrained tasks requesting different services arriving at the same time will inevitably lead to resource competition on an edge server. Task scheduling for these tasks faces new challenges [10]. It is worth noting that violating the deadlines of tasks will negatively impact the quality of experience(QoE) [8]. In other words, multiple delay-sensitive tasks have varying impacts on QoE. Thus, we introduce the concept of long-term satisfaction of system service called LTSS to reflect QoE and task completion rate to reflect the quality of service (QoS).

There are many studies on task scheduling in edge computing. Some studies consider both *task dispatching* and *resource allocation*, i.e., *task scheduling* [6,11,16]. Some studies also consider the heterogeneity of resources and deadlines to optimize different goals [14,18]. Some studies have proposed scheduling strategies to consider different task service quality requirements [12]. Other studies use existing optimization theories to optimize task scheduling strategies in edge computing, such as reinforcement learning, queuing theory, and genetic algorithm [3,16,17]. However, some of the above studies set the processing time of tasks to a certain value that overlooks the impact of resource heterogeneity on task scheduling. The majority of studies fail to consider the resource competition of tasks on resource-constrained edge computing. Thus, we consider a dynamic situation where multiple tasks arrive randomly and the information about these tasks is unknown until they arrive. At the same time, there is a resources competition for multiple tasks. Our work focus on problem of efficiently scheduling multiple deadline-constrained tasks in heterogeneous edge systems.

The rest of this paper is organized as follows: We introduce the heterogeneous edge computing system model and architecture in Sect. 2. The details of the task scheduling problem with deadline constraints are explained in Sect. 3. Then we introduce our proposed grouping-based multi-task scheduling strategy in Sect. 4. We present and discuss the simulation experiment results in Sect. 5. Finally, we draw conclusions in Sect. 6.

2   System Model and Architecture

This section describes the cloud-edge-end, a three-layer system architecture [5, 7,15] used in our work. The detailed architecture diagram is shown in Fig. 1.

2.1   Edge Network Model

The edge network consists of multiple edge servers. Every few edge servers in a nearby geographical area are connected to an access point (AP), denoted as $E \in \{E_1, E_2, \ldots, E_n\}$. In our model, an AP is defined as the summary node E^*. APs maintain a static routing table between each other, where $\alpha_{i,j}$ and $\beta_{i,j}$ represent the propagation delay and communication bandwidth between AP_i and AP_j. Edge computing nodes in nearby areas are interconnected by fiber optics, which have a low communication delay [4].

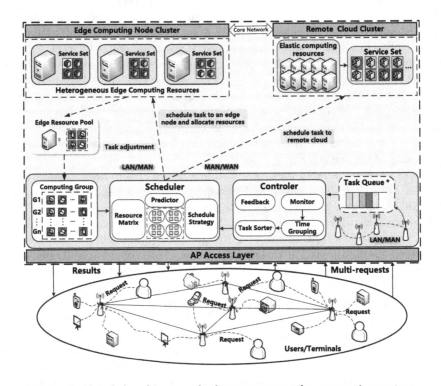

Fig. 1. thr detailed architecture for heterogeneous edge computing system.

2.2   Remote Cloud Model

In our system, the remote cloud E_{cloud} is regarded as a special server with sufficient computing resources. By flexibly launching relevant instances, all computing requests can be responded to by E_{cloud}. We assume that scheduling tasks to

E_{cloud} is non-blocking but incurs significant communication delays. The network propagation delay is denoted by α_{cloud}, the communication bandwidth between E^* and E_{cloud} is denoted by β_{cloud}, and the computational performance of the remote cloud is denoted by f_{cloud}.

2.3   Edge Resource Model

There are n edge servers managed by the summary node E^*. Each server is heterogeneous and configured with limited application services. For edge server E_i, we use E_i^c to represent the parallelism, E_i^m to represent the memory resources, f_i to represent the computational performance, and S_i to represent the set of application services. The computation resource of E_i is represented in the form of a triple:

$$Resource(E_i) = < E_i^c, E_i^m, S_i > \qquad (1)$$

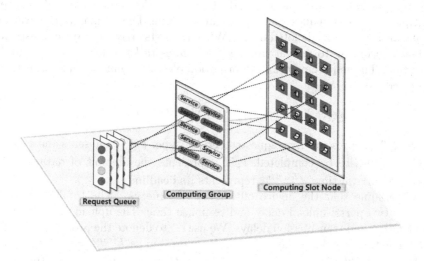

Fig. 2. One-to-one mapping relationship for request-service and one-to-many mapping relationship for service-computing resource.

We redefine the edge computing units in a logical way to manage heterogeneous edge computing resources and alleviate resource competition. Firstly, we define multiple slot nodes for the parallelism E_i^c of E_i. Furthermore, we classify computing nodes into multiple computing unit groups X_h, represented by an k-dimensional vector $X_h = \{x_{h,1}, x_{h,2}, ..., x_{h,k}\}$. The k-th computing slot in the h-th computing unit group is defined as $x_{h,k}$. Each group provides similar computing services. Heterogeneous edge resources can be defined as a resource matrix R^*, represented as

$$R^* = [X_1, X_2, ..., X_h] \qquad (2)$$

At time τ, let $Ava(x_{h,k}) \in \{0,1\}$ represent whether $x_{h,k}$ is idle or not. The available resources for the h-th application service are represented as

$$Xnum(h) = \sum_{x_{h,k} \in X_h, k=1,2,\dots} Ava(x_{h,k}) \tag{3}$$

By uniformly managing the granularity of computing units, we have achieved unified resource mapping, as shown in Fig. 2.

2.4 Task Arrival and Execution Model

Multiple user devices D access the edge network through WiFi or a 4G/5G mobile network and offload requests to the nearest edge summary node E^*. Each requests may include multiple tasks that request different types of services. These tasks are independent and indivisible module of applications, such as speech and face recognition, crowd sensing, and object detection applications. Moreover, multiple computing requests may arrive at any time. The E^* maintains a waiting queue for dispatching, denoted by Q^*. When requests arrive, the system responds to tasks according to arrival time and schedules tasks to computing units for execution. The requirement and information of t_j are unknown until it arrives, expressed as

$$Req(t_j) = <t_j^{kind}, t_j^{load}, t_j^{data}, t_j^{arrival}, t_j^{deadline}> \tag{4}$$

where, t_j^{kind} represents the requested service type, t_j^{load} represents the number of CPU cycles to be completed, t_j^{data} represents the amount of data, $t_j^{arrival}$ represents arrival time, $t_j^{deadline}$ represents its deadline.

We assume that the device offloads t_j at time rs_j. The total delay of t_j includes two parts: upload delay and response time. The upload delay includes propagation and transmission delays. We use α^* to denote the propagation delay of t_j. The transmission delay of t_j can be expressed as $T_j^{trans} = \alpha^* + t_j^{data}/\beta^*$.

We use T_j^{wait} and T_j^{back} to denote the waiting time and the back time of t_j for results. T_j^{wait} includes the waiting time of dispitch at E^* and the waiting time at $x_{h,k}$, which are denoted to $T_{j,*}^{wait}$ and $T_{j,h,k}^{wait}$. The process time of t_j at $x_{h,k}$ can be expressed as $T_{j,h,k}^{process} = t_j^{load}/f_{h,k}$. So, the response delay of t_j can be expressed as

$$T_j^{rsp} = T_{j,h,k}^{process} + T_j^{wait} + T_j^{back} \tag{5}$$

Here, $T_{j,*}^{wait}$ represents the delay generated by the scheduling process, taking a small average here. Since the amount of data in the computing result is usually small, T_j^{back} can be approximated to the propagation delay α^*. The total delay of t_j can be expressed as

$$T_j = T_j^{rsp} + T_j^{trans} = 2\alpha^* + \frac{t_j^{data}}{\beta^*} + \frac{t_j^{load}}{f_{h,k}} + T_j^{wait} \tag{6}$$

If current resource status of edge cannot meet deadline of task, task will be scheduled to the remote cloud for execution. We assume that scheduling task for the cloud will be executed immediately without waiting. Therefore, the total delay of the scheduling task in the cloud can be expressed as

$$T_j^{cloud} = 2\alpha_{cloud} + (1 + \rho)\frac{t_j^{data}}{\beta_{cloud}} + \frac{t_j^{load}}{f_{cloud}} + T_{j,*}^{wait} \qquad (7)$$

Here, ρ is the packet loss rate caused by network jitter during task transmission.

3   Problem Formulation

We introduce the concepts of "penalty" and "reward" to express the impact of scheduling multiple tasks with resource competition on service quality. Here, penalties are defined as the influencing factor for the long-term satisfaction of system services called LTSS. LTSS is used to reflect the quality of user experience (QoE). If the task is completed within its deadline, the system will receive a positive reward; otherwise, the it will receive a negative penalty.

Meanwhile, we introduce a variable $\vartheta_j \in \{-1, 0, 1\}$ to indicate the execution status of the task. If t_j is completed within the deadline, then $\vartheta_j = 1$; otherwise, if t_j responded but was not completed within the deadline, $\vartheta_j = -1$; otherwise, $\vartheta_j = 0$, represented as:

$$\vartheta_j = \begin{cases} 1, & rs_j + T_j \le t_j^{deadline}, \\ -1, & rs_j + T_j > t_j^{deadline}, \\ 0, & otherwise. \end{cases} \qquad (8)$$

We use CRD to denote task completion rate within deadlines and introduce a variable $y_{h,k}^j \in \{0, 1\}$ to represent the allocation of tasks. If t_j is assigned to computing node $x_{h,k}$, $y_{h,k}^j = 1$; otherwise, set $y_{h,k}^j = 0$. The slack time of t_j from current to deadline can be expressed as $T_j^{slack} = t_j^{deadline} - t_j^{arrival}$. We use $w \in \{w_1, w_2, \ldots, w_h\}$ to denote an initial weight factor for each service. The penalty of t_j is expressed as

$$\varphi_j = w_j \times exp(-\frac{T_j^{slack}}{t_j^{deadline} - rs_j}) \qquad (9)$$

The task completion reward of t_j is expressed as

$$rw_j = y_{h,k}^j \vartheta_j \varphi_j \qquad (10)$$

The total reward of system is expressed as

$$Rw = \sum_{j=1,2,\ldots,m} rw_j \qquad (11)$$

The main objective function is to maximize CRD, while the secondary objective is to maximize RW.

4   Grouping-Based Multi-task Scheduling Strategy

In this section, we propose a grouping-based multi-task scheduling strategy (GMTSS) to solve task scheduling problems with resource competition and dead-line constraints. Our strategy is divided into three main parts. Firstly, we develop models of task regrouping and priority sorting to balance efficiency and fair-ness in the response to tasks. Furthermore, we propose a resource-aware greedy scheduling algorithm. This algorithm aims to find a suitable node based on the state of the resource group. Then, we proposed a task adjusting method to achieve a better task scheduling solution, especially for tasks that cannot meet deadlines. Next, we will provide a detailed introduction to these three parts of the strategy. The process of scheduling is illustrated in Fig. 3.

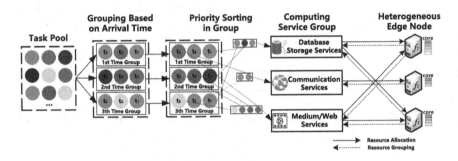

Fig. 3. The process of Grouping-Based Multi-task Scheduling Strategy.

4.1   Task Grouping and Priority Sorting Model

In our model, multiple requests for applications arrive randomly, and each request includes multiple tasks. The information about these tasks is unknown until they arrive. Prioritizing tasks is necessary to utilize more fragmented resources. Firstly, the **Monitor** component takes charge of determining the request information associated with each task and subsequently grouping them based on inherent priority and resource requirements.

Furthermore, the system assigns new priorities to the tasks within each group. The priority weight of t_j is represented as

$$p_j = \frac{\overline{rp_j}}{T_j^{slack}} \tag{12}$$

Due to the heterogeneity of computing nodes, we cannot directly and accu-rately calculate the remaining execution time of t_j, i.e. rp_j when t_j arrives online. So, we determine the priority of t_j by introducing the average execution time on m computing node, which deploys the services requested by t_j, represented as

$$\overline{rp_j} = \frac{\sum_{k=1}^m f_k}{k \times t_j^{load}} \tag{13}$$

According to Eq. (14), it is evident that for tasks with the same deadline, the larger the p_j, the less reserved time for task computation and return, and the more urgent the response.

Algorithm 1. Priority Sorting in Group

Input: $PreQueue$, node's performance table $fTable$
Output: $NewQueue$
 1: Initialize array of p, $NewQueue$ as $\emptyset$
 2: **for** each $i \in PreQueue$ **do**
 3:     $\bar{rp}_i \leftarrow$ Calculate $\bar{rp}_i$ of $PreQueue[i]$ by Eq.(14)
 4:     $p[i] \leftarrow$ Calculate $p[i]$ by Eq.(13)
 5: **end for**
 6: $NewQueue \leftarrow$ Sort tasks of $PreQueue$ by p
 7: **return** $NewQueue$

4.2 Resource-Aware Greedy Scheduling Algorithm

The greedy scheduling strategy tries to schedule tasks to computing nodes based on the state of resources. A scheduling decision considers three factors: the current status of resources within the computing group, the deadline of task, and the type of task. Generally, computing resources have two states: idle and tight. So, the scheduling strategy must consider utilizing different scheduling strategies for two states. The pseudocode of the algorithm is shown in Algorithm 2.

Firstly, **Scheduler** map t_j to the corresponding computing resource group X_h. Then, **Scheduler** searches in the computing group X_h. If there are idle computing nodes in the X_h, greedily dispatch tasks to the node with the shortest execution time. Otherwise, The state of computing resources within the group is tight. **Scheduler** preassigns t_j to each node within X_h and calculate the earliest completion time for the array of $EFT[]$. If there is a node $x_{h,k}$ that meets $t_j^{deadline}$, t_j is assigned a task-node pair$<t_j, x_{h,k}>$ at time ea_j. Initial strategy follows the FCFS for resource allocation. If there are no nodes meeting $t_j^{deadline}$, call Algorithm 2 to find whether there is a feasible solution. If there is a feasible solution, perform task adjusting; otherwise, schedule t_j to E_{cloud}.

4.3 Task Adjusting Method

Due to resource constraints, some tasks may inevitably miss their deadlines. However, the impact of missed deadlines on LTSS varies among different tasks. We try to adjust the execution order and location of tasks to achieve smaller resource utilization gaps. When we have to abandon some tasks due to resource competition, we prioritize those tasks that have less impact on LTSS. Here, we see the waiting queue for resource allocation as a schedule. As a result, task adjusting seeks to produce an optimal feasible schedule (*BestScheme*) rather than the original schedule (*PreScheme*). The *BestScheme* must satisfy the following two conditions, i.e., Eq. (15).

Algorithm 2. Resource-aware Greedy Scheduling Algorithm

Input: Task t_j, Resource Matrix R^*, Service Set S,
Output: Task-Computing node allocation pair $<t_j, x_{h,k}>$
1: Initialize array of $AvaNode$, EFT, EAT as $\emptyset$
2: **if** $t_j^{kind} \in S$ **then**
3:    mapping t_j into Computing Group X_h from R^*
4:    **for** each $k \in X_h$ **do**
5:        **if** $x_{h,k}$ is idle **then**
6:            add $x_{h,k}$ to array of $AvaNode$, $Xnum(h)$ ++
7:        **end if**
8:    **end for**
9: **end if**
10: **if** $Xnum(h) \mathrel{!}= 0$ **then**
11:    $x_{h,k} \leftarrow$ Select a node from $AvaNode$ with the shortest execution time
12:    Obtain allocation pair $<t_j, x_{h,k}>$, $y_{h,k}^j = 1$
13: **else**
14:    $EAT \leftarrow$ Update the earliest idle time of X_h
15:    $EFT \leftarrow$ Calculate the earliest completion time of computing nodes within X_h
16:    $x_{h,k} \leftarrow$ Search for the index of candidate node with the smallest in EFT
17:    **if** $EFT(x_{h,k}) <= t_j^{deadline}$ **then**
18:        Obtain allocation pair $<t_j, x_{h,k}>$, $y_{h,k}^j = 1$
19:        Scheduling t_j to computing node $x_{h,k}$
20:    **else**
21:        Preallocation t_j to candidate node $x_{h,k}$, calling Algorithm 3
22:        **if** Algorithm 3 exist a feasible solution **then**
23:            Obtain allocation pair $<t_j, x_{h,k}>$, $y_{h,k}^j = 1$
24:            Scheduling t_j to computing node $x_{h,k}$ and execute task adjusting
25:        **else**
26:            $y_{h,k}^j = 0$
27:            Scheduling t_j to E_{cloud}
28:        **end if**
29:    **end if**
30: **end if**

Condition 1: *BestScheme* must obtain the same or more tasks completed within the deadline than *PreScheme*.

Condition 2: *BestScheme* must achieve a higher reward of LTSS from completing tasks than *PreScheme*.

$$\begin{cases} RD_{BestScheme} \geq RD_{PreScheme} \\ RW_{BestScheme} > RW_{PreScheme} \end{cases} \tag{14}$$

We assume that all tasks in the current schedule can be completed within the deadline. The remaining slack time of t_j is defined as

$$RD(t_j) = t_j^{deadline} - (ea_j + rp_j) \tag{15}$$

The algorithm seeks feasible solutions through the following three methods in sequence. The pseudocode of task adjusting method is shown in Algorithm 3.

Algorithm 3. Task Adjusting Method

Input: Task t_j, Resource Matrix R^*, Original schedule $PreScheme$
Output: an optimal schedule $BestScheme$

 1: PreCpN ← Calculate number of tasks completed within deadlines of $PreScheme$
 2: PreReward ← Calculate RW of $PreScheme$ by Eq. (11)
 3: SlackTime ← Calculate $RD(t_j)$ of tasks in $PreScheme$ by Eq. (16)
 4: $Index$ ← Calculate $Index$ of inserting t_j into $PreScheme$
 5: Divide $PreScheme$ into $AQ(t_k)$ and $NAQ(t_k)$ by $Index$
 6: **if** $rp_j \leq min\left\{RD_{k \in AQ(t_k)}(t_k)\right\}$ **then**
 7:     $BestScheme$ ← Insert t_j into $Index$ of $PreScheme$
 8: **else**
 9:     CurReward,$NewScheme$ ← Calculate RW of inserting t_j into $PreScheme$
10:     CurCompleteNum ← Calculate number of tasks completed within deadlines of $NewScheme$
11:     **if** CurBenefits>PreReward && CurCompleteNum>=PreCpN **then**
12:         $BestScheme$ ← $NewScheme$
13:         Rescheduling the affected task
14:     **else**
15:         $NewScheme$ ← Select a appropriate task to replacet from $NAQ(t_k)$
16:         RlcReward ← Calculate RW of $NewScheme$
17:         **if** RlcReward > PreReward **then**
18:             $BestScheme$ ← $PreScheme$ after executing task replacement
19:             Rescheduling the replaced task
20:         **else**
21:             Scheduling t_j to E_{cloud}
22:         **end if**
23:     **end if**
24: **end if**

$DirectInsertion(DI)$: We first calculate the queue insertion position of t_j that meets the deadline, denoted as $Index$. Theoretically, inserting t_j into a certain position of $PreScheme$ only affects the tasks arranged behind t_j and delays their completion time. Therefore, the $PreScheme$ can be divided into two parts: the affected queue $AQ(t_k)$ and the unaffected queue $NAQ(t_k)$. Then, calculate the remaining slack time of $AQ(t_k)$.

$$rp_j \leq min\left\{RD_{k \in AQ(t_k)}(t_k)\right\} \tag{16}$$

If the remaining execution time of t_j is less than the minimum remaining slack time of $AQ(t_k)$, this means that it meets **Condition 1** and will definitely meet **Condition 2**, i.e., Eq. (16). So, t_j can directly insert into the position of $Index$ in $PreScheme$. Then, update the earliest allocation time ea_k of t_k, here $t_k \in AQ(t_k)$.

$HighestRewardInsertion(HRI)$: If there is no feasible solution for DI, the algorithm attempts to find a schedule for the highest reward insertion. We first calculate the completion reward of $NewScheme$ after inserting t_j into position of $Index$, and compare it with $PreScheme$. If the number of tasks completed within deadlines for $NewScheme$ is equal to $PreScheme$ but the total reward is greater than $PreScheme$, it's also a better schedule, i.e.,

$$\begin{cases} rp_j > min\left\{RD_{k \in AQ(t_k)}(t_k)\right\} \\ RW_{NewScheme} > RW_{PreScheme} \end{cases} \tag{17}$$

This indicates that $NewScheme$ satisfies both **Condition 1** and **Condition 2**, and that there exists a feasible solution for the highest reward insertion(HRI).

Inserting t_j into the position of $Index$ must result in a task not meeting the deadline in $AQ(t_k)$. However, affected tasks may still meet the deadline after being scheduled to another node of the computing group. So, we introduce a rescheduling mechanism to handle this task. Considering the heterogeneity of resources, the rescheduling of t_j from node $x_{h,k}$ to node $x_{h,k'}$ will lead to a change in execution time. The pseudocode of task rescheduling is shown in Algorithm 4. The change of rp_j is expressed as

$$rp_j = \frac{f_{h,k} * rp_j}{f_{h,k'}} \tag{18}$$

Algorithm 4. Task Rescheduling Algorithm

Input: Task t_j, Resource Matrix R^*, pre-allocation $y_{h,k}^j$
Output: New Task-Computing node allocation pair $<t_j,x_{h,k'}>$
1: Initialize array of EFT as $\emptyset$
2: mapping t_j into Computing Group X_h
3: **for** each $i \in X_h$ **do**
4:　　$EFT[i] \leftarrow$ Calculate the earliest completion time of $x_{h,i}$
5: **end for**
6: $x_{h,k'} \leftarrow$ Find the candidate node with $Min\{EFT\}$
7: **if** $EFT[k'] <= t_j^{deadline}$ **then**
8:　　Update allocation pair $<t_j,x_{h,k'}>$, $y_{h,k'}^j = 1$
9:　　$rp_j \leftarrow$ Calculate rp_j by by Eq. (19)
10:　　Rescheduling t_j to computing node $x_{h,k'}$
11: **else**
12:　　Scheduling t_j to E_{cloud}
13: **end if**

$MaximumRewardReplacement(MRR)$: If there is no feasible solution for both DI and HRI, the algorithm attempts to replace a task in the executing or $NAQ(t_k)$ to find a schedule that is better than $PreScheme$. Due to replacing a task to form $NewScheme$, it must meet **Condition 1**. A feasible schedule of MRR also needs to meet **Condition 2** and **Condition 3**. Finally, the replaced task will be rescheduled within the computing group.

Condition 3: the $NewScheme$ achieves a lower average completion time.

$$\begin{cases} ACT_{NewScheme} < ACT_{PreScheme} \\ RW_{NewScheme} > RW_{PreScheme} \end{cases} \tag{19}$$

5   Experimental Setup and Performance Evaluation

In this section, we evaluate the performance of our proposed GMTSS and compare it with the current state-of-the-art benchmark strategy. We develop an edge-cloud collaborative simulation system for deadline-constrained task scheduling using Python, named D-Edge-CloudSim. These simulation experiments were conducted by implementing the proposed GMTSS and benchmark strategy on a PC with a Core Ryzen 7 6800H 3.2 GHz, and 16 GB RAM, running Windows 11.

5.1   Experimental Setup

In our experiment, we set up 4 edge servers, 2 Aps as the summary node, 1 public remote cloud, and multiple edge devices. The edge servers are configured with 2–8 cores, computing performance of cores ranging from 1.2 GHz to 3.2 GHz, and configured with 3–5 application services. The computing performance for the public remote cloud is set to 4 GHz. The communication bandwidth for edge nodes is set within the range of 400 to 500 Mbps. The bandwidth of the core network between edge nodes and the Remote Cloud is set at 200 Mbps, and the propagation delay is set between 200 and 300 ms.

We simulated real applications and randomly generated a large number of tasks in container instances. These tasks are delay-sensitive, such as face recognition, machine tracking recognition, and sensor detection. Multiple tasks of the application request arrive randomly. We simulate a total of 50,000 tasks. We adjust the release time and deadline for these tasks based on real data. Meanwhile, we divide 50,000 tasks into 5 groups based on their arrival density, denoted as L1-L5, with the lowest density for L1 and the highest density for L5. Then, we introduce eight types of tasks based on real data. We set the workload to a range of 0.2 G to 0.5 G and the size of data to a range of 5 Mb to 20 Mb for each task. The minimum action time slot of the system is set to 50 ms.

Evaluation Parameters: We utilize three indicators to evaluate the performance of our proposed GMTSS: task completion rate within deadline (CRD), task completion reward (RW), and average completion time (ACT). We use three benchmark strategies for task dispatching and three benchmark strategies for resource allocation, and combine them into four combined task scheduling strategies to effectively compare the overall performance of GMTSS as follows:

Dispatching strategy:

Random: Randomly dispatch task to a node.

LeastLoad: Dispatch task to a node with the shortest queue length.

ShortestWait: Dispatch task to a node with Minimum waiting time.

Resource allocation strategy: EDF(Earliest Deadline First), SRPT (Shortest Remain Process Time) and FCFS(First Come First Service).

Four Combination benchmark strategies as follow.

- MLEDF: LeastLoad + EDF.
- RSPT: Random + SRPT.

- SWFF: ShortestWait + FCFS.
- MLSPT: LeastLoad + SRPT.

5.2   Evaluate Experimental Results

Firstly, we evaluate the completion rate of tasks within the deadline, i.e., CRD. As results shown in Fig. 4, our proposed GMTSS outperforms MLEDF by 41.6% ,and RSPT by 8.2%, and SWFF by 45.5%, and MLSPT by 14% on average. We can see that as the task density increases, other benchmark strategies have a serious performance decline.

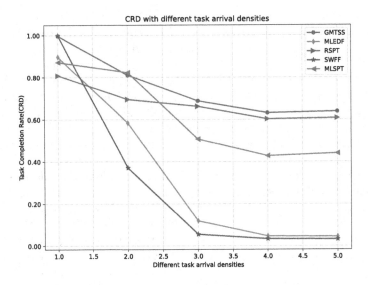

Fig. 4. The completion rate of tasks within the deadlines.

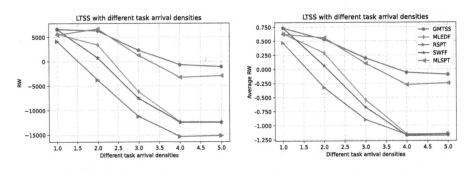

Fig. 5. The reward is used to reflect the quality of user experience(QoE).

Next, we evaluate the total and average task completion reward, i.e., RW. As shown in Fig. 5, the average rewards for GMTSS, MLEDF, RSPT, SWFF and MLSPT are 1.307, -0.384, -0.613, -0.447, and 0.165 respectively. Our proposed GMTSS has the highest overall task completion reward (RW) due to the following two points: a) GMTSS has a high task completion rate. b) GMTSS always generates a better schedule, ensuring higher RW. Those tasks generating higher RW are also more urgent.

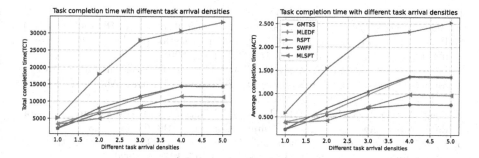

Fig. 6. The completion time of tasks.

Finally, we evaluate the completion time of tasks. From the results of Fig. 6, it shows that the average completion time for GMTSS, MLEDF, RSPT, SWFF and MLSPT are 0.6, 0.934, 1.562, 0.944 and 0.692. It is worth noting that RSPT has advantages in general scenarios, but the advantage disappears as the task density increases. Our proposed GMTSS is slightly superior in ACT for all benchmark scheduling strategies. The reason can be summarized as follows: a) GMTSS achieves full utilization of resource fragments by adjusting tasks within the computing group. b) GMTSS outperforms MLEDF and MLSPT in terms of completion time because they do not consider the impact of resource heterogeneity on task execution.

The above experimental results show that our proposed GMTSS has obvious advantages in a resource-constrained heterogeneous edge computing system with multi-tasks arriving randomly.

6   Conclusion

As a promising computing paradigm, edge computing has attracted more and more attention from researchers in various fields. It is a key challenge to achieve efficient task scheduling to improve QoS of heterogeneous edge systems with resource competition. In realizing this, we first build a model for resource management. Meanwhile, we consider the impact of task completion with different levels of urgency on QoS and QoE of the system, proposing a group-based multi-task scheduling strategy called GMTSS. Finally, we develop an edge-cloud collaborative simulation system to test our proposed GMTSS, named D-Edge-CloudSim. Experimental results show that our proposed GMTSS can effectively schedule multiple tasks with resource competition.

Acknowledgements. This work was supported in part by the National Natural Science Foundation of China (Grant Nos. 61972146, 62002032, 62372064), the Postgraduate Scientific Research Innovation Project of Hunan Province(CX20220942).

References

1. Abadi, Z.J.K., Mansouri, N., Khalouie, M.: Task scheduling in fog environment-challenges, tools & methodologies: a review. Comput. Sci. Rev. **48**, 100550 (2023)
2. Bellendorf, J., Mann, Z.Á.: Classification of optimization problems in fog computing. Futur. Gener. Comput. Syst. **107**, 158–176 (2020)
3. Fang, J., Zhang, J., Lu, S., Zhao, H., Zhang, D., Cui, Y.: Task scheduling strategy for heterogeneous multicore systems. IEEE Consumer Electron. Mag. **11**(1), 73–79 (2021)
4. Feng, A., Dong, D., Lei, F., Ma, J., Yu, E., Wang, R.: In-network aggregation for data center networks: a survey. Comput. Commun. **198**, 63–76 (2023)
5. Filali, A., Abouaomar, A., Cherkaoui, S., Kobbane, A., Guizani, M.: Multi-access edge computing: a survey. IEEE Access **8**, 197017–197046 (2020)
6. Han, Z., Tan, H., Li, X.Y., Jiang, S.H.C., Li, Y., Lau, F.C.: Ondisc: online latency-sensitive job dispatching and scheduling in heterogeneous edge-clouds. IEEE/ACM Trans. Netw. **27**(6), 2472–2485 (2019)
7. Hong, C.H., Varghese, B.: Resource management in fog/edge computing: a survey on architectures, infrastructure, and algorithms. ACM Comput. Surv. (CSUR) **52**(5), 1–37 (2019)
8. Jagadish, T., Apte, O., Pradeep, K.: Task scheduling algorithms in fog computing: a comparison and analysis. In: 2022 International Conference on Automation, Computing and Renewable Systems (ICACRS), pp. 483–488. IEEE (2022)
9. Li, J., et al.: Maximizing user service satisfaction for delay-sensitive Iot applications in edge computing. IEEE Trans. Parallel Distrib. Syst. **33**(5), 1199–1212 (2021)
10. Luo, Q., Hu, S., Li, C., Li, G., Shi, W.: Resource scheduling in edge computing: a survey. IEEE Commun. Surv. Tutorials **23**(4), 2131–2165 (2021)
11. Meng, J., Tan, H., Li, X.Y., Han, Z., Li, B.: Online deadline-aware task dispatching and scheduling in edge computing. IEEE Trans. Parallel Distrib. Syst. **31**(6), 1270–1286 (2019)
12. Oo, T., Ko, Y.B.: Application-aware task scheduling in heterogeneous edge cloud. In: 2019 International Conference on Information and Communication Technology Convergence (ICTC), pp. 1316–1320. IEEE (2019)
13. Tang, X., et al.: Cost-efficient workflow scheduling algorithm for applications with deadline constraint on heterogeneous clouds. IEEE Trans. Parallel Distrib. Syst. **33**(9), 2079–2092 (2022)
14. Xu, B., et al.: Fine-grained task scheduling based on priority for heterogeneous mobile edge computing. In: 2022 China Automation Congress (CAC), pp. 4889–4894. IEEE (2022)
15. Yu, W., et al.: A survey on the edge computing for the internet of things. IEEE access **6**, 6900–6919 (2017)
16. Yuan, H., Tang, G., Li, X., Guo, D., Luo, L., Luo, X.: Online dispatching and fair scheduling of edge computing tasks: a learning-based approach. IEEE Internet Things J. **8**(19), 14985–14998 (2021)

17. Yuchong, L., Jigang, W., Yalan, W., Long, C.: Task scheduling in mobile edge computing with stochastic requests and m/m/1 servers. In: 2019 IEEE 21st International Conference on High Performance Computing and Communications; IEEE 17th International Conference on Smart City; IEEE 5th International Conference on Data Science and Systems (HPCC/SmartCity/DSS), pp. 2379–2382. IEEE (2019)
18. Zhu, T., Shi, T., Li, J., Cai, Z., Zhou, X.: Task scheduling in deadline-aware mobile edge computing systems. IEEE Internet Things J. **6**(3), 4854–4866 (2018)

Real-Time Driver Fatigue Detection Method Based on Comprehensive Facial Features

Yihua Zheng[1], Shuhong Chen[1]([✉]), Jianming Wu[1], Kairen Chen[1], Tian Wang[2], and Tao Peng[1]

[1] School of Computer Science and Cyber Engineering, Guangzhou University, Guangzhou, Guangdong, China
shuhongchen@gzhu.edu.cn
[2] BNU-UIC Institute of Artificial Intelligence and Future Networks, Beijing Normal University, Zhuhai, Guangdong, China

Abstract. In recent years, there have been frequent cases of vehicle accidents caused by fatigued driving, leading to considerable economic losses and a high number of casualties. Accordingly, it has an important social significance for avoiding accident risks to remind tried drivers to take a break promptly. Fatigue driving detection based on facial feature recognition technology has attracted much attention due to its non-invasive, low-cost, and convenient detection advantages. However, the current fatigue driving technology faces the challenge of balancing real-time performance and accuracy in practical applications. Therefore, a fatigue driving detection model based on deep learning is proposed to address this issue. This model includes modules for object detection, head pose estimation, fatigue detection, and distraction detection. First, this paper proposes a facial feature detection algorithm based on YOLOv5 and FSA-Net, which can quickly detects the driver's eye state, mouth state, and head state. Second, in order to better avoid accident risks, the designed system detects driver distraction behaviors during driving process from both cognitive distraction and visual distraction perspectives based on the facial feature detection algorithm. Finally, a comprehensive dangerous driving behavior detection and warning system based on closed-eye detection, yawning detection, 3D head pose estimation, and object detection is designed by integrating multiple fatigue and distraction indicators. The experimental results show that the developed detection and warning system has high detection accuracy, which can provide timely warning when dangerous driving behavior occurs and helps ensure driving safety.

Keywords: object detection · computer vision · fatigue driving

1 Introduction

Fatigue driving is characterized by the driver's tiredness and drowsiness caused by long hours of driving or lack of sufficient rest. This situation can lead to

Z. Tari et al. (Eds.): ICA3PP 2023, LNCS 14488, pp. 484–501, 2024.
https://doi.org/10.1007/978-981-97-0801-7_28

serious traffic accidents, endangering the lives of the driver and other road users. In recent years, fatigue driving has become an important cause of traffic accidents [1–3]. If the driver is reminded in time when they are fatigued, 90% of such accidents can be avoided [4]. With the development of deep learning theory, deep learning algorithms have been widely applied in driver fatigue detection [5–8]. Dwivedi et al. [9] used a convolutional neural network to learn facial features and used the trained network model to classify the driver's state. Zhang et al. [12] proposed a yawning detection system consisting of a facial detector, a nose detector, a nose tracker, and a yawning detector. The system uses deep learning algorithms to detect the driver's facial area and nose position. Then, combined with a Kalman filter and TLD-based tracker, the system tracks the target in dynamic driving situations. Finally, a neural network is established using multiple features to detect yawning. Although this method has high accuracy, it is difficult to perform real-time detection on embedded devices due to its high requirements on the performance of the equipment. Liu et al. [11] proposed a real-time driver fatigue detection method based on Convolutional Neural Network and Long Short-Term Memory (CNN-LSTM). The method applies CNN to detect features of the driver's eyes and mouth. These parameters are then inputted into LSTM as continuous time series, including eye feature parameter Perclos, mouth feature parameter MClosed, and face orientation feature parameter Phdown, to output the level of fatigue. Ansari et al. [10] proposed the XSENS motion capture system to detect the driver's head movements. They designed a novel bidirectional long short-term memory deep neural network with rectified linear unit layers, which leverages 3D time-series head angular acceleration data. The network has demonstrated a high accuracy in recognizing the driver's active, fatigue, and transition states. Fatigue driving detection algorithms based on deep learning are not only non-contact, but they can also detect the driver's fatigue status directly by detecting facial features with high accuracy and speed. However, these algorithms still have some drawbacks:

1. The detection of driver's facial features are susceptible to the influence of lighting, background, angles, and individual differences, leading to inaccurate results.
2. The detection model for facial features is relatively complex, and it is difficult to balance real-time performance and accuracy.
3. Relying on a single judgment index can easily result in missed detection or false alarms.

To address the above issues, this paper proposes a fatigue detection algorithm based on YOLOv5 and FSA-Net. Compared to other advanced fatigue detection algorithms, the proposed algorithm in this paper achieves one-step completion of facial feature localization and real-time state detection of the driver. It encodes the state of the driver's eyes, mouth, and head in each frame of video for a comprehensive analysis of fatigue. According to the encoding sequence, it calculates the PERCLOS value, PERYAWN value, and PERNOD value. The main contributions of this paper are as follows:

1. To ensure real-time detection, this paper proposes a facial feature detection algorithm based on YOLOv5 and FSA-Net, which can quickly detect the driver's eye, mouth, and head posture.
2. To improve accident prevention, this paper proposes an algorithm that detects cognitive distraction using YOLOv5 and an algorithm that detects visual distraction using FSA-Net. These algorithms determine whether the driver's attention is on the road, reducing the risk of accidents.
3. In order to avoid the missed detection and false alarm problem of a single judgment index, a comprehensive dangerous driving system combining multiple feature indicators is developed.

2   Related Work

Research on fatigue driving mainly focuses on detecting the driver's fatigue level from three aspects: physiological characteristics, driving behavior, and facial features. The fatigue detection method based on physiological characteristics usually utilizes sensors to measure the driver's physiological parameters, such as electrocardiogram (ECG) signals, electroencephalogram (EEG) signals, and electromyogram (EMG) signals, to determine whether the driver is fatigued. Suganiya Murugan et al. [13] extracted fatigue features from electroencephalogram (EEG) signals for statistical analysis, and selected features based on principal component analysis. Machine learning methods such as support vector machines and ensemble classifiers were used to classify the features. The method based on physiological characteristics is the most accurate, reliable, and has strong anti-interference ability. However, it is usually invasive, requiring the driver to wear related equipment, which may affect the driving experience. In addition, the relevant equipment is often expensive, structurally complex, and difficult to implement for widespread use. Fatigue detection methods based on driving behavior data commonly involve using specialized equipment to collect data on various aspects of the vehicle's performance during operation, including steering wheel angle, vehicle speed, accelerator and brake pedal position, and steering wheel grip force. Afterwards, the collected data is typically analyzed to determine whether the driver is fatigued. Sha Chunfa et al. [14] collected data on steering wheel grip force and EEG signals, and established a fatigue detection model based on steering wheel grip force using neural networks. Schwarz C [15] input information on driving behavior, such as the degree of pedal opening and closing, vehicle speed, and turning angle, into a Bayesian algorithm for fatigue detection. Fatigue detection methods based on driving behavior characteristics are cost-effective, easy to collect information, and have minimal impact on drivers. However, these methods are easily affected by driving habits and external factors, so their accuracy are not high.

Fatigue detection methods based on facial features have gained research attention due to their advantages of non-contact detection and convenient application, making them highly promising for a wide range of applications. These methods have relatively high accuracy and real-time performance, without relying too much on fixed parameter thresholds. In early studies, Wang et al. [16]

proposed a fatigue detection algorithm that tracks mouth movements, classifies mouth states, and issues warnings when the driver is fatigued. Pan et al. [17] designed an adaboost-based face detector to segment and track the eyes. They used a proposed fuzzy comprehensive evaluation method to determine the open/closed state of the eyes and utilized the PERCLOS metric to calculate the frequency of eye closure for fatigue detection. Zhang et al. [18] designed a fatigue detection algorithm based on estimating head posture using facial landmarks. The method involves real-time capture of the face through a camera and utilizes a facial landmark detection algorithm based on the Dlib library to determine the state of the eyes, mouth, and head movement. Experimental results show that the proposed method has good real-time performance and high accuracy. Zhao et al. [19] proposed a fully automatic fatigue detection algorithm using driving images, which utilized the MTCNN network for face detection and key point detection, and introduced the EM-CNN network to detect the mouth and eye states from the region of interest (ROI). Finally, the fatigue state was determined by using two parameters, PERCLOS and mouth opening degree (POM). Shulei et al. [31] designed an adaBoost face detection algorithm based on Haar features, and used a facial landmarks localization algorithm to obtain facial landmarks. They then computed the eye aspect ratio based on the feature point data and set an appropriate threshold to detect the driver's fatigue level. Additionally, they introduced the Road Rage Extent Measure (RBEM) for the first time to assess whether the driver exhibited road rage characteristics. Akrout et al. [32] proposed a fusion system based on yawning detection, drowsiness detection, and 3D head pose estimation and evaluated it using three distinct databases. The system introduces four basic alertness levels: alert, slightly drowsy, fatigued, and inattentive.

Computer vision has found widespread applications in various fields, such as autonomous driving, medical diagnosis, smart home systems, and industrial manufacturing. This paper conducts research on fatigue detection technology based on computer vision, which determines the driver's fatigue state by detecting facial features. The main research content of this paper includes the detection of facial features, head posture estimation, and determination of fatigue and distraction states. The detection methods for facial features include methods based on feature point and methods based on deep learning. The method based on feature point is more stable, but the annotation of each feature point requires significant manual effort, leading to a higher workload. On the other hand, the method based on deep learning detects facial features by training deep neural networks, allowing for automatic feature extraction and high detection accuracy. This method has been widely applied in face recognition and detection.

Object detection methods based on deep learning can be classified into two-stage and single-stage detection methods. Two-stage object detection algorithms are comprised of two steps to complete the object detection task. Firstly, a region proposal network is used to generate candidate boxes. Then, each candidate box is fed separately into a neural network for more detailed feature extraction. Finally, the method outputs detection results based on the classification scores and position regression results. Examples of representative algorithms include R-CNN [20], SPP-Net [21], Fast R-CNN [22], and Faster R-CNN [23]. Two-stage object detection algorithms are generally more accurate than single-stage object detection algorithms, but they are notably slower in speed. Single-stage object detection algorithms do not require generating candidate boxes in advance. Instead, they directly classify and regress the position of the entire image to output the detection results. Therefore, when compared with two-stage object detection algorithms, single-stage object detection algorithms have better speed performance and are more applicable for real-time object detection tasks. Common single-stage object detection algorithms include the YOLO series [24–26] and SSD [27]. Because of the high demand for real-time detection speed in fatigue driving detection, this paper designs a facial feature detection algorithm that utilizes the YOLOv5 model and can be used for real-time monitoring in embedded devices.

Head pose estimation refers to the estimation of a person's head rotation angles in 3D space using computer vision techniques. Currently, the primary methods for head pose estimation include face key point detection technology, traditional machine learning techniques, and deep learning techniques. Deep learning-based head pose algorithms such as 3DDFA [28], FSA-Net [29], and Deep Head Pose [30] have shown promising results. FSA-Net, in particular, is a fast and highly-accurate method that treats regression tasks as classification tasks. By providing an RGB image as input, FSA-Net can estimate head pose in real-time. With a model size of around 5M, FSA-Net is lightweight and suitable for real-time detection applications. Thus, this paper uses FSA-Net for head pose estimation.

3   Dangerous Driving Detection Algorithms

3.1   Overall Design Scheme

Most existing fatigue detection algorithms are based on a single network. When drivers are fatigued, the symptoms exhibited by each individual may vary. Some may exhibit frequent blinking, involuntary nodding, or frequent yawning. The fatigue detection algorithm proposed in this paper comprises two networks with different functions to accurately detect all facial features of the driver. In case of facial obstruction, the proposed multi-feature fusion detection algorithm can effectively address the issue. Specifically, the algorithm can detect a driver's fatigue status by analyzing their mouth and head posture when wearing glasses, and by analyzing their eye status and head posture when wearing a mask.

The detection method proposed in this paper consists of three components: an object detection model that detects the driver's eyes, mouth, and hands; a head pose estimation model; and a fatigue and distraction detection model. The detailed flow of the model is shown in Fig. 1. First, real-time driving video frames are captured using a camera and inputted into the object detection model YOLOv5 for detecting the driver's eye and mouth states. At the same time, the head pose estimation model FSA-Net is used to obtain three Euler angles of the

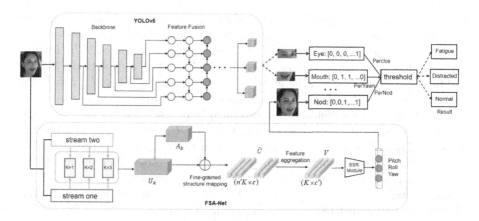

Fig. 1. Dangerous driving detection model

head pose, which are pitch, roll, and yaw. Then, the information detected are inputted into the fatigue and distraction detection model. The model encodes and stores the driver's eye features, mouth features, head pose state, and hand state in time sequence. The Perclos value, PerYawn value and PerNod of the data within a certain time sequence are calculated. At the same time, we count the number of times the driver is distracted and calculate the frequency of distraction by relying on hand state and head posture information. Finally, fatigue driving judgment and distraction driving judgment are made by setting multiple feature thresholds.

Facial Feature Detection Module. The primary function of the facial feature detection module is to rapidly and accurately capture the driver's eye and mouth features in complex driving environments. As fatigue detection technology demands real-time performance, an algorithm for facial feature extraction based on the YOLOv5 object detection model was designed. The YOLOv5 model comprises a backbone network, a neck network, and a head layer. The enhanced DarkNet-53 network is used as the backbone to extract features, which are then fused through the neck network. Finally, the head layer outputs the detection category and location information.

Head Pose Estimation Module. The Head Pose Estimation module's primary function is to estimate the driver's head posture angle and ascertain whether they are focusing on the road ahead. Facial keypoint-based algorithms are commonly used in head pose estimation. However, detecting facial keypoints can increase the model's computational load and affect the real-time detection speed. Therefore, some methods directly use an RGB image to estimate the head pose and provide missing 2D information on a 3D image. FSA-Net network is built on the SSR network and adopts a progressive soft regression approach to head pose estimation. The FSA-Net divides the input image into two streams for processing. As depicted in Fig. 1, there are three stages, and each stream extracts a feature map at one stage, followed by pairwise fusion. The feature maps are first multiplied pairwise, then fused with a 1×1 convolutional kernel, and finally pooled to obtain the $w \times h \times c$ feature map U_k. Then, U_k is fed into the aggregation module for feature refinement. In order to extract important features further, the FSA-Net network also introduces a scoring function to assign a pixel-level importance score to each feature map U_k, resulting in an attention map A_k. These feature maps U_k and attention maps A_k are then fed together into the fine-grained structure mapping module to extract representative features $\tilde{U}$, which are subsequently input into the feature aggregation module to generate a representative feature set V. Finally, the feature set V is fed into the SSR module to predict the three angles of the head poses.

3.2   Fatigue Detection Methods

The relevant experimental data indicates that when a driver is fatigued, they will involuntarily yawn, experience uncontrollable nodding of the head, and an increase in the duration of eye closure. The PERCLOS [33] represents the duration of eye closure during a unit of time, which is an indicator used to measure fatigue. This indicator usually has three standards: EM, which represents the proportion of time that the eyelid covers more than 50% of the eyeball area; P70, which represents the proportion of time that the eyelid covers more than 70% of the eyeball area; P80, which represents the proportion of time that the eyelid covers more than 80% of the eyeball area. The research data shows that the P80 standard has the strongest correlation with the degree of fatigue, so the fatigue detection indicator of the eye features used in this paper is based on the P80 standard. The formula for calculating the PERCLOS is as follows:

$$PERCLOS = \frac{\sum_{i=1}^{N} f_i}{N} \times 100\% \tag{1}$$

where f_i represents a frame with closed eyes, $\sum_{i=1}^{N} f_i$ represents the number of frames with closed eyes per unit time, and N represents the total number of frames per unit time.

When drivers are fatigued, they tend to yawn involuntarily, and the frequency of yawning increases with the degree of fatigue. Therefore, the frequency of yawning can be used to determine the level of fatigue. This paper utilizes the idea of PERCLOS to compute the frequency of yawning, which is determined as the ratio of yawning frames to the total number of detection frames captured within a specific time frame. To prevent normal mouth movements, such as talking or singing, with a small degree of mouth opening or short duration, from being mistakenly identified as yawning, this paper records the number of continuous frames f_{yawn} of a yawn. Only when the degree of mouth opening exceeds a certain threshold and f_{yawn} surpasses a specific threshold, it can be considered as a yawn. The formula for calculating the yawn frequency is as follows:

$$PERYAWN = \frac{\sum_{i=1}^{N} f_i}{N} \times 100\% \tag{2}$$

where f_i indicates a frame with the mouth open, $\sum_{i=1}^{N} f_i$ represents the number of consecutive frames in which the mouth is opened within a unit time, and N represents the overall number of frames within that unit time.

When a driver is excessively fatigued, they may involuntarily nod their head. Therefore, the fatigue level of the driver can be determined based on the frequency of nodding. When the head posture estimation algorithm detects that the head is tilted beyond a threshold angle and the eyes are closed, it is judged as a nodding state. The formula for calculating nodding frequency is:

$$PERNOD = \frac{\sum_{i=1}^{N} f_i}{N} \times 100\% \tag{3}$$

where f_i is a frame of nodding, $\sum_{i=1}^{N} f_i$ indicates the number of frames where the driver is nodding during a unit of time, and N represents the total number of frames per unit of time.

3.3 Distraction Behavior Judgment

While driving, the driver needs to continuously keep an eye on the road ahead, but external factors or unrelated behavior (such as using a mobile phone, looking at roadside billboards, turning their head to talk to a passenger, etc.) may cause the driver's gaze to deviate from the road ahead, leading to visual distraction. This deviation of the gaze is usually accompanied by a turn of the head. By monitoring the driver's head posture, we can determine whether there is visual distraction. The driver's image is input into the FSA-Net model, which outputs three dimensions of head pose: pitch, roll, and yaw. The range of normal head pose deviation is given in formula (4):

$$\begin{cases} P_1 \leq pitch \leq P_2 \\ R_1 \leq Roll \leq R_2 \\ Y_1 \leq Yaw \leq Y_2 \end{cases} \tag{4}$$

Considering the situation where the driver's head deviation is too large to detect facial features, this paper sets up a set of rules for detecting distracted driving:

1.If the driver's eye and mouth status cannot be detected at the same time, it is considered a severe distracted behavior.

2.If the driver's facial features can be detected, the head posture estimation is performed. If any of the estimated head posture angles exceed the normal range set and last for a long time, it is considered visual distraction.

However, not all distracted behaviors in reality involve head movement, such as using a phone, smoking, or eating. These behaviors can still divert the driver's attention, leading to a slow response to unexpected events. Therefore, this paper proposes the detection of cognitive distraction behavior based on object detection algorithms. The YOLOv5 model is utilized to detect whether the driver is holding an object. If the duration of holding the object exceeds a certain threshold, it is considered as cognitive distraction. This paper selects two common objects, phones and cigarettes, as detection targets for assessing cognitive distraction.

3.4   Algorithm for Detecting Dangerous Driving Behavior

The dangerous driving detection algorithm based on Yolov5 and FSA-Net is as shown above, with the following specific steps:

1) First, the algorithm Captures a driving image by the real-time camera and inputs it the pre-trained YOLOv5 model and FSA-Net model to perform target detection and head pose estimation tasks respectively.
2) Secondly, the algorithm utilizes YOLOv5 to detect the eye and mouth status. If the eyes are closed, the eye state is encoded as 1, otherwise 0. The PERCLOS value is calculated and updated. If PERCLOS value exceeds the threshold, a fatigue warning is triggered. Similarly, if the degree of mouth

---

Algorithm 1. Detection of dangerous driving behavior

Output: Totall frames $\mathcal{T}$, eye threshold $\mathcal{E}$, mouth threshold $\mathcal{M}$, head threshold $\mathcal{H}$, threshold for cognitive distraction $\mathcal{C}$, threshold of visual distraction $\mathcal{V}$.
Input: Fatigue $\mathcal{F}$, Distracted $\mathcal{D}$.
  for all $f \in \{1,..., \mathcal{T}\}$ **do**
    $F \leftarrow 0, D \leftarrow 0$
    $state_{eye}, state_{mouth}, phone, cigarette \leftarrow YOLOv5(f)$
    $state_{head} \leftarrow FSA - Net(f)$
    Calculate and update the PERCLOS value by Eq.(1).
    Calculate and update the PERYAWN value by Eq.(2).
    Calculate and update the PERNOD value by Eq.(3).
    Calculate and update the $count_c$ value.
    Calculate and update the $count_v$ value.
    if PERCLOS$\geq \mathcal{E}$ or PERYAWN$\geq \mathcal{M}$ or PERNOD$\geq \mathcal{H}$ **then**
      $F \leftarrow 1$
    else
      $F \leftarrow 0$
    end if
    if $count_c \geq \mathcal{C}$ or $count_v \geq \mathcal{V}$ **then**
      $D \leftarrow 1$
    else
      $D \leftarrow 0$
    end if
  end for

opening exceeds a certain threshold, the mouth state is encoded as 1, otherwise 0. If the coding continues to be 1 for a certain time period (about 1.5 s), the yawn count is increased. If the yawn count exceeds three times within one minute, a fatigue warning is triggered.

3) Then, the algorithm utilizes the FSA-Net model to obtain the pitch of the head posture. If the pitch is below a certain threshold and the eyes are closed, the head state is encoded as 1, otherwise 0. When the ratio of frames encoded as 1 within 10 s to the total detected frames exceeds 30%, a fatigue warning is triggered.

4) If the driver is holding a phone or a cigarette in the current frame, the cognitive distraction count will be incremented. Otherwise, the count will be reset and the algorithm will continue to detect the next frame of the video. If the cognitive distraction count equals the total number of detected frames within a certain period of time, the system will issue a warning.

5) The algorithm determines whether the driver's head deviates from the normal range. If so, the visual distraction count is incremented. Otherwise, the count is reset and the algorithm continues to detect the next frame. If the visual distraction count equals the total number of detected frames within a certain period of time, a warning will be issued.

6) If the driver's eyes and mouth are not detected, it is considered as severe distraction, and a warning will be issued immediately.

4   Experimental Results and Analysis

4.1   Construction of Experimental Dataset

This paper collected sample data from different genders, lighting conditions, and age groups, totaling over 4,000 images of real driving environments. The dataset is divided into seven categories, i.e.,open eyes, closed eyes, yawning, closed mouth, wearing a mask, using a phone call, and smoking. The details are shown in Table 1. The samples in the dataset include images with slightly opened eyes or mouths, and when annotating these images, this paper categorized them

Table 1. Training set and testing set

Category	Training set	Testing set
Closed	834	151
Open	1893	317
Yawn	561	138
Normal	1561	304
Mask	644	25
Phone	1256	232
Cigarette	962	270

into the closed eye and closed mouth categories. The purpose is to avoid errors caused by shooting angles or individual differences as much as possible. For example, there may be some samples in a talking or smiling state with their mouths open, but the degree of opening in these cases is very small.

4.2   Experimental Environment and Design

The experimental environment of this paper is based on Ubuntu 20.04 operating system, with NVIDIA GeForce RTX 3080 GPU, Intel Xeon Platinum 8255C CPU, 43 GB memory, and 10 GB graphics memory. The Python language used is version 3.8, the deep learning framework used is Pytorch 1.11.0, and the cuda version is 11.3. The software packages used include Pycharm and Anaconda.

4.3   Evaluation Metrics

To evaluate the performance of the trained Yolov5 model, this paper used common evaluation metrics for object detection models including mAP, Params, FLOPs, and FPS.

mAP is one of the common metrics to evaluate the performance of object detection algorithms, which represents the mean of the average precision (AP) for all categories. AP is the average precision for a single category. The mAP value is calculated by sorting the predicted boxes by confidence, calculating the corresponding precision values (AP) according to different IoU thresholds, and finally taking the average of the AP for all categories to obtain the mAP value. The calculation formula is shown in Equation (5).

$$mAP = \frac{1}{n} \sum_{i=1}^{n} AP(i) \tag{5}$$

Params: represents the size of the model's parameters which determines the hardware requirements of the model. Smaller parameter sizes are desirable since they require less computing resources, making them more practical to deploy. Therefore, smaller model parameters are often preferred.

FLOPs: represents the number of floating-point operations, used to measure the computational complexity and efficiency of the model. The larger the FLOPs, the more complex the model.

FPS: represents the number of frames detected per second and is used to measure the inference speed of the model. The higher the FPS, the better the real-time performance. Since this study requires high real-time detection speed of the model, the FPS indicator is particularly important for evaluating the model.

4.4   Training and Analysis of the Object Detection Model

At first, the model is initialized with pre-trained weights, and Mosaic data augmentation is applied to enhance the dataset's variability and improve the network's capacity for generalization. Due to the relatively small size of the self-built

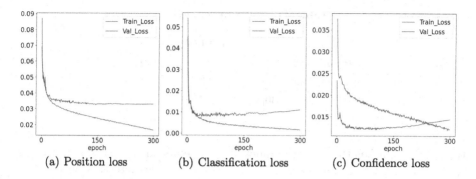

(a) Position loss   (b) Classification loss   (c) Confidence loss

Fig. 2. The loss rate of model training changes

dataset, the model is trained for only 300 epochs, using a batch size of 32 and the SGD optimizer. Moreover, the initial learning rate is set to 0.01, while the input image resolution is fixed at 640×640.

This paper discusses the training and testing of three models (YOLO5n, YOLOv5s, and YOLOv5m) and uses YOLOv5s as an example to illustrate how position loss, classification loss, and confidence loss change during training on the self-built dataset. Figure 2 show the trends in these loss functions. During training, the classification loss, position error, and confidence error of the training set initially decreased linearly, and then plateaued. However, during validation (as indicated in Fig. 2.b and Fig. 2.c), the loss and confidence loss of the validation set first decreased and then slightly increased. This may be due to the small size and unbalanced categories of the self-built dataset, which can lead to overfitting. The training process took 1.70 h, with an average time of 21.1 s per epoch. At the 189th epoch, the model's performance peaked, and this iteration was saved as the best model.

Table 2. models' AP value

version	AP						
	open	closed	yawn	normal	mask	phone	cigarette
Yolov5n	86.8	80.7	98.8	89.6	93.8	71.4	74.0
Yolov5s	87.8	83.2	99.1	89.3	94.9	77.5	75.5
Yolov5m	86.3	79.6	99.1	89.8	93.1	78.0	79.4

This paper uses metrics such as mAP, params, FLOPs, and FPS to evaluate and compare object detection models. The experimental results are shown in Tables 2 and 3

According to Tables 2 and 3, the YOLOv5n model has the smallest number of parameters and the fastest detection speed compared to other models, but at the cost of sacrificing some accuracy. Although YOLOv5m has a similar mAP

Table 3. Comparison of models

version	mAP	Params/M	FLOPs/G	FPS
Yolov5n	85.0	3.75	4.2	126.37
Yolov5s	86.8	13.8	15.8	117.87
Yolov5m	86.5	40.3	47.9	88.24

to YOLOv5s, it has the largest number of parameters and the slowest detection speed. In comparison, although YOLOv5s has a slower detection speed than YOLOv5n, it has the highest detection accuracy. Except for the categories of phones and cigarettes, the detection accuracy of other categories is above 80%, and the mAP reaches 86.8%. Therefore, in this study, the YOLOv5s version is adopted as the basic model for fatigue driving and distracted driving detection.

4.5   Detection Results of the Model

To verify the effectiveness of the proposed method in practical application scenarios, four testers simulated the normal, fatigued, and distracted driving state. The examples are shown in Fig. 3. Testers A and B are in well-lit conditions, while testers C and D are in dim light conditions. The four selected test videos were input into the detection model, and the eye status, mouth status, and head pitch angle of the four testers were obtained during the detection period. The results are as shown in Fig. 4. It can be seen from the Fig. 4 that the mouth state of tester A exhibited brief fluctuations, potentially indicating moments of speech. The eye-closed duration and blink frequency of testers A and C were both lower

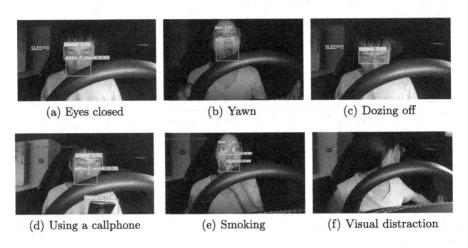

(a) Eyes closed	(b) Yawn	(c) Dozing off
(d) Using a callphone	(e) Smoking	(f) Visual distraction

Fig. 3. Real-time fatigue and distraction detection

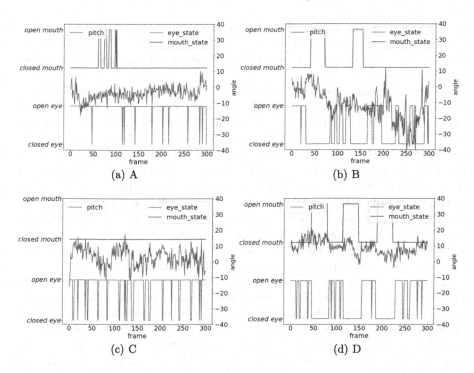

Fig. 4. Statistics on changes in the eyes, mouth and head states of 4 testers in one cycle

Table 4. Validation of fatigue driving detection

Tester	Frames	Closed-eyes	Perclos	Yawns	PerYawn	Nods	PerNod	Detected status	Actual status
A	300	11	0.037	0	0	0	0	awake	awake
B	300	193	0.643	2	0.177	3	0.103	drowsy	drowsy
C	300	40	0.133	0	0	0	0	awake	awake
D	300	145	0.483	3	0.343	0	0	drowsy	drowsy

Table 5. Validation of distracted driving detection

Tester	Time/min	Calling	Smoking	Visual distraction	Warning
A	2	3	1	0	4
B	2	2	2	1	5
C	2	1	1	1	4
D	2	3	1	2	5

compared to those of testers B and D. Furthermore, during the second half of the period, tester B exhibited more frequent changes in the head pitch angle. Both testers B and D had intervals where their mouths remained open. Finally, according to the fatigue judgment method in this paper, the fatigue judgment results of the four testers are shown in Table 4. It can be seen from the results in the table that the detection results of this model are consistent with the actual simulation results.

At the same time, in order to verify the effectiveness of the distraction detection algorithm, four testers individually conducted a two-minute distracted driving simulation. It was set that the tester was distracted for more than ten seconds as a distraction. The model counts the number of distractions, and the test results are shown in Table 5. Based on the data presented in the table, it is evident that Tester C experienced distraction only three times, whereas the distraction detection model issued four warnings. Furthermore, Tester D was distracted by using their mobile phone three times, whereas the model detected it only twice. This could potentially be attributed to the model's limitations in detecting distractions under dim light conditions, resulting in missed detections and false alarms. Nevertheless, apart from this issue, the model's overall detection results align reasonably well with the actual situation.

In addition, this paper sets the fatigue indicator thresholds based on numerous simulated experiments. These experimental data indicate that when a person is fatigued, the PERCLOS is usually greater than 0.3, and the mouth opening duration during yawning generally exceeds 1 s. Additionally, the deviation of pitch is typically more than 20 degrees. Therefore, this article sets the threshold of PERCLOS to 0.3 and the head deviation angle to 25 degrees, and sets the duration of one yawning to be not less than 1 s. In fact, the judgment threshold for fatigue driving actually varies due to different reaction and physiological states of individuals. Therefore, using a fixed threshold may not be optimal. To solve this problem, our developed system allows users to either choose the default threshold setting or set their own judgment threshold that is more appropriate for their individual needs.

5   Conclusion

To avoid traffic accidents and reduce economic losses and casualties, this paper focuses on the deficiencies of fatigue driving detection and distracted driving detection, and conducts research using deep learning methods. Using the YOLOv5 and FSA-Net models as frameworks, we have designed fatigue detection algorithms based on facial features, as well as distracted driving detection algorithms.

In this paper, we trained and tested the YOLOv5 model using a self-built dataset. The experimental results have demonstrated the high accuracy and speed of our model, which effectively solves the problem of reduced detection accuracy in fatigue detection algorithms that rely on facial landmarks, resulting from occlusion and inadequate lighting.

This paper combines the facial detection algorithm based on YOLOv5 and the head pose estimation algorithm based on FSA-Net to design a fatigue detection algorithm capable of identifying eye features, mouth features, and head features. Multiple feature indicators are used to determine the driver's fatigue state, ensuring real-time and high-precision detection.

Moreover, We uses the YOLOv5 model to determine driver's cognitive distractions and adopts the lightweight FSA-Net network for visual distraction detection. This approach effectively addresses the limitations of image-based classification methods which only detect limited distracted driving behavior categories. Besides, it resolves the issues with traditional head pose estimation methods, which may suffer from low accuracy, slow processing, and difficulty in deployment. By setting appropriate threshold values, real-time and effective distracted driving detection can be achieved.

Acknowledgements. This work was supported by the Guangdong Provincial Natural Science Foundation under Grant No. 2022A1515011386, the National Key Research and Development Program of China under Grant No. 2020YFB1005804, the National Natural Science Foundation of China under Grant 61632009, and the Guangdong Provincial Natural Science Foundation under Grant 2017A030308006.

References

1. Khunpisuth, O., Chotchinasri, T., Koschakosai, V., Hnoohom, N.: Driver drowsiness detection using eye-closeness detection. In: 2016 12th International Conference on Signal-Image Technology & Internet-Based Systems (SITIS), pp. 661–668 (2016)
2. Zhou, Z., Cai, Y., Ke, R., Yang, J.: A collision avoidance model for two-pedestrian groups: considering random avoidance patterns. Phys. A **475**, 142–154 (2017)
3. Zhou, Z., Zhou, Y., Pu, Z., Xu, Y.: Simulation of pedestrian behavior during the flashing green signal using a modified social force model: Transportmetrica A: Transport. Science **15**, 1019–1040 (2019)
4. Koh, S., et al.: Driver drowsiness detection via PPG biosignals by using multimodal head support. In: 2017 4th International Conference on Control, Decision and Information Technologies (CoDIT), pp. 0383–0388 (2017)
5. Anund, A., Fors, C., Ahlstrom, C.: The severity of driver fatigue in terms of line crossing: a pilot study comparing day- and night time driving in simulator. Eur. Transp. Res. Rev. **9**, 1–7 (2017)
6. Ravi, A., Phanigna, T.R., Lenina, Y., Ramcharan, P., Teja, P.S.: Real time driver fatigue detection and smart rescue system. In: 2020 International Conference on Electronics and Sustainable Communication Systems (ICESC), pp. 434–439 (2020)
7. Savas, B.K., Becerikli, Y.: Real time driver fatigue detection system based on multi-task ConNN. IEEE Access. **8**, 12491–12498 (2020)
8. Zhao, Y., Xie, K., Zou, Z., He, J.B.: Intelligent recognition of fatigue and sleepiness based on inceptionV3-LSTM via multi-feature fusion. IEEE Access. **8**, 144205–144217 (2020)
9. Dwivedi, K., Biswaranjan, K., Sethi, A.: Drowsy driver detection using representation learning. In: 2014 IEEE International Advance Computing Conference (IACC), pp. 995–999 (2014)
10. Ansari, S., Naghdy, F., Du, H., Pahnwar, Y.N.: Driver mental fatigue detection based on head posture using new modified reLU-BiLSTM deep neural network. IEEE Trans. Intell. Transp. Syst. **23**, 10957–10969 (2022)
11. Liu, M.-Z., Xu, X., Hu, J., Jiang, Q.N.: Real time detection of driver fatigue based on CNN-LSTM. IET Image Proc. **16**, 576–595 (2022)

12. Zhang, W., Murphey, Y.L., Wang, T., Xu, Q.: Driver yawning detection based on deep convolutional neural learning and robust nose tracking. In: 2015 International Joint Conference on Neural Networks (IJCNN), pp. 1–8 (2015)

13. Murugan, S., Selvaraj, J., Sahayadhas, A.: Detection and analysis: driver state with electrocardiogram (ECG). Phys Eng Sci Med. **43**, 525–537 (2020)

14. Sha, C.F., Li, R., Zhang, M. M.: Research on fatigue driving detection based on steering wheel grip force. Sci. Technol. Eng. Vol. 16, pp. 299–304(2016)

15. McDonald, A.D., Lee, J.D., Schwarz, C., Brown, T.L.: A contextual and temporal algorithm for driver drowsiness detection. Accident Anal. Prevent. **113**, 25–37 (2018)

16. Rongben, W., Lie, G., Bingliang, T., Lisheng, J.: Monitoring mouth movement for driver fatigue or distraction with one camera. In: Proceedings. The 7th International IEEE Conference on Intelligent Transportation Systems (IEEE Cat. No.04TH8749), pp. 314–319 (2004)

17. Pan, Z.G., Liu, R.F., Zhang, M.M.: Research on fatigue driving detection algorithm based on fuzzy comprehensive evaluation. J. Software **30**, 2954–2963 (2019)

18. Zhang, N., Zhang, H., Huang, J.: Driver fatigue state detection based on facial key points. In: 2019 6th International Conference on Systems and Informatics (ICSAI), pp. 144–149 (2019)

19. Zhao, Z., Zhou, N., Zhang, L., Yan, H., Xu, Y., Zhang, Z.: Driver fatigue detection based on convolutional neural networks using EM-CNN. Computational Intelligence and Neuroscience (2020)

20. Girshick, R., Donahue, J., Darrell, T., Malik, J.: Rich feature hierarchies for accurate object detection and semantic segmentation. Presented at the Proceedings of the IEEE Conference on Computer Vision and Pattern Recognition, pp. 580–587(2014)

21. He, K., Zhang, X., Ren, S., Sun, J.: Spatial pyramid pooling in deep convolutional networks for visual recognition. IEEE Trans. Pattern Anal. Mach. Intell. **37**, 1904–1916 (2015)

22. Girshick, R.: Fast R-CNN. In: Presented at the Proceedings of the IEEE International Conference on Computer Vision, pp. 1440–1448(2015)

23. Ren, S., He, K., Girshick, R., Sun, J.: Faster R-CNN: towards real-time object detection with region proposal networks. In: Advances in Neural Information Processing Systems. Curran Associates, Inc. (2015)

24. Redmon, J., Divvala, S., Girshick, R., Farhadi, A.: You only look once: unified, real-time object detection. Presented at the Proceedings of the IEEE Conference on Computer Vision and Pattern Recognition(CVPR), pp. 779–788(2016)

25. Redmon, J., Farhadi, A.: YOLO9000: better, faster, stronger. In: Presented at the Proceedings of the IEEE Conference on Computer Vision and Pattern Recognition(CVPR), pp. 7263–7271(2017)

26. Redmon, J., Farhadi, A.: YOLOv3: an incremental improvement. Computer Vision Pattern Recognition, Vol. 1804, pp. 1–6(2018)

27. Liu, W., Anguelov, D., Erhan, D., Szegedy, C., Reed, S., Fu, C.Y., Berg, A.C.: SSD: single shot multibox detector. In: Leibe, B., Matas, J., Sebe, N., Welling, M. (eds.) Computer Vision-ECCV 2016, pp. 21–37. Springer International Publishing, Cham (2016)

28. Zhu, X., Liu, X., Lei, Z., Li, S.Z.: Face alignment in full pose range: a 3D total solution. IEEE Trans. Pattern Anal. Mach. Intell. **41**, 78–92 (2019)

29. Yang, T.Y., Chen, Y.T., Lin, Y.Y., Chuang, Y.Y.: FSA-Net: learning fine-grained structure aggregation for head pose estimation from a single image. In: Presented

at the Proceedings of the IEEE/CVF Conference on Computer Vision and Pattern Recognition, pp. 1087–1096(2019)

30. Ruiz, N., Chong, E., Rehg, J.M.: Fine-grained head pose estimation without keypoints. In: Presented at the Proceedings of the IEEE Conference on Computer Vision and Pattern Recognition Workshops, pp. 2074–2083(2018)

31. Shulei, W., et al.: Road rage detection algorithm based on fatigue driving and facial feature point location. Neural Comput. Applic. **34**, 12361–12371 (2022)

32. Akrout, B., Mahdi, W.: A novel approach for driver fatigue detection based on visual characteristics analysis. J Ambient Intell Human Comput. **14**, 527–552 (2023)

33. Trutschel, U., Sirois, B., Sommer, D., Golz, M., Edwards, D.: PERCLOS: An Alertness Measure of the Past. Driving Assessment Conference. 6 (2011)

Author Index

Z. Tari et al. (Eds.): ICA3PP 2023, LNCS 14488, pp. 503–504, 2024.
https://doi.org/10.1007/978-981-97-0801-7

Printed in the United States
by Baker & Taylor Publisher Services